Jasmonic Acid Pathway in Plants

Jasmonic Acid Pathway in Plants

Special Issue Editor

Kenji Gomi

MDPI • Basel • Beijing • Wuhan • Barcelona • Belgrade • Manchester • Tokyo • Cluj • Tianjin

Special Issue Editor
Kenji Gomi
Kagawa University
Japan

Editorial Office
MDPI
St. Alban-Anlage 66
4052 Basel, Switzerland

This is a reprint of articles from the Special Issue published online in the open access journal *International Journal of Molecular Sciences* (ISSN 1422-0067) (available at: https://www.mdpi.com/journal/ijms/special_issues/plant_JA).

For citation purposes, cite each article independently as indicated on the article page online and as indicated below:

LastName, A.A.; LastName, B.B.; LastName, C.C. Article Title. *Journal Name* **Year**, *Article Number*, Page Range.

ISBN 978-3-03928-488-7 (Pbk)
ISBN 978-3-03928-489-4 (PDF)

© 2020 by the authors. Articles in this book are Open Access and distributed under the Creative Commons Attribution (CC BY) license, which allows users to download, copy and build upon published articles, as long as the author and publisher are properly credited, which ensures maximum dissemination and a wider impact of our publications.

The book as a whole is distributed by MDPI under the terms and conditions of the Creative Commons license CC BY-NC-ND.

Contents

About the Special Issue Editor

Kenji Gomi is an Associate Professor at Kagawa University, Japan. He completed his undergraduate studies and M.S. at Yokohama City University, Japan, in 1999 and received his Ph.D. at Kagawa University, Japan, in 2002. He studied the signal transduction regulated by a plant hormone, gibberellic acid, in rice as a postdoctoral researcher at Nagoya University, Japan (2002–2004). He next studied MAP kinase signaling pathways in tobacco as a postdoctoral researcher at the National Institute of Agrobiological Sciences, Japan (2004–2007). He again studied the signaling pathways regulated by plant hormones in rice as a lecturer at Kyoto University, Japan (2007–2008). He became an Assistant Professor in 2008 and an Associate Professor in 2011 at Kagawa University, Japan. He received the Young Scientist Award at the 2009 Annual Meeting of the Phytopathological Society of Japan. His current research program focuses on the signal transduction regulated by a plant hormone, jasmonic acid, under abiotic and biotic stresses in rice.

 International Journal of
Molecular Sciences

Editorial

Jasmonic Acid: An Essential Plant Hormone

Kenji Gomi

Plant Genome and Resource Research Center, Faculty of Agriculture, Kagawa University, Miki,
Kagawa 761-0795, Japan; gomiken@ag.kagawa-u.ac.jp

Received: 10 February 2020; Accepted: 12 February 2020; Published: 13 February 2020

The plant hormone jasmonic acid (JA) and its derivative, an amino acid conjugate of JA (jasmonoyl isoleucine: JA-Ile), are signaling compounds involved in the regulation of cellular defense and development in plants. The number of articles on JA has increased dramatically since the 1990s. JA was recognized as a stress hormone that regulates plant responses to biotic stresses, such as those elicited by hervivores and pathogens, as well as abiotic stresses, such as wounding and ultraviolet radiation. Recent studies have progressed remarkably in understanding the importance of JA in the life cycle of plants. It has been revealed that JA is directly involved in many physiological processes, including stamen growth, senescence, and root growth. Furthermore, it has been known to regulate the production of various metabolites, such as phytoalexins and terpenoids. Many active regulatory proteins involved in the JA signaling pathway have been identified by screening for *Arabidopsis* mutants. The discovery of the JA receptor, CORONATINE INSENSITIVE 1 (COI1), and the central repressors, jasmonate ZIM (JAZ)-domain proteins, further promotes the efforts to understand the JA signaling pathway in *Arabidopsis*. However, many aspects about the JA signaling pathway in other plant species remain to be elucidated.

This Special Issue, "Jasmonic Acid Pathway in Plants" contains five review articles published by field experts. Information available on the important role of JA in plant growth has been updated in these reviews [1–5]. These reviews will help in understanding the crucial roles of JA in its response to the several environmental stresses and developments in plants. In addition, this Special Issue also contains 15 original research articles. The noteworthy fact is that studies published in this Special Issue were performed using several plant species, including those belonging to *Arabidopsis*: *Camellia sinensis* [6], *Panax ginseng* [7], *Oryza sativa* L. [8,9], *Prunus avium* L. [10], *Brassica rapa* [11], *Sorghum bicolor* L. [12,13], *Nicotiana benthamiana* [14], *Pogostemon cablin* [15], *Zea mays* [16], and *Arabidopsis thaliana* [17–20]. This indicates that JA is an essential plant hormone across different plant species. Furthermore, these articles prove that JA has different roles during the vegetative and reproductive stages of plant growth. Gladman et al. [12] and Dampanabonia et al. [13] reported that JA negatively affects grain number in sorghum, suggesting that a new breeding approach that modifies JA-biosynthesis genes using genome editing can lead to increased grain yield in cereals. Below, I will focus on two studies published in this Special Issue as a plant pathologist.

Uji et al. demonstrated that the JA-induced VQ-motif-containing protein, OsVQ13, positively regulates the JA signaling pathway in rice [9]. Interestingly, OsVQ13 also activates the salicylic acid (SA) signaling pathway and confers resistance to rice bacterial blight, which is caused by the hemibiotrophic pathogen, *Xanthomonas oryzae* pv. *oryzae* (*Xoo*), in rice [9]. Generally, SA confers resistance to biotrophic and hemibiotrophic pathogens, whereas JA has been known to confer resistance to necrotrophic pathogens in plants. The relationship between JA and SA is antagonistic in many plant species. However, this antagonistic crosstalk between JA- and SA-dependent defense signaling pathways remains unclear in rice. It has been reported that a central repressor of the JA signaling pathway negatively affects *Xoo* resistance in rice [21]. JA-induced volatiles, such as monoterpenes, act as antibacterial or signaling compounds in the defense response against *Xoo* [22,23]. Uji et al. also demonstrated that the central positive regulators of the SA signaling pathway are induced by JA in rice [9]. It has also been suggested that OsVQ13 plays a critical role as an activator involved in both JA- and SA-induced resistance to *Xoo*. Taken together, these results strongly indicate that the JA and SA signaling pathways interact coordinately to yield induced defense responses in rice. Rice may develop a unique system to shield itself against pathogens. Recently, this unique system has been called "Common Defense System" [24].

Nakano and Mukaihara published an excellent study in this Special Issue [14]. The pathogen *Ralstonia solanacearum* is known to be hemibiotrophic and causes bacterial wilt disease in more than 200 plant species, such as tomato, potato, banana, and eggplant. Plants have developed a specialized defense system, the so-called pattern-triggered immunity (PTI), to inhibit or attenuate infection due to *R. solanacearum*. To suppress PTI, the pathogen injects approximately 70 type III effectors into the plant cells through the Hrp type III secretion system. Nakano and Mukaihara identified an effector, RipE1, which promoted the degradation of JAZ repressors and induced the expressions of JA-responsive genes in a cysteine–protease-activity-dependent manner [14]. JA and SA signaling pathways have been shown to antagonize each other in *N. benthamiana*. Thus, the activation of the JA signaling pathway causes the suppression of the SA signaling pathway, which is essential for the defense response against *R. solanacearum*. They also revealed that the effector RipAL induces JA production to activate its signaling pathway and simultaneously suppress the SA-mediated defense response in these plants [25]. These results indicate that *R. solanacearum* hijacks the JA signaling pathway and exploits antagonistic interactions between the JA and SA signaling pathways to promote a successful infection. They unraveled one of the survival strategies, which was previously unknown.

Finally, I would like to express my heartfelt gratitude to all of the authors and referees for their tremendous and relentless efforts in supporting this Special Issue. Without their valuable assistance, I would not have had an opportunity to publish this timely and successful publication, with its useful updates on the JA signaling pathway in plants. I also thank the assistant editor Sydney Tang for supporting my works.

Conflicts of Interest: The author declares no conflict of interest.

References

1. Ruan, J.; Zhou, Y.; Zhou, M.; Yan, J.; Khurshid, M.; Weng, W.; Cheng, J.; Zhang, K. Jasmonic Acid Signaling Pathway in Plants. *Int. J. Mol. Sci.* **2019**, *20*, 2479. [CrossRef]
2. Zhang, Y.; Bo, C.; Wang, L. Novel Crosstalks between Circadian Clock and Jasmonic Acid Pathway Finely Coordinate the Tradeoff among Plant Growth, Senescence and Defense. *Int. J. Mol. Sci.* **2019**, *20*, 5254. [CrossRef]
3. Jang, G.; Yoon, Y.; Choi, Y.D. Crosstalk with Jasmonic Acid Integrates Multiple Responses in Plant Development. *Int. J. Mol. Sci.* **2020**, *21*, 305. [CrossRef]
4. Ali, M.S.; Baek, K.-H. Jasmonic Acid Signaling Pathway in Response to Abiotic Stresses in Plants. *Int. J. Mol. Sci.* **2020**, *21*, 621. [CrossRef]
5. Ho, T.-T.; Murthy, H.N.; Park, S.-Y. Methyl Jasmonate Induced Oxidative Stress and Accumulation of Secondary Metabolites in Plant Cell and Organ Cultures. *Int. J. Mol. Sci.* **2020**, *21*, 716. [CrossRef]
6. Li, J.; Zeng, L.; Liao, Y.; Gu, D.; Tang, J.; Yang, Z. Influence of Chloroplast Defects on Formation of Jasmonic Acid and Characteristic Aroma Compounds in Tea (*Camellia sinensis*) Leaves Exposed to Postharvest Stresses. *Int. J. Mol. Sci.* **2019**, *20*, 1044. [CrossRef]
7. Liu, T.; Luo, T.; Guo, X.; Zou, X.; Zhou, D.; Afrin, S.; Li, G.; Zhang, Y.; Zhang, R.; Luo, Z. PgMYB2, a MeJA-Responsive Transcription Factor, Positively Regulates the Dammarenediol Synthase Gene Expression in *Panax Ginseng*. *Int. J. Mol. Sci.* **2019**, *20*, 2219. [CrossRef]
8. Bertini, L.; Palazzi, L.; Proietti, S.; Pollastri, S.; Arrigoni, G.; Polverino de Laureto, P.; Caruso, C. Proteomic Analysis of MeJa-Induced Defense Responses in Rice against Wounding. *Int. J. Mol. Sci.* **2019**, *20*, 2525. [CrossRef]

9. Uji, Y.; Kashihara, K.; Kiyama, H.; Mochizuki, S.; Akimitsu, K.; Gomi, K. Jasmonic Acid-Induced VQ-Motif-Containing Protein OsVQ13 Influences the OsWRKY45 Signaling Pathway and Grain Size by Associating with OsMPK6 in Rice. *Int. J. Mol. Sci.* **2019**, *20*, 2917. [CrossRef]

10. Berni, R.; Cai, G.; Xu, X.; Hausman, J.-F.; Guerriero, G. Identification of Jasmonic Acid Biosynthetic Genes in Sweet Cherry and Expression Analysis in Four Ancient Varieties from Tuscany. *Int. J. Mol. Sci.* **2019**, *20*, 3569. [CrossRef]

11. Xu, Y.-M.; Xiao, X.-M.; Zeng, Z.-X.; Tan, X.-L.; Liu, Z.-L.; Chen, J.-W.; Su, X.-G.; Chen, J.-Y. BrTCP7 Transcription Factor Is Associated with MeJA-Promoted Leaf Senescence by Activating the Expression of *BrOPR3* and *BrRCCR*. *Int. J. Mol. Sci.* **2019**, *20*, 3963. [CrossRef]

12. Gladman, N.; Jiao, Y.; Lee, Y.K.; Zhang, L.; Chopra, R.; Regulski, M.; Burow, G.; Hayes, C.; Christensen, S.A.; Dampanaboina, L.; et al. Fertility of Pedicellate Spikelets in Sorghum Is Controlled by a Jasmonic Acid Regulatory Module. *Int. J. Mol. Sci.* **2019**, *20*, 4951. [CrossRef]

13. Dampanaboina, L.; Jiao, Y.; Chen, J.; Gladman, N.; Chopra, R.; Burow, G.; Hayes, C.; Christensen, S.A.; Burke, J.; Ware, D.; et al. Sorghum *MSD3* Encodes an ω-3 Fatty Acid Desaturase that Increases Grain Number by Reducing Jasmonic Acid Levels. *Int. J. Mol. Sci.* **2019**, *20*, 5359. [CrossRef]

14. Nakano, M.; Mukaihara, T. Comprehensive Identification of PTI Suppressors in Type III Effector Repertoire Reveals that *Ralstonia solanacearum* Activates Jasmonate Signaling at Two Different Steps. *Int. J. Mol. Sci.* **2019**, *20*, 5992. [CrossRef]

15. Wang, X.; Chen, X.; Zhong, L.; Zhou, X.; Tang, Y.; Liu, Y.; Li, J.; Zheng, H.; Zhan, R.; Chen, L. PatJAZ6 Acts as a Repressor Regulating JA-Induced Biosynthesis of Patchouli Alcohol in *Pogostemon Cablin*. *Int. J. Mol. Sci.* **2019**, *20*, 6038. [CrossRef]

16. Ahmad, R.M.; Cheng, C.; Sheng, J.; Wang, W.; Ren, H.; Aslam, M.; Yan, Y. Interruption of Jasmonic Acid Biosynthesis Causes Differential Responses in the Roots and Shoots of Maize Seedlings against Salt Stress. *Int. J. Mol. Sci.* **2019**, *20*, 6202. [CrossRef]

17. Li, L.; Su, B.; Qi, X.; Zhang, X.; Song, S.; Shan, X. JA-Induced Endocytosis of AtRGS1 Is Involved in G-Protein Mediated JA Responses. *Int. J. Mol. Sci.* **2019**, *20*, 3779. [CrossRef]

18. Della Rovere, F.; Fattorini, L.; Ronzan, M.; Falasca, G.; Altamura, M.M.; Betti, C. Jasmonic Acid Methyl Ester Induces Xylogenesis and Modulates Auxin-Induced Xylary Cell Identity with NO Involvement. *Int. J. Mol. Sci.* **2019**, *20*, 4469. [CrossRef]

19. Garrido-Bigotes, A.; Valenzuela-Riffo, F.; Figueroa, C.R. Evolutionary Analysis of JAZ Proteins in Plants: An Approach in Search of the Ancestral Sequence. *Int. J. Mol. Sci.* **2019**, *20*, 5060. [CrossRef]

20. Qi, J.; Zhao, X.; Li, Z. iTRAQ-Based Quantitative Proteomic Analysis of the Arabidopsis Mutant *opr3-1* in Response to Exogenous MeJA. *Int. J. Mol. Sci.* **2020**, *21*, 571. [CrossRef]

21. Kashihara, K.; Onohata, T.; Okamoto, Y.; Uji, Y.; Mochizuki, S.; Akimitsu, K.; Gomi, K. Overexpression of *OsNINJA1* negatively affects a part of OsMYC2-mediated abiotic and biotic responses in rice. *J. Plant Physiol.* **2019**, *232*, 180–187. [CrossRef]

22. Taniguchi, S.; Hosokawa-Shinonaga, Y.; Tamaoki, D.; Yamada, S.; Akimitsu, K.; Gomi, K. Jasmonate induction of the monoterpene linalool confers resistance to rice bacterial blight and its biosynthesis is regulated by JAZ protein in rice. *Plant Cell Environ.* **2014**, *37*, 451–461. [CrossRef]

23. Yoshitomi, K.; Taniguchi, S.; Tanaka, K.; Uji, Y.; Akimitsu, K.; Gomi, K. Rice *terpene synthase 24* (*OsTPS24*) encodes a jasmonate-responsive monoterpene synthase that produces an antibacterial γ-terpinene against rice pathogen. *J. Plant Physiol.* **2016**, *191*, 120–126. [CrossRef]

24. Tamaoki, D.; Seo, S.; Yamada, S.; Kano, A.; Miyamoto, A.; Shishido, H.; Miyoshi, S.; Taniguchi, S.; Akimitsu, K.; Gomi, K. Jasmonic acid and salicylic acid activate a common defense system in rice. *Plant Signal Behav.* **2013**, *8*, e24260. [CrossRef]

25. Nakano, M.; Mukaihara, T. Ralstonia solanacearum type III e_ector RipAL targets chloroplasts and induces jasmonic acid production to suppress salicylic acid-mediated defense responses in plants. *Plant Cell Physiol.* **2018**, *59*, 2576–2589.

© 2020 by the author. Licensee MDPI, Basel, Switzerland. This article is an open access article distributed under the terms and conditions of the Creative Commons Attribution (CC BY) license (http://creativecommons.org/licenses/by/4.0/).

International Journal of
Molecular Sciences

Review

Jasmonic Acid Signaling Pathway in Plants

Jingjun Ruan [1,2], Yuexia Zhou [1], Meiliang Zhou [2], Jun Yan [3], Muhammad Khurshid [2,4], Wenfeng Weng [1], Jianping Cheng [1] and Kaixuan Zhang [2,*]

[1] College of Agriculture, Guizhou University, Guiyang 550025, China; jjruan@gzu.edu.cn (J.R.); zhouyuexiagui@hotmail.com (Y.Z.); wenfengweng@hotmail.com (W.W.); jpcheng@gzu.edu.cn (J.C.)
[2] Institute of Crop Sciences, Chinese Academy of Agricultural Sciences, Beijing 100081, China; zhoumeiliang@caas.cn (M.Z.); khurshid.ibb@pu.edu.pk (M.K.)
[3] Schools of Pharmacy and Bioengineering, Chengdu University, Chengdu 610106, China; yanjun62@cdu.edu.cn
[4] Institute of Biochemistry and Biotechnology, University of the Punjab, Lahore 54590, Pakistan
* Correspondence: zhangkaixuan@caas.cn; Tel./Fax: +86-10-8210-6368

Received: 17 April 2019; Accepted: 7 May 2019; Published: 20 May 2019

Abstract: Jasmonic acid (JA) and its precursors and dervatives, referred as jasmonates (JAs) are important molecules in the regulation of many physiological processes in plant growth and development, and especially the mediation of plant responses to biotic and abiotic stresses. JAs biosynthesis, perception, transport, signal transduction and action have been extensively investigated. In this review, we will discuss the initiation of JA signaling with a focus on environmental signal perception and transduction, JA biosynthesis and metabolism, transport of signaling molecules (local transmission, vascular bundle transmission, and airborne transportation), and biological function (JA signal receptors, regulated transcription factors, and biological processes involved).

Keywords: jasmonic acid; signaling pathway; environmental response; biological function

1. Introduction

Plants undergo many physiological changes to cope with biotic and abiotic stress. The survival of plants mainly depends on their ability to adapt in a varying environment through signaling networks [1]. These networks establish connections between the environmental signals and cell responses [2]. Plant hormones play major roles in the establishment of signaling networks to regulate plant growth and stress-related responses. Jasmonic acid (3-oxo-2-2′-*cis*-pentenyl-cyclopentane-1-acetic acid, abbreviated as JA) is an endogenous growth-regulating substance found in higher plants. JA and its methyl ester (MeJA) and isoleucine conjugate (JA-Ile) are derivatives of a class of fatty acids and are collectively known as jasmonates (JAs). Initially identified as a stress-related hormone, JAs are also involved in the regulation of important growth and developmental processes [3,4]. For example, JAs can induce stomatal opening, inhibit Rubisco biosynthesis, and affect the uptake of nitrogen and phosphorus and the transport of organic matter such as glucose. In particular, as a signaling molecule, JAs can effectively mediate responses against environmental stresses by inducing a series of genes expression [5]. JAs and salicylic acid (SA)-mediated signaling pathways are mainly related to plant resistance, prompting plant responses to external damage (mechanical, herbivore, and insect damage) and pathogen infection, thereby inducing resistance gene expression. In this review, the initiation, transmission, and biological functions of jasmonic acid signaling are introduced from the point of view of environmental signal molecules.

2. Initiation of Jasmonic Acid Signaling

2.1. Signal Perception and Transduction

In the last decades, a large number of studies have been conducted on how biotic and abiotic stress signals are perceived by plants and the biosynthesis of JA is initiated. In tomato (*Lycopersicon esculentum*), Pearce et al. in 1991 found a systemin that responded to mechanical damage such as insect damage [6]. Systemin is a polypeptide signal molecule consisting of 18 amino acids derived from a precursor protein consisting of 200 amino acids: prosystemin [7]. After the tomato is mechanically damaged, the prosystemin is hydrolyzed into a systemin, which can be transported to other cells via the apoplast and combined with the cell surface receptor SR160 (a protein rich in leucine repeat units) to finally activate the JA signaling pathway [8,9]. In addition to the traumatic signals, oligosaccharide signals induced by pathogens and fungal elicitors were also found in tomatoes, which ultimately activate the JA signaling pathway. It is speculated that the mechanism of action of oligosaccharides has a similar pathway to that of the systemin, but the specific mechanism of induction is still unclear [10,11].

Later, a polypeptide having the same function as a systemin, AtPEP1, consisting of 23 amino acids, was also found in *Arabidopsis thaliana*. Similar to the production of systemin, mechanical damage or pathogen infection promotes the hydrolysis of the precursor protein PROPEP1 (consisting of 92 amino acids) to AtPEP1, which binds to the receptor PEPR1 (an enzyme rich in leucine repeat units) on the plasma membrane, which ultimately activates the JA signaling pathway [12].

The process by which systemin and AtPEP1 bind to the receptor to activate the JA signaling pathway is complex. It is known that the most important process is to activate phospholipase on the plasma membrane, and then the phospholipase acts on the membrane to release linolenic acid (a precursor of JA synthesis) from the phospholipid [13]. However, the mechanism by which systemin activates phospholipase is unclear. So far, several phospholipases that can be induced by systemin and AtPEP1 have been identified, including PLA2 in tomato and DAD1, DGL, and PLD in *Arabidopsis*, and these phospholipases have similar mechanisms of action [14–16].

2.2. Synthesis and Metabolism of Jasmonate Compounds

Biosynthesis of JAs has been studied in a variety of monocotyledonous and dicotyledonous plants during the last decades. Most of the work is done in the model plants *Arabidopsis thaliana* and *L. esculentum* (tomato). So far, various enzymes in the JAs synthetic pathway have been identified, and our knowledge of the relationship between the JA synthesis pathway and other metabolic pathways is gradually improving (Figure 1). In *Arabidopsis*, there are three pathways for the synthesis of JAs, including the octadecane pathway starting from α-linolenic acid (18:3) and the hexadecane pathway starting from hexadecatrienoic acid (16:3) [17]. All three pathways require three reaction sites: the chloroplast, peroxisome, and cytoplasm. The synthesis of 12-oxo-phytodienoic acid (12-OPDA) or deoxymethylated vegetable dienic acid (dn-OPDA) from unsaturated fatty acid takes place in the chloroplast, which is then converted to JA in the peroxisome. In the cytoplasm, JA is metabolized into different structures by various chemical reactions, such as MeJA, JA-Ile, *cis*-jasmone (CJ), and 12-hydroxyjasmonic acid (12-OH-JA).

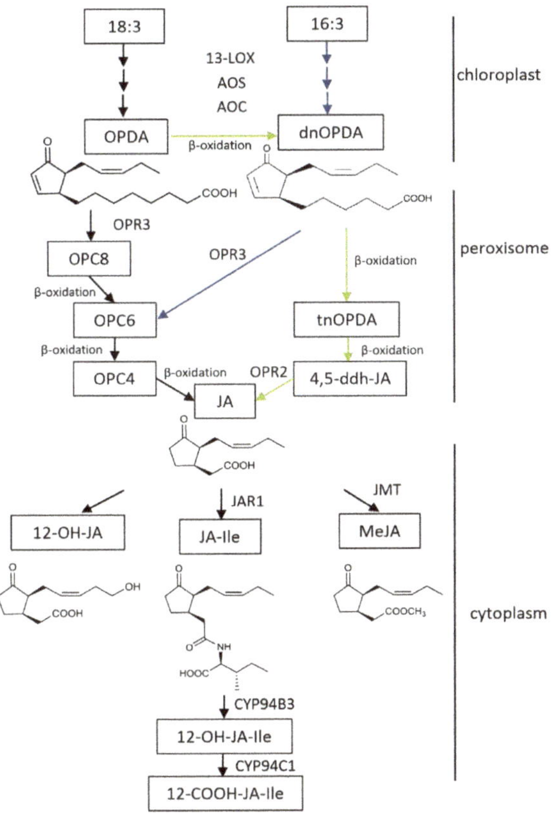

Figure 1. Scheme of the JAs biosynthesis pathway in *Arabidopsis thaliana*. The enzymes and the intermediates are indicated as follows: LOX for lipoxygenase, AOS for allene oxide synthase, AOC for allene oxide cyclase, OPR3 for OPDA reductase, JAR1 for jasmonate resistant 1, JMT for JA carboxyl methyltransferase; 18:3 for α-linolenic acid, 16:3 for hexadecatrienoic acid, OPDA for 12-oxo-phytodienoic acid, dnOPDA for dinor-12-oxo-phytodienoic acid, OPC8 for 8-(3-oxo-2-(pent-2-enyl)cyclopentyl) octanoic acid, OPC6 for 6-(3-oxo-2-(pent-2-enyl)cyclopentyl) hexanoic acid, OPC4 for 4-(3-oxo-2-(pent-2-enyl)cyclopentyl) butanoic acid, tnOPDA for tetranor-OPDA, 4,5-ddh-JA for 4,5-didehydrjasmonate, JA for jasmonic acid, JA-Ile for jasmonoyl-L-isoleucine, and MeJA for methyl jasmonate.

3. Transmission of the Jasmonic Acid Signal

The defense response triggered by a traumatic signal can result in a local defense response near the wound and/or a systemic acquired resistance (SAR) at the uninjured site, and/or even induced defense responses from adjacent plants. In these defense responses, short-distance transmission and long-distance transmission of JA signals are involved [18]. With the studies in the area of mechanisms of hormone signaling networks, it has been found that salicylic acid, ethylene, auxin, and other plant hormones interact with JA to regulate plant adaptation to the environment. At present, the understanding of complex regulatory networks and metabolic processes after plants perceive environmental signals is still very limited.

3.1. Short-Distance Signal Transmission

In plants, mechanical damage or insect feeding can cause rapid and transient accumulation of JA and JA-Ile at the site of injury, thereby activating the expression of defense genes surrounding the wound and producing a local defense response. In the local defense response, there are two ways of short-distance transmission of the JA signal. First, the systemin produced by the wounding acts as a signaling molecule, which is transmitted to the adjacent site through the apoplast and phloem to activate the JA cascade reaction pathway. Second, JA and JA-Ile induced by systemin act as signals and are transported to adjacent sites for defensive responses [19].

3.2. Long-Distance Signal Transmission

So far, it is known that the long-distance transmission of JA signals is via vascular bundle transmission and/or airborne transmission.

3.2.1. Vascular Bundle Transmission

Previously, many researchers believed that systemin functions directly in the long-distance signal transmission and is a mobile signal molecule. However, a series of grafting experiments using tomato jasmonate-insensitive mutant (*jai1*), systemin-insensitive mutant (*spr1*), and the JAs biosynthesis deletion mutants *spr2* and *acx1A* demonstrated that the systemin was not the systemically transmitted signal [19]. After the induction of the synthesis of JA, JA and MeJA are systemically transmitted in plants [20]. Thorpe et al. demonstrated by radioisotopic labeling experiments that MeJA can be transferred to phloem and xylem in vascular bundles [21]. Some work has also shown that JAs are not simply transported along the vascular bundle, but are accompanied by resynthesis of JAs during transport [20]. The localization of various JA synthetases (such as LOX, AOS, etc.) was also found in the companion cell–sieve element complex (CC-SE) of tomato vascular bundles [22], and the sieve molecules in the phloem have the ability to form the JA precursor OPDA [23]. Recently, Koo et al. [24] found that the systemic JA and JA-Ile caused by injury induction are not all transferred from the injured site, at least part of which is resynthesized and cascading cycles in the uninjured site produce more JA-Ile, which was later confirmed by Larrieu et al. [25].

3.2.2. Airborne Transmission

It was found that the flow rate of the tomato phloem signal is 1–5 cm per hour [26], but the accumulation of JA and JA-Ile can be detected in the whole plant within 15 min after mechanical damage [27]. In the 1990s, ring-cutting experiments demonstrated that although the vascular bundle transmission was blocked, there was still a rapid and strong defense gene expression in the distal leaves [27]. A large number of studies showed that in addition to vascular bundle transmission, there are other long-distance transmission routes for JA signals. Compared with JA, which has difficulty in penetrating the cell membrane without carrier assistance, MeJA easily penetrates the cell membrane and has strong volatility, and thus can be spread by airborne diffusion to distant leaves and adjacent plants [28]. It has been confirmed in a range of plants, such as *Arabidopsis thaliana* [20], *Nicotiana tabacum* [29], *Phaseolus lunatus* [30], and *Artemisia kawakamii* [31], that MeJA can be transmitted by air between damaged and undamaged leaves or between adjacent plants.

4. Perception of the Jasmonic Acid Signal and Induction of Response

4.1. Jasmonic Acid Receptor

The nuclear transport mechanism of JAs was systematically analyzed by means of molecular genetics, molecular biology, biochemistry, and cell biology. The ABC transporter AtJAT1/AtABCG16 with JAs transport ability was screened by a yeast system [32]. Radioactive isotope uptake experiments and autoradiography experiments showed that AtJAT1/AtABCG16 acts as a high-affinity transporter to regulate the subcellular distribution of JAs [32]. AtJAT1/AtABCG16 is localized on the nuclear and plasma membranes of plant cells and mediates the transport of JAs across the plasma membrane and the bioactive JA-Ile across the inner membrane of the nuclear membrane to activate JA responses at low concentration. When the concentration of JAs is high, the function of the JA transporter on the cytoplasmic membrane is dominant, which reduces the intracellular JA and JA-Ile concentrations to desensitize the JA signal. The JAs signaling pathway is activated in other cells by transporting JA to the apoplast. AtJAT1/AtABCG16 can rapidly regulate the dynamics of JA/JA-Ile in cells, which leads to the quick transport of JA-Ile into the nucleus when the plant is under stress, as well as the quick desensitization of the JA signal to avoid the inhibition of plant growth and development by the defense response (Figure 2).

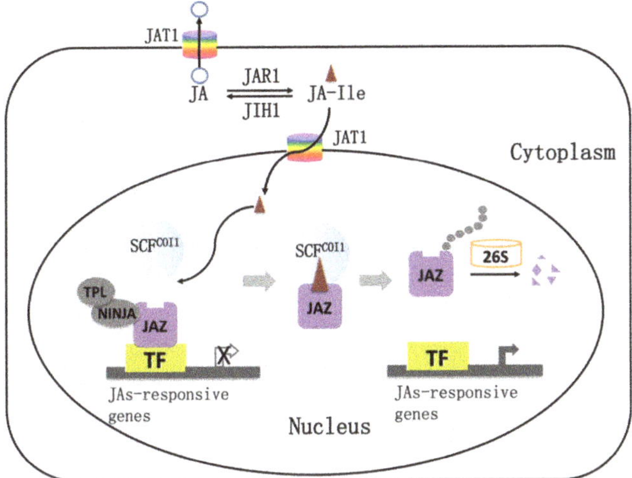

Figure 2. The working model of jasmonic acid transport and signaling pathway. JAT1: jasmonic acid transfer protein1; SCF: Skp1, Cullin and F-box proteins; COI1: coronatine insensitive1; JAZ: jasmonate ZIM-domain protein; TF: transcription factor; TPL: TOPLESS protein; NINJA: NOVEL INTERACTOR OF JAZ; 26S: 26S proteasome.

The understanding of JA receptors has undergone a complex process. In 1994, Feys first found that the *Arabidopsis coronatine insensitive1* (*coi1*) mutant lost all responses to JA [33], and further studies indicated that the *COI1* gene encodes an F-box protein that is a component of E3 ubiquitin ligase [34]. In this case, COI1 associates with the SKP1 protein and Cullin protein to form the SCF-type E3 ubiquitin ligase that is referred to as SCF^COI1, which targets the repressor proteins for degradation by ubiquitination [34,35]. The discovery of COI1 protein is of great significance for the study of theJA signaling pathway.

It was once thought that COI1 is the receptor for jasmonic acid signaling in cells, until the discovery of a jasmonate Zinc finger Inflorescence Meristem (ZIM)-domain (JAZ) protein family, which gave a new understanding of the jasmonic acid signal transduction pathway. In 2007, three research groups simultaneously found that JAZ proteins act as repressors in the JA signaling pathway [36–38]. To date, 13 JAZ proteins have been found in *Arabidopsis*, most of which have two conserved domains, Jas and ZIM [39]. The JAZ protein interacts with COI1 via the Jas domain and interacts with MYC2 via the ZIM domain [40]. Therefore, many researchers believe that JAZ proteins are the target protein of COI1 and the degradation of JAZ proteins is a key step to relieve the inhibition of the JAs pathway. However, in 2010, Sheard et al. proposed different views on JAs receptors through the analysis of crystal structure and confirmed that the COI1–JAZ complex is a high-affinity receptor for the bioactive JA-Ile; that is, COI1 and JAZ are coreceptors of JA signaling [41]. It is currently believed that plants perceive stimuli from the external environment to generate JA-Ile, which promotes the interaction between COI1 and JAZ proteins. Subsequently, JAZ proteins are degraded after being transferred to the 26S proteasome, and simultaneously, transcription factors (TFs) are released to activate the expression of downstream genes (Figure 2).

4.2. Jasmonic Acid Signal-Regulated Transcription Factor

JA-Ile activates the MYC transcription factors by directly binding to JAZ and COI1, which results in the degradation of JAZ through the 26S proteasome pathway (Figure 2). Recent studies have shown that the MYB transcription factors also bind with JAZ repressors and can be activated by the degradation of JAZ in the presence of JA-Ile. In addition, several other transcription factors (TFs) such as NAC, ERF, and WRKY are also involved in the JA signaling. These JA-responsive TFs regulate the expression of many genes involved in the growth and development of plants, and especially the responses and adaptation of plants to the environment (Figure 3). Studies have also shown that JA signaling can also induce the MAP kinase cascade pathway [42], calcium channel [43], and many processes that interact with signaling molecules such as ethylene, salicylic acid, and abscisic acid to regulate plant growth and development [44].

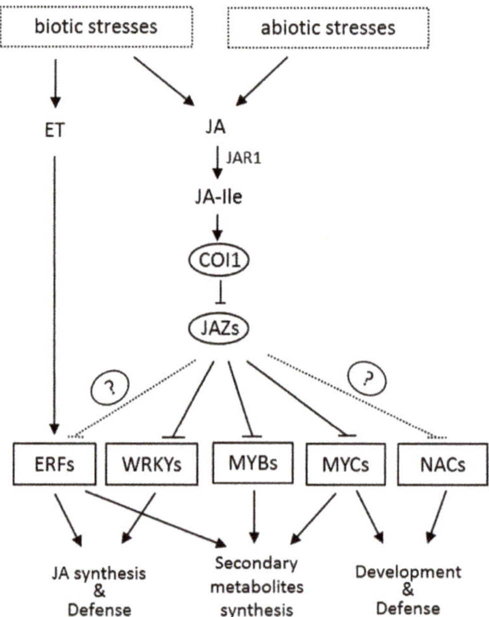

Figure 3. The regulation network of the jasmonic acid signaling pathway. Biotic and abiotic stresses induce the synthesis of JA, which can be converted to the biologically active JA-Ile by JAR1. Perception of JA-Ile by its receptor COI1 triggers the degradation of JAZ repressors, leading to the release of downstream transcription factors and the regulation of JAs-responsive genes in various processes. The question marker indicates an adaptor protein which is not identified yet.

4.2.1. MYC Transcription Factor

The basic helix–loop–helix (bHLH) transcription factor MYC2 is a well-known regulatory protein encoded by the *JIN1* gene. Most members of the JAZ protein family interact with MYC2 [45]. For a long time, it was believed that only the MYC2 protein can directly interact with the JAZ protein. In 2011, Fernandez-Calvo et al. identified that two other bHLH proteins, MYC3 and MYC4, share high sequence similarity with MYC2, suggesting they probably have similar functions. Indeed, MYC3 and MYC4 interact with JAZ proteins in vivo and in vitro, have similar DNA-binding specificity to MYC2, and act synergistically and distinctly with MYC2 [46]. A closely related TF, MYC5 (bHLH28), is induced by JAs and required for male fertility [47]. Besides transcriptional activators, JA-associated MYC2-like (JAM) proteins, JAM1, JAM2, and JAM3, were discovered as transcriptional repressors via forming protein–protein interactions with JAZs to regulate JAs responses [48].

4.2.2. MYB Transcription Factor

Most of the JAs-responsive MYB TFs belong to the R2R3-MYB family, which are widely distributed in plants and required for many processes. Schmiesing et al. showed that MYB51 and MYB34 regulate the synthesis of tryptophan and glucosinolates and act downstream of MYC2 [45]. However, many studies have found that MYB TFs can directly bind to JAZ proteins, indicating the release from JAZs to activate their target genes. For instance, in *Arabidopsis*, MYB21 and MYB24 are key factors in stamen and pollen maturation [49], and MYB75 can positively regulate the anthocyanin accumulation and trichome initiation [50]. Recently, a set of MYB TFs, MYB11, MYB13, MYB14, MYB15, and MYB16, were identified as repressors in the regulation of rutin biosynthesis in tartary buckwheat [51,52].

4.2.3. NAC Transcription Factor

ATAF1 and ATAF2 TFs in the *Arabidopsis* NAC family are both induced by JA signaling and involved in plant resistance to drought, salt stress, *Botrytis cinerea*, and other pathogens [53]. At the same time, ATAF1 and ATAF2 have an important regulatory effect on oxidative stress, flowering, and pod development of plants [54]. Two other NAC TFs in *Arabidopsis*, ANAC019 and ANAC055, are also present downstream of the MYC2 protein and regulate seed germination, cell division, and the synthesis of secondary walls of cells [55]. In addition, ATAF1, ATAF2, ANAC019, and ANAC055 are also involved in the crosstalk between JA and SA signaling pathways [53–55].

4.2.4. Ethylene-Responsive Factor (ERF) Transcription Factor

Microarray experiments at the genetic level have confirmed that JA signaling can induce the transcription of many *ERF* genes. The first evidence for a link between AP2/ERF TFs and JA signaling was found in *Catharanthus roseus*. The JAs-induced ORCA proteins, ORCA2 and ORCA3, belong to the AP2/ERF-domain family and can activate the expression of monoterpenoid indole alkaloid biosynthesis genes [56]. Based on the observation of ORCAs, the *Arabidopsis* ERF proteins, ERF1 and ORA59, function dependently on JAs and/or ET for the defenses against *Botrytis cinerea* [57,58]. Moreover, ORA59, rather than ERF1, acts as the integrator of JAs and ET signals [58] and regulates the biosynthesis of hydroxycinnamic acid amides [59]. The JAs-induced ORA47 can activate the expression of the JAs biosynthesis gene *AOC2*, indicating that ORA47 might act as an important regulator in the positive JAs-responsive feedback loop [60]. Moreover, JAs-responsive AtERF3 and AtERF4 act as repressors by not only down regulating their target genes' expression, but also interfering with the activity of other activators [61]. Interestingly, the activity of above TFs is not directly repressed by JAZ proteins, suggesting the presence of adaptors or corepressors in the JA signaling pathway.

4.2.5. WRKY Transcription Factor

WRKY transcription factors play an important regulatory role in plant development, senescence, and coping with environmental stress. In *Arabidopsis*, there are 89 members in the WRKY transcription factor family. It has been shown that some WRKY TFs are regulated by the JA signaling pathway, such as WRKY70 [62], WRKY22 [63], WRKY50 [64], WRKY57 [65], and WRKY89 [66]. These WRKY transcription factors are mostly associated with plant defense functions. In *Nicotiana attenuata*, two WRKY transcription factors, NaWRKY3 and NaWRKY6, regulate the expression of JAs biosynthesis-related genes (*LOX*, *AOS*, *AOC*, and *OPR*) to increase the levels of JA and JA-Ile [67]. In addition, *Arabidopsis* WRKY57 interacts with the inhibitor JAZ4/JAZ8 in the JA signaling pathway and the inhibitor IAA29 in the auxin signaling pathway, thereby regulating the interaction between JA and auxin-mediated signaling pathways and effects on plant leaf senescence [65].

5. Biological Processes Involved in Jasmonic Acid Signaling

5.1. Environmental Responses Affected by Jasmonic Acid Signaling

JA and its derivatives are plant signaling molecules closely related to plant defense and resistance to microbial pathogens, herbivorous insects, wounding, drought, salt stress, and low temperature. In addition to the traumatic signals, oligosaccharide signals induced by pathogens and fungal elicitors were also found in tomatoes, which ultimately activated the JA signaling pathway. Upon mechanical wounding in tomato, the prosystemin is hydrolyzed into a systemin, which can be transported to other cells via the apoplast and interacts with the cell surface receptor SR160 (a protein rich in leucine repeat units) to finally activate JAs responses. In this section, we will discuss the role of JA signaling in regulating plant responses in varying environments.

5.1.1. Effect of Light on Jasmonic Acid Signal Changes

To a large extent, the early development of plants is affected by light. It is observed that JA signaling mediates two aspects of the plant response to light: the photomorphogenesis of plants and the damage of plants by UVB. Red light/far-red light-mediated photomorphogenesis was observed in *Arabidopsis* and rice to be affected by JA signaling [68]. The involvement of the JA signaling pathway was also observed in blue light-mediated light morphogenesis in *Arabidopsis* and tomato [69,70]. Enhancement of UVB radiation induces the biosynthesis of JAs in the *Nicotiana* and *Brassica* genera to initiate the JA signaling pathway, and these processes are associated with chemical defenses of plants [68,71].

5.1.2. Effect of Temperature on Jasmonic Acid Signal Changes

The JA signaling pathway is involved in the response and adaptation process of plants to low temperatures. Recent studies on bananas have shown that the MeJA treatment can significantly induce the expression of MYC-family TFs and many cold-responsive genes (*MaCBF1*, *MaCBF2*, *MaCOR1*, *MaKIN2*, *MaRD2*, *MaRD5*, etc.) after cold storage, thereby reducing the damage of plants caused by cold [72]. It was also shown that the endogenous JA content of the banana (*Musa acuminata*) decreased slightly after low-temperature treatment, and the change was not significant [72]. However, in the cold, the MYC gene is activated rapidly in response to exogenous MeJA and resynthesizes a large amount of JAs in the plant body to protect against cold damage. In addition, studies in tomato (*L. esculentum*) [73], pomegranate (*Punica granatum*) [74], loquat (*Eribotrya japonica*) [75], mango (*Mangifera indica*) [76], and guava (*Psidium guajava*) [77] have shown that exogenous MeJA treatment can induce heat shock protein family transcription, increase antioxidant synthesis, reduce lipoxygenase activity, and increase plant resistance to cold damage (with 0 °C upper temperature). At present, there are relatively few reports on the JA signal involved in plant heat resistance, but JA is a key signal molecule for the formation of sesquiterpene induced by heat shock in eaglewood (*Aquilaria sinensis*) [1,2,78].

5.1.3. Effect of Drought on Jasmonic Acid Signal Changes

Many studies have found that JA signaling pathways are involved in drought stress. In *Arabidopsis thaliana* [79] and citrus (*Citrus paradisi × Poncirus trifoliate*) [80], it was found that the increase of endogenous JA content after drought stress was rapid and transient, and then gradually decreased to the basal level with the prolonged stress. On the other hand, the application of exogenous JA can also effectively alleviate the damage caused by drought to plants. Peanut (*Arachis hypogaea*) seedlings treated with MeJA showed enhanced drought resistance [81]. It was also observed that MeJA treatment can improve drought resistance in rice (*Oryza sativa*) [82], soybean (*Glycine max*) [83], and broccoli (*Brassica oleracea*) [84] by adjusting metabolism. Studies on broad bean (*Vicia faba*) and barley (*Hordeum vulgare*) have shown that MeJA may regulate stomatal closure through K^+ channels, thereby increasing the ability of plants to resist drought [85,86].

5.1.4. Effect of Salt on Jasmonic Acid Signal Changes

It was found that the endogenous JA content increased significantly after salt treatment in *Arabidopsis thaliana* [87], tomato (*L. esculentum*) [88], and potato (*Solanum tuberosum*) [89], among other plants. Moreover, JA content increased rapidly and persistently in salt-sensitive plants, while changes in JA content in salt-tolerant plants were not significant [89]. It has also been observed that exogenous JA can enhance the resistance of plants such as pepper (*Capsicum annuum*) [90] and verbena (*Rupestris riparia*) [91] to salt stress.

5.1.5. Effect of CO_2 Concentration on Jasmonic Acid Signal Changes

Ballhorn et al. found that high concentrations of CO_2 (500, 700, and 1000 µmol/mol) led to an increase in the release of JAs (MeJA and cis-JA) into the environment in lima bean (*P. lunatus*) [92]. The induced defense of plants infested with nematodes may be affected by elevated CO_2 concentrations, and CO_2-induced changes in plant resistance may result in genotypic-specific responses of plants to nematodes under elevated CO_2 conditions [93]. However, there are few studies on the JA signaling pathway in plants under the action of CO_2.

5.1.6. Effect of Ozone on Jasmonic Acid Signal Changes

The endogenous JA content of wild-type *Arabidopsis* increased significantly after ozone treatment. Experiments with the ozone-sensitive mutants *rcd1* (radical-induced cell death 1) [94] and *oji1* (ozone-sensitive and jasmonate-insensitive) [91,95] and the JA signal mutant *jar1* [96] have shown that exogenous MeJA can inhibit the spread of programmed cell death caused by ozone. Repression of the JA signaling pathway causes the plant to have a more intense response to ozone. Application of exogenous MeJA also resulted in reduced sensitivity of the hybrid poplar (*Populus maximowizii* × *P. trichocarpa*) [97] and tomato (*L. esculentum*) [96] to ozone. However, recent studies on cotton (*Gossypium hirsutum*) have shown that MeJA exhibits inhibition of ozone damage diffusion only at high concentrations (volume fraction: 685) of ozone, accompanied by antagonism with ethylene [98].

5.2. Gene Chip and Proteomics Studies on Jasmonic Acid Signal Changes

RNA-seq and proteomics studies further confirm the involvement of the JA signaling pathway in regulating the physiological processes of plants. Systematic biological approaches such as using gene chip and proteomics can study various physiological processes of plants at the whole genome level, and thus analyze the relationship between various metabolic pathways. Jung et al. identified 137 genes with altered expression levels in MeJA-treated *Arabidopsis* using gene chip technology. Among them, 74 genes were upregulated, including JAs biosynthesis genes, various defense genes (such as *pdf1.2*, encoding genes of myrosinase-binding protein), oxidative stress genes (oxidases, glutathione transfer), aging-related genes, cell wall modification-related genes, hormone metabolism-related (such as ACC oxidase, responsible for ethylene synthesis) genes, and genes involved in storage, signal transduction, and primary and secondary metabolism. The 63 downregulated genes included photosynthesis-related genes (Rubisco enzyme gene, chlorophyll protein gene, early photoinduced protein gene), cold regulatory genes, drought-responsive genes, defense-responsive genes, plant growth and development-related genes, cell wall-modification -related genes, and some other unknown functional genes, etc. [99]. Chen et al. compared the protein content of *Arabidopsis thaliana* before and after MeJA treatment by proteomics and found 186 differentially expressed proteins. These proteins are involved in plant photosynthesis, carbohydrate metabolism, hormone metabolism, secondary metabolism, product transport, stress and defense, and gene transcription [100].

6. Future Prospects

In the last decades, the JA signaling pathway has been studied extensively, but our knowledge about the role of JA signaling in response to different environmental stimuli is limited. Environmental signals usually result in a complex response network regulated by multiple signaling pathways. The mechanism of action of JA signaling in plant–environment interaction is still not clear. A series of signal transduction pathways related to JAs biosynthesis and transmission are well known, but it has not been systematically studied how different environmental signals are perceived by plants and initiate JAs synthesis. Due to the large number of receptors and kinases on the cell membrane, different biotic and abiotic signals stimulate the activation of different enzymes, accompanied by a series of complex reactions, such as calcium channel and potassium channel opening. Therefore, the perception of environmental signals has still a lot of research space. The studies on JAs receptors has made great progress, and the JA signal transduction pathway has also been established, but there are still many questions regarding the regulatory process which need to be answered.

Author Contributions: J.R., J.C. and M.Z. designed the review; J.R., J.Y., M.K. and Y.Z. access to information; J.R., K.Z. and M.K. revised the manuscript; M.K., W.W. and K.Z. designed the chart; J.R., Y.Z. and K.Z. wrote the paper.

Funding: Financial support was given by the National Key R&D Program of China "Intergovernmental International Science and Technology Innovation Cooperation/Hong Kong, Macao and Taiwan Science and Technology Innovation Cooperation" Key Project (No. 2017YFE0117600), the National Natural Science Foundation of China (No. 31660531 and 3180100249), and the Guizhou Science and Technology Support Project (No. 2018-2292).

Acknowledgments: We are very grateful to En-tang Tian (Guizhou University) for fruitful discussions on this work and Su-qin Zhang (National Wheat Improvement Center, Guizhou sub-center, University of Guizhou) for critical reading of the manuscript.

Conflicts of Interest: The authors declare no conflicts of interest.

References

1. Mur, L.A.; Kenton, P.; Atzorn, R.; Miersch, O.; Wasternack, C. The outcomes of concentration—Specific interactions between salicylate and jasmonate signaling include synergy, antagonism, and oxidative stress leading to cell death. *Plant Physiol.* **2006**, *140*, 249–262. [CrossRef] [PubMed]

2. Clarke, S.M.; Cristescu, S.M.; Miersch, O.; Harren, F.J.; Wasternack, C.; Mur, L.A. Jasmonates act with salicylic acid to confer basal thermotolerance in *Arabidopsis thaliana*. *New Phytol.* **2009**, *182*, 175–187. [CrossRef]

3. Wasternack, C.; Hause, B. Jasmonates: Biosynthesis, perception, signal transduction and action in plant stress response, growth and development. An update to the 2007 review in Annals of Botany. *Ann. Bot.* **2013**, *111*, 1021–1058. [CrossRef]

4. Campos, M.L.; Kang, J.H.; Howe, G.A. Jasmonate-triggered plant immunity. *J. Chem. Ecol.* **2014**, *40*, 657–675. [CrossRef] [PubMed]

5. Gupta, A.; Hisano, H.; Hojo, Y.; Matsuura, T.; Ikeda, Y.; Mori, I.C.; Kumar, M.S. Global profiling of phytohormone dynamics during combined drought and pathogen stress in *Arabidopsis thaliana* reveals ABA and JA as major regulators. *Sci. Rep-UK* **2017**, *7*, 4017. [CrossRef]

6. Pearce, G.; Strydom, D.; Johnson, S.; Ryan, C.A. A polypeptide from tomato leaves induces wound-inducible proteinase inhibitor proteins. *Science* **1991**, *253*, 895–897. [CrossRef]

7. Ryan, C.A.; Pearce, G. Systemins: A functionally defined family of peptide signals that regulate defensive genes in *Solanaceae* species. *Proc. Natl. Acad. Sci. USA* **2003**, *100*, 14577–14580. [CrossRef]

8. Li, C.; Liu, G.; Xu, C.; Lee, G.I.; Bauer, P.; Ling, H.Q.; Ganal, M.W.; Howe, G.A. The tomato suppressor of prosystemin-mediated responses2 gene encodes a fatty acid desaturase required for the biosynthesis of jasmonic acid and the production of a systemic wound signal for defense gene expression. *Plant Cell* **2003**, *15*, 1646–1661. [CrossRef]

9. Scheer, J.M.; Ryan, C.A. The systemin receptor SR160 from *Lycopersicon peruvianum* is a member of the LRR receptor kinase family. *Proc. Natl. Acad. Sci. USA* **2002**, *99*, 9585–9590. [CrossRef] [PubMed]

10. Stratmann, J.W.; Ryan, C.A. Myelin basic protein kinase activity in tomato leaves is induced systemically by wounding and increases in response to systemin and oligosaccharide elicitors. *Proc. Natl. Acad. Sci. USA* **1997**, *94*, 11085–11089. [CrossRef] [PubMed]

11. Narváez-Vásqueza, J.; Florin-Christensen, J.; Ryan, C.A. Positional specificity of a phospholipase A activity induced by wounding systemin oligosaccharide elicitors in tomato leaves. *Plant Cell* **1999**, *11*, 2249–2260. [CrossRef] [PubMed]

12. Yamaguchi, Y.; Huffaker, A.; Bryan, A.C.; Tax, F.E.; Ryan, C.A. PEPR2 is a second receptor for the Pep1 and Pep2 peptides and contributes to defense responses in Arabidopsis. *Plant Cell* **2010**, *22*, 508–522. [CrossRef]

13. Hind, S.; Malinowski, R.; Yalamanchili, R.; Stratmann, J.W. Tissue-type specific systemin perception and the elusive systemin receptor. *Plant Signal. Behav.* **2010**, *5*, 42–44. [CrossRef] [PubMed]

14. Yan, L.; Zhai, Q.; Wei, J.; Li, S.; Wang, B.; Huang, T.; Du, M.; Sun, J.; Kang, L.; Li, C.B.; et al. Role of tomato lipoxygenase D in wound-induced jasmonate biosynthesis and plant immunity to insect herbivores. *PLoS Genet.* **2013**, *9*, e1003964. [CrossRef] [PubMed]

15. Ellinger, D.; Sting, N.; Kubigsteltig, I.I.; Bals, T.; Juenger, M.; Pollmann, S.; Berger, S.; Schuenemann, D.; Mueller, M.J. DGL and DAD1 lipases are not essential for wound- and pathogen induced jasmonate biosynthesis: Redundant lipases contribute to jasmonate formation. *Plant Physiol.* **2010**, *153*, 114–127. [CrossRef] [PubMed]

16. Turner, J.G.; Ellis, C.; Devoto, A. The jasmonate signal pathway. *Plant Cell* **2002**, *14* (Suppl. 1), S153–S164. [CrossRef]

17. Chini, A.; Monte, I.; Zamarreño, A.M.; Hamberg, M.; Lassueur, S.; Reymond, P.; Weiss, S.; Stintzi, A.; Schaller, A.; Porzel, A.; et al. An OPR3-independent pathway uses 4,5-didehydrojasmonate for jasmonate synthesis. *Nat. Chem. Biol.* **2018**, *14*, 171–178. [CrossRef]

18. Ryan, C.A.; Moura, D.S. Systemic wound signaling in plants: A new perception. *Proc. Natl. Acad. Sci. USA* **2002**, *99*, 6519–6520. [CrossRef]

19. Truman, W.; Bennett, M.H.; Kubigsteltig, I.; Turnbull, C.; Grant, M. Arabidopsis systemic immunity uses conserved defense signaling pathways and is mediated by jasmonates. *Proc. Natl. Acad. Sci. USA* **2007**, *104*, 1075–1080. [CrossRef] [PubMed]

20. Heil, M.; Ton, J. Long-distance signaling in plant defense. *Trends Plant Sci.* **2008**, *13*, 264–272. [CrossRef]

21. Thorpe, M.R.; Ferrieri, A.P.; Herth, M.M.; Ferrieri, R.A. 11Cimaging: Methyl jasmonate moves in both phloem and xylem, promotes transport of jasmonate and of photoassimilate even after proton transport is decoupled. *Planta* **2007**, *226*, 541–551. [CrossRef]

22. Hause, B.; Stenzel, I.; Miersch, O.; Maucher, H.; Kramell, R.; Ziegler, J.; Wasternack, C. Tissue-specific oxylipin signature of tomato flowers: Allene oxide cyclase is highly expressed in distinct flower organs and vascular bundles. *Plant J.* **2000**, *24*, 113–126. [CrossRef] [PubMed]

23. Hause, B.; Hause, G.; Kutter, C.; Miersch, O.; Wasternack, C. Enzymes of jasmonate biosynthesis occur in tomato sieve elements. *Plant Cell Physiol.* **2003**, *44*, 643–648. [CrossRef] [PubMed]

24. Koo AJ, K.; Gao, X.; Jones, A.D.; Howe, G.A. A rapid wound signal activates systemic synthesis of bioactive jasmonates in Arabidopsis. *Plant J.* **2009**, *59*, 974–986. [CrossRef] [PubMed]

25. Larrieu, A.; Vernoux, T. Q&A: How does jasmonate signaling enable plants to adapt and survive? *BMC Boil.* **2016**, *14*, 79.

26. Nelson, C.E.; Walker-simmons, M.; Makus, D.; Zuroske, G.; Graham, J.; Ryan, C.A. Regulation of synthesis and accumulation of proteinase-inhibitors in leaves of wounded tomato plants. *ACS Symp. Ser.* **1983**, *208*, 103–122.

27. Malone, M. Rapid, long-distance signal transmission in higher plants. *Adv. Bot. Res.* **1996**, *22*, 163–228.

28. Farmer, E.E.; Ryan, C.A. Interplant communication: Airborne methyl jasmonate induces synthesis of proteinase inhibitors in plant leaves. *Proc. Natl. Acad. Sci. USA* **1990**, *87*, 7713–7716. [CrossRef]

29. Park, S.W.; Kaimoyo, E.; Kumar, D.; Mosher, S.; Klessig, D.F. Methyl salicylate is a critical mobile signal for plant systemic acquired resistance. *Science* **2007**, *318*, 113–116. [CrossRef]

30. Kost, C.; Heil, M. The defensive role of volatile emission and extra-floral nectar secretion for lima bean in nature. *J. Chem. Ecol.* **2008**, *34*, 2–13. [CrossRef]

31. Karban, R.; Baldwin, I.T.; Baxter, K.J.; Laue, G.; Felton, G.W. Communication between plants: Induced resistance in wild tobacco plants following clipping of neighboring sagebrush. *Oecologia* **2000**, *125*, 66–71. [CrossRef]

32. Li, Q.; Zheng, J.; Li, S.; Huang, G.; Skilling, S.J.; Wang, L.; Li, L.; Li, M.; Yuan, L.; Liu, P. Transporter-mediated nuclear entry of jasmonoyl-Isoleucine is essential for jasmonate signaling. *Mol. Plant.* **2017**, *10*, 695–708. [CrossRef] [PubMed]

33. Feys, B.; Benedetti, C.E.; Penford, C.N.; Turner, J.G. Arabidopsis mutants selected for resistance to the phytotoxin coronatine are male sterile, insensitive to methyl jasmonate, and resistant to a bacterial pathogen. *Plant Cell* **1994**, *6*, 751–759. [CrossRef]

34. Xie, D.X.; Feys, B.F.; James, S.; Nieto-Rostro, M.; Turner, J.G. COI1: An Arabidopsis gene required for jasmonate-regulated defense and fertility. *Science* **1998**, *280*, 1091–1094. [CrossRef] [PubMed]

35. Zhai, Q.; Zhang, X.; Wu, F.; Feng, H.; Deng, L.; Xu, L.; Zhang, M.; Wang, Q.; Li, C. Transcriptional mechanism of jasmonate receptor COI1-mediated delay of flowering time in Arabidopsis. *Plant Cell* **2015**, *27*, 2814–2828. [CrossRef] [PubMed]

36. Chini, A.; Fonseca, S.; Fernandez, G.; Adie, B.; Chico, J.; Lorenzo, O.; Garcia-Casado, G.; Lopez-Vidriero, I.; Lozano, F.; Ponce, M. The JAZ family of repressors is the missing link in jasmonate signalling. *Nature* **2007**, *448*, 666–671. [CrossRef] [PubMed]

37. Thines, B.; Katsir, L.; Melotto, M.; Niu, Y.; Mandaokar, A.; Liu, G.; Nomura, K.; He, S.Y.; Howe, G.A.; Browse, J. JAZ repressor proteins are targets of the SCF^{COI1} complex during jasmonate signaling. *Nature* **2007**, *448*, 661–665. [CrossRef] [PubMed]

38. Yan, Y.; Stolz, S.; Chételat, A.; Reymond, P.; Pagni, M.; Dubugnon, L.; Farmer, E.E. A downstream mediator in the growth repression limb of the jasmonate pathway. *Plant Cell* **2007**, *19*, 2470–2483. [CrossRef] [PubMed]

39. Melotto, M.; Mecey, C.; Niu, Y.; Chung, H.S.; Katsir, L.; Yao, J.; Zeng, W.; Thines, B.; Staswick, P.; Browse, J.; et al. A critical role of two positively charged amino acids in the Jas motif of Arabidopsis JAZ proteins in mediating coronatine- and jasmonoyl isoleucine-dependent interactions with the COI1 F-box protein. *Plant J.* **2008**, *55*, 979–988. [CrossRef]

40. Chini, A.; Fonseca, S.; Chico, J.M.; Fernández-Calvo, P.; Solano, R. The ZIM domain mediates homo- and heteromeric interactions between Arabidopsis JAZ proteins. *Plant J.* **2009**, *59*, 77–87. [CrossRef]

41. Sheard, L.B.; Tan, X.; Mao, H.; Withers, J.; Ben-Nissan, G.; Hinds, T.R.; Kobayashi, Y.; Hsu, F.F.; Sharon, M.; Browse, J.; et al. Jasmonate perception by inositolphosphate-potentiated COI1-JAZ co-receptor. *Nature* **2010**, *468*, 400–405. [CrossRef]

42. Li, Y.; Qin, L.; Zhao, J.; Muhammad, T.; Cao, H.; Li, H.; Zhang, Y.; Liang, Y. SlMAPK3 enhances tolerance to tomato yellow leaf curl virus (TYLCV) by regulating salicylic acid and jasmonic acid signaling in tomato (*Solanum lycopersicum*). *PLoS ONE* **2017**, *12*, e0172466. [CrossRef]

43. Kenton, P.; Mur LA, J.; Draper, J. A requirement for calcium and protein phosphatase in the jasmonate-induced increase in tobacco leaf acid phosphatase specific activity. *J. Exp. Bot.* **1999**, *50*, 1331–1341. [CrossRef]

44. Santner, A.; Estelle, M. Recent advances and emerging trends in plant hormone signaling. *Nature* **2009**, *459*, 1071–1078. [CrossRef]

45. Fernández-Calvo, P.; Chini, A.; Fernández-Barbero, G.; Chico, J.M.; Gimenez-Ibanez, S.; Geerinck, J.; Eeckhout, D.; Schweizerd, F.; Godoy, M.; Franco-Zorrillae, J.M.; et al. The Arabidopsis bHLH transcription factors MYC3 and MYC4 are targets of JAZ repressors and act additively with MYC2 in the activation of jasmonate responses. *Plant Cell* **2011**, *23*, 701–715. [CrossRef]

46. Schmiesing, A.; Emonet, A.; Gouhier-Darimont, C.; Reymond, P. Arabidopsis MYC transcription factors are the target of hormonal salicylic acid/jasmonic acid cross talk in response to *Pieris brassicae* egg extract. *Plant physiol.* **2016**, *170*, 2432–2443. [CrossRef]

47. Figueroa, P.; Browse, J. Male sterility in Arabidopsis induced by overexpression of a MYC5-SRDX chimeric repressor. *Plant J.* **2015**, *81*, 849–860. [CrossRef]

48. Sasaki-Sekimoto, Y.; Jikumaru, Y.; Obayashi, T.; Saito, H.; Masuda, S.; Kamiya, Y.; Ohta, H.; Shirasu, K. Basic helix-loop-helix transcription factors JASMONATE-ASSOCIATED MYC2-LIKE1 (JAM1), JAM2, and JAM3 are negative regulators of jasmonate responses in Arabidopsis. *Plant Physiol.* **2013**, *163*, 291–304. [CrossRef]

49. Song, S.; Qi, T.; Huang, H.; Ren, Q.; Wu, D.; Chang, C.; Peng, W.; Liu, Y.; Peng, J.; Xie, D. The jasmonate-ZIM domain proteins interact with the R2R3-MYB transcription factors MYB21 and MYB24 to affect jasmonate-regulated stamen development in Arabidopsis. *Plant Cell* **2011**, *23*, 1000–1013. [CrossRef]

50. Qi, T.; Song, S.; Ren, Q.; Wu, D.; Huang, H.; Chen, Y.; Fan, M.; Peng, W.; Ren, C.; Xie, D. The jasmonate-ZIM-domain proteins interact with the WD-Repeat/bHLH/MYB complexes to regulate jasmonate-mediated anthocyanin accumulation and trichome initiation in *Arabidopsis thaliana*. *Plant Cell* **2011**, *23*, 1795–1814. [CrossRef]

51. Zhou, M.; Sun, Z.; Ding, M.; Logacheva, M.D.; Kreft, I.; Wang, D.; Yan, M.; Shao, J.; Tang, Y.; Wu, Y.; et al. FtSAD2 and FtJAZ1 regulate the activity of FtMYB11 transcription repressor in phenylpropanoid pathway of *Fagopyrum tataricum*. *New Phytol.* **2017**, *216*, 814–828. [CrossRef] [PubMed]

52. Zhang, K.; Logacheva, M.D.; Meng, Y.; Hu, J.; Wan, D.; Li, L.; Dagmar, J.; Wang, Z.; Georgiev, M.I.; Yu, Z.; et al. Jasmonate-responsive MYB factors spatially repress rutin biosynthesis in *Fagopyrum tataricum*. *J. Exp. Bot.* **2018**, *69*, 1955–1966. [CrossRef] [PubMed]

53. Delessert, C.; Kazan, K.; Wilson, I.W.; van der Straeten, D.; Manners, J.; Dennis, E.S.; Dolferus, R. The transcription factor ATAF2 represses the expression of pathogenesis-related genes in Arabidopsis. *Plant J.* **2005**, *43*, 745–757. [CrossRef]

54. Nuruzzaman, M.; Sharoni, A.M.; Kikuchi, S. Roles of NAC transcription factors in the regulation of biotic and abiotic stress responses in plants. *Front. Microbiol.* **2013**, *4*, 248. [CrossRef] [PubMed]

55. Bu, Q.; Jiang, H.; Li, C.B.; Zhai, J.; Wu, X.; Sun, J.; Xie, Q.; Li, C. Role of the *Arabidopsis thaliana* NAC transcription factors ANAC019 and ANAC055 in regulating jasmonic acid-signaled defense responses. *Cell Res.* **2008**, *18*, 756–767. [CrossRef]

56. van der Fits, L.; Memelink, J. ORCA3, a jasmonate-responsive transcriptional regulator of plant primary and secondary metabolism. *Science* **2001**, *289*, 295–297. [CrossRef]

57. Lorenzo, O.; Piqueras, R.; Sánchez-Serrano, J.J.; Solano, R. ETHYLENE RESPONSE FACTOR1 integrates signals from ethylene and jasmonate pathways in plant defense. *Plant Cell* **2003**, *15*, 165–178. [CrossRef]

58. Pré, M.; Atallah, M.; Champion, A.; de Vos, M.; Pieterse, C.M.J.; Memelink, J. The AP2 /ERF domain transcription factor ORA59 integrates jasmonic acid and ethylene signals in plant defense. *Plant Physiol.* **2008**, *147*, 1347–1357. [CrossRef]

59. Li, J.; Zhang, K.; Meng, Y.; Hu, J.; Ding, M.; Bian, J.; Yan, M.; Han, J.; Zhou, M. Jasmonic acid/ethylene signaling coordinates hydroxycinnamic acid amides biosynthesis through ORA59 transcription factor. *Plant J.* **2018**, *95*, 444–457. [CrossRef]

60. Chen, H.Y.; Hsieh, E.J.; Cheng, M.C.; Chen, C.Y.; Hwang, S.Y.; Lin, T.P. ORA47 (octade-canoid-responsive AP2/ERF-domain transcription factor 47) regulates jasmonic acid and abscisic acid biosynthesis and signaling through binding to a novel *cis*-element. *New Phytol.* **2016**, *211*, 599–613. [CrossRef] [PubMed]

61. Fujimoto, S.Y.; Ohta, M.; Usui, A.; Shinshi, H.; Ohme-Takagi, M. Arabidopsis ethylene-responsive element binding factors act as transcriptional activators or repressors of GCC box–mediated gene expression. *Plant Cell* **2000**, *12*, 393–404. [CrossRef]

62. Li, J.; Zhong, R.; Palva, E.T. WRKY70 and its homolog WRKY54 negatively modulate the cell wall-associated defenses to necrotrophic pathogens in Arabidopsis. *PLoS ONE* **2017**, *12*, e0183731. [CrossRef]

63. Kloth, K.J.; Wiegers, G.L.; Busscher-Lange, J.; van Haarst, J.C.; Kruijer, M.; Bouwmeester, H.J.; Dicke, M.; Jongsma, M.A. AtWRKY22 promotes susceptibility to aphids and modulates salicylic acid and jasmonic acid signalling. *J. Exp. Bot.* **2016**, *67*, 3383–3396. [CrossRef]

64. Gao, Q.M.; Venugopal, S.; Navarre, D.; Kachroo, A. Low oleic acid-derived repression of jasmonic acid-inducible defense responses requires the WRKY50 and WRKY51 proteins. *Plant Physiol.* **2011**, *155*, 464–476. [CrossRef]

65. Jiang, Y.; Liang, G.; Yang, S.; Yu, D. Arabidopsis WRKY57 functions as a node of convergence for jasmonic acid–and auxin-mediated signaling in jasmonic acid–induced leaf senescence. *Plant Cell* **2014**, *26*, 230–245. [CrossRef]

66. Jiang, Y.; Guo, L.; Liu, R.; Jiao, B.; Ling, Z.; Luo, K. Overexpression of poplar PtrWRKY89 in transgenic Arabidopsis leads to a reduction of disease resistance by regulating defense-related genes in salicylate-and jasmonate-dependent signaling. *PLoS ONE* **2016**, *11*, e0149137. [CrossRef]

67. Skibbe, M.; Qu, N.; Galis, I.; Balawin, I.T. Induced plant defenses in the natural environment: *Nicotiana attenuala* WRKY3 and WRKY6 coordinate responses to herbivory. *Plant Cell* **2008**, *20*, 1984–2000. [CrossRef]

68. Mewis, I.; Schreiner, M.; Nguyen, C.N.; Krumbein, A.; Ulrichs, C.; Lohse, M.; Zrenner, R. UV-B irradiation changes specifically the secondary metabolite profile in broccoli sprouts: Induced signaling overlaps with defense response to biotic stressors. *Plant Cell Physiol.* **2012**, *53*, 1546–1560. [CrossRef]

69. Cerrudo, I.; Keller, M.M.; Cargnel, M.D.; Demkura, P.V.; Wit, M.; Patitucci, M.S.; Pierik, R.; Pieterse CM, J.; Ballaré, C.L. Low Red/Far-Red ratios reduce Arabidopsis resistance to Botrytis cinerea and jasmonate responses via a COI1-JAZ10-dependent, salicylic acid independent mechanism. *Plant Physiol.* **2012**, *158*, 2042–2052. [CrossRef]

70. Gupta, N.; Prasad, V.B.R.; Chattopadhyay, S. LeMYC2 acts as a negative regulator of blue light mediated photomorphogenic growth, and promotes the growth of adult tomato plants. *BMC Plant Boil.* **2014**, *14*, 38. [CrossRef]

71. Svyatyna, K.; Riemann, M. Light-dependent regulation of the jasmonate pathway. *Protoplasma* **2012**, *249*, 137–145. [CrossRef]

72. Zhao, M.L.; Wang, J.N.; Shan, W.; Fan, J.G.; Kuang, J.F.; Wu, K.Q.; Li, X.P.; Chen, W.X.; He, F.Y.; Chen, J.Y.; et al. Induction of jasmonate signaling regulators MaMYC2s and their physical interactions with MaICE1 in methyl jasmonate-induced chilling tolerance in banana fruit. *Plant Cell Environ.* **2012**, *36*, 30–51. [CrossRef]

73. Li, S.; Yang, Y.; Zhang, Q.; Liu, N.; Xu, Q.; Hu, L. Differential physiological and metabolic response to low temperature in two zoysiagrass genotypes native to high and low latitude. *PLoS ONE* **2018**, *13*, e0198885. [CrossRef]

74. Sayyari, M.; Babalar, M.; Kalantari, S.; Martínez-Romero, D.; Guillén, F.; Serrano, M.; Valero, D. Vapour treatments with methyl salicylate or methyl jasmonate alleviated chilling injury and enhanced antioxidant potential during postharvest storage of pomegranates. *Food Chem.* **2011**, *124*, 964–970. [CrossRef]

75. Cao, S.F.; Zheng, Y.H.; Wang, K.T.; Jin, P.; Rui, H.J. Methyl jasmonate reduces chilling injury and enhances antioxidant enzyme activity in postharvest loquat fruit. *Food Chem.* **2009**, *115*, 1458–1463. [CrossRef]

76. González-Aguilar, G.A.; Fortiz, J.; Cruz, R.; Baez, R.; Wang, C.Y. Methyl jasmonate reduces chilling injury and maintains postharvest quality of mango fruit. *J. Agric. Food Chem.* **2000**, *48*, 515–519. [CrossRef] [PubMed]

77. González-Aguilar, G.A.; Tiznado-Hernández, M.E.; Zavaleta-Gatica, R.; Martínez-Téllez, M.A. Methyl jasmonate treatments reduce chilling injury and activate the defense response of guava fruits. *Biochem. Bioph. Res. Commun.* **2003**, *313*, 694–701.

78. Xu, Y.H.; Liao, Y.C.; Zhang, Z.; Liu, J.; Sun, P.W.; Gao, Z.H.; Sui, C.; Wei, J.H. Jasmonic acid is a crucial signal transducer in heat shock induced sesquiterpene formation in *Aquilaria sinensis*. *Sci. Rep-UK* **2016**, *6*, 21843. [CrossRef]

79. Balbi, V.; Devoto, A. Jasmonate signalling network in Arabidopsis thaliana: Crucial regulatory nodes and new physiological scenarios. *New Phytologist.* **2008**, *177*, 301–318. [CrossRef]

80. Ollas, C.; Hernando, B.; Arbona, V.; Gómez-Cadenas, A. Jasmonic acid transient accumulation is needed for abscisic acid increase in citrus roots under drought stress conditions. *Physiol. Plantarum.* **2013**, *147*, 296–306. [CrossRef] [PubMed]

81. Todaka, D.; Shinozaki, K.; Yamaguchi-Shinozaki, K. Recent advances in the dissection of drought-stress regulatory networks and strategies for development of drought-tolerant transgenic rice plants. *Front. Plant Sci.* **2015**, *6*, 84. [CrossRef] [PubMed]

82. Fu, J.; Wu, H.; Ma, S.; Xiang, D.; Liu, R.; Xiong, L. OsJAZ1 attenuates drought resistance by regulating JA and ABA signaling in rice. *Front. Plant Sci.* **2017**, *8*, 2108. [CrossRef]

83. Mohamed, H.I.; Latif, H.H. Improvement of drought tolerance of soybean plants by using methyl jasmonate. *Physiol. Mol. Biol. Plants* **2017**, *23*, 545–556. [CrossRef]

84. Wu, H.L.; Wu, X.L.; Li, Z.H.; Duan, L.S.; Zhang, M.C. Physiological evaluation of drought stress tolerance and recovery in cauliflower (*Brassica oleracea* L.) seedlings treated with methyl jasmonate and coronatine. *J. Plant Growth Regul.* **2012**, *31*, 113–123. [CrossRef]

85. Evans, N.H. Modulation of guard cell plasma membrane potassium currents by methyl jasmonate. *Plant Physiol.* **2003**, *131*, 8–11. [CrossRef]

86. Horton, R.F. Methyl jasmonate and transpiration in barley. *Plant Physiol.* **1991**, *96*, 1376–1378. [CrossRef] [PubMed]

87. Ellouzi, H.; Hamed, K.B.; Cela, J.; Müller, M.; Abdelly, C.; Munné-Bosch, S. Increased sensitivity to salt stress in tocopherol-deficient Arabidopsis mutants growing in a hydroponic system. *Plant Signal Behav.* **2013**, *8*, e23136. [CrossRef]

88. Pedranzani, H.; Racagni, G.; Alemano, S.; Miersch, O.; Ramírez, I.; Pea-Cortés, H.; Taleisnik, E.; Machado-Domenech, E.; Abdala, G. Salt tolerant tomato plants show increased levels of jasmonic acid. *J. Plant Growth Regul.* **2003**, *41*, 149–158. [CrossRef]

89. De Domenico, S.; Taurino, M.; Gallo, A.; Poltronieri, P.; Pastor, V.; Flors, V.; Santino, A. Oxylipin dynamics in *Medicago truncatula* in response to salt and wounding stresses. *Physiol. Plant.* **2019**, *165*, 198–208. [CrossRef]

90. Molina, C.; Zaman-Allah, M.; Khan, F.; Fatnassi, N.; Horres, R.; Rotter, B.; Steinhauer, D.; Amenc, L.; Drevon, J.J.; Winter, P.; et al. The salt-responsive transcriptome of chickpea roots and nodules via deepSuperSAGE. *BMC Plant Biol.* **2011**, *11*, 31. [CrossRef]

91. Ismail, A.; Riemann, M.; Nick, P. The jasmonate pathway mediates salt tolerance in grapevines. *J. Exp. Bot.* **2012**, *63*, 2127–2139. [CrossRef]

92. Ballhorn, D.J.; Reisdorff, C.; Pfanz, H. Quantitative effects of enhanced CO_2 on jasmonic acid induced plant volatiles of lima bean (*Phaseolus lunatus* L.). *J. Appl. Bot. Food Qual.* **2011**, *84*, 65–71.

93. Sun, Y.; Yin, J.; Cao, H.; Li, C.; Kang, L.; Ge, F. Elevated CO_2 influences nematode-induced defense responses of tomato genotypes differing in the JA pathway. *PLoS ONE* **2011**, *6*, e19751. [CrossRef]

94. Overmyer, K.; Tuominen, H.; Kettunen, R.; Betz, C.; Langebartels, C.; Sandermann, H.J.; Kangasjärvi, J. Ozone-sensitive Arabidopsis rcd1 mutant reveals opposite roles for ethylene and jasmonate signaling pathways in regulating superoxide-dependent cell death. *Plant Cell* **2000**, *12*, 1849–1862. [CrossRef]

95. Kanna, M.; Tamaoki, M.; Kubo, A.; Nakajima, N.; Rakwal, R.; Agrawal, G.K.; Tamogami, S.; Ioki, M.; Ogawa, D.; Saji, H.; et al. Isolation of an ozone-sensitive and jasmonate-semi-insensitive Arabidopsis mutant (*oji1*). *Plant Cell Physiol.* **2003**, *44*, 1301–1310. [CrossRef]

96. Rao, M.V.; Lee, H.; Creelman, R.A.; Mullet, J.E.; Davis, K.R. Jasmonic acid signaling modulates ozone-induced hypersensitive cell death. *Plant Cell* **2000**, *12*, 1633–1646. [CrossRef]

97. Koch, J.R.; Creelman, R.A.; Eshita, S.M.; Seskar, M.; Mullet, J.E.; Davis, K.R. Ozone sensitivity in hybrid poplar correlates with insensitivity to both salicylic acid and jasmonic acid. The role of programmed cell death in lesion formation. *Plant Physiol.* **2000**, *123*, 487–496. [CrossRef]

98. Grantz, D.A.; Vu, H.B. Root and shoot gas exchange respond additively to moderate ozone and methyl jasmonate without induction of ethylene: Ethylene is induced at higher O3 concentrations. *J. Exp. Bot.* **2012**, *63*, 4303–4313. [CrossRef]

99. Devoto, A.; Ellis, C.; Magusin, A.; Chang, H.S.; Chilcott, C.; Zhu, T.; Turnter, J.G. Expression profiling reveals COI1 to be a key regulator of genes involved in wound- and methyl jasmonate induced secondary metabolism, defense and hormone interactions. *Plant Mol. Biol.* **2005**, *58*, 497–513. [CrossRef]

100. Chen Chen, Y.Z.; Pang, Q.Y.; Dai, S.J.; Wang, Y.; Chen, S.X.; Yan, X.F. Proteomic identification of differentially expressed proteins in Arabidopsis in response to methyl jasmonate. *Plant Physiol.* **2011**, *168*, 995–1008. [CrossRef]

© 2019 by the authors. Licensee MDPI, Basel, Switzerland. This article is an open access article distributed under the terms and conditions of the Creative Commons Attribution (CC BY) license (http://creativecommons.org/licenses/by/4.0/).

International Journal of
Molecular Sciences

Review

Novel Crosstalks between Circadian Clock and Jasmonic Acid Pathway Finely Coordinate the Tradeoff among Plant Growth, Senescence and Defense

Yuanyuan Zhang [1], Cunpei Bo [1,2] and Lei Wang [1,2,*]

[1] Key Laboratory of Plant Molecular Physiology, CAS Center for Excellence in Molecular Plant Sciences, Institute of Botany, Chinese Academy of Sciences, Beijing 100093, China; zhangyy@ibcas.ac.cn (Y.Z.); bcp1984@163.com (C.B.)
[2] University of Chinese Academy of Sciences, Beijing 100049, China
* Correspondence: wanglei@ibcas.ac.cn

Received: 4 September 2019; Accepted: 17 October 2019; Published: 23 October 2019

Abstract: Circadian clock not only functions as a cellular time-keeping mechanism, but also acts as a master regulator to coordinate the tradeoff between plant growth and defense in higher plants by timing a few kinds of phytohormone biosynthesis and signaling, including jasmonic acid (JA). Notably, circadian clock and JA pathway have recently been shown to intertwine with each other to ensure and optimize the plant fitness in an ever-changing environment. It has clearly demonstrated that there are multiple crosstalk pathways between circadian clock and JA at both transcriptional and post-transcriptional levels. In this scenario, circadian clock temporally modulates JA-mediated plant development events, herbivory resistance and susceptibility to pathogen. By contrast, the JA signaling regulates clock activity in a feedback manner. In this review, we summarized the cross networks between circadian clock and JA pathway at both transcriptional and post-transcriptional levels. We proposed that the novel crosstalks between circadian clock and JA pathway not only benefit for the understanding the JA-associated circadian outputs including leaf senescence, biotic, and abiotic defenses, but also put timing as a new key factor to investigate JA pathway in the future.

Keywords: circadian clock; jasmonic acid; crosstalk

---

1. Introduction

Circadian clock, an internal timekeeping mechanism, regulates plant growth and development by synchronizing the internal biological and physiological events with the external daily light-dark cycle, thus to enhance fitness of plants [1,2]. The circadian clock molecular system is majorly composed of three parts, namely the input pathways, core oscillator, and output pathways. The input pathways can perceive and recognize the rhythmic environmental cues, then transfer the external timing information to core oscillator through entraining mechanism. Conceptually, the self-sustained central oscillator is based on a series of transcriptional-translational feedback loops. The central loop of core oscillator is formed by the reciprocally repression between TIMING OF CAB EXPRESSION 1 (TOC1) which is the founding member of the PSEUDO-RESPONSE REGULATOR (PRR) family, and CIRCADIAN CLOCK-ASSOCIATED 1 (CCA1)/LATE ELONGATED HYPOCOTYL (LHY), two MYB-domain containing transcriptional repressors, with the expression peak at dawn [3,4]. In the morning loop, CCA1/LHY can repress the expression of *PRR7* and *PRR9*, by directly binding to the evening element (EE) within their respective promoters [4]. Reciprocally, PRR9/7/5 proteins also sequentially suppress the expression of *CCA1/LHY* from dawn to dusk [5,6]. In the evening, another important component is evening complex (EC), composed by EARLY FLOWERING 3 (ELF3),

EARLY FLOWERING 4 (ELF4), and LUX ARRHYTHMO (LUX), which is able to act as transcriptional repression complex. ELF3 and ELF4 proteins localize in the nucleus and do not contain any of the identified functional domains so far [7,8]. LUX is a GARP transcription factor with a single MYB domain [9]. EC confers the nighttime repression to the clock by repressing the expression of *TOC1*, *GIGANTEA* (*GI*), and *PRR9* [9–11]. Loss of function any of the individual EC components will lead to circadian arrhythmia, indicating that EC plays a crucial role in maintaining the proper circadian clock activity [1,9,10]. F-box protein ZEITLUPE (ZTL), has been identified as a blue light receptor, containing light, oxygen, and voltage (LOV) domain at its N-terminus and tandem KELCH domain at its C-terminus, respectively [12]. ZTL plays an essential role in mediating the degradation of TOC1 and PRR5 at post-translation level [13,14]. Intriguingly, GI serves as a chaperone and interacts directly with both HEAT SHOCK PROTEIN (HSP90) and ZTL to form a ternary complex, thus, specifically facilitates the maturation of ZTL [15]. Additionally, GI recruits the deubiquitylases, UBP12 and UBP13, to regulate the accumulation of ZTL photoreceptor complex [16] (Figure 1). Circadian output pathways represent a plethora of downstream events regulated by, including the temporal regulation on plant growth and development, timing biotic and abiotic stresses, and modulation on multiple phytohormone signaling pathways. For instance, EC directly binds and represses the expression of PHYTOCHROME INTERACTING FACTOR 4 (*PIF4*) and *PIF5* to gate the hypocotyls growth in late night [10]. In the other hand, independent of EC, ELF3 alone can also regulate hypocotyl elongation by physically interacting with PIF4 protein to inhibit its transcriptional activity [17]. Importantly, circadian clock regulates biological processes mediated by hormones through affecting hormone biosynthesis, signaling, and response pathways, such as the defense hormones, salicylic acid (SA) and jasmonic acid (JA) [18,19].

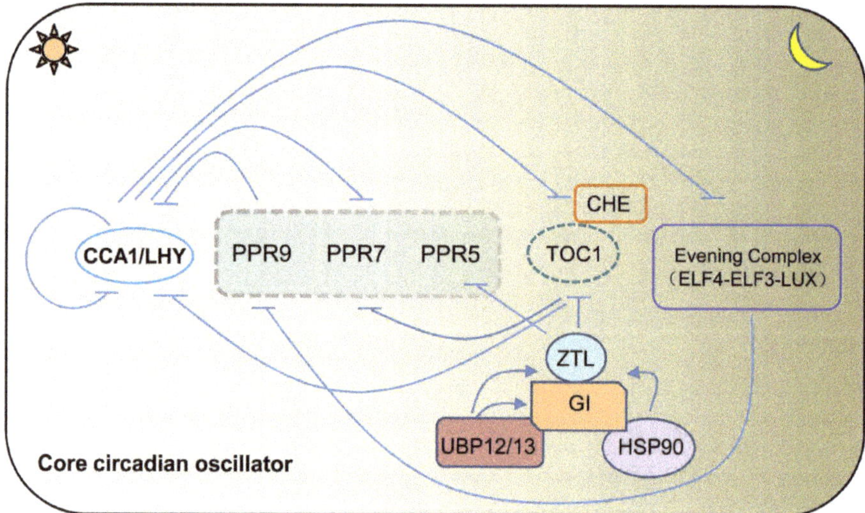

Figure 1. A simplified model for the circadian clock in *Arabidopsis thaliana*. Morning expressed CIRCADIAN CLOCK-ASSOCIATED1 (CCA1) and LATE ELONGATED HYPOCOTYL (LHY) repress the expression of all of the *PRR* family members and evening complex (EC). All the PSEUDO-RESPONSE REGULATOR (PRRs) reciprocally repress the expression of *CCA1* and *LHY*. EC is composed by EARLY FLOWERING 4 (ELF4), EARLY FLOWERING 3 (ELF3), and LUX ARRHYTHMO (LUX), and acts as a negative regulator of *PRR9*. GIGANTEA (GI) acts as a co-chaperone, recruiting HSP90 for the maturation of the ZEITLUPE (ZTL) protein. Additionally, GI recruits the deubiquitylases, UBP12 and UBP13, to regulate the accumulation of ZTL photoreceptor complex.

JA has been well recognized as a plant defense related hormone, which mainly regulates plant response to biotic stresses, including herbivore and pathogen attack. JA also plays crucial roles in mediating various biological events, such as photomorphogenesis, root growth, leaf senescence, wounding response, regeneration, abiotic stress responses, herbivory, and pathogen infections [20–25]. The biosynthesis of JA has been well characterized [26–28]. In brief, bioactive JA, (+)-7-*iso*-JA-Ile (JA-Ile), is generated from a trienoic fatty acid through the octadecanoid pathways (Figure 2). JA metabolism pathways convert JA into active and inactive compounds (Figure 2). The JA perception shares canonical ubiquitin-proteasome system with other hormones, such as gibberellin (GA) and auxin. CORONATINE INSENSITIVE 1 (COI1), a F-box protein, acts as JA receptor, which can bind bioactive JA and trigger the formation of receptor complex COI1-JA-Ile-JAZ, to promote the ubiquitination and degradation of JASMONATE ZIM DOMAIN (JAZ) proteins [29–32]. JAZ family consists of 13 members in *Arabidopsis*, and most of them possess two conserved domains, namely Zn-finger protein expressed in inflorescence meristem (ZIM) and Jas domains [33]. ZIM domain is responsible to mediate the interaction with NOVEL INTERACTOR OF JAZ (NINJA) or dimerization of JAZ proteins themselves, while Jas domain facilitates its interaction with COI1 and other transcription factors [25,34,35]. To date, lots of JAZ targets have been identified, including MYC, MYB, NAC, ERF, and WRKY family transcriptional factors, which mediate various downstream JA responses. MYC2, a basic-helix-loop-helix (bHLH) transcription factor, has been considered as a master downstream regulator of JA signaling pathway. A recent study showed that MEDIATOR 25 (MED25), a subunit of the mediator coactivator complex, could bridge COI1 to RNA Polymerase II and chromatin, thus, to trigger JA signaling. MED25 physically interacts with COI1 and histone acetyltransferase 1 (HAC1), and cooperatively mediates the histone (H) 3 lysine (K) 9 acetylation (H3K9ac) modification within the promoters of MYC2 target genes [36] (Figure 2).

MYC2, together with MYC3 and MYC4, belongs to the basic-helix-loop-helix IIIe transcription factor family. They are direct targets of JAZ repressors to play critical roles in mediating various aspects of the JA response in *Arabidopsis* [37]. They redundantly regulate the activation of JA-induced leaf senescence, by binding *SENESCENCE-ASSOCIATED GENE 29* (*SAG29*)promoter thus to activate its expression [38]. Interestingly, members of bHLH subgroup IIId transcriptional factors including bHLH03, bHLH13, bHLH14, and bHLH17, can inhibit the functions of MYC2, MYC3, and MYC4 in regulating JA-induced leaf senescence [38]. In addition, MYC2, MYC3, and MYC4 can also directly bind and promote the transcription of *PHEOPHORBIDE A OXYGENASE* (*PAO*), *NON-YELLOWING 1* (*NYE1*), and *NON-YELLOW COLORING 1* (*NYC1*), which are associated with chlorophyll degradation during leaf senescence [39]. Altogether, these evidences demonstrated that the JA signaling pathway was involved in JA-mediated senescence. Furthermore, it is also reported that exogenous JA treatment repressed flowering time by inhibiting *FT* expression, partially through MYC2/3/4 [40]. Recently, it has been implicated that both JA homeostasis and signaling pathway are regulated by circadian clock, while JA is capable of regulating circadian speed in a feedback manner. Here we summarize the emerging crosstalks between circadian clock and JA pathway, and list out the perspective for future investigation on their crosstalk network, which might shed light on the tradeoff among plant growth, development, and defense to optimize plant growth and reproductive behavior.

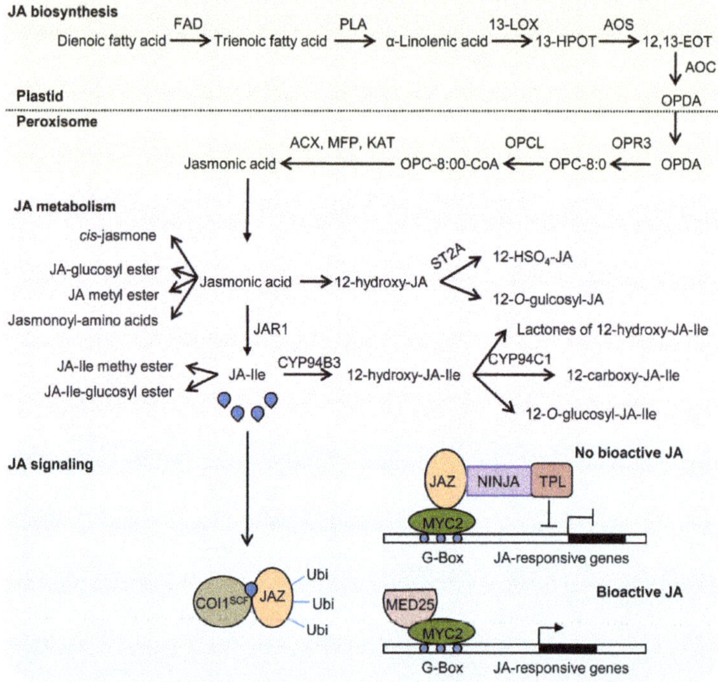

Figure 2. A model for jasmonic acid (JA) biosynthesis, metabolism and signaling pathways. JA-Ile is generated from trienoic fatty acid through the octadecanoid pathways. JA metabolism pathways can convert JA into active or inactive compounds. Coronatine insensitive1 (COI1), a F-box protein, acts as JA receptor, which can bind bioactive JA to trigger the formation of receptor complex COI1-JA-Ile-JAZ, hence to promote the ubiquitination and degradation of jasmonate zim domain (JAZ) proteins, then resulting in the release of downstream transcription factors, such as MYC2, and activation of JA responsive genes. ST2a, 12-OH-JA sulfotransferase; FAD, fatty acid desaturase; PLA, phospholipase A1; 13-LOX, 13-lipoxygenase; 13-HPOT, 13-hydroperoxyoctadecatrienoic acid; AOS, allene oxide synthase; 12,13-EOT, 12,13(S)-epoxyoctadecatrienoic acid; AOC, allene oxide cyclase; OPDA, (9S,13S)-12-oxo-phytodienoic acid; OPR, OPDA reductase. OPC-8:0, 3-oxo-2(cis-2′-pentenyl)-cyclopentane-1-octanoic acid; OPCL, OPC-8:0 CoA ligase; ACX, acyl-CoA oxidase; KAT, 3-ketoacyl-CoA thiolase; MFP, multifunctional protein; JAR1, jasmonate resistant 1; JA-Ile, jasmonoyl-L-isoleucine.

2. Biosynthesis and Metabolism of JA are Regulated by Circadian Clock

Transcriptomic studies have shown many of the defense-associated genes are regulated by circadian clock [41–44]. Consistently, the accumulation of JA content displayed a well rhythmic oscillation pattern with the peak at middle of the subjective day time and the trough level at around the middle night, indicating that JA biosynthesis and homoeostasis might be regulated by circadian clock [45]. As expected, the transcriptional profile of *SULFOTRANSFERASE 2A* (*ST2A*) which encodes a sulfotransferase family protein to involve the metabolism of JA, is significantly up-regulated at the end of dark phase under short day, and controlled by circadian clock [46]. Moreover, transcriptomic profiling analysis displayed that *LIPOXYGENASE 3* (*LOX3*) and *LIPOXYGENASE 4* (*LOX4*), encoding two 13-lipoxygenases which directly catalyze the biosynthesis of JA, were significantly up-regulated in *lux arrhythmo* (*lux*) mutant, which further indicated the biosynthesis of JA may be regulated by Evening

Complex [47]. Collectively, both JA biosynthesis and metabolism genes are controlled by circadian clock with peak at specific time of the day, hence to cause the rhythmic JA accumulation pattern.

3. Circadian Clock Regulates JA-Mediated Plant Development Events

In animals, the circadian clock has been reported to be tightly associated with aging process, and the dysfunction or disruption of circadian clock will dramatically accelerate the aging process. Nevertheless, whether the circadian clock regulates the aging or senescence process in higher plants is still largely unknown. Intriguingly, circadian stress from the changed regime of light-dark duration results in the lesser transcript levels of *CCA1* and *LHY*. Strikingly, circadian stress, which changed the regime of light-dark duration, also causes dramatically JA-dependent cell death process in cytokinin deficient plants and clock related mutants [48]. Very recently, transcriptomic profiling analysis revealed that evening complex is involved in the regulation of JA signaling and response by timing *MYC2* transcription (Figure 3). Meanwhile, the JA content is decreased in EC mutant examined at midday, the peak time for JA accumulation, which may result from the feedback regulation of activated JA signaling pathway [47]. Time for coffee (TIC), a component of circadian clock, functions as a negative regulator of JA signaling pathway, and the JA responses to root length inhibition is defective in *tic* mutant. In this scenario, TIC protein can directly interact with MYC2 and inhibit its protein turnover specifically in the evening phase [49] (Figure 3). Thus, we concluded that, not only JA content, but also its signaling pathway are modulated by circadian clock, indicating the complex cross network between circadian clock and JA regulated cellular events.

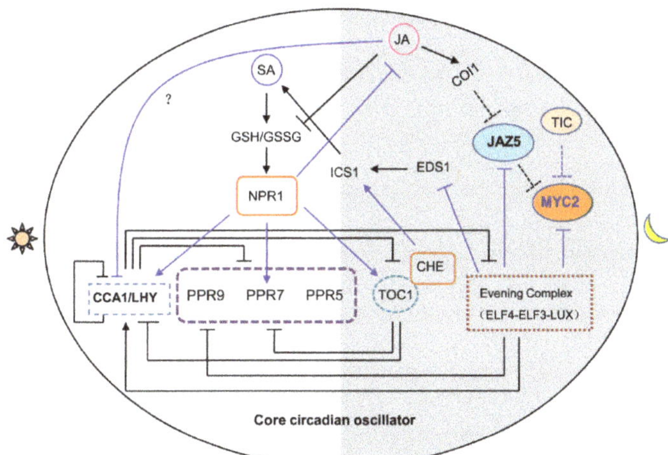

Figure 3. A proposed model for crosstalks between circadian clock and JA pathway in *Arabidopsis*. Evening complex (EC) transcriptionally represses the expression of *MYC2*, an essential master of JA signaling pathway to mediate leaf senescence. *EDS1* and *JAZ5* are the direct targets of LUX in regulating salicylic acid (SA) and JA signaling. Time for coffee (TIC) directly interacts with MYC2 and inhibits MYC2 protein turnover by timing the transcriptional level of CORONATINE INSENSITIVE 1 (COI1) in an evening-phase-specific manner. CCA1 hiking expedition (CHE) serves as an activation of ISOCHORISMATE SYNTHASE 1 (*ICS1*), an enzyme essential for SA biosynthesis. SA modulates the amplitude of circadian clock accompany by the redox rhythm and NON-EXPRESSOR OF PATHOGENESIS-RELATED GENE 1 (NPR1). NPR1 regulates the expression of circadian genes, *CCA1*, *LHY*, *PRR7*, and *TIMING OF CAB EXPRESSION 1* (*TOC1*). Reciprocally, JA signaling affects clock activity through unknown mechanisms. Novel molecular links between circadian clock and JA signaling pathway await to be further investigated.

4. Circadian Clock Gates JA Regulated Herbivory Resistance

Circadian clock confers the ability of plant to anticipate diel abiotic threats including herbivory and pathogen resistance, while JA is one of the major hormones during this process. The feeding behavior of cabbage loopers, *Trichoplusia ni* (*T. ni*), is rhythmic under constant conditions [45]. Plants entrained in-phase with the insects display much more resistance to the insect-attacking, with less tissue damage [45]. These phenotypes indicated that the feeding behavior of insects could be under the control of circadian clock. Furthermore, the circadian arrhythmic plants, *lux* mutant and *CCA1-OX* plants, did not display drastic difference in plant tissue loss compared with wild type, when challenged with in- and out-of-phase entrained *T. ni*. This finding implies that circadian clock is required for plant defense against herbivory. The plant tissue loss of *allene oxide synthase* (*aos*)and *jasmonate resistant 1* (*jar1*) mutants, in which the biosynthesis of JA is defective, also show no obvious difference when treated with in- and out-phase entrained *T. ni.* Both the circadian and JA mutants fail to display enhanced *T. ni* resistance, even entrained in-phase to insect, implying that both circadian clock and JA pathway are required for the in-phase-dependent enhanced herbivory resistance [45]. When plants encounter herbivores, it will emit complex volatile compounds to against this biotic stress, of which green leaf volatiles (GLVs) are a kind of fatty acid-derived compounds emitted upon plant damage [50]. *HYDROPEROXIDE LYASE* (*HPL*), encoding a GLV biosynthetic enzyme, is regulated by circadian clock at transcriptional level in *Nicotiana attenuata*. Accordingly, the emission of GLV is also rhythmic, with a peak at midday while its trough level at night. Moreover, JA signaling increases the basal turnover of *NaHPL* transcripts. Taken together, the GLV emissions are co-regulated by damage, JA signaling and circadian clock [51]. A study showed that the internal floral rhythm is abrogated in *NaLHY* and *NaZTL* RNA interference transgenic lines in *Nicotiana attenuata* [52]. The *NaZTL* RNAi transgenic lines are more susceptible to generalist herbivore *Spodoptera littoralis* compared to wild type. Plants usually produce various secondary metabolites to defense herbivores [53]. Nicotine is one of the most efficient defense-related metabolites in *Nicotiana attenuata*. To investigate whether the accumulation of nicotine confers to the attenuated *Spodoptera littoralis* resistance in *NaZTL* RNAi line, the secondary metabolites levels were measured. The results displayed that the nicotine level was significantly decreased in *NaZTL* RNAi transgenic lines than in the control plants. Furthermore, exogenous supplementation of nicotine could rescue the attenuated resistance to *Spodoptera littoralis* in *NaZTL* RNAi transgenic lines [54]. These findings suggested that the nicotine levels mediate the resistance against to *Spodoptera littoralis* in *NaZTL* RNAi transgenic lines. Moreover, they also found that the transcriptional levels of nicotine biosynthesis genes were significantly decreased in *NaZTL* RNAi plants. Meanwhile, the biosynthesis of nicotine is also mediated by JA signaling. NaZTL interacts with JASMONATE ZIM domain (JAZ) protein in the COI1 dependent manner, thus mediating JA signaling to gate plant defense response. All these evidences showed that *NaZTL* RNAi plants were more susceptible to *Spodoptera littoralis*, partially due to the reduced JA-regulated accumulation of nicotine in *Nicotiana attenuata* [54]. Taken together, all these evidences implied the essential roles of circadian clock in gating JA regulated herbivory resistance.

5. Circadian Clock Gates Temporal Variation Susceptibility to Pathogen by Jasmonates

Besides of gating herbivory resistance, circadian clock is also playing a vital role in gating JA mediated-defense against to pathogen. CCA1, a single MYB-domain containing transcriptional repressor with peak expression at dawn, has been identified as a regulator of plant defense [41]. TIC has also been reported to involve the circadian clock gated pathogen defense [49]. To address whether there is a temporal variation susceptibility to pathogen in *Arabidopsis*, Col-0 plants were challenged with pathogen *Pseudomonas syringae* pv. *tomato* (*Pst*) DC3000 at different circadian times under constant light conditions to avoid the effect of light/dark photoperiod. It turned out that Col-0 is more susceptible to *Pst* DC3000 at subjective midnight than at subjective morning [55]. However, the temporal variation susceptibility to pathogen was vanished in two circadian arrhythmic plants, namely *CCA1*-overexpressing line (*CCA1-ox*) and *elf3* mutant. Expression profiling analysis further revealed that a series of known defense related genes, including components of SA-related signaling pathway, namely *ISOCHORISMATE SYNTHASE 1* (*ICS1*), *ENHANCED DISEASE SUSCEPTIBILITY 1* (*EDS1*), *ENHANCED DISEASE SUSCEPTIBILITY 5* (*EDS5*), and CONSTITUTIVE EXPRESSION OF PR GENE 5 (*CPR5*) and the receptor of jasmonic acid (*COI1*), are transcriptionally regulated by circadian clock in *Arabidopsis* [55]. These lines of evidences suggest that the temporal variation susceptibility to pathogen is indeed under the control of circadian clock [55]. In addition, Korneli and colleagues show that circadian clock controls the discrepancy of pre-invasive and post-invasive defense responses against pathogens. They found that the distinct time of day responses to pathogen in another arrhythmic mutant *lux* (*lux arrhythmo*), such as oxidative burst and cell death, is different to those in wild type [56]. Very recently, the interplay between circadian clock and JA signaling was further elaborated. *LUX* was found to be induced when challenging with pathogen, and *LUX* is partially involved into the stomata-dependent defense response. Transcriptomic profiling and ChIP-seq analysis characterized *EDS1* and *JAZ5* as the novel targets of LUX in the regulation of JA signaling (Figure 3) [57]. Taken together, the above evidences clearly indicated the prominent role of circadian clock in gating plant defense, especially in a JA signaling dependent manner.

6. Circadian Clock is Associated with the Crosstalk between JA and SA Signaling

JA, together with SA, has been characterized as defense associated hormones. Intriguingly, the JA and SA signaling display antagonistic roles to each other in many defense processes, and their abundance are usually reverse to each other [1,58]. So far, several studies have indicated that the circadian clock delicately primes the JA and SA signaling. The peak of SA is at the nighttime while JA is at the midday, which is associated with the defense against morning biotrophic and dusk herbivore attacks respectively [58]. *PHT4;1* has been previously shown to affect SA-mediated defense. Wang and colleagues found that the expression of phosphate transporter gene *PHT4;1* was under the control of CCA1, and CCA1 could directly bind the promoter of *PHT4;1* to shape its diel transcription pattern, thus mediating SA dependent defense resistance [59]. ICS1 is a central enzyme for biosynthesis of SA, its expression can be activated upon various pathogens challenges. Yeast one hybrid assay found that CCA1 HIKING EXPEDITION (CHE) protein could directly bind *ICS1* promoter. Further study found that the circadian expression patterns of *CHE* and *ICS1* were similar. Moreover, the expression of *ICS1* was reduced in *che-2* mutant. Consistently, the SA accumulation was also decreased in *che-2*. These findings suggested that CHE served as an activator of *ICS1* and SA contents (Figure 3) [60]. It seems like that the timing SA and JA oscillations at different time window by circadian clock confers plant temporal variation susceptibility to specific invaders. Meanwhile, this separation of JA and SA may be able to avoid their potential antagonism [58]. Plant immunity usually causes the alteration of the cellular redox state including the total level of glutathione, the ratio of reduced (GSH) and oxidized (GSSG) forms of glutathione. SA increases the ratio of GSH/GSSG, by contrast, JA decreases the ratio of GSH/GSSG and the accumulation of glutathione. Unsurprisingly, glutathione is also involved in the regulatory crosstalk between JA and SA [58,61,62]. *Arabidopsis* NON-EXPRESSOR OF PATHOGENESIS-RELATED GENE 1 (NPR1) functions as a redox state sensor and regulates transcription of core circadian clock genes such as *PRR7* and *TOC1* [58]. SA triggers the enhanced redox status, increased glutathione accumulation and GSH/GSSG ratio, which resulted in the reinforcement of circadian clock [58] (Figure 3). Altogether, this balanced regulatory loop may maximize plants adaptability to external environment.

7. JA Signaling Reciprocally Affects Clock Activity

It has been shown that many of plant hormones biosynthesis and signaling are under the control of circadian clock to exert their function in specific time of day. Nonetheless, whether and how phytohormones regulate clock activity such as circadian phase and period, amplitude had been rarely explored. In higher plants, there are nine kinds of well-known hormones including auxin, gibberellins, cytokinin, abscisic acid (ABA), ethylene, SA, JA, brassinosteroide (BR), and strigolactones (SL) [63–66].

Years ago, the phytohormone effects on circadian clock have been examined using pharmaceutical treatment and hormone related mutants. They found that auxin could regulate circadian amplitude and clock precision, cytokinins could delay circadian phase, while brassinosteroid and abscisic acid could modulate circadian period. In contrast, gibberellins and ethylene had no effects on circadian clock [67]. SA has been experimentally shown to modulate the amplitude of circadian clock accompanied by the redox rhythm [68]. As for JA, recent study suggested that JA signaling also could reciprocally affect clock activity, and the expression of a few core components of circadian clock, such as *CCA1*, *LUX*, and *GRP7*, were reduced upon JA treatment. Further, JA treatment also resulted in a dampened amplitude of *CCA1:LUC* [57]. Moreover, the circadian period could be significantly lengthened in Col-0 by treated with JA-isoleucine (JA-Ile), a kind of bioactive JA derivative [57]. However, the underlying mechanism of JA treatment caused the lengthened circadian period still remains unclear (Figure 3). In the future, it would be required to systematically investigate the circadian phenotypes of JA signaling related mutants to pinpoint the exact molecular links from JA signaling to circadian clock, which might be occurred at transcriptional and post-transcriptional levels, or both.

8. Perspectives

To battle against various pathogens and pests, sessile plants had evolved many conserved and sophisticated defense mechanisms. Circadian clock is an internal time-keeping mechanism, which confers the anticipation of plants to the surrounding environmental cues such as daily changing light and temperature information. The emerging studies indicate that circadian clock play crucial roles in mediating leaf senescence. Evening complex, constituted by ELF4-ELF3-LUX, can directly bind and repress the expression of *MYC2*, which is a master regulator for JA signaling pathway, thus gating the circadian output regulation on JA-induced leaf senescence [47]. Another recent study confirmed that Evening Complex mutants showed hypersensitivity to dark-induced leaf senescence [69]. By contrast, the dark-induced leaf senescence is significantly delayed in *prr9* mutant, as PRR9 can directly bind the promoter of *ORESARA1* (*ORE1*) and activate its transcription. Genetically, *ORE1* overexpression can rescue the delayed leaf senescence of *prr9*. Collectively, circadian component PRR9 serves as a novel regulator of leaf senescence via positively activation of *ORE1* [69]. Reciprocally, circadian period is also feedback regulated by the leaf aging process. The circadian periods are about 1 h shorter in older leaves than in younger leaves, implying that aging process is associated with the regulation of circadian period [69]. Further study showed that TOC1 plays a central role in linking age to circadian clock period regulation [69]. Intriguingly, as aging, the JA response decays in *Arabidopsis*, which is regulated by the interaction between SQUAMOSA PROMOTER BINDING PROTEIN-LIKE 9 (SPL9) and JAZ protein [70]. Whether SPL directly mediate the circadian period needs to be further investigated in the future. Recent study shows that JA signaling reciprocally affects clock activity, however, its elaborated mechanism still remains elusive. The mechanism of JA signaling coupled to circadian clock is warranted to be further investigated by using JA signaling and circadian clock related mutants. Moreover, whether the regulation of JA on circadian activity is direct or indirect is still unknown. The network and the links between JA and circadian clock need to be further disclosed.

Besides of regulation on leaf senescence and defense resistance, JA also plays essential roles in regulating photomorphogenesis, root growth, photoperiodic dependent flowering time control, abiotic stress response and sterility as well [26]. Whether circadian clock gates these outputs and orchestrates the interaction between various JA responses, especially the tradeoff between growth and defense, awaits to be future investigated.

Furthermore, it has been known the biosynthesis of JA is circadian clock regulated and with a peak at midday [45], however, its underlying molecular mechanism is still unknown yet. The biosynthesis of JA is extremely complex, and a series of enzymes are involved in. Grundy and colleagues proposed that TOC1, PRR5, and PRR7 may be the novel regulators of both JA signaling and biosynthesis, due to the binding of these proteins to the promoters of JA responses genes by ChIP-seq analysis, including *LOX2*, *LOX3*, *LOX4*, *JAZ1*, *JAZ9*, *OCP3*, *PFT1*, *WRKY40*, *MYB108*, and *ANAC019* [71]. It would be interesting to know whether these genes are regulated by circadian clock. Further investigation with big-data driven approaches including transcriptomics, together with classical biochemical assays including protein-protein interaction, protein-DNA interaction, and genetic tools might be able to reveal this mask.

JA is usually produced when attacked and regulates inducible defenses. However, multiple lines of evidence showed that JA was regulated by circadian clock, which strongly suggested a more constitutive role of circadian clock in gating plant abiotic stress. Circadian clock confers the ability of plant to anticipate diel abiotic threats including herbivory and pathogen resistance at specific time of day, thus to enhance fitness of plants. However, the reciprocal regulation mechanisms among circadian clock, JA signaling, and abiotic stress are still not clear, which need to be further disclosed.

Author Contributions: Y.Z., C.B. and L.W. wrote the manuscript; Y.Z., C.B. and L.W. contributed to the discussion and approved the final manuscript.

Funding: The work was supported by the National Natural Sciences Foundation of China (31670290 and 31570292), Youth Innovation Promotion Association CAS (NO.2017110) and Young Elite Scientists Sponsorship Program CAST (2017QNRC001) to Y.Z., and Strategic Priority Research Program of the Chinese Academy of Sciences (XDB27030206) and National Key Research and Development Program of Ministry of Science and Technology of China (2016YFD0100604) to L.W.

Acknowledgments: We apologize to those authors whose excellent relevant work could not be cited due to space limitations.

Conflicts of Interest: The authors declare that the research was conducted in the absence of any commercial or financial relationships that could be construed as a potential conflict of interest.

References

1. Sanchez, S.E.; Kay, S.A. The plant circadian clock: From a simple timekeeper to a complex developmental manager. *Cold Spring Harb. Perspect. Biol.* **2016**, *8*. [CrossRef]

2. Greenham, K.; McClung, C.R. Integrating circadian dynamics with physiological processes in plants. *Nat. Rev. Genet.* **2015**, *16*, 598–610. [CrossRef]

3. Nagel, D.H.; Doherty, C.J.; Pruneda-Paz, J.L.; Schmitz, R.J.; Ecker, J.R.; Kay, S.A. Genome-wide identification of CCA1 targets uncovers an expanded clock network in *Arabidopsis*. *Proc. Natl Acad. Sci. USA* **2015**, *112*, E4802–E4810. [CrossRef]

4. Harmer, S.L.; Hogenesch, J.B.; Straume, M.; Chang, H.S.; Han, B.; Zhu, T.; Wang, X.; Kreps, J.A.; Kay, S.A. Orchestrated transcription of key pathways in *Arabidopsis* by the circadian clock. *Science* **2000**, *290*, 2110–2113. [CrossRef]

5. Pokhilko, A.; Fernandez, A.P.; Edwards, K.D.; Southern, M.M.; Halliday, K.J.; Millar, A.J. The clock gene circuit in *Arabidopsis* includes a repressilator with additional feedback loops. *Mol. Syst. Biol.* **2012**, *8*. [CrossRef]

6. Wang, L.; Kim, J.; Somers, D.E. Transcriptional corepressor TOPLESS complexes with pseudoresponse regulator proteins and histone deacetylases to regulate circadian transcription. *Proc. Natl. Acad. Sci. USA* **2013**, *110*, 761–766. [CrossRef]

7. Hicks, K.A.; Albertson, T.M.; Wagner, D.R. *EARLY FLOWERING3* encodes a novel protein that regulates circadian clock function and flowering in *Arabidopsis*. *Plant Cell* **2001**, *13*, 1281–1292. [CrossRef]

8. Doyle, M.R.; Davis, S.J.; Bastow, R.M.; McWatters, H.G.; Kozma-Bognar, L.; Nagy, F.; Millar, A.J.; Amasino, R.M. The *ELF4* gene controls circadian rhythms and flowering time in *Arabidopsis thaliana*. *Nature* **2002**, *419*, 74–77. [CrossRef]

9. Helfer, A.; Nusinow, D.A.; Chow, B.Y.; Gehrke, A.R.; Bulyk, M.L.; Kay, S.A. *LUX ARRHYTHMO* encodes a nighttime repressor of circadian gene expression in the *Arabidopsis* core clock. *Curr. Biol.* **2011**, *21*, 126–133. [CrossRef]

10. Nusinow, D.A.; Helfer, A.; Hamilton, E.E.; King, J.J.; Imaizumi, T.; Schultz, T.F.; Farre, E.M.; Kay, S.A. The ELF4-ELF3-LUX complex links the circadian clock to diurnal control of hypocotyl growth. *Nature* **2011**, *475*, 398–402. [CrossRef]

11. Herrero, E.; Kolmos, E.; Bujdoso, N.; Yuan, Y.; Wang, M.; Berns, M.C.; Uhlworm, H.; Coupland, G.; Saini, R.; Jaskolski, M.; et al. EARLY FLOWERING4 recruitment of EARLY FLOWERING3 in the nucleus sustains the *Arabidopsis* circadian clock. *Plant. Cell* **2012**, *24*, 428–443. [CrossRef]

12. Kim, W.Y.; Fujiwara, S.; Suh, S.S.; Kim, J.; Kim, Y.; Han, L.; David, K.; Putterill, J.; Nam, H.G.; Somers, D.E. ZEITLUPE is a circadian photoreceptor stabilized by GIGANTEA in blue light. *Nature* **2007**, *449*, 356–360. [CrossRef]

13. Baudry, A.; Ito, S.; Song, Y.H.; Strait, A.A.; Kiba, T.; Lu, S.; Henriques, R.; Pruneda-Paz, J.L.; Chua, N.H.; Tobin, E.M.; et al. F-box proteins FKF1 and LKP2 act in concert with ZEITLUPE to control *Arabidopsis* clock progression. *Plant. Cell* **2010**, *22*, 606–622. [CrossRef]

14. Mas, P.; Kim, W.Y.; Somers, D.E.; Kay, S.A. Targeted degradation of TOC1 by ZTL modulates circadian function in *Arabidopsis thaliana*. *Nature* **2003**, *426*, 567–570. [CrossRef]

15. Cha, J.Y.; Kim, J.; Kim, T.S.; Zeng, Q.; Wang, L.; Lee, S.Y.; Kim, W.Y.; Somers, D.E. GIGANTEA is a co-chaperone which facilitates maturation of ZEITLUPE in the *Arabidopsis* circadian clock. *Nat. Commun.* **2017**, *8*. [CrossRef]

16. Lee, C.M.; Li, M.W.; Feke, A.; Liu, W.; Saffer, A.M.; Gendron, J.M. GIGANTEA recruits the UBP12 and UBP13 deubiquitylases to regulate accumulation of the ZTL photoreceptor complex. *Nat. Commun.* **2019**, *10*. [CrossRef]

17. Nieto, C.; Lopez-Salmeron, V.; Daviere, J.M.; Prat, S. ELF3-PIF4 interaction regulates plant growth independently of the Evening Complex. *Curr. Biol.* **2015**, *25*, 187–193. [CrossRef]

18. Rawat, R.; Schwartz, J.; Jones, M.A.; Sairanen, I.; Cheng, Y.; Andersson, C.R.; Zhao, Y.; Ljung, K.; Harmer, S.L. REVEILLE1, a Myb-like transcription factor, integrates the circadian clock and auxin pathways. *Proc. Natl. Acad. Sci. USA* **2009**, *106*, 16883–16888. [CrossRef]

19. Legnaioli, T.; Cuevas, J.; Mas, P. TOC1 functions as a molecular switch connecting the circadian clock with plant responses to drought. *EMBO J.* **2009**, *28*, 3745–3757. [CrossRef]

20. Goossens, J.; Fernandez-Calvo, P.; Schweizer, F.; Goossens, A. Jasmonates: Signal transduction components and their roles in environmental stress responses. *Plant Mol. Biol.* **2016**, *91*, 673–689. [CrossRef]

21. Zhang, G.; Zhao, F.; Chen, L.; Pan, Y.; Sun, L.; Bao, N.; Zhang, T.; Cui, C.X.; Qiu, Z.; Zhang, Y.; et al. Jasmonate-mediated wound signalling promotes plant regeneration. *Nat. Plants* **2019**, *5*, 491–497. [CrossRef] [PubMed]

22. Kazan, K. Diverse roles of jasmonates and ethylene in abiotic stress tolerance. *Trends Plant. Sci.* **2015**, *20*, 219–229. [CrossRef] [PubMed]

23. Zhu, Z.; Lee, B. Friends or foes: New insights in jasmonate and ethylene co-actions. *Plant. Cell Physiol.* **2015**, *56*, 414–420. [CrossRef]

24. Chini, A.; Gimenez-Ibanez, S.; Goossens, A.; Solano, R. Redundancy and specificity in jasmonate signalling. *Curr. Opin. Plant. Biol.* **2016**, *33*, 147–156. [CrossRef] [PubMed]

25. Sharma, M.; Laxmi, A. Jasmonates: Emerging players in controlling temperature stress tolerance. *Front. Plant. Sci.* **2015**, *6*. [CrossRef] [PubMed]

26. Huang, H.; Liu, B.; Liu, L.; Song, S. Jasmonate action in plant growth and development. *J. Exp. Bot.* **2017**, *68*, 1349–1359. [CrossRef] [PubMed]

27. Wasternack, C.; Hause, B. Jasmonates: Biosynthesis, perception, signal transduction and action in plant stress response, growth and development. An update to the 2007 review in Annals of Botany. *Ann. Bot.* **2013**, *111*, 1021–1058. [CrossRef]

28. Ruan, J.; Zhou, Y.; Zhou, M.; Yan, J.; Khurshid, M.; Weng, W.; Cheng, J.; Zhang, K. Jasmonic acid signaling pathway in plants. *Int. J. Mol. Sci.* **2019**, *20*. [CrossRef]

29. Fonseca, S.; Chini, A.; Hamberg, M.; Adie, B.; Porzel, A.; Kramell, R.; Miersch, O.; Wasternack, C.; Solano, R. (+)-7-iso-Jasmonoyl-L-isoleucine is the endogenous bioactive jasmonate. *Nat. Chem. Biol.* **2009**, *5*, 344–350. [CrossRef]

30. Fonseca, S.; Chico, J.M.; Solano, R. The jasmonate pathway: The ligand, the receptor and the core signalling module. *Curr. Opin. Plant. Biol.* **2009**, *12*, 539–547. [CrossRef]

31. Xie, D.X.; Feys, B.F.; James, S.; Nieto-Rostro, M.; Turner, J.G. COI1: An *Arabidopsis* gene required for jasmonate-regulated defense and fertility. *Science* **1998**, *280*, 1091–1094. [CrossRef] [PubMed]

32. Yan, J.; Zhang, C.; Gu, M.; Bai, Z.; Zhang, W.; Qi, T.; Cheng, Z.; Peng, W.; Luo, H.; Nan, F.; et al. The *Arabidopsis* CORONATINE INSENSITIVE1 protein is a jasmonate receptor. *Plant. Cell* **2009**, *21*, 2220–2236. [CrossRef] [PubMed]

33. Melotto, M.; Mecey, C.; Niu, Y.; Chung, H.S.; Katsir, L.; Yao, J.; Zeng, W.; Thines, B.; Staswick, P.; Browse, J.; et al. A critical role of two positively charged amino acids in the Jas motif of *Arabidopsis* JAZ proteins in mediating coronatine- and jasmonoyl isoleucine-dependent interactions with the COI1 F-box protein. *Plant. J.* **2008**, *55*, 979–988. [CrossRef] [PubMed]

34. Chung, H.S.; Niu, Y.; Browse, J.; Howe, G.A. Top hits in contemporary JAZ: An update on jasmonate signaling. *Phytochemistry* **2009**, *70*, 1547–1559. [CrossRef]

35. Pauwels, L.; Goossens, A. The JAZ proteins: A crucial interface in the jasmonate signaling cascade. *Plant. Cell* **2011**, *23*, 3089–3100. [CrossRef]

36. An, C.; Li, L.; Zhai, Q.; You, Y.; Deng, L.; Wu, F.; Chen, R.; Jiang, H.; Wang, H.; Chen, Q.; et al. Mediator subunit MED25 links the jasmonate receptor to transcriptionally active chromatin. *Proc. Natl. Acad. Sci. USA* **2017**, *114*, E8930–E8939. [CrossRef]

37. Goossens, J.; Mertens, J.; Goossens, A. Role and functioning of bHLH transcription factors in jasmonate signalling. *J. Exp. Bot.* **2017**, *68*, 1333–1347. [CrossRef]

38. Qi, T.; Wang, J.; Huang, H.; Liu, B.; Gao, H.; Liu, Y.; Song, S.; Xie, D. Regulation of jasmonate-induced leaf senescence by antagonism between bHLH subgroup IIIe and IIId factors in *Arabidopsis*. *Plant. Cell* **2015**, *27*, 1634–1649. [CrossRef]

39. Zhu, X.; Chen, J.; Xie, Z.; Gao, J.; Ren, G.; Gao, S.; Zhou, X.; Kuai, B. Jasmonic acid promotes degreening via MYC2/3/4- and ANAC019/055/072-mediated regulation of major chlorophyll catabolic genes. *Plant. J.* **2015**, *84*, 597–610. [CrossRef]

40. Wang, H.; Li, Y.; Pan, J.; Lou, D.; Hu, Y.; Yu, D. The bHLH transcription factors MYC2, MYC3, and MYC4 are required for jasmonate-mediated inhibition of flowering in *Arabidopsis*. *Mol. Plant.* **2017**, *10*, 1461–1464. [CrossRef]

41. Wang, W.; Barnaby, J.Y.; Tada, Y.; Li, H.; Tor, M.; Caldelari, D.; Lee, D.U.; Fu, X.D.; Dong, X. Timing of plant immune responses by a central circadian regulator. *Nature* **2011**, *470*, 110–114. [CrossRef] [PubMed]

42. Covington, M.F.; Maloof, J.N.; Straume, M.; Kay, S.A.; Harmer, S.L. Global transcriptome analysis reveals circadian regulation of key pathways in plant growth and development. *Genome Biol.* **2008**, *9*. [CrossRef]

43. Michael, T.P.; Mockler, T.C.; Breton, G.; McEntee, C.; Byer, A.; Trout, J.D.; Hazen, S.P.; Shen, R.; Priest, H.D.; Sullivan, C.M.; et al. Network discovery pipeline elucidates conserved time-of-day-specific cis-regulatory modules. *PLoS Genet.* **2008**, *4*. [CrossRef] [PubMed]

44. Sauerbrunn, N.; Schlaich, N.L. PCC1: A merging point for pathogen defence and circadian signalling in *Arabidopsis*. *Planta* **2004**, *218*, 552–561. [CrossRef] [PubMed]

45. Goodspeed, D.; Chehab, E.W.; Min-Venditti, A.; Braam, J.; Covington, M.F. *Arabidopsis* synchronizes jasmonate-mediated defense with insect circadian behavior. *Proc. Natl. Acad. Sci. USA* **2012**, *109*, 4674–4677. [CrossRef] [PubMed]

46. Yamashino, T.; Kitayama, M.; Mizuno, T. Transcription of *ST2A* encoding a sulfotransferase family protein that is involved in jasmonic acid metabolism is controlled according to the circadian clock- and PIF4/PIF5-mediated external coincidence mechanism in *Arabidopsis thaliana*. *Biosci. Biotech. Bioch.* **2013**, *77*, 2454–2460. [CrossRef] [PubMed]

47. Zhang, Y.; Wang, Y.; Wei, H.; Li, N.; Tian, W.; Chong, K.; Wang, L. Circadian evening complex represses jasmonate-induced leaf senescence in *Arabidopsis*. *Mol. Plant.* **2018**, *11*, 326–337. [CrossRef]

48. Nitschke, S.; Cortleven, A.; Iven, T.; Feussner, I.; Havaux, M.; Riefler, M.; Schmulling, T. Circadian stress regimes affect the circadian clock and cause jasmonic acid-dependent cell death in cytokinin-deficient *Arabidopsis* plants. *Plant. Cell* **2016**, *28*, 1616–1639. [CrossRef]

49. Shin, J.; Heidrich, K.; Sanchez-Villarreal, A.; Parker, J.E.; Davis, S.J. TIME FOR COFFEE represses accumulation of the MYC2 transcription factor to provide time-of-day regulation of jasmonate signaling in *Arabidopsis*. *Plant Cell* **2012**, *24*, 2470–2482. [CrossRef]

50. Ameye, M.; Allmann, S.; Verwaeren, J.; Smagghe, G.; Haesaert, G.; Schuurink, R.C.; Audenaert, K. Green leaf volatile production by plants: A meta-analysis. *New Phytol* **2018**, *220*, 666–683. [CrossRef]

51. Joo, Y.; Schuman, M.C.; Goldberg, J.K.; Wissgott, A.; Kim, S.G.; Baldwin, I.T. Herbivory elicits changes in green leaf volatile production via jasmonate signaling and the circadian clock. *Plant. Cell Environ.* **2019**, *42*, 972–982. [CrossRef] [PubMed]

52. Yon, F.; Joo, Y.; Cortes Llorca, L.; Rothe, E.; Baldwin, I.T.; Kim, S.G. Silencing Nicotiana attenuata *LHY* and *ZTL* alters circadian rhythms in flowers. *New Phytol.* **2016**, *209*, 1058–1066. [CrossRef]

53. Mithofer, A.; Boland, W. Plant defense against herbivores: Chemical aspects. *Annu. Rev. Plant. Biol.* **2012**, *63*, 431–450. [CrossRef] [PubMed]

54. Li, R.; Llorca, L.C.; Schuman, M.C.; Wang, Y.; Wang, L.; Joo, Y.; Wang, M.; Vassao, D.G.; Baldwin, I.T. ZEITLUPE in the roots of wild tobacco regulates jasmonate-mediated nicotine biosynthesis and resistance to a generalist herbivore. *Plant. Physiol.* **2018**, *177*, 833–846. [CrossRef]

55. Bhardwaj, V.; Meier, S.; Petersen, L.N.; Ingle, R.A.; Roden, L.C. Defence responses of *Arabidopsis thaliana* to infection by Pseudomonas syringae are regulated by the circadian clock. *PLoS ONE* **2011**, *6*. [CrossRef] [PubMed]

56. Korneli, C.; Danisman, S.; Staiger, D. Differential control of pre-invasive and post-invasive antibacterial defense by the *Arabidopsis* circadian clock. *Plant. Cell Physiol.* **2014**, *55*, 1613–1622. [CrossRef]

57. Zhang, C.; Gao, M.; Seitz, N.C.; Angel, W.; Hallworth, A.; Wiratan, L.; Darwish, O.; Alkharouf, N.; Dawit, T.; Lin, D.; et al. LUX ARRHYTHMO mediates crosstalk between the circadian clock and defense in *Arabidopsis*. *Nat. Commun.* **2019**, *10*. [CrossRef] [PubMed]

58. Karapetyan, S.; Dong, X. Redox and the circadian clock in plant immunity: A balancing act. *Free Radical Biol. Med.* **2018**, *119*, 56–61. [CrossRef]

59. Wang, G.; Zhang, C.; Battle, S.; Lu, H. The phosphate transporter PHT4;1 is a salicylic acid regulator likely controlled by the circadian clock protein CCA1. *Front. Plant. Sci.* **2014**, *5*. [CrossRef]

60. Zheng, X.Y.; Zhou, M.; Yoo, H.; Pruneda-Paz, J.L.; Spivey, N.W.; Kay, S.A.; Dong, X. Spatial and temporal regulation of biosynthesis of the plant immune signal salicylic acid. *Proc. Natl. Acad. Sci. USA* **2015**, *112*, 9166–9173. [CrossRef]

61. Tada, Y.; Spoel, S.H.; Pajerowska-Mukhtar, K.; Mou, Z.; Song, J.; Wang, C.; Zuo, J.; Dong, X. Plant immunity requires conformational changes of NPR1 via S-nitrosylation and thioredoxins. *Science* **2008**, *321*, 952–956. [CrossRef] [PubMed]

62. Mou, Z.; Fan, W.; Dong, X. Inducers of plant systemic acquired resistance regulate NPR1 function through redox changes. *Cell* **2003**, *113*, 935–944. [CrossRef]

63. Santner, A.; Calderon-Villalobos, L.I.; Estelle, M. Plant hormones are versatile chemical regulators of plant growth. *Nat. Chem. Biol.* **2009**, *5*, 301–307. [CrossRef] [PubMed]

64. Gomez-Roldan, V.; Fermas, S.; Brewer, P.B.; Puech-Pages, V.; Dun, E.A.; Pillot, J.P.; Letisse, F.; Matusova, R.; Danoun, S.; Portais, J.C.; et al. Strigolactone inhibition of shoot branching. *Nature* **2008**, *455*, 189–194. [CrossRef] [PubMed]

65. Hagihara, S.; Yamada, R.; Itami, K.; Torii, K.U. Dissecting plant hormone signaling with synthetic molecules: Perspective from the chemists. *Curr. Opin. Plant. Biol.* **2019**, *47*, 32–37. [CrossRef] [PubMed]

66. Larrieu, A.; Vernoux, T. Comparison of plant hormone signalling systems. *Essays Biochem.* **2015**, *58*, 165–181. [CrossRef] [PubMed]

67. Hanano, S.; Domagalska, M.A.; Nagy, F.; Davis, S.J. Multiple phytohormones influence distinct parameters of the plant circadian clock. *Genes Cells* **2006**, *11*, 1381–1392. [CrossRef]

68. Zhou, M.; Wang, W.; Karapetyan, S.; Mwimba, M.; Marques, J.; Buchler, N.E.; Dong, X. Redox rhythm reinforces the circadian clock to gate immune response. *Nature* **2015**, *523*, 472–476. [CrossRef]

69. Kim, H.; Kim, H.J.; Vu, Q.T.; Jung, S.; McClung, C.R.; Hong, S.; Nam, H.G. Circadian control of *ORE1* by PRR9 positively regulates leaf senescence in *Arabidopsis*. *Proc. Natl. Acad. Sci. USA* **2018**, *115*, 8448–8453. [CrossRef]
70. Mao, Y.B.; Liu, Y.Q.; Chen, D.Y.; Chen, F.Y.; Fang, X.; Hong, G.J.; Wang, L.J.; Wang, J.W.; Chen, X.Y. Jasmonate response decay and defense metabolite accumulation contributes to age-regulated dynamics of plant insect resistance. *Nat. Commun.* **2017**, *8*. [CrossRef]
71. Grundy, J.; Stoker, C.; Carre, I.A. Circadian regulation of abiotic stress tolerance in plants. *Front. Plant. Sci.* **2015**, *6*. [CrossRef] [PubMed]

 © 2019 by the authors. Licensee MDPI, Basel, Switzerland. This article is an open access article distributed under the terms and conditions of the Creative Commons Attribution (CC BY) license (http://creativecommons.org/licenses/by/4.0/).

International Journal of
Molecular Sciences

Review

Crosstalk with Jasmonic Acid Integrates Multiple Responses in Plant Development

Geupil Jang [1], Youngdae Yoon [2] and Yang Do Choi [3],*

[1] School of Biological Sciences and Technology, Chonnam National University, Gwangju 61186, Korea; yk3@jnu.ac.kr
[2] Department of Environmental Health Science, Konkuk University, Seoul 05029, Korea; yyoon21@gmail.com
[3] The National Academy of Sciences, Seoul 06579, Korea
* Correspondence: choiyngd@snu.ac.kr

Received: 25 November 2019; Accepted: 20 December 2019; Published: 2 January 2020

Abstract: To date, extensive studies have identified many classes of hormones in plants and revealed the specific, nonredundant signaling pathways for each hormone. However, plant hormone functions largely overlap in many aspects of plant development and environmental responses, suggesting that studying the crosstalk among plant hormones is key to understanding hormonal responses in plants. The phytohormone jasmonic acid (JA) is deeply involved in the regulation of plant responses to biotic and abiotic stresses. In addition, a growing number of studies suggest that JA plays an essential role in the modulation of plant growth and development under stress conditions, and crosstalk between JA and other phytohormones involved in growth and development, such as gibberellic acid (GA), cytokinin, and auxin modulate various developmental processes. This review summarizes recent findings of JA crosstalk in the modulation of plant growth and development, focusing on JA–GA, JA–cytokinin, and JA–auxin crosstalk. The molecular mechanisms underlying this crosstalk are also discussed.

Keywords: jasmonic acid; crosstalk; gibberellic acid; cytokinin; auxin

---

1. Introduction

Plant growth and physiology are regulated by endogenous processes and environmental signals; phytohormones govern these processes by controlling transcriptional and translational networks. Jasmonates, including jasmonic acid (JA) and its derivatives, were initially isolated as a methyl ester form of JA in *Jasminum grandiflorum*. JA is classified as a cyclopentane fatty acid and is biosynthesized from linolenic acid, a major fatty acid of membranes in plant cells. Details of the JA biosynthetic pathway have been well reviewed [1,2]. Briefly, JA biosynthesis is regulated by enzymes such as lipoxygenase, allene oxide synthase, and allene oxide cyclase, which mediate the octadecanoid pathway. The free acid JA can be further metabolized into methyl jasmonate or the JA-isoleucine conjugate (JA-Ile) via the activity of jasmonate methyl transferase and jasmonate-amido synthetase, respectively. In response to environmental signals, the expression of the genes involved in JA metabolism is dynamically regulated, leading to changes in endogenous JA levels and stress responses, supporting the view that JA is a key hormone mediating plant responses to environmental stresses [3].

Early studies on JA showed that JA treatment rapidly and dynamically regulates genes involved in plant defense, suggesting the existence of a JA-specific signaling pathway and the integral role of JA in regulating gene expression networks [4,5]. In 1994, *Arabidopsis thaliana coronatine insensitive 1 (coi1)* mutants, in which the JA response is blocked, were identified [6] and a series of studies on *coi1* extended our understanding of the JA signaling pathway. *COI1* encodes an F-box protein that acts as the JA receptor and functions in E3-ubiquitin ligase-mediated proteolysis of target proteins [7–9] such as the JASMONATE ZIM-DOMAIN (JAZ) proteins. Further identification of JA signaling components, including JA-responsive MYC transcription factors, revealed a JA signaling pathway that includes JA perception and JA-dependent gene regulation. Briefly, the expression of JA-dependent genes and activation of the JA response are inhibited in plant cells with low JA levels. In these cells, the MYC2 transcription factors, which are responsible for the expression of JA-responsive genes, stay inactive through the direct interaction with JAZ proteins, which are JA signaling repressors. JAZ proteins contain two domains, ZIM and Jas, and these domains mediate the interaction of JAZs with other proteins. The ZIM domain is responsible for its dimerization and interaction with NINJA, which connects the transcriptional suppressor TOPLESS to JA signaling, and the Jas domain mediates the JAZ–COI1 interaction [10,11].

When JA biosynthesis is activated in response to endogenous or environmental signals, and JA, especially JA-Ile, accumulates in cells, JA-Ile activates JA signaling through interaction with the COI1 receptor. This direct interaction induces proteolysis of the JAZ proteins and activates the expression of JA-responsive genes by releasing the MYC2 transcription factor from the JAZ–MYC2 complex [8]. Unlike the JAZ repressors, the MYC2 transcription factor activates the transcription of JA-responsive genes and promotes the JA response. As JAZs and MYC2 are key factors in plant growth and development as positive and negative regulators, respectively, they may mediate JA-dependent growth inhibition under stress conditions [12–14].

Plant hormones have their own specific biosynthetic and signaling pathways, but their roles in plant development and physiology overlap. This suggests that plant hormones modulate plant growth and physiology through interactions with other hormones, and the extensive interplay between auxin and cytokinin in the regulation of all aspects of plant growth and development supports this idea [15,16]. JA mediates the plant response to biotic and abiotic stresses through interaction with salicylic acid, ethylene, and abscisic acid (ABA), and details of this crosstalk and its underlying molecular mechanisms have been well reported in previous studies [3,17–19]. JA also modulates plant development, such as root, stamen, hypocotyl, chloroplast, and xylem development, and increasing evidence suggests that JA-dependent modulation of plant growth and development largely depends on the interaction of JA with other phytohormones such as gibberellins (GAs), cytokinin, and auxin that govern endogenous developmental programs. Many studies have revealed that the crosstalk between phytohormones is mediated through regulatory proteins controlling phytohormone metabolic and signaling pathways [3,20]. This review briefly describes the metabolism and signaling pathways of the phytohormones GA, cytokinin, and auxin that interact with JA in the modulation of plant growth and development, and recent findings on JA crosstalk, focusing on the JA–GA, JA–cytokinin, and JA–auxin interactions. The molecular mechanisms underlying the JA–GA, JA–cytokinin, and JA–auxin interactions are also discussed in this review.

2. The JA–GA Interaction

2.1. GA Metabolism and Signaling Pathway

GAs regulate plant growth and development, such as stem elongation, seed germination, leaf expansion, root development, and stamen and flower development [21]. Due to the essential role of GAs in plant growth, the GA response affects plant growth and productivity [22], and many studies suggest that GA is fundamental to stress-related growth inhibition through interactions with stress-response hormones [23–30].

GAs are a large class of tetracyclic diterpenoid compounds, and approximately 136 forms have been identified in higher plants and fungi. However, only a few of them, including GA_1, GA_3, GA_4, and GA_7, are biologically active, while other GAs are intermediate forms in the GA biosynthetic process or inactive forms of GAs. Therefore, GA metabolism, including its biosynthesis, is integral to GA homeostasis and the GA response in plants [31,32]. The biosynthetic pathway of GAs includes the biosynthesis of *ent*-kaurene, the conversion of *ent*-kaurene to GA_{12}, and the formation of C_{20}- and C_{19}-GAs in the cytosol, and three different classes of enzymes, terpene synthases, cytochrome

P450 monooxygenases, and 2-oxoglutarate-dependent dioxygenases, mediate this process [20,33,34]. Further metabolic processes are required for the formation of active GAs and the deactivation of bioactive GAs, and GA 20-oxidase, GA 3-oxidase, and GA 2-oxidase mediate these metabolic process [35–38].

GA signaling is another key step controlling the transcription of GA-dependent genes and the regulation of the GA response, and, similar to other plant hormones such as JA, auxin, and strigolactone, the GA signaling process is based on E3 ubiquitin ligase-mediated proteolysis of DELLAs. The Arabidopsis genome encodes five DELLAs, including REPRESSOR OF GA1-3 (RGA), which functions as an intracellular negative regulator of GA signaling [39,40]. In Arabidopsis, direct interaction between GAs and the GA INSENSITIVE DWARF1 (GID1) receptor induces the interaction between GID1 and DELLAs, and provokes the degradation of DELLAs through E3 ubiquitin ligase-mediated ubiquitinylation and 26S proteasome-mediated proteolysis [39,41,42]. The proteolysis of DELLAs leads to the release of GA-responsive transcription factors such as PHYTOCHROME INTERACTING FACTORS (PIFs) in Arabidopsis and PIF-LIKE (PIL) proteins in rice (*Oryza sativa*), and triggers the transcription of GA-responsive genes and the GA response [43–45]. The finding that *RGA*-overexpressing plants displayed a reduced GA response while mutants lacking *RGA* expression showed an enhanced GA response indicates a crucial role of DELLAs in GA signaling pathways [46–48].

2.2. The JA–GA Interaction and Its Underlying Molecular Mechanism

Environmental stresses strongly affect plant growth. To survive under stress conditions, plants activate defense programs and suppress developmental programs, leading to growth inhibition. By contrast, in the proper conditions for growth, plants activate developmental programs while suppressing defense programs, leading to vigorous growth. This indicates that plants dynamically coordinate growth and defense strategies in response to environmental stresses. The essential role that GAs play in the regulation of plant growth suggests that GAs have key roles in this coordination, and the finding that environmental stresses, such as salinity, promote the accumulation of DELLAs but reduce endogenous levels of bioactive GAs supports this idea [24–26]. In addition, stress-induced growth reduction was attenuated in quadruple-*della* mutants, while plants with reduced GA levels, such as the GA biosynthesis *mutant ga1-3*, exhibited enhanced tolerance to salt stress [24]. These findings indicated that GA plays an essential role in the coordination of plant growth and defense, and further analysis of *della* mutant plants suggested that DELLAs are deeply involved in GA-dependent coordination process [49].

Many studies reported that developmental flexibility under stress conditions largely depends on the interplay between stress-related hormones and growth-related hormones, and increasing evidence indicates that JA and GA antagonistically interact to coordinate plant growth and defense [50–52]. Extensive studies on the JAZ JA signaling repressor proteins, and the DELLA GA signaling repressor proteins revealed that direct interaction between JAZs and DELLAs mediates the antagonistic interaction between JA and GA (Figure 1) [27,28]. In the "relief of repression" model, the JAZ–DELLA interaction attenuates the functions of JAZs and DELLAs as signaling repressors. For example, in GA-free conditions, DELLAs directly interact with JAZs and allow MYC2 to promote the JA response, while in the presence of GA, JAZs are released from the DELLA–JAZ complex by degradation of DELLAs, and the free JAZs attenuate the JA response through direct interact with MYC2. The model explained the DELLA-mediated upregulation of the JA response and the antagonistic interaction between JA and GA [27]. This model was supported by studies showing that JA promotes transcription of *RGA3*, and the JA-responsive MYC2 transcription factor directly binds to the promoter of *RGA3* [29].

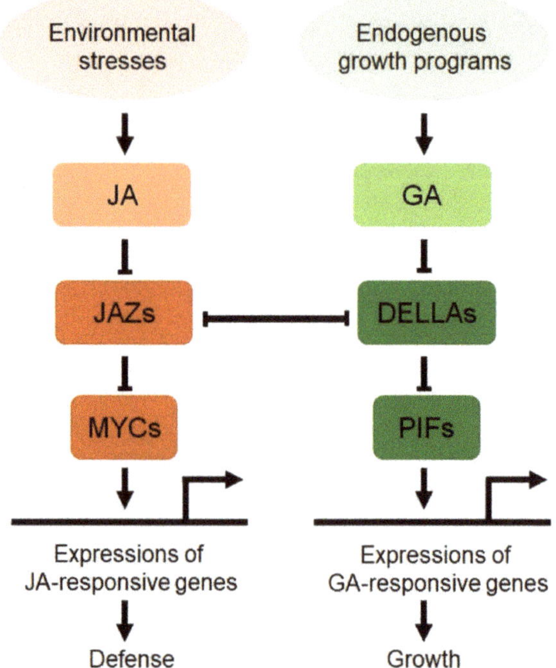

Figure 1. A schematic of crosstalk between jasmonic acid (JA) and gibberellic acid (GA) in coordination between plant growth and defense. JA and GA antagonistically interact to coordinate plant growth and defense, and the crosstalk is mediated by direct interaction between JA signaling repressors, JASMONATE ZIM-DOMAIN (JAZs), and GA signaling repressors, DELLAs. MYCs and PHYTOCHROME INTERACTING FACTORS (PIFs) indicates transcription factors responsible for transcription of JA-responsive and GA-responsive genes, respectively. The arrows and T bars indicate positive and negative regulation, respectively.

A recent study using overexpression plants and knock-out mutants of *OsJAZ9* revealed that OsJAZ9 is a key JAZ protein that mediates the antagonistic interaction of JA and GA [21]. In this study, they identified OsJAZ9 proteins that directly interact with the rice DELLA protein SLENDER RICE 1 (SLR1), and demonstrated that the OsJAZ9–SLR1 interaction mediates the antagonistic interaction of JA and GA in rice by showing that overexpression of *OsJAZ9* promotes the GA response while knock-out of *OsJAZ9* reduces the GA response. Together, these data suggest that JA is an essential hormone that modulates plant growth under stress conditions, and its antagonistic interaction with GA mediates this process.

3. The JA–Cytokinin Interaction

3.1. Cytokinin Metabolism and Signaling

Cytokinin regulates the maintenance of stem cell identity and cell proliferation; therefore, cytokinin affects most aspects of plant growth and development [53]. The expression of genes involved in cytokinin responses is largely affected by the stress-response hormone JA or JA-dependent stress responses [54–56]. Furthermore, cytokinin-deficient mutant plants displayed increased tolerance to stresses, similar to transgenic plants with higher JA responses [57–60]. These studies suggested that the cytokinin response is integral to the JA-dependent stress response and growth modulation.

Most naturally occurring cytokinins are derivatives of isopentenyladenine, and zeatin is the ubiquitous form of cytokinins in higher plants [53,61]. Zeatin occurs as two isomers, *trans*-zeatin (*tZ*) and *cis*-zeatin (*cZ*); *tZ* is the active form of cytokinin in all plant species and *cZ* is less active than *tZ* [62,63]. Isopentenyl transferases (IPTs), and cytochrome P450 CYP735A1 and CYP735A2 mediate the production of *tZ* cytokinin [53]. The IPT-catalyzed reaction is the rate limiting step in cytokinin biosynthesis process, and the results showing that overexpression of *AtIPT1, 3, 4, 5, 7,* or *8* promoted cytokinin production and shoot growth support this [61,64,65]. The biological activity and homeostasis of cytokinins can be regulated by conjugation with glucose or amino acids, or by degradation. For example, glucosyl-conjugated cytokinins, which do not interact with cytokinin receptors, are inactive, and overexpression of cytokinin oxidase, which is responsible for cytokinin cleavage, reduces endogenous levels of cytokinins [66,67].

The cytokinin signaling pathway, which is composed of cytokinin receptors, histidine phosphotransfer proteins, and transcription factors, regulates cytokinin responses in plants. In Arabidopsis, three histidine kinases (AHK2, AHK3, and AHK4/WOODEN LEG) function as cytokinin receptors [53,68]. Direct interaction between cytokinins and the histidine kinase receptors activates the kinase activity of the receptors, leading to autophosphorylation on the conserved histidine residue. The phosphate is transferred to the histidine phosphotransfer proteins (AHPs) via the conserved aspartate residue of the receptors. In Arabidopsis, five genes encode AHPs that normally function as histidine phosphotransferases and one gene (*AHP6*) encodes a pseudo-AHP that negatively regulates cytokinin signaling. The AHPs activated by phosphorylation move into the nucleus and sequentially activate B-type ARABIDOPSIS RESPONSE REGULATOR (ARR) transcription factors responsible for the transcription of cytokinin-responsive genes [69,70]. Genes encoding components of the cytokinin signaling pathways, such as *AHKs, AHPs,* and *ARRs,* are affected by JA or environmental stresses such as drought, salt, and cold, suggesting that the cytokinin response is involved in plant stress responses [50,56,71,72].

3.2. The JA–Cytokinin Interaction and Its Underlying Molecular Mechanism

Previous studies have proposed that JA antagonistically interacts with cytokinin in various aspects of plant development. For example, JA inhibits cytokinin-induced soybean (*Glycine max*) callus growth [73], and nullifies the effect of cytokinin on chlorophyll development [74,75]. Furthermore, JA and cytokinin differently regulate the expression of the genes involved in the chlorophyll development, indicating the existence of an antagonistic interaction between JA and cytokinin. A recent study revealed that xylem differentiation is regulated by JA in Arabidopsis roots, and an antagonistic interaction between JA and cytokinin is fundamentally important for JA-dependent xylem development [50]. Xylem is responsible for water and nutrient transport and it develops from procambial/cambial cells, which are stem cells of the vascular system [76,77]. In Arabidopsis roots, cytokinin maintains stem cell identity and functions as a negative regulator of xylem differentiation. The role of cytokinin in xylem differentiation was demonstrated by showing that exogenous cytokinin treatment inhibits xylem development, and the *wooden leg* mutants with defects in cytokinin signaling strongly exhibit an all-xylem phenotype and lack procambial cells in their roots. Additionally, mutants that lack transcription of Type-B *ARRs,* such as *ARR1, ARR10,* and *ARR12,* or transgenic plants overexpressing *AHP6,* a negative regulator of cytokinin signaling, form extra xylem [50,78]. Similar to the cytokinin signaling mutants, the wild-type plants or JA-deficient *OPDA reductase 3* (*opr3*) mutants treated with exogenous JA showed an extra xylem phenotype, whereas JA signaling mutants, such as *coi1* and *jasmonate resistant 1* (*jar1*), did not [50]. Together with the results that JA suppresses the procambium-specific cytokinin response, and that the effect of JA on extra xylem formation is nullified by cytokinin, suggest that the stress hormone JA antagonistically interacts with cytokinin in xylem development in Arabidopsis roots.

These findings were supported by the results that JA reduces the expression of the cytokinin-responsive *PIN-FORMED 7* (*PIN7*) gene, which is responsible for xylem development, and the finding that drought stress induces the formation of extra xylem in Arabidopsis roots further supported this idea [50,79]. Furthermore, *myc2* mutant did not form extra xylem in response to exogenous JA, and the expression of *AHP6*, encoding a cytokinin signaling inhibitor, was reduced in *myc2* mutant, suggesting that the JA-responsive MYC2 transcription factor mediates this process by promoting *AHP6* expression (Figure 2). It is likely that an antagonistic interaction between JA and cytokinin is also involved in the regulation of JA-dependent stress responses. A recent study by Nitschke et al. (2016) showed that plants with reduced cytokinin levels or defective cytokinin signaling exhibited a JA-dependent cell death phenotype in response to circadian stress, unlike wild-type plants [80], suggesting that JA and cytokinin antagonistically interact in the plant response to circadian stress.

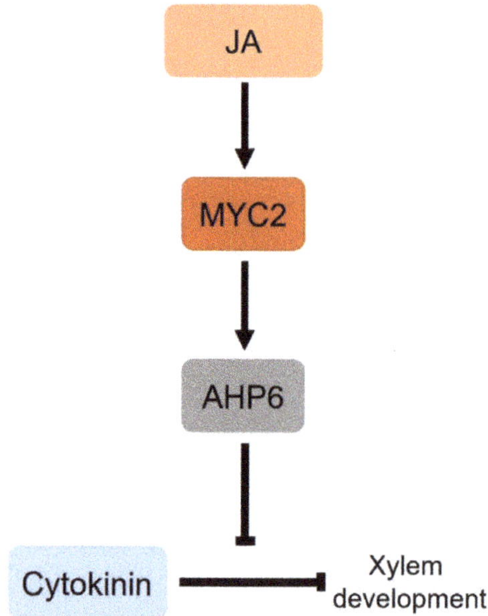

Figure 2. A schematic of crosstalk between JA and cytokinin in xylem development. JA antagonistically interacts with cytokinin in xylem development and the JA-responsive MYC2 transcription factor mediates this process. MYC2 negatively regulates cytokinin response by promoting expression of *AHP6*, a cytokinin signaling inhibitor. The arrows and T bars indicate positive and negative regulation, respectively.

Despite many studies supporting an antagonistic interaction between JA and cytokinin in modulation of plant development and physiology, the molecular mechanisms underlying the JA–GA crosstalk remain largely unknown. Regulation of *AHP6* expression by MYC2 transcription suggests that components of JA and cytokinin signaling pathways might mediate the interaction between JA and cytokinin. However, the observation that cytokinin levels were affected by stress conditions also suggests that regulation of JA and cytokinin metabolism might also be involved in the JA–cytokinin interaction [60,81,82].

4. The JA–Auxin Interaction

4.1. Auxin Metabolism and Signaling

Auxin has essential functions in cell fate determination and cell division, thus mediating most aspects of plant growth and development [83]. Indole-3-acetic acid (IAA) is the predominant form of natural auxins in plants. IAA can be produced through tryptophan-dependent and -independent pathways, and the tryptophan-dependent pathway is currently the best understood auxin biosynthetic pathway in plants [84,85]. The tryptophan-dependent pathway is mediated by tryptophan aminotransferase (TAA), and the flavin monooxygenase YUCCA (YUC). TAA and YUC are responsible for the conversion of tryptophan to indole-3-pyruvate (IPA) and the conversion of IPA to IAA, respectively. Similar to other plant hormones, IAA can be deactivated by conjugation with amino acids or sugars, and by oxidation [86].

Auxin-mediated regulation of gene expression is crucial for auxin-dependent regulation of plant growth and development, and this process is regulated by the auxin signaling pathway. Similar to JA and GA, auxin signaling is based on E3 ubiquitin ligase-mediated proteolysis of signaling repressor proteins [87]. AUXIN RESPONSE FACTORS (ARFs) are transcription factors responsible for the transcription of auxin-responsive genes, and they regulate the transcription of auxin-responsive genes by directly binding auxin responsive elements through their B3-like DNA binding motif. The transcriptional activity of ARFs depends on the interaction with the auxin signaling repressor Aux/IAAs, and degradation of Aux/IAAs induces the release of ARFs with transcriptional activity and activates the transcription of auxin-responsive genes. The degradation of Aux/IAAs is provoked by the SCF^{TIR1} E3 ubiquitin ligase, and the direct interaction between auxin and the TIR1 auxin receptor enhances the physical association between TIR1 and Aux/IAAs and sequential ubiquitination of Aux/IAAs [88].

4.2. Interaction of JA and Auxin and the Underlying Molecular Mechanism

The interaction of JA and auxin in plant development and physiology plays a role in processes such as cell elongation, tendril coiling, and the production of secondary metabolites [89,90], but this interaction has not been elucidated at the molecular level. The identification of genes involved in JA and auxin metabolism and signaling pathways have revealed that JA and auxin interact to modulate plant development.

The interaction between JA and auxin has been well demonstrated in the regulation of root development. JA inhibits apical growth of roots; JA-treated wild-type plants form much shorter roots than untreated wild-type plants, while mutant plants with defects in JA signaling form similar roots in length to the roots of wild-type plants even in JA-treated conditions [50]. By contrast, auxin is essential for root growth and auxin deficiency or signaling mutants, such as (*trp2-12*) and *auxin resistant 3 (arx3-1)*, develop very short roots compared to wild-type plants [91,92]. This suggests that JA-induced inhibition of root growth might be mediated by an interaction with auxin, and a study by Chen et al. (2011) demonstrated this [93]. In the study, they showed that the JA-mediated inhibition of root growth is caused by a reduction of root meristem activity, and exogenous JA treatment suppresses the expression of the auxin-responsive transcription factors *PLETHORAs* (*PLTs*), which are responsible for maintenance of the stem cell niche and cell proliferation [94]. However, the expression levels of *PLTs* was not suppressed in JA-signaling mutants, such as *coi1-1* and *myc2*, suggesting that COI1-dependent JA signaling mediates the JA-induced root phenotype, and the MYC2 transcription factor suppresses the expression of *PLTs*. Together with the result that MYC2 directly binds to the promoters of *PLTs*, indicate that JA-responsive MYC2 mediates JA-induced inhibition of root growth by directly repressing the expression of auxin-responsive *PLTs* (Figure 3), and suggest that JA and auxin antagonistically interact in the regulation of apical root growth.

The JA–auxin interaction is involved in various aspects of plant development as well as root development. The development of floral organs, such as petals and stamens, is coordinated as flowers mature, and a study by Reeves et al. (2012) showed that an interaction between auxin-responsive transcription factors and JA biosynthesis modulates this process [95]. The R2R3 MYB transcription factors MYB21 and MYB24 are key regulators of petal and stamen growth, and the auxin-responsive transcription factors ARF6 and ARF8 regulate the expression of JA-responsive *MYB21* and *MYB24* by controlling JA biosynthesis, indicating that auxin interacts with JA to regulate the development of floral organs [95,96]. The JA-auxin interaction was also observed in the regulation of leaf senescence.

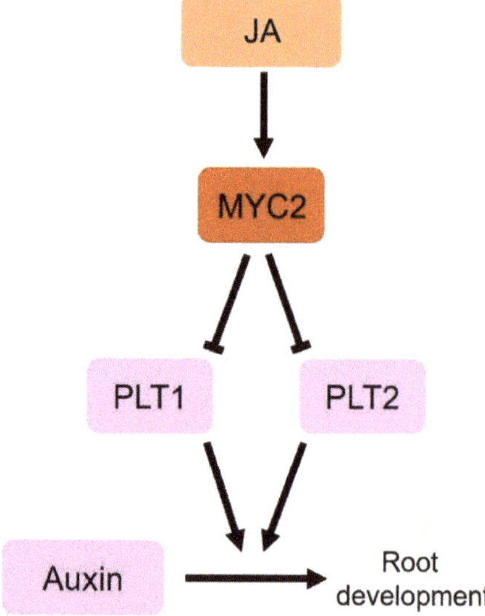

Figure 3. A schematic of crosstalk between JA and auxin in root development. PLT1 and 2 transcription factors are key regulators of root development downstream auxin. JA inhibits root growth, and MYC2 transcription factor mediates this development process by reducing expression of *PLT1* and 2. The arrows and T bars indicate positive and negative regulation, respectively.

JA plays an essential role as a positive regulator of leaf senescence. JAZ7 suppresses dark-induced leaf senescence, while MYCs, including MYC2, promote senescence by activating the expression of senescence-associated genes and chlorophyll degradation-related genes, indicating that JA activates leaf senescence through a COI1-dependent JA signaling pathway [97–99]. In JA-dependent leaf senescence, the JA signaling repressors JAZ4 and JAZ8 function as negative regulators while the auxin signaling repressor IAA29 functions as a positive regulator. In JA-dependent leaf senescence, WRKY57 is another negative regulator that negatively affects the expression of senescence-associated genes. More importantly, WRKY57 interacts with JAZ4/8 and IAA29. These results suggest that competition between the WRKY57–JAZ4/8 and WRKY57–IAA29 interactions mediates JA-dependent leaf senescence, suggesting that an antagonistic interaction of JA and auxin is involved in leaf senescence [100].

5. Complexity of JA Crosstalk

Hormonal interactions are a critical component of plant growth and physiology [3]. This review described the role of JA crosstalk with other phytohormones in the modulation of plant growth and development, focusing on JA–GA, JA–cytokinin, and JA–auxin interactions and the molecular mechanisms underlying these processes. JA interacts with most plant hormones, and as shown in previous studies, JA extensively interacts with salicylic acid to modulate plant defenses against pathogen attacks. The interaction between JA, which mediates disease resistance to necrotrophic pathogens, and salicylic acid, which mediates broad-spectrum resistance to biotrophic pathogens, allows plants to establish an efficient defense system against a variety of pathogen attacks, and ethylene is also involved in this process [17,101]. In addition, ABA interacts with JA to regulate cellular metabolic processes, and ABA receptor PYRABACTIN RESISTANCE1-Like proteins with the ability to interact with JAZs mediate this process by modulating JA signaling [102].

Brassinosteroids (BR) mediate various aspects of plant growth and development and modulate JA signaling and JA-dependent growth inhibition. For example, *DWARF4* encodes a key enzyme responsible for BR biosynthesis and a leaky mutation of *DWARF4* restored JA sensitivity in the *coi1* mutant background and showed JA hypersensitivity in the wild-type background. Furthermore, expression of *DWARF4* was downregulated by JA in a COI1-dependent manner, and exogenous BR treatment attenuated the effects of JA on root growth inhibition [103]. These results indicate that a BR–JA interaction is involved in the modulation of JA signaling.

As described in this review, JA interacts with a variety of hormones involved in growth regulation, such as GA, cytokinin, auxin, and BR, to modulate plant growth and development, and the nature of the interaction is generally antagonistic. These interactions may help optimize plant growth and development under stress conditions. However, the nature of the interaction appears to differ depending on the type of cell and tissue. For example, JA and GA antagonistically interact in stem elongation, while they interact synergistically in stamen development [21,104]. JA antagonistically interacts with auxin to modulate apical growth of roots, but synergistically to promote lateral root growth [93,105].

6. Future Perspectives

Identification and characterization of the components involved in plant hormone metabolism and signaling have provided important clues to understand the hormonal interactions underlying the regulation of plant growth and physiology in response to endogenous and exogenous signals. JA is a key hormone that mediates the plant response to biotic and abiotic stresses, and is deeply involved in stress-induced modulation of plant growth and development. Increasing evidence indicates that JA-dependent growth regulation largely depends on the crosstalk of JA with other growth-related hormones such as auxin, cytokinin, GA, and BR. Although some of the molecular mechanisms underlying these processes have been revealed, including protein–protein interactions between hormone signaling components, many of the questions about the complexity and dynamics of hormonal interactions still remain unanswered. Further molecular and genetic studies will expand our understanding of the mechanisms underlying JA crosstalk in the modulation of plant growth and development under stress conditions.

Author Contributions: G.J. and Y.D.C. designed the review; G.J. and Y.Y. access information; G.J. and Y.D.C. wrote the article with contributions of Y.Y. All authors have read and agreed to the published version of the manuscript.

Funding: This work was carried out with the support of the Cooperative Research Program for Agriculture Science and Technology Development (Project No. PJ01323901 and PJ01364301) Rural Development Administration, Republic of Korea, and the National Research Foundation of Korea Grant funded by the Korean Government (MOE) [NRF-2019R1A2C1007103].

Conflicts of Interest: The authors declare no conflict of interest.

References

1. Wasternack, C.; Strnad, M. Jasmonates: News on occurrence, biosynthesis, metabolism and action of an ancient group of signaling compounds. *Int. J. Mol. Sci.* **2018**, *19*, 2539. [CrossRef] [PubMed]
2. Ruan, J.; Zhou, Y.; Zhou, M.; Yan, J.; Khurshid, M.; Weng, W.; Cheng, J.; Zhang, K. Jasmonic acid signaling pathway in plants. *Int. J. Mol. Sci.* **2019**, *20*, 2479. [CrossRef] [PubMed]
3. Yang, J.; Duan, G.; Li, C.; Liu, L.; Han, G.; Zhang, Y.; Wang, C. The Crosstalks Between Jasmonic Acid and Other Plant Hormone Signaling Highlight the Involvement of Jasmonic Acid as a Core Component in Plant Response to Biotic and Abiotic Stresses. *Front. Plant Sci.* **2019**, *10*, 1349. [CrossRef] [PubMed]
4. Wang, J.; Wu, D.; Wang, Y.; Xie, D. Jasmonate action in plant defense against insects. *J. Exp. Bot.* **2019**, *70*, 3391–3400. [CrossRef]

5. Genva, M.; Akong, F.O.; Andersson, M.X.; Deleu, M.; Lins, L.; Fauconnier, M.-L. New insights into the biosynthesis of esterified oxylipins and their involvement in plant defense and developmental mechanisms. *Phytochem. Rev.* **2019**, *18*, 343–358. [CrossRef]

6. Feys, B.J.; Benedetti, C.E.; Penfold, C.N.; Turner, J.G. Arabidopsis mutants selected for resistance to the phytotoxin coronatine are male sterile, insensitive to methyl jasmonate, and resistant to a bacterial pathogen. *Plant Cell* **1994**, *6*, 751–759. [CrossRef]

7. Thines, B.; Katsir, L.; Melotto, M.; Niu, Y.; Mandaokar, A.; Liu, G.; Nomura, K.; He, S.Y.; Howe, G.A.; Browse, J. JAZ repressor proteins are targets of the SCF COI1 complex during jasmonate signalling. *Nature* **2007**, *448*, 661–665. [CrossRef]

8. Chini, A.; Fonseca, S.; Fernandez, G.; Adie, B.; Chico, J.; Lorenzo, O.; Garcia-Casado, G.; López-Vidriero, I.; Lozano, F.; Ponce, M. The JAZ family of repressors is the missing link in jasmonate signalling. *Nature* **2007**, *448*, 666–671. [CrossRef]

9. Xu, L.; Liu, F.; Lechner, E.; Genschik, P.; Crosby, W.L.; Ma, H.; Peng, W.; Huang, D.; Xie, D. The SCFCOI1 ubiquitin-ligase complexes are required for jasmonate response in Arabidopsis. *Plant Cell* **2002**, *14*, 1919–1935. [CrossRef]

10. Pauwels, L.; Barbero, G.F.; Geerinck, J.; Tilleman, S.; Grunewald, W.; Pérez, A.C.; Chico, J.M.; Bossche, R.V.; Sewell, J.; Gil, E. NINJA connects the co-repressor TOPLESS to jasmonate signalling. *Nature* **2010**, *464*, 788–791. [CrossRef]

11. Melotto, M.; Mecey, C.; Niu, Y.; Chung, H.S.; Katsir, L.; Yao, J.; Zeng, W.; Thines, B.; Staswick, P.; Browse, J. A critical role of two positively charged amino acids in the Jas motif of Arabidopsis JAZ proteins in mediating coronatine-and jasmonoyl isoleucine-dependent interactions with the COI1 F-box protein. *Plant J.* **2008**, *55*, 979–988. [CrossRef] [PubMed]

12. Gasperini, D.; Chételat, A.; Acosta, I.F.; Goossens, J.; Pauwels, L.; Goossens, A.; Dreos, R.; Alfonso, E.; Farmer, E.E. Multilayered organization of jasmonate signalling in the regulation of root growth. *PLoS Genet.* **2015**, *11*, e1005300. [CrossRef] [PubMed]

13. Major, I.T.; Yoshida, Y.; Campos, M.L.; Kapali, G.; Xin, X.F.; Sugimoto, K.; de Oliveira Ferreira, D.; He, S.Y.; Howe, G.A. Regulation of growth–defense balance by the JASMONATE ZIM-DOMAIN (JAZ)-MYC transcriptional module. *New Phytol.* **2017**, *215*, 1533–1547. [CrossRef] [PubMed]

14. Campos, M.L.; Yoshida, Y.; Major, I.T.; de Oliveira Ferreira, D.; Weraduwage, S.M.; Froehlich, J.E.; Johnson, B.F.; Kramer, D.M.; Jander, G.; Sharkey, T.D. Rewiring of jasmonate and phytochrome B signalling uncouples plant growth-defense tradeoffs. *Nat. Commun.* **2016**, *7*, 12570. [CrossRef] [PubMed]

15. Chandler, J.W.; Werr, W. Cytokinin–auxin crosstalk in cell type specification. *Trends Plant Sci.* **2015**, *20*, 291–300. [CrossRef] [PubMed]

16. Liu, J.; Moore, S.; Chen, C.; Lindsey, K. Crosstalk complexities between auxin, cytokinin, and ethylene in Arabidopsis root development: From experiments to systems modeling, and back again. *Mol. Plant* **2017**, *10*, 1480–1496. [CrossRef] [PubMed]

17. Li, N.; Han, X.; Feng, D.; Yuan, D.; Huang, L.-J. Signaling crosstalk between salicylic acid and ethylene/jasmonate in plant defense: Do we understand what they are whispering? *Int. J. Mol. Sci.* **2019**, *20*, 671. [CrossRef]

18. Liu, L.; Sonbol, F.-M.; Huot, B.; Gu, Y.; Withers, J.; Mwimba, M.; Yao, J.; He, S.Y.; Dong, X. Salicylic acid receptors activate jasmonic acid signalling through a non-canonical pathway to promote effector-triggered immunity. *Nat. Commun.* **2016**, *7*, 13099. [CrossRef]

19. De Ollas, C.; Dodd, I.C. Physiological impacts of ABA–JA interactions under water-limitation. *Plant Mol. Biol.* **2016**, *91*, 641–650. [CrossRef]

20. Pacifici, E.; Polverari, L.; Sabatini, S. Plant hormone cross-talk: The pivot of root growth. *J. Exp. Bot.* **2015**, *66*, 1113–1121. [CrossRef]

21. Um, T.Y.; Lee, H.Y.; Lee, S.; Chang, S.H.; Chung, P.J.; Oh, K.-B.; Kim, J.-K.; Jang, G.; Choi, Y.D. JASMONATE ZIM-DOMAIN PROTEIN 9 interacts with SLENDER RICE 1 to mediate the antagonistic interaction between jasmonic and gibberellic acid signals in rice. *Front. Plant Sci.* **2018**, *9*, 1866. [CrossRef] [PubMed]

22. Miceli, A.; Moncada, A.; Sabatino, L.; Vetrano, F. Effect of Gibberellic Acid on Growth, Yield, and Quality of Leaf Lettuce and Rocket Grown in a Floating System. *Agronomy* **2019**, *9*, 382. [CrossRef]

23. Liu, X.; Hou, X. Antagonistic regulation of ABA and GA in metabolism and signaling pathways. *Front. Plant Sci.* **2018**, *9*, 251. [CrossRef] [PubMed]
24. Achard, P.; Cheng, H.; De Grauwe, L.; Decat, J.; Schoutteten, H.; Moritz, T.; Van Der Straeten, D.; Peng, J.; Harberd, N.P. Integration of plant responses to environmentally activated phytohormonal signals. *Science* **2006**, *311*, 91–94. [CrossRef] [PubMed]
25. Magome, H.; Yamaguchi, S.; Hanada, A.; Kamiya, Y.; Oda, K. The DDF1 transcriptional activator upregulates expression of a gibberellin-deactivating gene, GA2ox7, under high-salinity stress in Arabidopsis. *Plant J.* **2008**, *56*, 613–626. [CrossRef] [PubMed]
26. Colebrook, E.H.; Thomas, S.G.; Phillips, A.L.; Hedden, P. The role of gibberellin signalling in plant responses to abiotic stress. *J. Exp. Biol.* **2014**, *217*, 67–75. [CrossRef]
27. Hou, X.; Lee, L.Y.C.; Xia, K.; Yan, Y.; Yu, H. DELLAs modulate jasmonate signaling via competitive binding to JAZs. *Dev. Cell* **2010**, *19*, 884–894. [CrossRef]
28. Yang, D.-L.; Yao, J.; Mei, C.-S.; Tong, X.-H.; Zeng, L.-J.; Li, Q.; Xiao, L.-T.; Sun, T.-P.; Li, J.; Deng, X.-W. Plant hormone jasmonate prioritizes defense over growth by interfering with gibberellin signaling cascade. *Proc. Natl. Acad. Sci. USA* **2012**, *109*, E1192–E1200. [CrossRef]
29. Wild, M.; Davière, J.-M.; Cheminant, S.; Regnault, T.; Baumberger, N.; Heintz, D.; Baltz, R.; Genschik, P.; Achard, P. The Arabidopsis DELLA RGA-LIKE3 is a direct target of MYC2 and modulates jasmonate signaling responses. *Plant Cell* **2012**, *24*, 3307–3319. [CrossRef]
30. Galvão, V.C.; Collani, S.; Horrer, D.; Schmid, M. Gibberellic acid signaling is required for ambient temperature-mediated induction of flowering in Arabidopsis thaliana. *Plant J.* **2015**, *84*, 949–962. [CrossRef]
31. Hedden, P.; Phillips, A.L. Gibberellin metabolism: New insights revealed by the genes. *Trends Plant Sci.* **2000**, *5*, 523–530. [CrossRef]
32. Thomas, S.G.; Hedden, P. Gibberellin metabolism and signal transduction. *Annu. Plant Rev. Online* **2018**, *2018*, 147–184.
33. Sun, T.-p. Gibberellin metabolism, perception and signaling pathways in Arabidopsis. *Arab. Book Am. Soc. Plant Biol.* **2008**, *6*, e0103. [CrossRef] [PubMed]
34. Yamaguchi, S. Gibberellin metabolism and its regulation. *Annu. Rev. Plant Biol.* **2008**, *59*, 225–251. [CrossRef] [PubMed]
35. Hedden, P.; Sponsel, V. A century of gibberellin research. *J. Plant Growth Regul.* **2015**, *34*, 740–760. [CrossRef]
36. Phillips, A.L.; Ward, D.A.; Uknes, S.; Appleford, N.E.; Lange, T.; Huttly, A.K.; Gaskin, P.; Graebe, J.E.; Hedden, P. Isolation and expression of three gibberellin 20-oxidase cDNA clones from Arabidopsis. *Plant Physiol.* **1995**, *108*, 1049–1057. [CrossRef]
37. Schomburg, F.M.; Bizzell, C.M.; Lee, D.J.; Zeevaart, J.A.; Amasino, R.M. Overexpression of a novel class of gibberellin 2-oxidases decreases gibberellin levels and creates dwarf plants. *Plant Cell* **2003**, *15*, 151–163. [CrossRef]
38. Rieu, I.; Eriksson, S.; Powers, S.J.; Gong, F.; Griffiths, J.; Woolley, L.; Benlloch, R.; Nilsson, O.; Thomas, S.G.; Hedden, P. Genetic analysis reveals that C19-GA 2-oxidation is a major gibberellin inactivation pathway in Arabidopsis. *Plant Cell* **2008**, *20*, 2420–2436. [CrossRef]
39. Daviere, J.-M.; Achard, P. A pivotal role of DELLAs in regulating multiple hormone signals. *Mol. Plant* **2016**, *9*, 10–20. [CrossRef]
40. Silverstone, A.L.; Chang, C.W.; Krol, E.; Sun, T.P. Developmental regulation of the gibberellin biosynthetic gene GA1 in Arabidopsis thaliana. *Plant J.* **1997**, *12*, 9–19. [CrossRef]
41. Sun, T.-p. The molecular mechanism and evolution of the GA–GID1–DELLA signaling module in plants. *Curr. Biol.* **2011**, *21*, R338–R345. [CrossRef] [PubMed]
42. Willige, B.C.; Ghosh, S.; Nill, C.; Zourelidou, M.; Dohmann, E.M.; Maier, A.; Schwechheimer, C. The DELLA domain of GA INSENSITIVE mediates the interaction with the GA INSENSITIVE DWARF1A gibberellin receptor of Arabidopsis. *Plant Cell* **2007**, *19*, 1209–1220. [CrossRef] [PubMed]
43. Todaka, D.; Nakashima, K.; Maruyama, K.; Kidokoro, S.; Osakabe, Y.; Ito, Y.; Matsukura, S.; Fujita, Y.; Yoshiwara, K.; Ohme-Takagi, M. Rice phytochrome-interacting factor-like protein OsPIL1 functions as a key regulator of internode elongation and induces a morphological response to drought stress. *Proc. Natl. Acad. Sci. USA* **2012**, *109*, 15947–15952. [CrossRef] [PubMed]

44. Feng, S.; Martinez, C.; Gusmaroli, G.; Wang, Y.; Zhou, J.; Wang, F.; Chen, L.; Yu, L.; Iglesias-Pedraz, J.M.; Kircher, S. Coordinated regulation of Arabidopsisthaliana development by light and gibberellins. *Nature* **2008**, *451*, 475–479. [CrossRef] [PubMed]

45. Sultan, S.E. Plant developmental responses to the environment: Eco-devo insights. *Curr. Opin. Plant Biol.* **2010**, *13*, 96–101. [CrossRef] [PubMed]

46. Cheng, H.; Qin, L.; Lee, S.; Fu, X.; Richards, D.E.; Cao, D.; Luo, D.; Harberd, N.P.; Peng, J. Gibberellin regulates Arabidopsis floral development via suppression of DELLA protein function. *Development* **2004**, *131*, 1055–1064. [CrossRef] [PubMed]

47. Dill, A.; Sun, T.-p. Synergistic derepression of gibberellin signaling by removing RGA and GAI function in Arabidopsis thaliana. *Genetics* **2001**, *159*, 777–785.

48. King, K.E.; Moritz, T.; Harberd, N.P. Gibberellins are not required for normal stem growth in Arabidopsis thaliana in the absence of GAI and RGA. *Genetics* **2001**, *159*, 767–776.

49. Achard, P.; Renou, J.-P.; Berthomé, R.; Harberd, N.P.; Genschik, P. Plant DELLAs restrain growth and promote survival of adversity by reducing the levels of reactive oxygen species. *Curr. Biol.* **2008**, *18*, 656–660. [CrossRef]

50. Jang, G.; Chang, S.H.; Um, T.Y.; Lee, S.; Kim, J.-K.; Do Choi, Y. Antagonistic interaction between jasmonic acid and cytokinin in xylem development. *Sci. Rep.* **2017**, *7*, 10212. [CrossRef]

51. Huot, B.; Yao, J.; Montgomery, B.L.; He, S.Y. Growth–defense tradeoffs in plants: A balancing act to optimize fitness. *Mol. Plant* **2014**, *7*, 1267–1287. [CrossRef] [PubMed]

52. Hu, Y.; Jiang, Y.; Han, X.; Wang, H.; Pan, J.; Yu, D. Jasmonate regulates leaf senescence and tolerance to cold stress: Crosstalk with other phytohormones. *J. Exp. Bot.* **2017**, *68*, 1361–1369. [CrossRef] [PubMed]

53. Kieber, J.J.; Schaller, G.E. Cytokinin signaling in plant development. *Development* **2018**, *145*, dev149344. [CrossRef] [PubMed]

54. Hare, P.; Cress, W.; Van Staden, J. The involvement of cytokinins in plant responses to environmental stress. *Plant Growth Regul.* **1997**, *23*, 79–103. [CrossRef]

55. Howe, G.A.; Schilmiller, A.L. Oxylipin metabolism in response to stress. *Curr. Opin. Plant Biol.* **2002**, *5*, 230–236. [CrossRef]

56. Argueso, C.T.; Ferreira, F.J.; Kieber, J.J. Environmental perception avenues: The interaction of cytokinin and environmental response pathways. *Plant Cell Environ.* **2009**, *32*, 1147–1160. [CrossRef]

57. Seo, J.S.; Joo, J.; Kim, M.J.; Kim, Y.K.; Nahm, B.H.; Song, S.I.; Cheong, J.J.; Lee, J.S.; Kim, J.K.; Choi, Y.D. OsbHLH148, a basic helix-loop-helix protein, interacts with OsJAZ proteins in a jasmonate signaling pathway leading to drought tolerance in rice. *Plant J.* **2011**, *65*, 907–921. [CrossRef]

58. Bandurska, H.; Stroiński, A.; Kubiś, J. The effect of jasmonic acid on the accumulation of ABA, proline and spermidine and its influence on membrane injury under water deficit in two barley genotypes. *Acta Physiol. Plant.* **2003**, *25*, 279–285. [CrossRef]

59. Qiu, Z.; Guo, J.; Zhu, A.; Zhang, L.; Zhang, M. Exogenous jasmonic acid can enhance tolerance of wheat seedlings to salt stress. *Ecotoxicol. Environ. Saf.* **2014**, *104*, 202–208. [CrossRef]

60. Nishiyama, R.; Watanabe, Y.; Fujita, Y.; Le, D.T.; Kojima, M.; Werner, T.; Vankova, R.; Yamaguchi-Shinozaki, K.; Shinozaki, K.; Kakimoto, T. Analysis of cytokinin mutants and regulation of cytokinin metabolic genes reveals important regulatory roles of cytokinins in drought, salt and abscisic acid responses, and abscisic acid biosynthesis. *Plant Cell* **2011**, *23*, 2169–2183. [CrossRef]

61. Miyawaki, K.; Tarkowski, P.; Matsumoto-Kitano, M.; Kato, T.; Sato, S.; Tarkowska, D.; Tabata, S.; Sandberg, G.; Kakimoto, T. Roles of Arabidopsis ATP/ADP isopentenyltransferases and tRNA isopentenyltransferases in cytokinin biosynthesis. *Proc. Natl. Acad. Sci. USA* **2006**, *103*, 16598–16603. [CrossRef] [PubMed]

62. Schäfer, M.; Brütting, C.; Meza-Canales, I.D.; Großkinsky, D.K.; Vankova, R.; Baldwin, I.T.; Meldau, S. The role of cis-zeatin-type cytokinins in plant growth regulation and mediating responses to environmental interactions. *J. Exp. Bot.* **2015**, *66*, 4873–4884. [CrossRef] [PubMed]

63. Gajdošová, S.; Spíchal, L.; Kamínek, M.; Hoyerová, K.; Novák, O.; Dobrev, P.I.; Galuszka, P.; Klíma, P.; Gaudinová, A.; Žižková, E. Distribution, biological activities, metabolism, and the conceivable function of cis-zeatin-type cytokinins in plants. *J. Exp. Bot.* **2011**, *62*, 2827–2840. [CrossRef] [PubMed]

64. Sun, J.; Niu, Q.-W.; Tarkowski, P.; Zheng, B.; Tarkowska, D.; Sandberg, G.; Chua, N.-H.; Zuo, J. The Arabidopsis AtIPT8/PGA22 gene encodes an isopentenyl transferase that is involved in de novo cytokinin biosynthesis. *Plant Physiol.* **2003**, *131*, 167–176. [CrossRef] [PubMed]

65. Kakimoto, T. Identification of plant cytokinin biosynthetic enzymes as dimethylallyl diphosphate: ATP/ADP isopentenyltransferases. *Plant Cell Physiol.* **2001**, *42*, 677–685. [CrossRef]

66. Spíchal, L.; Rakova, N.Y.; Riefler, M.; Mizuno, T.; Romanov, G.A.; Strnad, M.; Schmülling, T. Two cytokinin receptors of Arabidopsis thaliana, CRE1/AHK4 and AHK3, differ in their ligand specificity in a bacterial assay. *Plant Cell Physiol.* **2004**, *45*, 1299–1305. [CrossRef]

67. Werner, T.; Motyka, V.; Laucou, V.; Smets, R.; Van Onckelen, H.; Schmülling, T. Cytokinin-deficient transgenic Arabidopsis plants show multiple developmental alterations indicating opposite functions of cytokinins in the regulation of shoot and root meristem activity. *Plant Cell* **2003**, *15*, 2532–2550. [CrossRef]

68. Xie, M.; Chen, H.; Huang, L.; O'Neil, R.C.; Shokhirev, M.N.; Ecker, J.R. A B-ARR-mediated cytokinin transcriptional network directs hormone cross-regulation and shoot development. *Nat. Commun.* **2018**, *9*, 1604. [CrossRef]

69. Hutchison, C.E.; Li, J.; Argueso, C.; Gonzalez, M.; Lee, E.; Lewis, M.W.; Maxwell, B.B.; Perdue, T.D.; Schaller, G.E.; Alonso, J.M. The Arabidopsis histidine phosphotransfer proteins are redundant positive regulators of cytokinin signaling. *Plant Cell* **2006**, *18*, 3073–3087. [CrossRef]

70. Yan, Z.; Liu, X.; Ljung, K.; Li, S.; Zhao, W.; Yang, F.; Wang, M.; Tao, Y. Type B response regulators act as central integrators in transcriptional control of the auxin biosynthesis enzyme TAA1. *Plant Physiol.* **2017**, *175*, 1438–1454. [CrossRef]

71. Cortleven, A.; Leuendorf, J.E.; Frank, M.; Pezzetta, D.; Bolt, S.; Schmülling, T. Cytokinin action in response to abiotic and biotic stresses in plants. *Plant Cell Environ.* **2019**, *42*, 998–1018. [CrossRef] [PubMed]

72. Tran, L.-S.P.; Urao, T.; Qin, F.; Maruyama, K.; Kakimoto, T.; Shinozaki, K.; Yamaguchi-Shinozaki, K. Functional analysis of AHK1/ATHK1 and cytokinin receptor histidine kinases in response to abscisic acid, drought, and salt stress in Arabidopsis. *Proc. Natl. Acad. Sci. USA* **2007**, *104*, 20623–20628. [CrossRef] [PubMed]

73. Ueda, J.; Kato, J. Inhibition of cytokinin-induced plant growth by jasmonic acid and its methyl ester. *Physiol. Plant.* **1982**, *54*, 249–252. [CrossRef]

74. Liu, L.; Li, H.; Zeng, H.; Cai, Q.; Zhou, X.; Yin, C. Exogenous jasmonic acid and cytokinin antagonistically regulate rice flag leaf senescence by mediating chlorophyll degradation, membrane deterioration, and senescence-associated genes expression. *J. Plant Growth Regul.* **2016**, *35*, 366–376. [CrossRef]

75. Mukherjee, I.; Reid, D.; Naik, G. Influence of cytokinins on the methyl jasmonate-promoted senescence in Helianthus annuus cotyledons. *Plant Growth Regul.* **2002**, *38*, 61–68.

76. Bishopp, A.; Help, H.; El-Showk, S.; Weijers, D.; Scheres, B.; Friml, J.; Benková, E.; Mähönen, A.P.; Helariutta, Y. A mutually inhibitory interaction between auxin and cytokinin specifies vascular pattern in roots. *Curr. Biol.* **2011**, *21*, 917–926. [CrossRef]

77. De Rybel, B.; Mähönen, A.P.; Helariutta, Y.; Weijers, D. Plant vascular development: From early specification to differentiation. *Nat. Rev. Mol. Cell Biol.* **2016**, *17*, 30. [CrossRef]

78. Yokoyama, A.; Yamashino, T.; Amano, Y.-I.; Tajima, Y.; Imamura, A.; Sakakibara, H.; Mizuno, T. Type-B ARR transcription factors, ARR10 and ARR12, are implicated in cytokinin-mediated regulation of protoxylem differentiation in roots of Arabidopsis thaliana. *Plant Cell Physiol.* **2007**, *48*, 84–96. [CrossRef]

79. Jang, G.; Choi, Y.D. Drought stress promotes xylem differentiation by modulating the interaction between cytokinin and jasmonic acid. *Plant Signal. Behav.* **2018**, *13*, e1451707. [CrossRef]

80. Nitschke, S.; Cortleven, A.; Iven, T.; Feussner, I.; Havaux, M.; Riefler, M.; Schmülling, T. Circadian stress regimes affect the circadian clock and cause jasmonic acid-dependent cell death in cytokinin-deficient Arabidopsis plants. *Plant Cell* **2016**, *28*, 1616–1639. [CrossRef]

81. Pavlů, J.; Novak, J.; Koukalová, V.; Luklova, M.; Brzobohatý, B.; Černý, M. Cytokinin at the crossroads of abiotic stress signalling pathways. *Int. J. Mol. Sci.* **2018**, *19*, 2450. [CrossRef] [PubMed]

82. Le, D.T.; Nishiyama, R.; Watanabe, Y.; Vankova, R.; Tanaka, M.; Seki, M.; Yamaguchi-Shinozaki, K.; Shinozaki, K.; Tran, L.-S.P. Identification and expression analysis of cytokinin metabolic genes in soybean under normal and drought conditions in relation to cytokinin levels. *PLoS ONE* **2012**, *7*, e42411. [CrossRef] [PubMed]

83. Figueiredo, D.D.; Köhler, C. Auxin: A molecular trigger of seed development. *Genes Dev.* **2018**, *32*, 479–490. [CrossRef] [PubMed]

84. Wang, Y.; Yang, W.; Zuo, Y.; Zhu, L.; Hastwell, A.H.; Chen, L.; Tian, Y.; Su, C.; Ferguson, B.J.; Li, X. GmYUC2a mediates auxin biosynthesis during root development and nodulation in soybean. *J. Exp. Bot.* **2019**, *70*, 3165–3176. [CrossRef]

85. Wang, B.; Chu, J.; Yu, T.; Xu, Q.; Sun, X.; Yuan, J.; Xiong, G.; Wang, G.; Wang, Y.; Li, J. Tryptophan-independent auxin biosynthesis contributes to early embryogenesis in Arabidopsis. *Proc. Natl. Acad. Sci. USA* **2015**, *112*, 4821–4826. [CrossRef]

86. Casanova-Sáez, R.; Voß, U. Auxin metabolism controls developmental decisions in land plants. *Trends Plant Sci.* **2019**, *24*, 741–754. [CrossRef]

87. Leyser, O. Auxin signaling. *Plant Physiol.* **2017**, *176*, 465–479. [CrossRef]

88. Dharmasiri, N.; Dharmasiri, S.; Estelle, M. The F-box protein TIR1 is an auxin receptor. *Nature* **2005**, *435*, 441–445. [CrossRef]

89. Miyamoto, K.; Oka, M.; Ueda, J. Update on the possible mode of action of the jasmonates: Focus on the metabolism of cell wall polysaccharides in relation to growth and development. *Physiol. Plant.* **1997**, *100*, 631–638. [CrossRef]

90. Saniewski, M.; Ueda, J.; Miyamoto, K. Relationships between jasmonates and auxin in regulation of some physiological processes in higher plants. *Acta Physiol. Plant.* **2002**, *24*, 211. [CrossRef]

91. Ursache, R.; Miyashima, S.; Chen, Q.; Vatén, A.; Nakajima, K.; Carlsbecker, A.; Zhao, Y.; Helariutta, Y.; Dettmer, J. Tryptophan-dependent auxin biosynthesis is required for HD-ZIP III-mediated xylem patterning. *Development* **2014**, *141*, 1250–1259. [CrossRef] [PubMed]

92. Zhang, Y.; He, P.; Ma, X.; Yang, Z.; Pang, C.; Yu, J.; Wang, G.; Friml, J.; Xiao, G. Auxin-mediated statolith production for root gravitropism. *New Phytol.* **2019**, *224*, 761–774. [CrossRef] [PubMed]

93. Chen, Q.; Sun, J.; Zhai, Q.; Zhou, W.; Qi, L.; Xu, L.; Wang, B.; Chen, R.; Jiang, H.; Qi, J. The basic helix-loop-helix transcription factor MYC2 directly represses PLETHORA expression during jasmonate-mediated modulation of the root stem cell niche in Arabidopsis. *Plant Cell* **2011**, *23*, 3335–3352. [CrossRef] [PubMed]

94. Mähönen, A.P.; Ten Tusscher, K.; Siligato, R.; Smetana, O.; Díaz-Triviño, S.; Salojärvi, J.; Wachsman, G.; Prasad, K.; Heidstra, R.; Scheres, B. PLETHORA gradient formation mechanism separates auxin responses. *Nature* **2014**, *515*, 125–129. [CrossRef] [PubMed]

95. Reeves, P.H.; Ellis, C.M.; Ploense, S.E.; Wu, M.-F.; Yadav, V.; Tholl, D.; Chételat, A.; Haupt, I.; Kennerley, B.J.; Hodgens, C. A regulatory network for coordinated flower maturation. *PLoS Genet.* **2012**, *8*, e1002506. [CrossRef] [PubMed]

96. Huang, H.; Gao, H.; Liu, B.; Qi, T.; Tong, J.; Xiao, L.; Xie, D.; Song, S. Arabidopsis MYB24 regulates jasmonate-mediated stamen development. *Front. Plant Sci.* **2017**, *8*, 1525. [CrossRef]

97. Qi, T.; Wang, J.; Huang, H.; Liu, B.; Gao, H.; Liu, Y.; Song, S.; Xie, D. Regulation of jasmonate-induced leaf senescence by antagonism between bHLH subgroup IIIe and IIId factors in Arabidopsis. *Plant Cell* **2015**, *27*, 1634–1649. [CrossRef]

98. Yu, J.; Zhang, Y.; Di, C.; Zhang, Q.; Zhang, K.; Wang, C.; You, Q.; Yan, H.; Dai, S.Y.; Yuan, J.S. JAZ7 negatively regulates dark-induced leaf senescence in Arabidopsis. *J. Exp. Bot.* **2015**, *67*, 751–762. [CrossRef]

99. Zhu, X.; Chen, J.; Xie, Z.; Gao, J.; Ren, G.; Gao, S.; Zhou, X.; Kuai, B. Jasmonic acid promotes degreening via MYC 2/3/4-and ANAC 019/055/072-mediated regulation of major chlorophyll catabolic genes. *Plant J.* **2015**, *84*, 597–610. [CrossRef]

100. Jiang, Y.; Liang, G.; Yang, S.; Yu, D. Arabidopsis WRKY57 functions as a node of convergence for jasmonic acid–and auxin-mediated signaling in jasmonic acid–induced leaf senescence. *Plant Cell* **2014**, *26*, 230–245. [CrossRef]

101. He, X.; Jiang, J.; Wang, C.Q.; Dehesh, K. ORA59 and EIN3 interaction couples jasmonate-ethylene synergistic action to antagonistic salicylic acid regulation of PDF expression. *J. Integr. Plant Biol.* **2017**, *59*, 275–287. [CrossRef] [PubMed]

102. Per, T.S.; Khan, M.I.R.; Anjum, N.A.; Masood, A.; Hussain, S.J.; Khan, N.A. Jasmonates in plants under abiotic stresses: Crosstalk with other phytohormones matters. *Environ. Exp. Bot.* **2018**, *145*, 104–120. [CrossRef]

103. Ren, C.; Han, C.; Peng, W.; Huang, Y.; Peng, Z.; Xiong, X.; Zhu, Q.; Gao, B.; Xie, D. A leaky mutation in DWARF4 reveals an antagonistic role of brassinosteroid in the inhibition of root growth by jasmonate in Arabidopsis. *Plant Physiol.* **2009**, *151*, 1412–1420. [CrossRef] [PubMed]

104. Qi, T.; Huang, H.; Wu, D.; Yan, J.; Qi, Y.; Song, S.; Xie, D. Arabidopsis DELLA and JAZ proteins bind the WD-repeat/bHLH/MYB complex to modulate gibberellin and jasmonate signaling synergy. *Plant Cell* **2014**, *26*, 1118–1133. [CrossRef] [PubMed]
105. Cai, X.-T.; Xu, P.; Zhao, P.-X.; Liu, R.; Yu, L.-H.; Xiang, C.-B. Arabidopsis ERF109 mediates cross-talk between jasmonic acid and auxin biosynthesis during lateral root formation. *Nat. Commun.* **2014**, *5*, 5833. [CrossRef] [PubMed]

 © 2020 by the authors. Licensee MDPI, Basel, Switzerland. This article is an open access article distributed under the terms and conditions of the Creative Commons Attribution (CC BY) license (http://creativecommons.org/licenses/by/4.0/).

International Journal of
Molecular Sciences

Review

Jasmonic Acid Signaling Pathway in Response to Abiotic Stresses in Plants

Md. Sarafat Ali and Kwang-Hyun Baek *

Department of Biotechnology, Yeungnam University, Gyeongsan, Gyeongbuk 38541, Korea;
sarafatbiotech@ynu.ac.kr
* Correspondence: khbaek@ynu.ac.kr; Tel.: +82-53-810-3029

Received: 26 December 2019; Accepted: 16 January 2020; Published: 17 January 2020

Abstract: Plants as immovable organisms sense the stressors in their environment and respond to them by means of dedicated stress response pathways. In response to stress, jasmonates (jasmonic acid, its precursors and derivatives), a class of polyunsaturated fatty acid-derived phytohormones, play crucial roles in several biotic and abiotic stresses. As the major immunity hormone, jasmonates participate in numerous signal transduction pathways, including those of gene networks, regulatory proteins, signaling intermediates, and proteins, enzymes, and molecules that act to protect cells from the toxic effects of abiotic stresses. As cellular hubs for integrating informational cues from the environment, jasmonates play significant roles in alleviating salt stress, drought stress, heavy metal toxicity, micronutrient toxicity, freezing stress, ozone stress, CO_2 stress, and light stress. Besides these, jasmonates are involved in several developmental and physiological processes throughout the plant life. In this review, we discuss the biosynthesis and signal transduction pathways of the JAs and the roles of these molecules in the plant responses to abiotic stresses.

Keywords: abiotic stresses; jasmonates; JA-Ile; JAZ repressors; transcription factor; signaling

---

1. Introduction

Plants grow in environments that impose a variety of biotic and abiotic stresses. The primary abiotic stresses that influence plant growth include light, temperature, salt, carbon dioxide, water, ozone, and soil nutrient content and availability [1], where the fluctuation of any of these can hamper the normal physiological processes. Being static organisms, plants are unable to avoid abiotic stresses simply by moving to a suitable environment. Consequently, they have evolved mechanisms to compensate for the unwanted stressful conditions by altering their own developmental and physiological processes.

The growth, development, and survival of plants depend on complex biological networks coupled with anabolic and catabolic pathways [2]. Abiotic stresses can disrupt these network pathways, resulting in their uncoupling. For example, extremely high or low temperatures might inhibit a subset of enzymes in the same or connected pathways [3], and hence various intermediate compounds might accumulate as a result of this functional uncoupling of metabolic pathways [4]. These intermediate compounds could be converted to toxic by-products that might affect the cell's survival or longevity [5]. Reactive oxygen species (ROS) are one of the most common groups of toxic intermediates produced by abiotic stresses.

Phytohormones, the regulators of plant development, are central players in sensing and signaling diverse environmental conditions, such as drought, osmotic stress, chilling injury, heavy metal toxicity, etc. [6]. There are currently nine known major classes of naturally occurring phytohormones (viz., auxins, gibberellins, cytokinins, abscisic acid (ABA), ethylene (ET), brassinosteroids, jasmonic acid (JA), salicylic acid (SA), and strigolactones), all of which evoke many different responses.

Specifically, JA and its derivatives (e.g., jasmonyl isoleucine (JA-Ile), *cis*-jasmone, JA-glucosyl ester, methyl jasmonate (MeJA), jasmonoyl-amino acid, 12-hydroxyjasmonic acid sulfate (12-HSO$_4$-JA), 12-*O*-glucosyl-JA, JA-Ile methyl ester, JA-Ile glucosyl ester, 12-carboxy-JA-Ile, 12-*O*-glucosyl-JA-Ile, and lactones of 12-hydroxy-JA-Ile), which are collectively known as jasmonates (JAs), are fatty acids derived from cyclopentanones and belong to the family of oxidized lipids that are collectively known as oxylipins [7]. These oxylipins are biologically active signaling molecules that are produced either enzymatically by lipoxygenases or alpha-dioxygenases, or nonenzymatically through the autoxidation of polyunsaturated fatty acids [8].

The JAs are ubiquitous in higher plant species, where their levels are high in the reproductive tissues and flowers, but very low in the mature leaves and roots [9,10]. JAs modulate many crucial processes in plant growth and development, such as vegetative growth, cell cycle regulation, anthocyanin biosynthesis, stamen and trichome development, fruit ripening, senescence, rubisco biosynthesis inhibition, stomatal opening, nitrogen and phosphorus uptake, and glucose transport [10–25]. As signaling molecules, JAs regulate the expression of numerous genes in response to abiotic stresses (e.g., salt, drought, heavy metals, micronutrient toxicity, low temperature, etc.) and promote specific protective mechanisms (Figure 1) [26]. In this review, we focus on the biosynthesis and signaling of JA, *cis*-jasmone, MeJA, and JA-Ile in response to abiotic stresses because of the high bioactivity of these compounds.

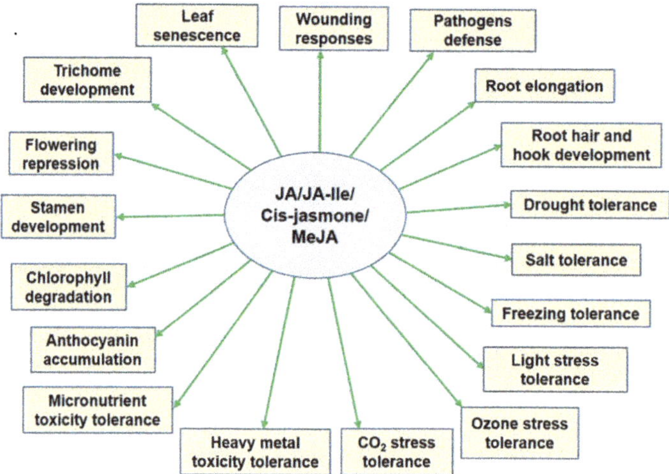

Figure 1. Various plant processes modulated by jasmonic acid and its isoleucine conjugate in response to abiotic stresses. JA, jasmonic acid; JA-Ile, jasmonyl isoleucine; MeJA, methyl jasmonate.

2. Abiotic Stress-Sensing Mechanisms in Plants

Abiotic stresses alter the physiological processes in plants by affecting gene expression, RNA or protein stability, the coupling of reactions, ion transport, or other cellular functions [27]. Any of these alterations could be a signal to the plant that a change in environmental conditions has occurred and that it is the optimum time to respond by either activating the stress-response pathways or altering existing ones. Some of the mechanisms used by plants to sense the abiotic stresses are as follows [28]: (i) Physical sensing, involving mechanical effects of the stress on the plant or cell structure, such as contraction of the plasma membrane from the cell wall during drought stress; (ii) biophysical sensing, involving changes of the protein structure or enzymatic activity, such as the inhibition of different enzymes during heat stress; (iii) metabolic sensing, involving the detection of by-product accumulation due to the uncoupling of electron transfer or enzymatic reactions, such ROS accumulation due to high light intensity; (iv) biochemical sensing, involving the presence of specialized proteins to sense a particular stress, such as calcium channels that can alter the Ca^{2+} homeostasis and sense changes in the temperature; and (v) epigenetic sensing, involving modifications of the DNA or RNA structure without altering the genetic sequences, such as the changes in chromatin that occur during temperature stress [28–30]. These stress-sensing mechanisms can activate downstream signal transduction pathways individually or in combination. Consequently, plants activate various anti-stress mechanisms to acclimate or adapt to the various stresses.

3. Biosynthesis and Metabolism of Jasmonic Acid during Abiotic Stress

During the last decades, the biosynthesis of JA has been well characterized in a variety of monocotyledonous and dicotyledonous plants [10,31,32]. To summarize, JA is biosynthesized through the consecutive action of enzymes present in the plastid, peroxisome, and cytoplasm (Figure 2) [33]. Abiotic (and biotic) stimuli activate phospholipases in the plastid membrane, promoting the synthesis of linolenic acid (18:3) in the plant [10,34]. Linolenic acid, a precursor in the JA biosynthesis process, is converted to 12-oxo-phytodienoic acid (12-oxo-PDA) through oxygenation with lipoxygenase (LOX), allene oxide synthase (AOS), and allene oxide cyclase (AOC). JA is then synthesized from 12-oxo-PDA by the activity of 12-oxo-phytodienoic acid reductase (OPR) and 3 cycles of beta-oxidation. Therefore, the JA biosynthetic pathway is known as the octadecanoid pathway [32,34,35].

In the cytosol, JA metabolic pathways convert the phytohormone into more than 30 distinct active and inactive derivatives, depending on the chemical modification of the carboxylic acid group, the pentenyl side chain, or the pentanone ring (Figure 2) [36–40]. Among the series of metabolites, free JA, *cis*-jasmone, MeJA, and JA-Ile are considered to be the major forms of bioactive JA in plants [10,41]. *cis*-jasmone is produced through the decar-boxylation of JA (Figure 3) [42]. The volatile MeJA is produced from JA through the activity of JA carboxyl methyltransferase (Figure 3) [26]. Jasmonate amino acid synthetase 1 (JAR1) catalyzes the reversible conversion between JA and JA-Ile (Figure 3) [41]. Evidence suggests that JA-Ile is an important compound in the JA signal transduction pathway [43].

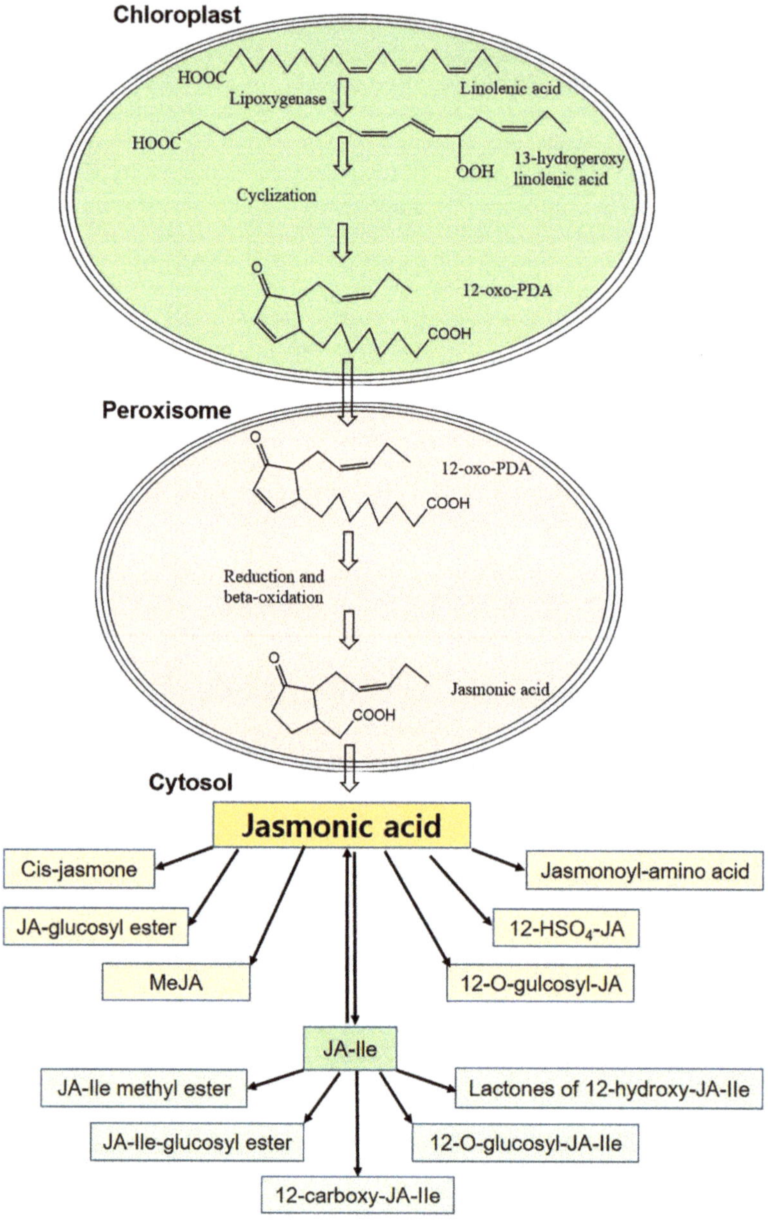

Figure 2. Schematic diagram of jasmonic acid biosynthesis and metabolism in response to abiotic stresses. In the chloroplast, JA biosynthesis begins with the chloroplast membrane release of linolenic acid, which is finally converted to 12-oxo-PDA. Upon transport of 12-oxo-PDA into the peroxisome, a series of enzymes work to convert it to JA, which is then exported to the cytoplasm. JA may be metabolized into different compounds depending on the chemical modification of the carboxylic acid group, the pentenyl side chain, or the pentanone ring. JA, jasmonic acid; JA-Ile, jasmonyl isoleucine; MeJA, methyl jasmonate; 12-HSO$_4$-JA, 12-hydroxyjasmonic acid sulfate; 12-oxo-PDA, 12-oxo-phytodienoic acid.

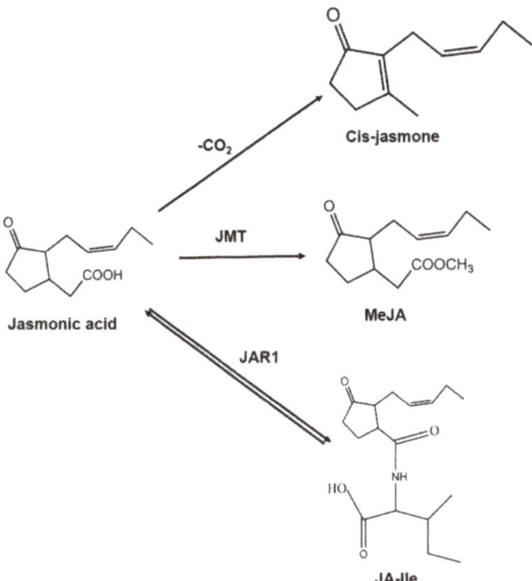

Figure 3. Major bioactive jasmonates in plants and their bioconversion. $-CO_2$, decar-boxylation; JMT, jasmonic acid carboxyl methyltransferase; MeJA, methyl jasmonate; JAR1, jasmonate amino acid synthetase 1; JA-Ile, jasmonyl isoleucine.

4. Jasmonic Acid Signaling during Abiotic Stress

In the plant cell cytoplasm, the most bioactive JA is JA-Ile, the level of which is very low under normal conditions [41]. Upon stress stimulation, JA undergoes epimerization to form JA-Ile, which accumulates in the cytoplasm of the stressed leaves. JA-Ile is transported to the nucleus and adjacent sites of the leaves for defensive responses [44,45]. In *Arabidopsis thaliana* (At), the subcellular localization of JAs are regulated by a high-affinity transporter, jasmonic acid transfer protein 1 (AtJAT1, also known as AtABCG16) [46]. Both the plasma membrane and nuclear membrane of plant cells contain JAT1, through which JA or JA-Ile is exported from the cytoplasm to the nucleus and apoplast [46]. Therefore, the dynamics of JA or JA-Ile in the cytoplasm, nucleus, and apoplast is regulated by JAT1 during abiotic stress.

JA or JA-Ile in the apoplast activates the JA signaling pathways in other cells. JA signals can transmit long distances via vascular bundles and/or air transmission. After their synthesis, JA and MeJA are transmitted in plants systemically [47]; that is, they can transfer to different parts of the plant via the vascular bundles [48]. During such transportation, JAs are not only transported but are also resynthesized [47], a fact that has been proven by the localization of various JA synthetases in the companion cell–sieve element complex of the vascular bundles in the tomato plant [49]. The JA precursor 12-oxo-PDA is formed in the sieve elements of the phloem, which is another indication of the resynthesis of JAs transported through the vascular bundles [50]. Compared with JA, MeJA can diffuse easily to distant leaves and adjacent plants owing its strong volatility and high capability of penetrating the cell membrane [40].

Under normal conditions, the promoters of jasmonate-responsive genes are not activated by the different types of transcription factors (TFs) due to the low level of JA-Ile (Figure 4). The various TFs [51] are repressed by a series of jasmonate-zinc finger inflorescence meristem (ZIM) domain (JAZ) proteins that act as transcriptional repressors (Table 1). The JAZ repressors recruit the protein topless (TPL) and the interactor/adaptor protein novel interactor of JAZ (NINJA); together, they form an effective transcriptional repression complex that acts to inhibit the expression of jasmonate-responsive genes by changing the open complex to a closed one through the further recruitment of histone deacetylase 6 (HDA6) and HDA19 [43,52–55].

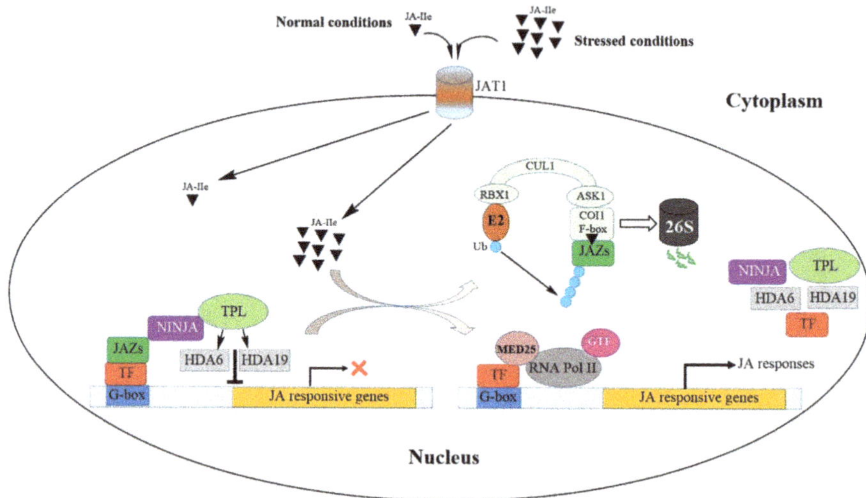

Figure 4. Jasmonic acid perception and signal transduction during abiotic stress. In the absence of abiotic stimuli or at a low level of JA-Ile, the transcription factors are repressed by JAZ proteins, thereby preventing their activation of the promoters of jasmonate-responsive genes. JAZ proteins recruit TPL and adaptor protein NINJA to form an active transcriptional repression complex that inhibits JA responses by changing the open complex to a closed one through the further recruitment of HDA6 and HDA19. Abiotic stresses elevate JA synthesis, which is readily epimerized to JA-Ile. The latter is then transported to the nucleus by the JAT1 transporter. JA-Ile facilitates the interaction of JAZ with the F-box protein COI1 within the SCF complex, leading to the proteasomal degradation of JAZ. The derepressed TF binds to the G-box element, whereupon MED25, RNA Pol II, and GTF are recruited, resulting in the expression of jasmonate-responsive genes. JA, jasmonic acid; JA-Ile, jasmonyl isoleucine; JAT1, jasmonic acid transfer protein 1; TF, transcription factor; JAZ, jasmonate ZIM domain; NINJA, novel interactor of JAZ; TPL, topless; HDA6, HDA19, histone deacetylase 6, 19; Ub, ubiquitin; E2, ubiquitin-conjugating enzymes; RBX1, ring box 1; CUL1, cullin 1; ASK1, *Arabidopsis* SKP1 homolog 1; COI1, coronatine insensitive 1; MED25, mediator 25; RNA Pol II, RNA polymerase II; GTF, general transcription factor.

To date, 13 JAZ proteins have been identified in *Arabidopsis*, most of which have two conserved domains: the central domain known as the ZIM domain [56–59], and the C-terminal JA-associated (Jas) domain [56]. The various domains present in the JAZ proteins facilitate their protein-protein interactions [60]. The JAZ proteins interact with the TFs via the ZIM domain, interacting with NINJA (which contains an ethylene-responsive element binding factor-associated amphiphilic repression (EAR) motif) and recruiting TPL to form the JAZ–NINJA–TPL repressor complex [54,55]. Among the 13 JAZ proteins of *Arabidopsis*, JAZ5, JAZ6, JAZ7, JAZ8, and JAZ13 contain an additional EAR motif that can interact directly with TPL in the absence of NINJA [57,59]. Within the Jas domain, the minimal amino acid sequence that can bind the coronatine or JA-Ile is termed the JAZ degron, the bipartite structure of which contains a loop and an amphipathic alpha-helix that binds to coronatine or JA-Ile and coronatine insensitive 1 (COI1), respectively [61].

Abiotic stresses elevate the processes that lead to JA-Ile formation in the cytosol and its transportation to the nucleus. JA-Ile is the natural bioactive ligand of *A. thaliana*, as affirmed by gas chromatography-mass spectrometry and high-performance liquid chromatography analyses [41]. Among JA, JA-Ile, MeJA, and 12-oxo-PDA, only JA-Ile can promote *COI1*-JAZ binding [58].

The ubiquitin–proteasome complex comprises suppressor of kinetochore protein 1 (SKP1)–cullin–F-box (SCF). The *Arabidopsis COI1* mutant lacks all responses to JA [62]. The *COI1* gene encodes an F-box protein, which associates with SKP1 and cullin to form SCF-type E3 ubiquitin ligase [63]. During abiotic stress, the JA-Ile that is formed and transported to the nucleus is recognized by the F-box protein COI1. JA-Ile facilitates the interaction of JAZ with COI1 within the SCF complex [63,64], with inositol pentakisphosphate serving as a cofactor in the formation of the COI1–JAZ co-receptor complex [61,65]. Ubiquitination of the JAZ protein leads to its proteasomal degradation and the release of the TFs to modulate the expression of jasmonate-responsive genes, thereby regulating the jasmonate-regulated defenses and growth. Mediator 25 (MED25), a subunit of the *Arabidopsis* mediator complex [66], bridges the communication between the gene-specific TF, RNA polymerase II, and the general transcription machinery [67]. Several lines of evidence have indicated that every aspect of JA function is due to the matching pairs of TFs with a subset of JAZ repressors to orchestrate the expression of jasmonate-responsive genes [64,68–71].

Table 1. Transcription factors that interact with the jasmonate-ZIM domain proteins and their corresponding JA-regulated plant responses (adapted from Zhai et al. [72]; Zhu and Lee [73]).

JAZ Domains	JAZ-Interacting DNA-Binding Transcription Factors	Physiological Functions
JAZs	MYC2/3/4/5	Root elongation, wounding responses, defense, metabolism, hook development [58,74–77]
JAZ1/8/10/11	MYB21/24	Stamen development and fertility [71]
JAZ1/2/5/6/8/9/10/11	TT8/GL3/EGL3 /MYB75/GL1	Trichome development and anthocyanin synthesis [70]
JAZ1/3/4/9	FIL/YAB1	Chlorophyll degradation and anthocyanin accumulation [78]
JAZ9/11	OsRSS3/OsbHLH148	Confer drought and salt tolerance [79,80]
JAZ1/4/9	ICE1/2	Increase freezing tolerance [68]
JAZ4/8	WRKY57	Promote leaf senescence [69]
JAZ1/3/9	EIN3/EIL1	Root elongation, defense, root hair and hook development [81]
JAZ1/3/4/9	TOE1/2	Repression of flowering during early vegetative development [82]
JAZs except JAZ7/12	bHLH03/13/14/17	Root elongation, fertility, defense, anthocyanin synthesis [83–86]

JA, jasmonic acid; JAZ, jasmonate ZIM domain.

5. Regulation of Diverse Jasmonic Acid Responses by Transcription Factors during Abiotic Stress

Abiotic stresses induce JA signaling through the derepression of TFs. JAZ proteins interact with the MYC and MYB TFs and suppress the expression of jasmonate-responsive genes [56]. JAZ proteins are stimulated for proteosomal degradation in the presence of the bioactive ligand JA-Ile [56]. Studies have revealed that several other TFs (e.g., NAC, ERF, and WRKY) are also involved in JA signaling [87–89]. In addition to the TFs, JA signaling also activates the calcium channel [90], mitogen-activated protein kinase cascade [45], and various other processes that interact with SA, ABA, and ET to govern plant growth and development in response to abiotic stresses [91].

MYC2, encoded by the *JIN1* gene, is a basic helix-loop-helix (bHLH) TF and a key regulator of JA signaling. MYC2 binds to the G-box (CACGTG) and G-box-related hexamers [76,92–95], and can interact with most members of the JAZ repressors [76]. However, it is the only MYC subtype that is not the target of JAZ repressors. A number of other TFs can interact with JAZ repressors and remodel the JA signals into specific context-dependent responses (Table 1). MYC3 and MYC4 have similar DNA-binding specificity as MYC2 and can interact with JAZ proteins [76]. MYC5 (bHLH28), which is closely related to MYC2, is activated by the JAs and is required for stamen development and seed production [96,97]. Besides the MYC TFs, the JA-associated MYC2-like (JAM) proteins bHLH3/JAM3, bHLH13/JAM2, bHLH14, and bHLH17/JAM1 regulate JA-mediated anthocyanin accumu-lation, chlorophyll loss, root growth, resistance to bacterial pathogens, and leaf senescence [83–85,98]. Inducer of CBF expression 1 (ICE1) and ICE2, which are bHLH-type TFs, interact with JAZ4 and JAZ9 for the regulation of JA-dependent freezing tolerance [68]. Rice salt sensitive 3 (RSS3) interacts with JAZ9 and JAZ11 and non-R/B-like bHLH TFs, forming the RSS3–JAZ–bHLH complex that regulates the JA-mediated salt stress response [79].

The MYB TFs, which belong to the R2R3-MYB family, show considerable response to JA signaling. They control many processes in plants; for example, the synthesis of tryptophan and glucosinolates is regulated by MYB51 and MYB34, which also play an important role downstream of MYC2 [76]. A subset of JAZ proteins repress the transcriptional activities of MYB21 and MYB24 through their N-terminal R2R3 domain [71]. Evidence suggests that MYB21 and MYB24 are crucial factors for regulating stamen development and pollen maturation in *Arabidopsis* [71]. Anthocyanin accumulation and trichome initiation are positively regulated by MYB75 [70]. MYB21 and MYB24 also interact with MYC2, MYC3, MYC4, and MYC5 to form an MYC–MYB transcription complex that regulates stamen development [97].

The NAC family of TFs is also activated by JA signaling. For example, the JA signal-activated proteins ATAF1 and ATAF2 are involved in the development of plant resistance to salt stress, drought, and plant pathogens like *Botrytis cinerea* [99]. ATAF1 and ATAF2 also play crucial regulatory roles in the oxidative stress caused by abiotic stresses. The NAC TF ANAC019 and ANAC055 work downstream of MYC2 to regulate cell division, secondary cell wall synthesis, and seed germination [100].

The TFs ORCA2 and ORCA3 belong to the AP2/ERF-domain family activated by JA signaling and regulate the expression of genes related to monoterpenoid indole alkaloid biosynthesis [101]. ORA59 regulates the biosynthesis of hydroxycinnamic acid amides and acts as the integrator of JA and ET signals [26,102]. ORA47 is a crucial regulator in the positive jasmonate-responsive feedback loop owing to the activation of the JA biosynthesis gene *AOC2* [103]. Jasmonate-responsive AtERF3 and AtERF4 act as repressors to downregulate the expression of their respective target genes and interfere with the activity of other activators [104]. JAZ repressors cannot repress the activity of the TFs directly, indicating the existence of adaptors or co-repressors in the JA signaling pathway.

WRKY TFs play a critical regulatory role in confronting environmental stresses, as well as in plant development and senescence. In *Arabidopsis*, WRKY70 [105], WRKY22 [106], WRKY50 [107], WRKY57 [69], and WRKY89 [108], which are regulated by the JA signaling pathway, are particularly associated with plant defense functions. In the *Nicotiana attenuata*, WRKY3 and WRKY6 increase the levels of JA and JA-Ile by regulating the expression of jasmonate biosynthesis-related genes (*LOX*, *AOS*, *AOC*, and *OPR*) [109]. In the *Arabidopsis* plant, WRKY57 combines with JAZ4 and JAZ8 to regulate JA-induced leaf senescence [69].

Filamentous flower (FIL), a YABBY family TF, interacts with JAZ3 to regulate JA-mediated responses, such as chlorophyll loss and anthocyanin accumulation [78]. Trichome initiation and anthocyanin accumulation in plants are regulated by the WD-repeat–bHLH–MYB protein complexes. JAZ1, JAZ8, and JAZ11 interact with these complexes and repress their transcriptional activity, leading to the inhibition of anthocyanin accumulation and trichome initiation [70]. Plants biosynthesize JA-Ile in response to environmental cues and induce the degradation of the JAZ proteins, thereby freeing the WD-repeat–bHLH–MYB complexes and allowing them to regulate the expression of genes essential for anthocyanin accumulation and trichome initiation [70,78].

6. Roles of Jasmonic Acid in Alleviating Abiotic Stresses in Plants

6.1. Jasmonic Acid Signaling under Salt Stress

Salinity stress has both osmotic and cytotoxic effects on plant growth and development. The endogenous JA content was increased in *A. thaliana* [110], tomato (*Lycopersicon esculentum*) [111], and potato (*Solanum tuberosum*) [112] after salt treatment. Transcript profile analysis of stressed sweet potato revealed that during salt stress JA level was significantly increased to cope with the effect of salt stress [113]. The JA content increased immediately and persistently in the salt-sensitive plants, whereas the changes were not significant in the salt-tolerant ones [112]. Exogenous MeJA increased the tolerance of the black locust tree (*Robinia pseudoacacia*) to salt stress by increasing the activities of superoxide dismutase (SOD) and ascorbate peroxidase (APX) [108]. These finding were similar to those of Faghih et al. [114], who found that MeJA enhanced the activities of the APX, peroxidase (POD), and SOD enzymes. These lines of evidence suggest that JAs can alleviate salt stress by increasing the endogenous hormones and the antioxidative system.

6.2. Jasmonic Acid Signaling under Drought Stress

Drought stress or water deficit decreases turgor pressure, increases ion toxicity, and inhibits photosynthesis. It has been reported in several studies that JA signaling pathways are associated with the alleviation of drought stress. The increase in the endogenous JA content was rapid and transient in *A. thaliana* [21] and citrus (*Citrus paradisi* × *Poncirus trifoliata*) [115] immediately after drought stress, but the content decreased to the basal level with prolongation of the stress. MeJA treatment could improve the drought resistance in peanut (*Arachis hypogaea*) [116], rice (*Oryza sativa*) [117], soybean (*Glycine max*) [118], and broccoli (*Brassica oleracea*) plants [119]. The application of exogenous MeJA not only increased the total carbohydrate, polysaccharide, total soluble sugar, free amino acid, total proline, and protein contents, but also the activities of catalase (CAT), POD, and SOD in maize plants (*Zea mays*) [120]. In the broad bean (*Vicia faba*) and barley (*Hordeum vulgare*) plants, MeJA increased their abilities to resist drought by regulating stomatal closure [121,122]. MeJA also increased the drought resistance of cauliflower (*B. oleracea*) by activating the enzymatic (SOD, POD, CAT, APX, and glutathione reductase) and nonenzymatic (proline and soluble sugar) antioxidative systems [119]. Therefore, MeJA effectively improves the drought tolerance of plants by increasing the organic osmoprotectants and antioxidative enzyme activity [123].

6.3. Jasmonic Acid Signaling under Heavy Metals Toxicity

Heavy metals can mimic the essential mineral nutrients and generate ROS. Several studies have revealed that JA signaling pathways are associated with heavy metal toxicity. Exogenous MeJA could alleviate the cadmium-induced damage in soybean (*G. max*) [124], *A. thaliana* [125], European black nightshade (*Solanum nigrum*) [126], chili pepper (*Capsicum frutescens*) [127], and mangrove (*Kandelia obovata*) plants by increasing the activities of SOD, APX, and CAT. MeJA mitigated the toxicity of boron in the sweet wormwood (*Artemisia annua*) by reducing the amount of lipid peroxidation and stimulating the synthesis of antioxidative enzymes [128]. In *B. napus*, oxidative stress was minimized by MeJA through the induction of the expression of genes encoding antioxidants and secondary metabolites [129]. Therefore, the exogenous application of MeJA effectively alleviates heavy metal damage by increasing the levels of antioxidative enzyme activity and secondary metabolites.

6.4. Jasmonic Acid Signaling under Micronutrient Toxicity

Several reports have suggested that JAs can protect plants from the effects of micronutrient toxicity. A high boron concentration is detrimental to plant growth and development [130,131] as reported in the apple (*Malus domestica*) root stock [132], wheat (*Triticum aestivum*) [133], barley (*H. vulgare*) [134], and tomato plants [135]. Treatment with exogenous MeJA could counter the boron toxicity in plants by activating the antioxidative defense enzymes (CAT, POD, and SOD) and inhibiting lipid peroxidation [9,128]. JAs also play a crucial role in plant defense responses against lead (Pb) stress. JA showed a reduction in Pb uptake and increased the growth of tomato plants when seeds were primed with JA [136].

6.5. Jasmonic Acid Signaling under Freezing Stress

Low temperature or cold stress causes extracellular ice crystal formation and cell dehydration. JA signaling plays a prominent role in the adaptation of plants to cold stress. The expression of the MYC TFs and several cold-responsive genes (*MaCBF1, MaCBF2, MaKIN2, MaCOR1, MaRD2, MaRD5,* etc.) was induced after the cold storage of bananas (*Musa acuminata*) [137]. MeJA could alleviate the cold stress in the tomato [138], loquat (*Eriobotrya japonica*) [139], pomegranate (*Punica granatum*) [140], mango (*Mangifera indica*) [141], guava (*Psidium guajava*) [142], cowpea (*Vigna sinensis*) plant [143], and peach (*Prunus persica*) [144] by increasing the synthesis of antioxidants and the activation of some defense compounds (e.g., phenolic compounds and heat shock proteins). These results suggest that JAs can mitigate cold injury through their promotion of the active defense compounds and the antioxidative system.

6.6. Jasmonic Acid Signaling under Ozone Stress

Ozone generates ROS that cause lesions and induce programmed cell death in plants. In wild-type *Arabidopsis*, the JA content was found to be significantly increased after ozone treatment [145]. The spread of programmed cell death caused by ozone could be inhibited by exogenous treatment with MeJA [145–148]. Moreover, the hybrid poplar (*Populus maximowizii* × *P. trichocarpa*) and tomato (*L. esculentum*) showed reduced sensitivity to ozone after exogenous MeJA treatment [145,149]. Elevated ozone activated the JA pathway in tomato plants which significantly up-regulated the emission rates of volatile compounds for the protection of plants from natural enemies [150].

6.7. Jasmonic Acid Signaling under Light Stress

Fewer reports are available about the effects of light and the JA signal on plant growth and development. In several studies, the JA signaling pathways in *Nicotiana* and *Brassica* species were initiated by the JA biosynthesis induced by UVB treatment, which increased the defensive mechanisms of the plants [151,152]. JA signaling had an effect on blue light-mediated light morphogenesis in *A. thaliana* and tomato (*L. esculentum*) [153,154] and on red light/far-red light-mediated photomorphogenesis in *A. thaliana* and rice (*O. sativa*) [152].

6.8. Jasmonic Acid Signaling under CO$_2$ Stress

There are few reports about the JA signal transduction pathway in plants under CO$_2$ stress, however, these reports have varied for various plant and insect species [155–157]. Ballhorn et al. reported that in lima bean (*Phaseolus lunatus*), the concentration of MeJA and cis-JA was increased at a high concentration of CO$_2$ (500, 700, and 1000 ppm) [158]. An elevated level of CO$_2$ (750 ppm) increased the defense mechanism of tomato plants against nematode by activating the JA- and SA-signaling pathway [159]. The elevated level of CO$_2$ also increased the JA and main defense-related metabolites in tobacco but decreased in rice [157].

7. Roles of Jasmonic Acid in Plant Species other than Angiosperms

The information herein regarding the biosynthesis and activities of JA and its derivatives is related to angiosperms. Aside from the angiosperms, the bryophytes, lycophytes, fern (lycophytes and ferns/horsetails, together known as pteridophytes), and gymnosperms have all been shown to contain JA compounds, including the precursor 12-oxo-PDA. Among the multicellular sporophytes (consisting of bryophytes and vascular plants), bryophytes such as the moss (*Physcomitrella patens*) and the liverwort (*Marchantia polymorpha*) produce 12-oxo-PDA but not JA [160,161], suggesting that only the first half of the octadecanoid pathway in chloroplasts remains in the bryophytes.

Among the vascular plants, lycophytes (seedless vascular plants) such as the spikemoss (*Selaginella moellendorffii*) have been shown to possess 12-oxo-PDA, JA, and JA-Ile, and the endogenous concentrations of 12-oxo-PDA and JA were also transiently increased within 10 min after wounding [162]. Therefore, the evolution of the JA biosynthetic pathway after that of 12-oxo-PDA is related to the plant acquisition of a vascular system. JA biosynthesis and its signal transduction pathway were also observed in the fern (*Pteridium aquilinum*), where wounding stimulated 12-oxo-PDA and JA in the plant [163], suggesting that JA and JA-Ile biosynthesis first emerged after the emergence of the bryophytes in plant evolution.

Jasmonates also act as cellular signaling compounds in gymnosperms [164,165]. As shown in several studies, the application of MeJA increased the resistance of the Norway spruce (*Picea abies*) to the root pathogen *Pythium ultimum* Trow [166], induced the expression of the 14-3-3 gene in the spruce plant [*Picea glauca* (Muench) Voss] [167], and accumulated a high amount of paclitaxel in several *Taxus* species [168]. The accumulation of JA in response to wounding is a common physiological feedback among all vascular plant species [1]. Therefore, JA has evolved as a plant hormone for stress adaptation, beginning with the emergence of vascular plants.

8. Conclusions and Future Perspectives

JA and its derivatives play crucial roles in the defense and resistance of plants in response to biotic and abiotic stresses. The roles of JAs in the plant defense responses and in growth protection provide a direct way of alleviating the stresses. In the presence of abiotic stresses, JAs induce tolerance chiefly by activating the plant's defense mechanisms, which mainly involve the antioxidative enzymes and other defensive compounds. Future studies will pinpoint how different environmental signals are perceived by plants in the various components in the signaling pathways and the biosynthesis of the JAs, especially in the initiation and establishment of cooperation between the TFs and JAZ repressors during JA signal transduction. Future studies will also elucidate the molecular mechanisms of JA movement through the transporter, resource allocation between growth- and defense-related processes, synergistic or antagonistic interactions between JA and other hormonal signaling pathways. Such works will expand our understanding of the molecular mechanisms underlying the actions of JA against biotic and abiotic stresses.

Author Contributions: M.S.A. and K.-H.B. designed and prepared the review. All authors have read and agreed to the published version of the manuscript.

Funding: This work was supported by the grant from NRF, Korea.

Acknowledgments: This research was supported by Basic Science Research Program through the National Research Foundation of Korea (NRF) funded by the Ministry of Education (NRF-2019R1F1A1052625).

Conflicts of Interest: The authors declare no conflicts of interest.

References

1. Isah, T. Stress and defense responses in plant secondary metabolites production. *Biol. Res.* **2019**, *52*, 1–25. [CrossRef]

2. Altaf-Ul-Amin, M.; Katsuragi, T.; Sato, T.; Kanaya, S. A glimpse to background and characteristics of major molecular biological networks. *Biomed Res. Int.* **2015**, *2015*, 1–14. [CrossRef]

3. Sulpice, R.; Trenkamp, S.; Steinfath, M.; Usadel, B.; Gibon, Y.; Witucka-Wall, H.; Pyl, E.T.; Tschoep, H.; Steinhauser, M.C.; Guenther, M.; et al. Network analysis of enzyme activities and metabolite levels and their relationship to biomass in a large panel of *Arabidopsis* accessions. *Plant Cell* **2010**, *22*, 2872–2893. [CrossRef] [PubMed]

4.   Demine, S.; Reddy, N.; Renard, P.; Raes, M.; Arnould, T. Unraveling biochemical pathways affected by mitochondrial dysfunctions using metabolomic approaches. *Metabolites* **2014**, *4*, 831–878. [CrossRef] [PubMed]

5.   Ayala, A.; Muñoz, M.F.; Argüelles, S. Lipid peroxidation: Production, metabolism, and signaling mechanisms of malondialdehyde and 4-Hydroxy-2-Nonenal. *Oxid. Med. Cell. Longev.* **2014**, *2014*, 1–31. [CrossRef] [PubMed]

6.   Lymperopoulos, P.; Msanne, J.; Rabara, R. Phytochrome and phytohormones: Working in tandem for plant growth and development. *Front. Plant Sci.* **2018**, *9*, 1–14. [CrossRef]

7.   Wasternack, C.; Feussner, I. The oxylipin pathways: Biochemistry and function. *Annu. Rev. Plant Biol.* **2018**, *69*, 363–386. [CrossRef]

8.   Göbel, C.; Feussner, I. Methods for the analysis of oxylipins in plants. *Phytochemistry* **2009**, *70*, 1485–1503. [CrossRef]

9.   Dar, T.A.; Uddin, M.; Khan, M.M.A.; Hakeem, K.R.; Jaleel, H. Jasmonates counter plant stress: A review. *Environ. Exp. Bot.* **2015**, *115*, 49–57. [CrossRef]

10.   Wasternack, C.; Hause, B. Jasmonates: Biosynthesis, perception, signal transduction and action in plant stress response, growth and development. An update to the 2007 review in Annals of Botany. *Ann. Bot.* **2013**, *111*, 1021–1058. [CrossRef]

11.   Camposa, M.L.; Kanga, J.-H.; Howea, G.A. Jasmonate-triggered plan immunity. *J. Chem. Ecol.* **2014**, *40*, 657–675. [CrossRef] [PubMed]

12.   Parthier, B. Jasmonates, new regulators of plant growth and development: Many facts and few hypotheses on their actions. *Bot. Acta* **1991**, *104*, 446–454. [CrossRef]

13.   Koda, Y.; Takahashi, K.; Kikuta, Y. Potato tuber-inducing activities of salicylic acid and related compounds. *J. Plant Growth Regul.* **1992**, *11*, 215–219. [CrossRef]

14.   Sembdner, G.; Parthier, B. The biochemistry and the physiological and molecular actions of jasmonates. *Annu. Rev. Plant Physiol. Plant Mol. Biol.* **1993**, *44*, 569–589. [CrossRef]

15.   Creelman, R.A.; Mullet, J.E. Jasmonic acid distribution and action in plants: Regulation during development and response to biotic and abiotic stress. *Proc. Natl. Acad. Sci. USA* **1995**, *92*, 4114–4119. [CrossRef] [PubMed]

16.   Creelman, R.A.; Mullet, J.E. Biosynthesis and action of jasmonates in plants. *Annu. Rev. Plant Physiol. Plant Mol. Biol.* **1997**, *48*, 355–381. [CrossRef]

17.   Koda, Y. Possible involvement of jasmonates in various morphogenic events. *Physiol. Plant.* **1997**, *100*, 639–646. [CrossRef]

18.   Wasternack, C.; Hause, B. Jasmonates and octadecanoids: Signals in plant stress responses and development. *Prog. Nucleic Acid Res. Mol. Biol.* **2002**, *72*, 165–221.

19.   Browse, J. Jasmonate: An oxylipin signal with many roles in plants. *Vitam. Horm.* **2005**, *72*, 431–456.

20.   Wasternack, C. Jasmonates: An update on biosynthesis, signal transduction and action in plant stress response, growth and development. *Ann. Bot.* **2007**, *100*, 681–697. [CrossRef]

21.   Balbi, V.; Devoto, A. Jasmonate signalling network in *Arabidopsis thaliana*: Crucial regulatory nodes and new physiological scenarios. *New Phytol.* **2008**, *177*, 301–318. [CrossRef] [PubMed]

22.   Pauwels, L.; Morreel, K.; De Witte, E.; Lammertyn, F.; Van Montagu, M.; Boerjan, W.; Inze, D.; Goossens, A. Mapping methyl jasmonate-mediated transcriptional reprogramming of metabolism and cell cycle progression in cultured *Arabidopsis* cells. *Proc. Natl. Acad. Sci. USA* **2008**, *105*, 1380–1385. [CrossRef] [PubMed]

23.   Zhang, Y.; Turner, J.G. Wound-induced endogenous jasmonates stunt plant growth by inhibiting mitosis. *PLoS ONE* **2008**, *3*, e3699. [CrossRef] [PubMed]

24.   Reinbothe, C.; Springer, A.; Samol, I.; Reinbothe, S. Plant oxylipins: Role of jasmonic acid during programmed cell death, defence and leaf senescence. *FEBS J.* **2009**, *276*, 4666–4681. [CrossRef] [PubMed]

25.   Yoshida, Y.; Sano, R.; Wada, T.; Takabayashi, J.; Okada, K. Jasmonic acid control of GLABRA3 links inducible defense and trichome patterning in *Arabidopsis*. *Development* **2009**, *136*, 1039–1048. [CrossRef]

26.   Li, J.; Zhang, K.; Meng, Y.; Hu, J.; Ding, M.; Bian, J.; Yan, M.; Han, J.; Zhou, M. Jasmonic acid/ethylene signaling coordinates hydroxycinnamic acid amides biosynthesis through ORA59 transcription factor. *Plant J.* **2018**, *95*, 444–457. [CrossRef]

27. Kosová, K.; Vítámvás, P.; Urban, M.O.; Klíma, M.; Roy, A.; Tom Prášil, I. Biological networks underlying abiotic stress tolerance in temperate crops-a proteomic perspective. *Int. J. Mol. Sci.* **2015**, *16*, 20913–20942. [CrossRef]

28. Hamant, O.; Haswell, E.S. Life behind the wall: Sensing mechanical cues in plants. *BMC Biol.* **2017**, *15*, 1–9. [CrossRef]

29. Kudla, J.; Becker, D.; Grill, E.; Hedrich, R.; Hippler, M.; Kummer, U.; Parniske, M.; Romeis, T.; Schumacher, K. Advances and current challenges in calcium signaling. *New Phytol.* **2018**, *218*, 414–431. [CrossRef]

30. Avramova, Z. Transcriptional "memory" of a stress: Transient chromatin and memory (epigenetic) marks at stress-response genes. *Plant J.* **2015**, *83*, 149–159. [CrossRef]

31. Huang, H.; Liu, B.; Liu, L.; Song, S. Jasmonate action in plant growth and development. *J. Exp. Bot.* **2017**, *68*, 1349–1359. [CrossRef] [PubMed]

32. Ruan, J.; Zhou, Y.; Zhou, M.; Yan, J.; Khurshid, M.; Weng, W.; Cheng, J.; Zhang, K. Jasmonic acid signaling pathway in plants. *Int. J. Mol. Sci.* **2019**, *20*, 2479. [CrossRef] [PubMed]

33. Feussner, I.; Wasternack, C. The lipoxygenase pathway. *Annu. Rev. Plant Biol.* **2002**, *53*, 275–297. [CrossRef]

34. Hou, Q.; Ufer, G.; Bartels, D. Lipid signalling in plant responses to abiotic stress. *Plant Cell Environ.* **2016**, *39*, 1029–1048. [CrossRef] [PubMed]

35. Han, G.Z. Evolution of jasmonate biosynthesis and signalling mechanisms. *J. Exp. Bot.* **2017**, *68*, 1323–1331.

36. Wasternack, C.; Strnad, M. Jasmonate signaling in plant stress responses and development—Active and inactive compounds. *N. Biotechnol.* **2016**, *33*, 604–613. [CrossRef]

37. Matthes, M.C.; Bruce, T.J.A.; Ton, J.; Verrier, P.J.; Pickett, J.A.; Napier, J.A. The transcriptome of cis-jasmone-induced resistance in *Arabidopsis thaliana* and its role in indirect defence. *Planta* **2010**, *232*, 1163–1180. [CrossRef]

38. Taki, N.; Sasaki-Sekimoto, Y.; Obayashi, T.; Kikuta, A.; Kobayashi, K.; Ainai, T.; Yagi, K.; Sakurai, N.; Suzuki, H.; Masuda, T.; et al. 12-Oxo-phytodienoic acid triggers expression of a distinct set of genes and plays a role in wound-induced gene expression in *Arabidopsis*. *Plant Physiol.* **2005**, *139*, 1268–1283. [CrossRef]

39. Heitz, T.; Smirnova, E.; Widemann, E.; Aubert, Y.; Pinot, F.; Ménard, R. The rise and fall of jasmonate biological activities. In *Lipids in Plant and Algae Development*; Nakamura, Y., Li-Beisson, Y., Eds.; Springer: Cham, Switzerland, 2016; pp. 405–426.

40. Farmer, E.E.; Ryan, C.A. Interplant communication: Airborne methyl jasmonate induces synthesis of proteinase inhibitors in plant leaves. *Proc. Natl. Acad. Sci. USA* **1990**, *87*, 7713–7716. [CrossRef]

41. Fonseca, S.; Chini, A.; Hamberg, M.; Adie, B.; Porzel, A.; Kramell, R.; Miersch, O.; Wasternack, C.; Solano, R. (+)-7-iso-Jasmonoyl-L-isoleucine is the endogenous bioactive jasmonate. *Nat. Chem. Biol.* **2009**, *5*, 344–350. [CrossRef]

42. Koch, T.; Bandemer, K.; Boland, W. Biosynthesis of cis-Jasmone: A pathway for the inactivation and the disposal of the plant stress hormone jasmonic acid to the gas phase? *Helv. Chim. Acta* **1997**, *80*, 838–850. [CrossRef]

43. Wasternack, C.; Song, S. Jasmonates: Biosynthesis, metabolism, and signaling by proteins activating and repressing transcription. *J. Exp. Bot.* **2017**, *68*, 1303–1321. [CrossRef] [PubMed]

44. Truman, W.; Bennet, M.H.; Kubigsteltig, I.; Turnbull, C.; Grant, M. *Arabidopsis* systemic immunity uses conserved defense signaling pathways and is mediated by jasmonates. *Proc. Natl. Acad. Sci. USA* **2007**, *104*, 1075–1080. [CrossRef]

45. Li, Y.; Qin, L.; Zhao, J.; Muhammad, T.; Cao, H.; Li, H.; Zhang, Y.; Liang, Y. SlMAPK3 enhances tolerance to tomato yellow leaf curl virus (TYLCV) by regulating salicylic acid and jasmonic acid signaling in tomato (*Solanum lycopersicum*). *PLoS ONE* **2017**, *12*, e0172466. [CrossRef] [PubMed]

46. Wang, F.; Yu, G.; Liu, P. Transporter-mediated subcellular distribution in the metabolism and signaling of jasmonates. *Front. Plant Sci.* **2019**, *10*. [CrossRef] [PubMed]

47. Heil, M.; Ton, J. Long-distance signalling in plant defence. *Trends Plant Sci.* **2008**, *13*, 264–272. [CrossRef]

48. Thorpe, M.R.; Ferrieri, A.P.; Herth, M.M.; Ferrieri, R.A. 11C-imaging: Methyl jasmonate moves in both phloem and xylem, promotes transport of jasmonate, and of photoassimilate even after proton transport is decoupled. *Planta* **2007**, *226*, 541–551. [CrossRef]

49. Hause, B.; Stenzel, I.; Miersch, O.; Maucher, H.; Kramell, R.; Ziegler, J.; Wasternack, C. Tissue-specific oxylipin signature of tomato flowers: Allene oxide cyclase is highly expressed in distinct flower organs and vascular bundles. *Plant J.* **2000**, *24*, 113–126. [CrossRef]

50. Hause, B.; Hause, G.; Kutter, C.; Miersch, O.; Wasternack, C. Enzymes of jasmonate biosynthesis occur in tomato sieve elements. *Plant Cell Physiol.* **2003**, *44*, 643–648. [CrossRef]

51. Zhou, M.; Memelink, J. Jasmonate-responsive transcription factors regulating plant secondary metabolism. *Biotechnol. Adv.* **2016**, *34*, 441–449. [CrossRef]

52. Chini, A.; Gimenez-Ibanez, S.; Goossens, A.; Solano, R. Redundancy and specificity in jasmonate signalling. *Curr. Opin. Plant Biol.* **2016**, *33*, 147–156. [CrossRef] [PubMed]

53. Causier, B.; Ashworth, M.; Guo, W.; Davies, B. The TOPLESS interactome: A framework for gene repression in *Arabidopsis*. *Plant Physiol.* **2012**, *158*, 423–438. [CrossRef] [PubMed]

54. Acosta, I.F.; Gasperini, D.; Chételat, A.; Stolz, S.; Santuari, L.; Farmer, E.E. Role of NINJA in root jasmonate signaling. *Proc. Natl. Acad. Sci. USA* **2013**, *110*, 15473–15478. [CrossRef] [PubMed]

55. Pauwels, L.; Barbero, G.F.; Geerinck, J.; Tilleman, S.; Grunewald, W.; Pérez, A.C.; Chico, J.M.; Vanden, R.; Sewell, J.; Gil, E.; et al. NINJA connects the co-repressor TOPLESS to jasmonate signalling. *Nature* **2010**, *464*, 788–791. [CrossRef]

56. Pauwels, L.; Goossens, A. The JAZ proteins: A crucial interface in the jasmonate signaling cascade. *Plant Cell* **2011**, *23*, 3089–3100. [CrossRef]

57. Shyu, C.; Figueroa, P.; de Pew, C.L.; Cooke, T.F.; Sheard, L.B.; Moreno, J.E.; Katsir, L.; Zheng, N.; Browse, J.; Howea, G.A. JAZ8 lacks a canonical degron and has an EAR motif that mediates transcriptional repression of jasmonate responses in *Arabidopsis*. *Plant Cell* **2012**, *24*, 536–550. [CrossRef]

58. Thines, B.; Katsir, L.; Melotto, M.; Niu, Y.; Mandaokar, A.; Liu, G.; Nomura, K.; He, S.Y.; Howe, G.A.; Browse, J. JAZ repressor proteins are targets of the SCFCOI1 complex during jasmonate signalling. *Nature* **2007**, *448*, 661–666. [CrossRef]

59. Thireault, C.; Shyu, C.; Yoshida, Y.; St. Aubin, B.; Campos, M.L.; Howe, G.A. Repression of jasmonate signaling by a non-TIFY JAZ protein in *Arabidopsis*. *Plant J.* **2015**, *82*, 669–679. [CrossRef]

60. Gimenez-Ibanez, S.; Boter, M.; Solano, R. Novel players fine-tune plant trade-offs. *Essays Biochem.* **2015**, *58*, 83–100.

61. Sheard, L.B.; Tan, X.; Mao, H.; Withers, J.; Ben-nissan, G.; Hinds, T.R.; Kobayashi, Y.; Hsu, F.; Sharon, M.; Browse, J.; et al. Jasmonate perception by inositol phosphate-potentiated COI1-JAZ co-receptor. *Nature* **2010**, *468*, 400–405. [CrossRef]

62. Feys, B.J.F.; Benedetti, C.E.; Penfold, C.N.; Turner, J.G. *Arabidopsis* mutants selected for resistance to the phytotoxin coronatine are male sterile, insensitive to methyl jasmonate, and resistant to a bacterial pathogen. *Plant Cell* **1994**, *6*, 751–759. [CrossRef] [PubMed]

63. Xie, D.; Feys, B.F.; James, S.; Nieto-Rostro, M.; Turner, J.G. COI1: An *Arabidopsis* gene required for jasmonate-regulated defense and fertility. *Science* **1998**, *280*, 1091–1094. [CrossRef] [PubMed]

64. Zhai, Q.; Zhang, X.; Wu, F.; Feng, H.; Deng, L.; Xu, L.; Zhang, M.; Wang, Q.; Li, C. Transcriptional mechanism of jasmonate receptor COI1-mediated delay of flowering time in *Arabidopsis*. *Plant Cell* **2015**, *27*, 2814–2828. [CrossRef] [PubMed]

65. Mosblech, A.; Thurow, C.; Gatz, C.; Feussner, I.; Heilmann, I. Jasmonic acid perception by COI1 involves inositol polyphosphates in *Arabidopsis thaliana*. *Plant J.* **2011**, *65*, 949–957. [CrossRef] [PubMed]

66. Bäckström, S.; Elfving, N.; Nilsson, R.; Wingsle, G.; Björklund, S. Purification of a plant mediator from *Arabidopsis thaliana* identifies PFT1 as the Med25 subunit. *Mol. Cell* **2007**, *26*, 717–729. [CrossRef]

67. Chen, R.; Jiang, H.; Li, L.; Zhai, Q.; Qi, L.; Zhou, W.; Liu, X.; Li, H.; Zheng, W.; Sun, J.; et al. The arabidopsis mediator subunit MED25 differentially regulates jasmonate and abscisic acid signaling through interacting with the MYC2 and ABI5 transcription factors. *Plant Cell* **2012**, *24*, 2898–2916. [CrossRef]

68. Hu, Y.; Jiang, L.; Wang, F.; Yu, D. Jasmonate regulates the INDUCER OF CBF expression-C-repeat binding factor/dre binding factor1 cascade and freezing tolerance in *Arabidopsis*. *Plant Cell* **2013**, *25*, 2907–2924. [CrossRef]

69. Jiang, Y.; Liang, G.; Yang, S.; Yu, D. *Arabidopsis* WRKY57 functions as a node of convergence for jasmonic acid- and auxin-mediated signaling in jasmonic acid-induced leaf senescence. *Plant Cell* **2014**, *26*, 230–245. [CrossRef]

70. Qi, T.; Song, S.; Ren, Q.; Wu, D.; Huang, H.; Chen, Y.; Fan, M.; Peng, W.; Ren, C.; Xiea, D. The jasmonate-ZIM-domain proteins interact with the WD-repeat/bHLH/MYB complexes to regulate jasmonate-mediated anthocyanin accumulation and trichome initiation in *Arabidopsis thaliana*. *Plant Cell* **2011**, *23*, 1795–1814. [CrossRef]

71. Song, S.; Qi, T.; Huang, H.; Ren, Q.; Wu, D.; Chang, C.; Peng, W.; Liu, Y.; Peng, J.; Xie, D. The jasmonate-ZIM domain proteins interact with the R2R3-MYB transcription factors MYB21 and MYB24 to affect jasmonate-regulated stamen development in *Arabidopsis*. *Plant Cell* **2011**, *23*, 1000–1013. [CrossRef]

72. Zhai, Q.; Yan, C.; Li, L.; Xie, D.; Li, C. Jasmonates. In *Hormone Metabolism and Signaling in Plants*; Li, J., Li, C., Smith, S.M., Eds.; Elsevier Ltd.: Amsterdam, The Netherlands, 2017; pp. 243–272.

73. Zhu, Z.; Lee, B. Friends or foes: New insights in jasmonate and ethylene co-actions. *Plant Cell Physiol.* **2015**, *56*, 414–420. [CrossRef] [PubMed]

74. Chini, A.; Fonseca, S.; Fernández, G.; Adie, B.; Chico, J.M.; Lorenzo, O.; García-Casado, G.; López-Vidriero, I.; Lozano, F.M.; Ponce, M.R.; et al. The JAZ family of repressors is the missing link in jasmonate signalling. *Nature* **2007**, *448*, 666–671. [CrossRef] [PubMed]

75. Cheng, Z.; Sun, L.; Qi, T.; Zhang, B.; Peng, W.; Liu, Y.; Xie, D. The bHLH transcription factor MYC3 interacts with the jasmonate ZIM-domain proteins to mediate jasmonate response in *Arabidopsis*. *Mol. Plant* **2011**, *4*, 279–288. [CrossRef] [PubMed]

76. Fernández-Calvo, P.; Chini, A.; Fernández-Barbero, G.; Chico, J.M.; Gimenez-Ibanez, S.; Geerinck, J.; Eeckhout, D.; Schweizer, F.; Godoy, M.; Franco-Zorrilla, J.M.; et al. The *Arabidopsis* bHLH transcription factors MYC3 and MYC4 are targets of JAZ repressors and act additively with MYC2 in the activation of jasmonate responses. *Plant Cell* **2011**, *23*, 701–715. [CrossRef] [PubMed]

77. Niu, Y.; Figueroa, P.; Browse, J. Characterization of JAZ-interacting bHLH transcription factors that regulate jasmonate responses in *Arabidopsis*. *J. Exp. Bot.* **2011**, *62*, 2143–2154. [CrossRef]

78. Boter, M.; Golz, J.F.; Giménez-Ibañeza, S.; Fernandez-Barbero, G.; Franco-Zorrilla, J.M.; Solano, R. Filamentous flower is a direct target of JAZ3 and modulates responses to jasmonate. *Plant Cell* **2015**, *27*, 3160–3174. [CrossRef]

79. Toda, Y.; Tanaka, M.; Ogawa, D.; Kurata, K.; Kurotani, K.I.; Habu, Y.; Ando, T.; Sugimoto, K.; Mitsuda, N.; Katoh, E.; et al. RICE SALT SENSITIVE3 forms a ternary complex with JAZ and class-C bHLH factors and regulates JASMONATE-induced gene expression and root cell elongation. *Plant Cell* **2013**, *25*, 1709–1725. [CrossRef]

80. Seo, J.S.; Joo, J.; Kim, M.J.; Kim, Y.K.; Nahm, B.H.; Song, S.I.; Cheong, J.J.; Lee, J.S.; Kim, J.K.; Choi, Y. Do OsbHLH148, a basic helix-loop-helix protein, interacts with OsJAZ proteins in a jasmonate signaling pathway leading to drought tolerance in rice. *Plant J.* **2011**, *65*, 907–921. [CrossRef]

81. Zhu, Z.; An, F.; Feng, Y.; Li, P.; Xue, L.; A, M.; Jiang, Z.; Kim, J.M.; To, T.K.; Li, W.; et al. Derepression of ethylene-stabilized transcription factors (EIN3/EIL1) mediates jasmonate and ethylene signaling synergy in *Arabidopsis*. *Proc. Natl. Acad. Sci. USA* **2011**, *108*, 12539–12544. [CrossRef]

82. Zhang, B.; Wang, L.; Zeng, L.; Zhang, C.; Ma, H. *Arabidopsis* TOE proteins convey a photoperiodic signal to antagonize CONSTANS and regulate flowering time. *Genes Dev.* **2015**, *29*, 975–987. [CrossRef]

83. Nakata, M.; Mitsuda, N.; Herde, M.; Koo, A.J.K.; Moreno, J.E.; Suzuki, K.; Howe, G.A.; Ohme-Takagi, M. A bHLH-type transcription factor, ABA-INDUCIBLE BHLH-TYPE TRANSCRIPTION FACTOR/JA-ASSOCIATED MYC2-LIKE1, acts as a repressor to negatively regulate jasmonate signaling in *Arabidopsis*. *Plant Cell* **2013**, *25*, 1641–1656. [CrossRef] [PubMed]

84. Sasaki-Sekimoto, Y.; Jikumaru, Y.; Obayashi, T.; Saito, H.; Masuda, S.; Kamiya, Y.; Ohta, H.; Shirasu, K. Basic helix-loop-helix transcription factors JASMONATE-ASSOCIATED MYC2-LIKE1 (JAM1), JAM2, and JAM3 are negative regulators of jasmonate responses in *Arabidopsis*. *Plant Physiol.* **2013**, *163*, 291–304. [CrossRef] [PubMed]

85. Song, S.; Qi, T.; Fan, M.; Zhang, X.; Gao, H.; Huang, H.; Wu, D.; Guo, H.; Xie, D. The bHLH subgroup IIId factors negatively regulate jasmonate-mediated plant defense and development. *PLoS Genet.* **2013**, *9*, e1003653. [CrossRef] [PubMed]

86. Fonseca, S.; Fernández-Calvo, P.; Fernández, G.M.; Díez-Díaz, M.; Gimenez-Ibanez, S.; López-Vidriero, I.; Godoy, M.; Fernández-Barbero, G.; Van Leene, J.; De Jaeger, G.; et al. bHLH003, bHLH013 and bHLH017 are new targets of JAZ repressors negatively regulating JA responses. *PLoS ONE* **2014**, *9*, e86182. [CrossRef] [PubMed]

87. Eulgem, T.; Somssich, I.E. Networks of WRKY transcription factors in defense signaling. *Curr. Opin. Plant Biol.* **2007**, *10*, 366–371. [CrossRef]

88. Gutterson, N.; Reuber, T.L. Regulation of disease resistance pathways by AP2/ERF transcription factors. *Curr. Opin. Plant Biol.* **2004**, *7*, 465–471. [CrossRef]

89. Nuruzzaman, M.; Sharoni, A.M.; Kikuchi, S. Roles of NAC transcription factors in the regulation of biotic and abiotic stress responses in plants. *Front. Microbiol.* **2013**, *4*, 1–16. [CrossRef]

90. Kenton, P.; Mur, L.A.J.; Draper, J. A requirement for calcium and protein phosphatase in the jasmonate-induced increase in tobacco leaf acid phosphatase specific activity. *J. Exp. Bot.* **1999**, *50*, 1331–1341. [CrossRef]

91. Santner, A.; Estelle, M. Recent advances and emerging trends in plant hormone signalling. *Nature* **2009**, *459*, 1071–1078. [CrossRef]

92. Abe, H.; Yamaguchi-Shinozaki, K.; Urao, T.; Iwasaki, T.; Hosokawa, D.; Shinozaki, K. Role of Arabidopsis MYC and MYB homologs in drought- and abscisic acid-regulated gene expression. *Plant Cell* **1997**, *9*, 1859–1868.

93. Boter, M.; Ruíz-Rivero, O.; Abdeen, A.; Prat, S. Conserved MYC transcription factors play a key role in jasmonate signaling both in tomato and *Arabidopsis*. *Genes Dev.* **2004**, *18*, 1577–1591. [CrossRef] [PubMed]

94. Dombrecht, B.; Gang, P.X.; Sprague, S.J.; Kirkegaard, J.A.; Ross, J.J.; Reid, J.B.; Fitt, G.P.; Sewelam, N.; Schenk, P.M.; Manners, J.M.; et al. MYC2 differentially modulates diverse jasmonate-dependent functions in *Arabidopsis*. *Plant Cell* **2007**, *19*, 2225–2245. [CrossRef] [PubMed]

95. Saijo, Y.; Uchiyama, B.; Abe, T.; Satoh, K.; Nukiwa, T. Contiguous four-guanosine sequence in c-myc antisense phosphorothioate oligonucleotides inhibits cell growth on human lung cancer cells: Possible involvement of cell adhesion inhibition. *Japanese J. Cancer Res.* **1997**, *88*, 26–33. [CrossRef] [PubMed]

96. Figueroa, P.; Browse, J. Male sterility in *Arabidopsis* induced by overexpression of a MYC5-SRDX chimeric repressor. *Plant J.* **2015**, *81*, 849–860. [CrossRef]

97. Qi, T.; Huang, H.; Song, S.; Xie, D. Regulation of jasmonate-mediated stamen development and seed production by a bHLH-MYB complex in *Arabidopsis*. *Plant Cell* **2015**, *27*, 1620–1633. [CrossRef]

98. Qi, T.; Wang, J.; Huang, H.; Liu, B.; Gao, H.; Liu, Y.; Song, S.; Xie, D. Regulation of jasmonate-induced leaf senescence by antagonism between bHLH subgroup IIIe and IIId factors in *Arabidopsis*. *Plant Cell* **2015**, *27*, 1634–1649. [CrossRef]

99. Delessert, C.; Kazan, K.; Wilson, I.W.; Van Der Straeten, D.; Manners, J.; Dennis, E.S.; Dolferus, R. The transcription factor ATAF2 represses the expression of pathogenesis-related genes in *Arabidopsis*. *Plant J.* **2005**, *43*, 745–757. [CrossRef]

100. Bu, Q.; Jiang, H.; Li, C.B.; Zhai, Q.; Zhang, J.; Wu, X.; Sun, J.; Xie, Q.; Li, C. Role of the *Arabidopsis thaliana* NAC transcription factors ANAC019 and ANAC055 in regulating jasmonic acid-signaled defense responses. *Cell Res.* **2008**, *18*, 756–767. [CrossRef]

101. Van Der Fits, L.; Memelink, J. ORCA3, a fasmonate-responsive transcriptional regulator of plant primary and secondary metabolism. *Science* **2000**, *289*, 295–297. [CrossRef]

102. Pré, M.; Atallah, M.; Champion, A.; De Vos, M.; Pieterse, C.M.J.; Memelink, J. The AP2/ERF domain transcription factor ORA59 integrates jasmonic acid and ethylene signals in plant defense. *Plant Physiol.* **2008**, *147*, 1347–1357. [CrossRef]

103. Saxena, I.; Srikanth, S.; Chen, Z. Cross talk between H_2O_2 and interacting signal molecules under plant stress response. *Front. Plant Sci.* **2016**, *7*, 1–16. [CrossRef] [PubMed]

104. Fujimoto, S.Y.; Ohta, M.; Usui, A.; Shinshi, H.; Ohme-Takagi, M. *Arabidopsis* ethylene-responsive element binding factors act as transcriptional activators or repressors of GCC box-mediated gene expression. *Plant Cell* **2000**, *12*, 393–404.

105. Li, J.; Zhong, R.; Palva, E.T. WRKY70 and its homolog WRKY54 negatively modulate the cell wall-associated defenses to necrotrophic pathogens in *Arabidopsis*. *PLoS ONE* **2017**, *12*, e0183731. [CrossRef] [PubMed]

106. Kloth, K.J.; Wiegers, G.L.; Busscher-Lange, J.; Van Haarst, J.C.; Kruijer, W.; Bouwmeester, H.J.; Dicke, M.; Jongsma, M.A. AtWRKY22 promotes susceptibility to aphids and modulates salicylic acid and jasmonic acid signalling. *J. Exp. Bot.* **2016**, *67*, 3383–3396. [CrossRef] [PubMed]

107. Gao, Q.M.; Venugopal, S.; Navarre, D.; Kachroo, A. Low oleic acid-derived repression of jasmonic acid-inducible defense responses requires the WRKY50 and WRKY51 proteins. *Plant Physiol.* **2011**, *155*, 464–476. [CrossRef] [PubMed]

108. Jiang, M.; Xu, F.; Peng, M.; Huang, F.; Meng, F. Methyl jasmonate regulated diploid and tetraploid black locust (*Robinia pseudoacacia* L.) tolerance to salt stress. *Acta Physiol. Plant.* **2016**, *38*, 1–13. [CrossRef]

109. Skibbe, M.; Qu, N.; Galis, I.; Baldwin, I.T. Induced plant defenses in the natural environment: *Nicotiana attenuata* WRKY3 and WRKY6 coordinate responses to herbivory. *Plant Cell* **2008**, *20*, 1984–2000. [CrossRef]

110. Ellouzi, H.; Ben Hamed, K.; Cela, J.; Müller, M.; Abdelly, C.; Munné-bosch, S. Increased sensitivity to salt stress in tocopherol-deficient *Arabidopsis* mutants growing in a hydroponic system. *Plant Signal. Behav.* **2013**, *8*, 1–13. [CrossRef]

111. Pedranzani, H.; Racagni, G.; Alemano, S.; Miersch, O.; Ramírez, I.; Peña-Cortés, H.; Taleisnik, E.; Machado-Domenech, E.; Abdala, G. Salt tolerant tomato plants show increased levels of jasmonic acid. *Plant Growth Regul.* **2003**, *41*, 149–158. [CrossRef]

112. De Domenico, S.; Taurino, M.; Gallo, A.; Poltronieri, P.; Pastor, V.; Flors, V.; Santino, A. Oxylipin dynamics in Medicago truncatula in response to salt and wounding stresses. *Physiol. Plant.* **2019**, *165*, 198–208. [CrossRef]

113. Zhang, H.; Zhang, Q.; Zhai, H.; Li, Y.; Wang, X.; Liu, Q.; He, S. Transcript profile analysis reveals important roles of jasmonic acid signalling pathway in the response of sweet potato to salt stress. *Sci. Rep.* **2017**, *7*, 1–12. [CrossRef] [PubMed]

114. Faghih, S.; Ghobadi, C.; Zarei, A. Response of Strawberry plant cv. 'Camarosa' to salicylic acid and methyl jasmonate application under salt stress condition. *J. Plant Growth Regul.* **2017**, *36*, 651–659. [CrossRef]

115. De Ollas, C.; Hernando, B.; Arbona, V.; Gómez-Cadenas, A. Jasmonic acid transient accumulation is needed for abscisic acid increase in citrus roots under drought stress conditions. *Physiol. Plant.* **2013**, *147*, 296–306. [CrossRef] [PubMed]

116. Todaka, D.; Shinozaki, K.; Yamaguchi-Shinozaki, K. Recent advances in the dissection of drought-stress regulatory networks and strategies for development of drought-tolerant transgenic rice plants. *Front. Plant Sci.* **2015**, *6*, 1–20. [CrossRef] [PubMed]

117. Fu, J.; Wu, H.; Ma, S.; Xiang, D.; Liu, R.; Xiong, L. OSJAZ1 attenuates drought resistance by regulating JA and ABA signaling in rice. *Front. Plant Sci.* **2017**, *8*, 1–13. [CrossRef] [PubMed]

118. Mohamed, H.I.; Latif, H.H. Improvement of drought tolerance of soybean plants by using methyl jasmonate. *Physiol. Mol. Biol. Plants* **2017**, *23*, 545–556. [CrossRef]

119. Wu, H.; Wu, X.; Li, Z.; Duan, L.; Zhang, M. Physiological evaluation of drought stress tolerance and recovery in cauliflower (*Brassica oleracea* L.) seedlings treated with methyl jasmonate and coronatine. *J. Plant Growth Regul.* **2012**, *31*, 113–123. [CrossRef]

120. Abdelgawad, Z.A.; Khalafaallah, A.A.; Abdallah, M.M. Impact of methyl jasmonate on antioxidant activity and some biochemical aspects of maize plant grown under water stress condition. *Agric. Sci.* **2014**, *05*, 1077–1088. [CrossRef]

121. Evans, N.H. Modulation of guard cell plasma membrane potassium currents by methyl jasmonate. *Plant Physiol.* **2003**, *131*, 8–11. [CrossRef]

122. Horton, R.F. Methyl jasmonate and transpiration in Barley. *Plant Physiol.* **1991**, *96*, 1376–1378. [CrossRef]

123. Qiu, Z.; Guo, J.; Zhu, A.; Zhang, L.; Zhang, M. Exogenous jasmonic acid can enhance tolerance of wheat seedlings to salt stress. *Ecotoxicol. Environ. Saf.* **2014**, *104*, 202–208. [CrossRef] [PubMed]

124. Poonam, S.; Kaur, H.; Geetika, S. Effect of jasmonic acid on photosynthetic pigments and stress markers in *Cajanus cajan* (L.) Millsp. seedlings under copper stress. *Am. J. Plant Sci.* **2013**, *04*, 817–823. [CrossRef]

125. Maksymiec, W.; Krupa, Z. Effects of methyl jasmonate and excess copper on root and leaf growth. *Biol. Plant.* **2007**, *51*, 322–326. [CrossRef]

126. Yan, Z.; Zhang, W.; Chen, J.; Li, X. Methyl jasmonate alleviates cadmium toxicity in *Solanum nigrum* by regulating metal uptake and antioxidative capacity. *Biol. Plant.* **2015**, *59*, 373–381. [CrossRef]

127. Yan, Z.; Chen, J.; Li, X. Methyl jasmonate as modulator of Cd toxicity in *Capsicum frutescens* var. fasciculatum seedlings. *Ecotoxicol. Environ. Saf.* **2013**, *98*, 203–209. [CrossRef]

128. Aftab, T.; Khan, M.M.A.; Idrees, M.; Naeem, M.; Moinuddin; Hashmi, N. Methyl jasmonate counteracts boron toxicity by preventing oxidative stress and regulating antioxidant enzyme activities and artemisinin biosynthesis in *Artemisia annua* L. *Protoplasma* **2011**, *248*, 601–612. [CrossRef]

129. Farooq, M.A.; Gill, R.A.; Islam, F.; Ali, B.; Liu, H.; Xu, J.; He, S.; Zhou, W. Methyl jasmonate regulates antioxidant defense and suppresses arsenic uptake in *Brassica napus* L. *Front. Plant Sci.* **2016**, *7*, 1–16. [CrossRef]

130. Karabal, E.; Yücel, M.; Öktem, H.A. Antioxidant responses of tolerant and sensitive barley cultivars to boron toxicity. *Plant Sci.* **2003**, *164*, 925–933. [CrossRef]

131. Papadakis, I.E.; Dimassi, K.N.; Bosabalidis, A.M.; Therios, I.N.; Patakas, A.; Giannakoula, A. Effects of B excess on some physiological and anatomical parameters of "Navelina" orange plants grafted on two rootstocks. *Environ. Exp. Bot.* **2004**, *51*, 247–257. [CrossRef]

132. Molassiotis, A.; Sotiropoulos, T.; Tanou, G.; Diamantidis, G.; Therios, I. Boron-induced oxidative damage and antioxidant and nucleolytic responses in shoot tips culture of the apple rootstock EM 9 (*Malus domestica* Borkh). *Environ. Exp. Bot.* **2006**, *56*, 54–62. [CrossRef]

133. Cervilla, L.M.; Blasco, B.; Ríos, J.J.; Romero, L.; Ruiz, J.M. Oxidative stress and antioxidants in tomato (*Solanum lycopersicum*) plants subjected to boron toxicity. *Ann. Bot.* **2007**, *100*, 747–756. [CrossRef] [PubMed]

134. Gunes, A.; Inal, A.; Bagci, E.G.; Coban, S.; Sahin, O. Silicon increases boron tolerance and reduces oxidative damage of wheat grown in soil with excess boron. *Biol. Plant.* **2007**, *51*, 571–574. [CrossRef]

135. Inal, A.; Pilbeam, D.J.; Gunes, A. Silicon increases tolerance to boron toxicity and reduces oxidative damage in barley. *J. Plant Nutr.* **2009**, *32*, 112–128. [CrossRef]

136. Bali, S.; Jamwal, V.L.; Kaur, P.; Kohli, S.K.; Ohri, P.; Gandhi, S.G.; Bhardwaj, R.; Al-Huqail, A.A.; Siddiqui, M.H.; Ahmad, P. Role of P-type ATPase metal transporters and plant immunity induced by jasmonic acid against lead (Pb) toxicity in tomato. *Ecotoxicol. Environ. Saf.* **2019**, *174*, 283–294. [CrossRef]

137. Zhao, M.L.; Wang, J.N.; Shan, W.; Fan, J.G.; Kuang, J.F.; Wu, K.Q.; Li, X.P.; Chen, W.X.; He, F.Y.; Chen, J.Y.; et al. Induction of jasmonate signalling regulators MaMYC2s and their physical interactions with MaICE1 in methyl jasmonate-induced chilling tolerance in banana fruit. *Plant Cell Environ.* **2013**, *36*, 30–51. [CrossRef]

138. Zhang, X.; Sheng, J.; Li, F.; Meng, D.; Shen, L. Methyl jasmonate alters arginine catabolism and improves postharvest chilling tolerance in cherry tomato fruit. *Postharvest Biol. Technol.* **2012**, *64*, 160–167. [CrossRef]

139. Jin, P.; Duan, Y.; Wang, L.; Wang, J.; Zheng, Y. Reducing chilling injury of Loquat fruit by combined treatment with hot air and methyl jasmonate. *Food Bioprocess Technol.* **2014**, *7*, 2259–2266. [CrossRef]

140. Sayyari, M.; Babalar, M.; Kalantari, S.; Martínez-Romero, D.; Guillén, F.; Serrano, M.; Valero, D. Vapour treatments with methyl salicylate or methyl jasmonate alleviated chilling injury and enhanced antioxidant potential during postharvest storage of pomegranates. *Food Chem.* **2011**, *124*, 964–970. [CrossRef]

141. González-Aguilar, G.A.; Fortiz, J.; Cruz, R.; Baez, R.; Wang, C.Y. Methyl jasmonate reduces chilling injury and maintains postharvest quality of mango fruit. *J. Agric. Food Chem.* **2000**, *48*, 515–519. [CrossRef]

142. González-Aguilar, G.A.; Tiznado-Hernández, M.E.; Zavaleta-Gatica, R.; Martínez-Téllez, M.A. Methyl jasmonate treatments reduce chilling injury and activate the defense response of guava fruits. *Biochem. Biophys. Res. Commun.* **2004**, *313*, 694–701. [CrossRef]

143. Fan, L.; Wang, Q.; Lv, J.; Gao, L.; Zuo, J.; Shi, J. Amelioration of postharvest chilling injury in cowpea (*Vigna sinensis*) by methyl jasmonate (MeJA) treatments. *Sci. Hortic.* **2016**, *203*, 95–101. [CrossRef]

144. Jin, P.; Zheng, Y.; Tang, S.; Rui, H.; Wang, C.Y. A combination of hot air and methyl jasmonate vapor treatment alleviates chilling injury of peach fruit. *Postharvest Biol. Technol.* **2009**, *52*, 24–29. [CrossRef]

145. Rao, M.V.; Lee, H.; Creelman, R.A.; Mullet, J.E.; Davis, K.R. Jasmonic acid signaling modulates ozone-induced hypersensitive cell death. *Plant Cell* **2000**, *12*, 1633. [CrossRef] [PubMed]

146. Ismail, A.; Riemann, M.; Nick, P. The jasmonate pathway mediates salt tolerance in grapevines. *J. Exp. Bot.* **2012**, *63*, 2127–2139. [CrossRef]

147. Overmyer, K.; Tuominen, H.; Kettunen, R.; Betz, C.; Langebartels, C.; Sandermann H., J.; Kangasjarvi, J. Ozone-sensitive *Arabidopsis rcd1* mutant reveals opposite roles for ethylene and jasmonate signaling pathways in regulating superoxide-dependent cell death. *Plant Cell* **2000**, *12*, 1849–1862. [CrossRef]

148. Kanna, M.; Tamaoki, M.; Kubo, A.; Nakajima, N.; Rakwal, R.; Agrawal, G.K.; Tamogami, S.; Ioki, M.; Ogawa, D.; Saji, H.; et al. Isolation of an ozone-sensitive and jasmonate-semi-insensitive *Arabidopsis mutant* (*oji1*). *Plant Cell Physiol.* **2003**, *44*, 1301–1310. [CrossRef]

149. Koch, J.R.; Creelman, R.A.; Eshita, S.M.; Seskar, M.; Mullet, J.E.; Davis, K.R. Ozone sensitivity in hybrid poplar correlates with insensitivity to both salicylic acid and jasmonic acid. The role of programmed cell death in lesion formation. *Plant Physiol.* **2000**, *123*, 487–496. [CrossRef]

150. Cui, H.; Wei, J.; Su, J.; Li, C.; Ge, F. Elevated O_3 increases volatile organic compounds via jasmonic acid pathway that promote the preference of parasitoid Encarsia formosa for tomato plants. *Plant Sci.* **2016**, *253*, 243–250. [CrossRef]

151. Svyatyna, K.; Riemann, M. Light-dependent regulation of the jasmonate pathway. *Protoplasma* **2012**, *249*, 137–145. [CrossRef]

152. Mewis, I.; Schreiner, M.; Nguyen, C.N.; Krumbein, A.; Ulrichs, C.; Lohse, M.; Zrenner, R. UV-B irradiation changes specifically the secondary metabolite profile in broccoli sprouts: Induced signaling overlaps with defense response to biotic stressors. *Plant Cell Physiol.* **2012**, *53*, 1546–1560. [CrossRef]

153. Cerrudo, I.; Keller, M.M.; Cargnel, M.D.; Demkura, P.V.; de Wit, M.; Patitucci, M.S.; Pierik, R.; Pieterse, C.M.J.; Ballaré, C.L. Low red/far-red ratios reduce arabidopsis resistance to *Botrytis cinerea* and jasmonate responses via a COI1-JAZ10-dependent, salicylic acid-independent mechanism. *Plant Physiol.* **2012**, *158*, 2042–2052. [CrossRef] [PubMed]

154. Gupta, N.; Prasad, V.B.R.; Chattopadhyay, S. LeMYC2 acts as a negative regulator of blue light mediated photomorphogenic growth, and promotes the growth of adult tomato plants. *BMC Plant Biol.* **2014**, *14*, 1–14. [CrossRef] [PubMed]

155. Casteel, C.L.; O'Neill, B.F.; Zavala, J.A.; Bilgin, D.D.; Berenbaum, M.R.; DeLucia, E.H. Transcriptional profiling reveals elevated CO_2 and elevated O_3 alter resistance of soybean (*Glycine max*) to Japanese beetles (*Popillia japonica*). *Plant Cell Environ.* **2008**, *31*, 419–434. [CrossRef] [PubMed]

156. Zavala, J.A.; Casteel, C.L.; DeLucia, E.H.; Berenbaum, M.R. Anthropogenic increase in carbon dioxide compromises plant defense against invasive insects. *Proc. Natl. Acad. Sci. USA* **2008**, *105*, 5129–5133. [CrossRef] [PubMed]

157. Lu, C.; Qi, J.; Hettenhausen, C.; Lei, Y.; Zhang, J.; Zhang, M.; Zhang, C.; Song, J.; Li, J.; Cao, G.; et al. Elevated CO_2 differentially affects tobacco and rice defense against lepidopteran larvae via the jasmonic acid signaling pathway. *J. Integr. Plant Biol.* **2018**, *60*, 412–431. [CrossRef] [PubMed]

158. Ballhorn, D.J.; Reisdorff, C.; Pfanz, H. Quantitative effects of enhanced CO_2 on jasmonic acid induced plant volatiles of lima bean (*Phaseolus lunatus* L.). *J. Appl. Bot. Food Qual.* **2011**, *84*, 65–71.

159. Sun, Y.; Yin, J.; Cao, H.; Li, C.; Kang, L.; Ge, F. Elevated CO_2 influences nematode-induced defense responses of tomato genotypes differing in the JA pathway. *PLoS ONE* **2011**, *6*, e19751. [CrossRef]

160. Stumpe, M.; Göbel, C.; Faltin, B.; Beike, A.K.; Hause, B.; Himmelsbach, K.; Bode, J.; Kramell, R.; Wasternack, C.; Frank, W.; et al. The moss *Physcomitrella patens* contains cyclopentenones but no jasmonates: Mutations in allene oxide cyclase lead to reduced fertility and altered sporophyte morphology. *New Phytol.* **2010**, *188*, 740–749. [CrossRef]

161. Yamamoto, Y.; Ohshika, J.; Takahashi, T.; Ishizaki, K.; Kohchi, T.; Matsuura, H.; Takahashi, K. Functional analysis of allene oxide cyclase, MpAOC, in the liverwort *Marchantia polymorpha*. *Phytochemistry* **2015**, *116*, 48–56. [CrossRef]

162. Pratiwi, P.; Tanaka, G.; Takahashi, T.; Xie, X.; Yoneyama, K.; Matsuura, H.; Takahashi, K. Identification of jasmonic acid and jasmonoyl-isoleucine, and characterization of AOS, AOC, OPR and JAR1 in the model lycophyte *Selaginella moellendorffii*. *Plant Cell Physiol.* **2017**, *58*, 789–801. [CrossRef]

163. Radhika, V.; Kost, C.; Bonaventure, G.; David, A.; Boland, W. Volatile emission in bracken fern is induced by jasmonates but not by *Spodoptera littoralis* or *Strongylogaster multifasciata* herbivory. *PLoS ONE* **2012**, *7*, e48050. [CrossRef] [PubMed]

164. Thaler, J.S.; Stout, M.J.; Karban, R.; Duffey, S.S. Jasmonate-mediated induced plant resistance affects a community of herbivores. *Ecol. Entomol.* **2001**, *26*, 312–324. [CrossRef]

165. Franceschi, V.R.; Krekling, T.; Christiansen, E. Application of methyl jasmonate on *Picea abies* (Pinaceae) stems induces defense-related responses in phloem and xylem. *Am. J. Bot.* **2002**, *89*, 578–586. [CrossRef] [PubMed]

166. Kozlowski, G.; Buchala, A.; Métraux, J.P. Methyl jasmonate protects Norway spruce [*Picea abies* (L.) Karst.] seedlings against Pythium ultimum Trow. *Physiol. Mol. Plant Pathol.* **1999**, *55*, 53–58. [CrossRef]

167. Lapointe, G.; Luckevich, M.D.; Séguin, A. Investigation on the induction of 14-3-3 in white spruce. *Plant Cell Rep.* **2001**, *20*, 79–84. [CrossRef] [PubMed]

168. Ketchum, R.E.B.; Gibson, D.M.; Croteau, R.B.; Shuler, M.L. The kinetics of taxoid accumulation in cell suspension cultures of Taxus following elicitation with methyl jasmonate. *Biotechnol. Bioeng.* **1999**, *62*, 97–105. [CrossRef]

© 2020 by the authors. Licensee MDPI, Basel, Switzerland. This article is an open access article distributed under the terms and conditions of the Creative Commons Attribution (CC BY) license (http://creativecommons.org/licenses/by/4.0/).

International Journal of
Molecular Sciences

Review

Methyl Jasmonate Induced Oxidative Stress and Accumulation of Secondary Metabolites in Plant Cell and Organ Cultures

Thanh-Tam Ho [1], Hosakatte Niranjana Murthy [2] and So-Young Park [3,*

[1]   Institute for Global Health Innovations, Duy Tan University, Danang 550000, Vietnam; hotamqn@gmail.com
[2]   Department of Botany, Karnatak University, Dharwad 580003, India; hnmurthy60@gmail.com
[3]   Department of Horticultural Science, Chungbuk National University, Cheongju 28644, Korea
*   Correspondence: soypark7@cbnu.ac.kr; Tel.: +82-432-612-531

Received: 16 December 2019; Accepted: 19 January 2020; Published: 22 January 2020

Abstract: Recently, plant secondary metabolites are considered as important sources of pharmaceuticals, food additives, flavours, cosmetics, and other industrial products. The accumulation of secondary metabolites in plant cell and organ cultures often occurs when cultures are subjected to varied kinds of stresses including elicitors or signal molecules. Application of exogenous jasmonic acid (JA) and methyl jasmonate (MJ) is responsible for the induction of reactive oxygen species (ROS) and subsequent defence mechanisms in cultured cells and organs. It is also responsible for the induction of signal transduction, the expression of many defence genes followed by the accumulation of secondary metabolites. In this review, the application of exogenous MJ elicitation strategies on the induction of defence mechanism and secondary metabolite accumulation in cell and organ cultures is introduced and discussed. The information presented here is useful for efficient large-scale production of plant secondary metabolites by the plant cell and organ cultures.

Keywords: antioxidant enzyme activity; elicitor; methyl jasmonate; secondary metabolite; signal molecules

---

1. Introduction

In plant defence systems, each cell has acquired the capability to respond to pathogens and environmental stresses and to build up a defence response. A plant defence mechanism is determined by several factors, mainly depending on their genetic characteristics and physiological state [1,2]. The supplementation of exogenous methyl jasmonate (MJ) to in vitro cultures is responsible for the induction of reactive oxygen species (ROS), regulation of defence response by an accumulation of antioxidant enzyme activity [3,4]. On the other hand, MJ also stimulates molecular signal transduction, regulation of gene expression which leads to accumulation of secondary metabolites [5,6].

Free radicals such as superoxide anion ($O_2^{\bullet-}$), hydroxyl radical ($^{\bullet}OH$), as well as non-radical molecules like hydrogen peroxide (H_2O_2), singlet oxygen ($^1O_2$), are accumulated in plant cells in response to stress mechanism [6,7]. Elicitation with MJ, salicylic acid (SA), environmental stresses, as well as pathogens attack lead to enhanced generation of ROS in plants due to disruption of cellular homeostasis [8,9]. When the level of ROS exceeds the defence mechanisms, a cell is said to be in a state of "oxidative stress" [10]. Although all ROS are extremely harmful to organisms at high concentrations, ROS are well-described second messengers in a variety of cellular processes including tolerance to environmental stresses at low concentration [11]. ROS when present in lower concentration, control many different processes in cells. However, higher concentration ROS in the cell is toxic and detrimental. Therefore, it is necessary for the cells to control the level of ROS tightly to avoid any oxidative injury and not to eliminate them completely [10]. The antioxidants enzymes include superoxide dismutase (SOD), catalase (CAT), guaiacol peroxidase (G-POD), enzymes of ascorbate-glutahione (AsA-GSH) cycle such as ascorbate peroxidase (APX), monodehydroascorbate reductase (MDHAR), dehydroascorbate reductase (DHAR), and glutathione reductase (GR). While ascorbate (AsA), glutathione (GSH), carotenoids, tocopherols, and phenolics serve as potent nonenzymic antioxidants within the cell, were achieved scavenging or detoxification of excess ROS [10].

Application of exogenous MJ to in vitro cultures has emerged as a novel technique for a hyperaccumulation of secondary metabolites. It has been demonstrated that application of MJ to in vitro cultures has induced the antioxidant enzyme activity, expression of defence-related genes, and over-production of secondary metabolites [3,4,12–14]. The process or methodology for enhancement of secondary metabolites in *Polygonum multiflorum* adventitious and hairy roots cultures by application of MJ has been presented in Figure 1. The first step is to establish an in vitro culture system, the second step is the optimization of culture conditions for biomass accumulation in the bioreactor culture system. Then, concentration and exposure time of MJ were determined in the third step, thereafter, scale-up of culture process could be achieved up to pilot-scale. HPLC and Fourier-transform infrared spectroscopy (FT-IR) analysis were applied to investigate the quality of bioactive compounds in the culture [15–17]. In this review, the role of MJ on the induction of oxidative stress, antioxidant enzyme activity, signal molecules transduction, and secondary metabolite accumulation in plant cell and organ culture system are introduced and discussed.

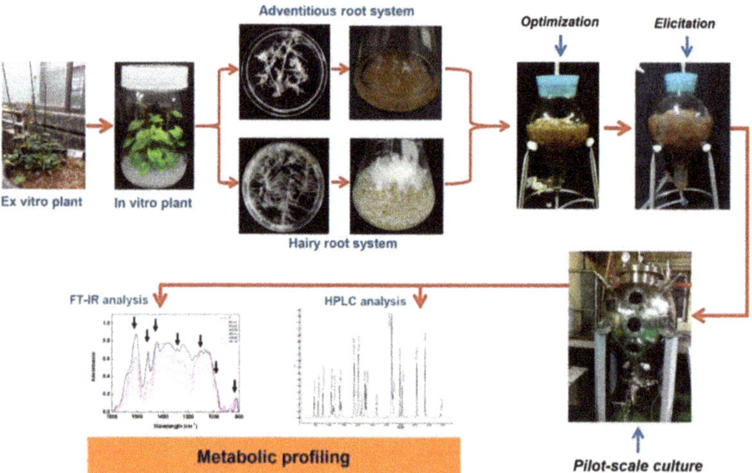

Figure 1. The experimental process of *Polygonum multiflorum* for enhanced production of secondary metabolites.

2. Jasmonic Acid (JA) and Methyl Jasmonate (MJ)

The Jasmonic acid (JA) and its derivatives were collectively called Jasmonates (JAs). JAs are cyclopentanone compounds or plant hormones derived from α-linolenic acid. It includes a group of oxygenated fatty acids, and JA is the main precursor of different compounds to this group like methyl jasmonate (MJ), which was first isolated from the essential oil of *Jasminum grandiflorum* [18], and the free acid was isolated subsequently from the culture filtrates of the fungus *Lasiodiplodia theobromae* [19]. JAs was synthesized from fatty acid including three steps. The first step occurs in the membrane chloroplast, where α-linolenic acid and hexadecatrienoic acid were released from membrane phospholipids [20,21]. Lipoxygenase (LOX), allene oxide synthase (AOS), allene oxide cyclase (AOC) are key enzyme of JA biosynthesis in chloroplast and they form (cis-(+)-12-oxophytodienoic acid (OPDA) and dn-OPDA (dinor-OPDA) [22,23]. The second step of JAs synthesis takes place in peroxisome by the process of β-oxidation to give finally jasmonic acid [21,23]. In the final step, JA is exported to the cytoplasm for further modification like MJ, JA-Ile (Figure 2) [21,24,25].

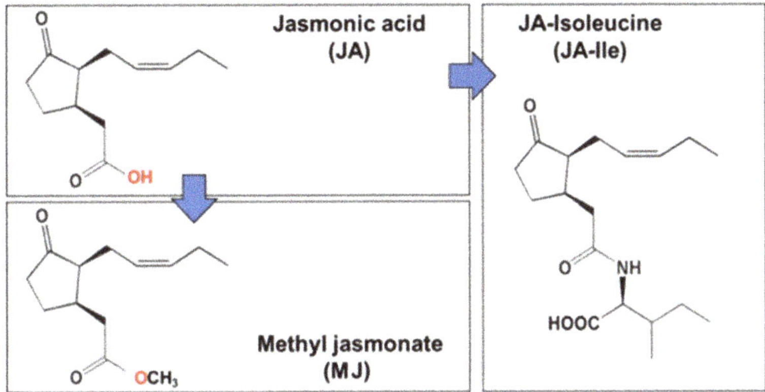

Figure 2. Jasmonic acid and its derivatives.

Jasmonic acid (JA) and its precursors and derivatives are known to possess many physiological processes in plant growth and development, and especially the mediation of plant responses to biotic and abiotic stresses [26–28]. They have been used extensively for elicitation studies within in vitro culture systems [29]. The elicitation process leads to crosstalk between jasmonates with their receptors which are present in the plasma membrane, which leads an array of defence responses with the cells, including the production of reactive oxygen species (ROS), reactive nitrogen species (RNS), and induction of enzymes of oxidative stress protection [29]. This also leads to synthesis and accumulation of signaling molecules such as JA, salicylic acid (SA), nitric oxide (NO), ethylene (ET) inside the cell and subsequent regulation of gene expression involved in secondary metabolite production [5,30,31]. Therefore, a large number of chemicals such as JA, MJ, SA, ET, are used for elicitation studies in vitro. However, MJ is the most commonly used elicitor which has shown a pronounced effect of accumulation secondary metabolites in plant cell and organ cultures [29,30]. Giri and Zaheer et al. [29] reported that MJ was most abundantly used chemical elicitor (60% of reports) in in vitro culture system, followed by SA and JA (approximately 15 and 10%, respectively). Cell suspension, adventitious roots and hairy roots are commonly used as culture system followed by multiple shoots and embryos for elicitation experiments [4,29,32].

3. Usage of Methyl Jasmonate (MJ) on Oxidative Stress and Antioxidant Response

3.1. Summary of MJ Elicitation Mechanism

Summary of MJ-treated elicitation mechanism is presented in Figure 3. In the first step, the elicitors are perceived by the specific receptors localized in the plasma membrane, which initiates signalling processes that activate plant defence mechanism [31]. Majority of the studies have revealed that binding of elicitors to receptors leads to induction of pathogenesis-related proteins and the production of ROS and RNS, enzymes of oxidative stress protection, the activation of defence-related genes [30–35]. During the process of signal transduction several processes such as protein phosphorylation, lipid oxidation, enhanced antioxidant enzyme activity (SOD, G-POD, APX, CAT), and the activation and the de novo biosynthesis of transcription factors and subsequent expression secondary metabolite biosynthetic genes have been reported by several researchers [5,21].

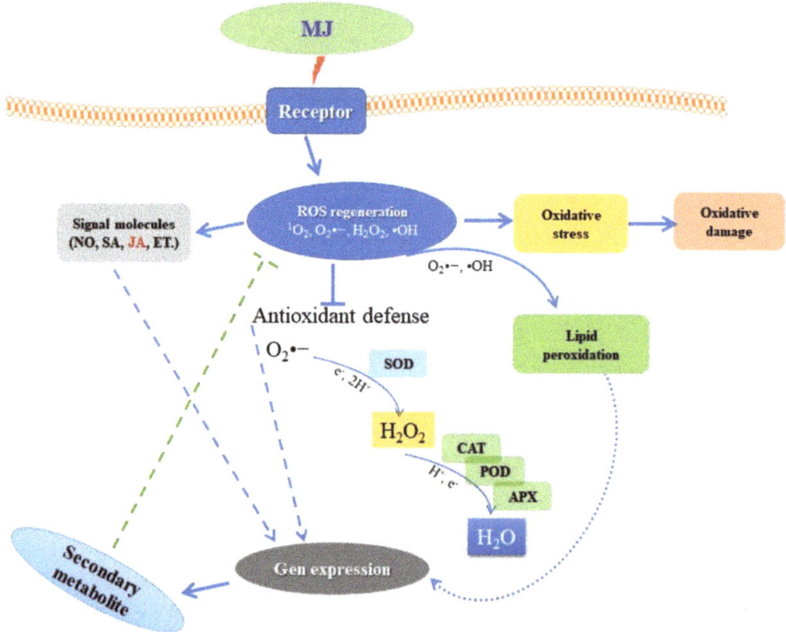

Figure 3. General mechanism after MJ perception. Abbreviations: *ROS* reactive oxygen species, SA salicylic acid, JA jasmonic acid, ET ethylene, O_2•$^-$ superoxide anion, •OH hydroxyl radical, H_2O_2 hydrogen peroxide, SOD superoxide dismutase, CAT catalase, G-POD Guaiacol peroxidase, APX ascorbate peroxidase.

3.2. Oxidative Stress-Induced by Methyl Jasmonate

Recent studies suggest that environmental stresses can increase the oxygen-induced damage to cells due to increased generation of ROS. ROS brings about the peroxidation of membrane lipids, which leads to membrane damage [13,27]. Some studies report that induction of oxidative stress has been observed in plants exposed to MJ and SA [36–38]. In agreement with these early studies, exposure to MJ significantly increased the MDA content, an index of lipid peroxidation. $O2^-$ is known to be harmful to all membrane constituents (Figures 3 and 4A). The accumulation of MDA content increased significantly in the MJ exposed roots of *Cnidium officinale* compared to the control (Figure 4A). This suggests that MJ lead to the production of O_2•$^-$, •OH resulting in increased lipid peroxidation. In other studies, the better protection in *P. ginseng* seems to result from the more efficient antioxidative system while a significant increase in MDA level in MJ-treated roots appeared to be derived from lower to increased activities of the enzymes [13,14].

3.3. Antioxidant Response

MJ is generally considered to modulate many physiological events in higher plants, such as defence responses, flowering and senescence [13,39]. Plants respond to a variety of environmental stresses (biotic and abiotic) through induction of antioxidant defence enzymes that provide protection against further damage [8,40]. Stress tolerance is closely related to the efficiency of antioxidant enzymes and these antioxidant enzymes and metabolites are reported to increase under various environmental stresses [8,12,41]. Numerous studies have demonstrated that MJ plays a key role in the enhancement of antioxidant enzymes activity such as APX, DHAR, MDHAR, AO, GR, GST, GPX, G-POD, CAT, SOD. Ali et al. [12] achieved that induced activity of SOD, G-POD and reduced ascorbate (ASC) contents indicated that antioxidants played an important role for protecting cells from MJ elicitation in *Panax ginseng and Panax quinquefolium* adventitious root cultures. In the apoplast and symplast of roots of sunflower (*Helianthus annuus* L.) seedlings, MJ-elicited roots showed a fast increase in ROS content, followed by a marked increase in the activity of H_2O_2-scavenging enzymes, G-POD, APX and CAT. The mechanisms responsible for MJ-induced H_2O_2 accumulation was investigated further by studying both the production and scavenging of H_2O_2 in the extracellular matrix [42].

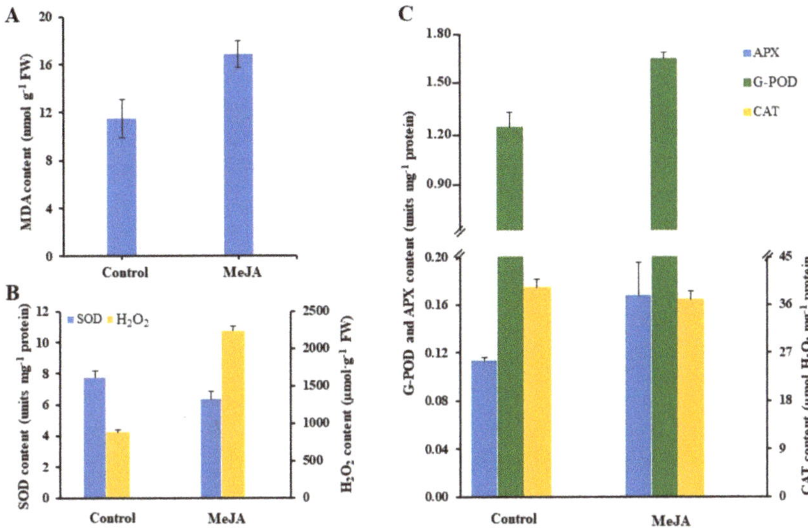

Figure 4. Effect of MJ on malondialdehyde (MDA) content (**A**), and antioxidant enzyme activity (**B,C**) in adventitious root cultures of *Cnidium officinale*. *SOD* superoxide dismutase, *CAT* catalase, *G-POD* guaiacol peroxidase, *APX* ascorbate peroxidase. Values represent mean ± SE (*n* = 3).

The effects of MJ on the accumulation of antioxidant enzyme activity in adventitious root culture of *Cnidium officinale* has been presented in Figure 4B,C. The activity of SOD was significantly decreased in MJ-treated roots, while the accumulation of H_2O_2 was increased compared to control (un-treated roots). MJ treatments resulted in a significant increase in G-POD and APX activity which were approximately 1.33 and 1.48-fold higher than the control, respectively. However, the CAT activity was slightly decreased in *C. officinale* adventitious roots treated with MJ. Similar results were also reported by Ali et al. [13] in *P. ginseng* adventitious root cultures.

The main detoxifying enzymes in oxidative stress are SOD, APX, CAT and G-POD, monodehydroascorbate reductase (MDHAR), dehydroascorbate reductase (DHAR), and glutathione reductase (GR) also showed an important role in scavenging stress-induced ROS generated in plants [10,13,14]. These enzymes operate in different subcellular compartments and respond in concert when cells are exposed to oxidative stress [10]. SOD plays a central role in defence against oxidative stress in all aerobic organisms [37]. The enzyme SOD belongs to the group of metalloenzymes and catalyses the dismutation of $O_2\bullet^-$ to O_2 and H_2O_2, which in turn is detoxified by CAT or G-POD or APX reactions (Figure 3). Therefore, the control of the steady-state $O_2\bullet^-$ levels by SOD is an important factor for protecting the cells against oxidative damage. Therefore, SOD is usually considered the first line of defence against oxidative stress [13,14,40,43]. CAT was the first enzyme to be discovered and characterized among antioxidant enzymes. It is a ubiquitous tetrameric heme-containing enzyme that catalyses the dismutation of two molecules of H_2O_2 into H_2O and O_2. G-POD is a heme-containing protein, preferably oxidizes aromatic electron donor such as guaiacol and pyrogallol at the expense of H_2O_2. APX is a central component of AsA-GSH cycle and plays an essential role in the control of intracellular ROS levels. APX uses two molecules of AsA to reduce H_2O_2 to water with a concomitant generation of two molecules of MDHA [10,11,44–46]. Farooq et al. [47] reported that the application of MJ minimized the oxidative stress, as revealed via a lower level of ROS synthesis in leaves of *Brassica napus*. This study also indicates that MJ plays an effective role in the regulation of multiple transcriptional pathways which were involved in oxidative stress responses, therefore enhanced enzymatic activities and gene expression of important antioxidants (SOD, APX, CAT, POD), secondary metabolites.

3.4. Signaling Pathway-Mediated Secondary Metabolite Accumulation in Plant Cell and Organ Cultures

Polyunsaturated fatty acids can produce signals, such as oxylipins, which include JA, JA methyl ester, JA amino acid conjugates, and further JA metabolites [48,49]. JA and its derivatives are known as the signalling molecules that can induce the biosynthetic enzymes involved in the formation of secondary metabolites in ginseng as shown in Figure 5 [5,50]. Zhao et al. [31] reported that exogenous application of MJ could induce endogenous JA biosynthesis, which stimulates expression of saponin biosynthetic genes in ginseng. Kim et al. [51] reported that the expression of farnesyl pyrophosphate synthase (FPS) was induced by MJ and produced the farnesyl diphosphate precursor for squalene biosynthesis. The transcription level of squalene synthase (SS) and squalene epoxidase (SE), key genes in ginseng biosynthesis was enhanced by MJ treatment in *P. notoginseng* and *P. ginseng* [5,51].

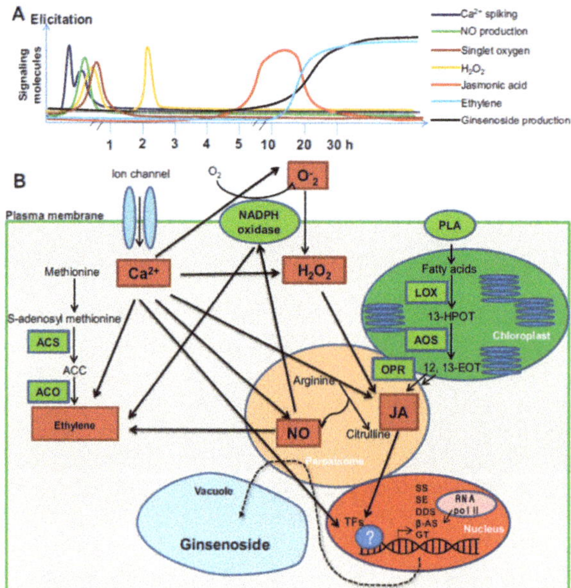

Figure 5. Schematic illustration of the sequential signalling pathways activated in elicited ginseng (**A**). Model of cross-talk between different signal transductions (**B**). Cross talk between different signalling molecules is shown by bold arrows. *ACC* 1-aminocyclopropane-1-carboxylic acid, *ACS* 1-aminocyclopropane-1-carboxylic acid synthase, *ACO* 1-aminocyclopropane-1-carboxylic acid, *AOS* allene oxide synthase, *β-AS* beta-amyrin synthase, *DDS* dammarenediol synthase, *12,13-EOT* 12,13(S)-epoxyoctadecatrienoic acid, H_2O_2 hydrogen peroxide, *13-HPOT* (13S)-hydroperoxyoctadecatrienoic acid, *JA* jasmonic acid, *LOX* lipoxygenase, *NO* nitric oxide, *NOS* nitric oxide synthase, O_2^- superoxide radical, *OPR* oxophytodienoate reductase, *PLA* phospholipase, *SS* squalene synthase, *SE* squalene epoxidase, *TFs* transcription factors, *UGRdGT* UDPG/ginsenoside Rd glucosyltransferase (Adapted from Rahimi et al. [5]).

The treatment of in vitro cultures with MJ may induce cross-talks between different signal transduction (auxin, ET, ABA, SA, GA, Ca^{2+}, O_2^-, H_2O_2, NO) as in the case in vivo systems. JA signalling pathway triggers *COL1* (*coronatine insensitive1*) gene in *Arabidopsis*, which encode F-box protein and involved in ubiquitination and removal of repressors of the JA signalling pathways [52]. Similarly, two other genes *JAR1* (Jasmonate resistant1) and *JIN1* (Jasmonate insentive1 and *JIN1* are also known as MYC2) were identified in *Arabidopsis*, which are involved in conjugation of jasmonic acid to Ile [53] and transcription regulation of JA-responsive genes, respectively [54]. Another protein *JAZ* (Jasmonate zim domain) mediates the interaction between *JAZ* and *COL1* or other transcriptional factors. COL1 protein, JAZ, and MYC2 constitute the core signal transduction mechanism of JA signalling and also responsible interaction with other transduction pathways [55]. For example, *NtCOL1*, *NtJAZ*, *NtMYC2a/2b* are responsible for nicotine biosynthesis in *Nicotiana tabacum* [55]. JA and auxin singling pathways participate in crosstalk and regulate various plant responses via COL1, MYC2 and JAZ components. When plants are activated with exogenous auxin, the auxin-TIR-AUX/IAA-ARF signalling is activated, which leads to JA synthesis. Contrarily, the endogenous JA induces the expression of auxin synthase gene (*ASA1*) and auxin levels, which leads the regulation of *JAZ1* [55]. During JA and ET crosstalk *ET3* (Ethylene insensitive3, which is involved in ET signalling pathway) and *JAZ-MYC2* (which is involved in the JA signalling pathway) will interact and regulate the stress responses [55]. During JA and ET crosstalk JA and ET may antagonise or coordinately regulated the plant stress responses. At the time of ABA and JA signalling pathways, JAZ-MYC2 participates in the crosstalk between JA and ABA and they regulate the plant responses coordinately. ABA receptor PYL (Pyrabactin resistance1-like proteins) interacts with the JA signalling pathway in many plants [55]. The crosstalk between JA and SA signalling pathways involves many components such as MAPK (mitogen-activated protein kinase), GRX (redox regulators of glutathione) and TRX (thioredoxin). It was demonstrated that JA signalling inhibits SA biosynthesis and accumulation [55]. JA and GA singling pathways regulate plant responses either coordinately or antagonistically. The interaction between JA and SA was brought about by the C-terminus of JAZs with that of MYC2 or DELLA proteins. DELLA interacts with JAZ to release MYC2, leads to activation of MYC2 genes. Concurrently, DELLA interacts with JAZ to inhibit the expression of JA biosynthetic genes via MYB21 and MYB24. Within vitro cultures of ginseng, elicitation with MJ induces crosstalk between MJ and Ca^{2+}, O_2^- (superoxide radical), H_2O_2, NO, ethylene and JA have been reported [5]. In the beginning, calcium signalling flux is required for further signalling, then both H_2O_2 and JA mediate early responses, whereas ethylene production is the late response in elicitor-induced for saponin biosynthesis (Figure 5) [5,56,57].

4. Usage of Methyl Jasmonate (MJ) in Plant Cell and Organ Cultures

4.1. Application of MJ for Enhancing Secondary Metabolite in In Vitro Culture System

The metabolic profiling of a plant is influenced by various micro-environmental and macro-environmental conditions. The plant responds to its surrounding environmental factors (as biotic and abiotic stress) through the stimulation of secondary metabolism to produce the desired compounds needed for its survival, the process is known as elicitation [58–62]. In recent days, MJ has been used extensively for elicitation studies involving in vitro culture systems such as cell suspension, adventitious root, hairy root and multiple shoot culture system (Table 1). Figure 6 shows the effect of MJ on phenolic compound production in adventitious root culture of *Cnidium officinale*. The adventitious roots treated with 100 µM MJ led to significantly higher yields (two-fold increment) of total phenolic compounds compared with the control treatments. These results corroborate the data for antioxidant enzyme activity (Figures 3 and 4). Elicitor treatment might have resulted in the production of ROS due to stress. In order to mitigate the effects of ROS, plant tissues exhibit a stimulated antioxidant enzyme activity leading to an increased production of secondary metabolites [14,63,64]. Han and Yuan [65] indicated that activation of phenylalanine ammonia-lyase (PAL) activity and accumulation of phenolic compounds is regulated by the oxidative burst in suspension culture of *Taxus cuspidata*, that is believed to change the membrane permeability and lead to the induction of secondary metabolism.

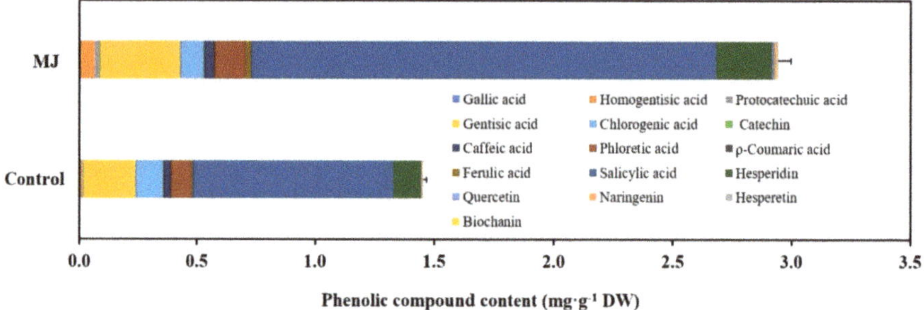

Figure 6. Effect of MJ on phenolic compounds production in *Cnidium officinale* adventitious root culture.

MJ treatment stimulates the biosynthesis of secondary metabolites in plant cell cultures via a large number of control points and triggers the expression of key genes that increase cellular activities at biochemical and molecular levels through the involvement of signal compounds [29]. It also plays a role in the signal transduction, which speeds up enzyme catalysis, thereby leading to the formation of specific compounds such as polyphenol, terpenoids, flavonoids, and alkaloids [4,30]. Ho et al. [62] reported that phenolic levels increased approximately 2-fold in adventitious root samples treated with 50 µM MJ (22.08 mg·g^{-1} DW) versus the control (10.08 mg·g^{-1} DW) in *Polygonum multiflorum*. In addition, hairy root cultures of *P. multiflorum*, the highest total phenolic content (52.87 mg·g^{-1} DW) was 3.4-fold higher than the control, especially, MJ treatment led to significantly higher levels of almost all individual phenolics, such as 1.13-fold increase of physcion, 3.83-fold of quercetin, 1.58-fold of kaempferol, 5.48-fold of p-hydroxybenzoic acid, and 4.3-fold of salicylic acid [62]. Level of ginsenoside accumulation showed a seven-fold enhancement in the adventitious root cultures treated with 100 µM MJ when compared to the control [66]. MJ at 100 µM enhanced the maximum production of xanthotoxin and bergapten at 1.1-fold and 39.6-fold higher than control in *Changium smyrnioides*, respectively [67].

Optimization of MJ concentration, the growth stage and exposure time of cultures are important critical factors for improving secondary metabolite synthesis [29,66]. Elicitation with 100 µM MJ influenced silymarin pigment accumulation in cell culture of *Silybum marimum* after 3 days of treatment [68]. In another report, elicitation with 4 µM MJ for 2 weeks resulted in a 6.5-fold enhancement of solasodine content (9.33 mg g^{-1} DW) than un-elicited hairy root cultures of *Solanum trilobatum* [69]. In addition, increasing of withanolide derivatives accumulation (14-fold) was achieved in 40-day-old hairy roots elicited with 15 µM MJ for 4 h exposure time [70]. In the previous studies on *P. multiflorum*, the phenolic compound in adventitious root culture reached to 2-fold higher than the control after 7-days treatment with 50 µM MJ, whereas 3.4-fold higher was observed in hairy roots after 5-day of exposure time [34,62]. Besides, an increase in 4.5-fold of triterpenoid in hairy root cultures of *Centella asiatica* was reported [71]. Treatment of cultures with 100 µM MJ for 6 days has been reported to influence the accumulation of both phenolic and tanshinones in *Salvia miltiorrhiza* hairy root cultures [72].

Table 1. Usage of MJ for enhance secondary metabolite within in vitro culture system.

Plant	Material [z]	Culture System	MJ	Elicitor Duration	Biomass	Target Compound	Fold Increase Compound	Reference
Panax notoginseng	CS	Flask	200 μM	7 day	Not significant	Rb, Rg ginsenosides	3.38	[73]
Silybum marimum	CS	Flask	100 μM	3 day	Decreased	Silymarin	4	[68]
Morinda citrifolia	CS	Flask	150 μM	2 day	Slightly decreased	Anthraquinone	4	[74]
Mentha x piperita	CS	Flask	100 μM	2 day	Decreased	Rosmarinic acid	1.8	[75]
Changium smyrnioides	CS	Flask	100 μM	5 day	Decreased	Xanthotoxin, bergapten	1.1 and 39.6	[67]
Scopolia parviflora	AR	Flask	1 mM	1 day	Slightly decreased	Scopolamine and hyoscyamine	2–3	[76]
Perovskia abrotanoides	AR	Flask	10 μM	7 day	Not significant	Tanshinone	2.3	[77]
Polygonum multiflorum	AR	Flask	50 μM	7 day	Decreased	Phenolic	2	[62]
Withania somnifera	HR	Flask	15 μM	4 h	Slightly decreased	Withanolide derivatives	14	[71]
Solanum trilobatum	HR	Flask	4 μM	14 day	Slightly decreased	Solasodine	6.5	[70]
Salvia miltiorrhiza	HR	Flask	100 μM	6 day	Decreased	Phenolic and tanshinones	3.3 and 2.5	[71]
Polygonum multiflorum	HR	Flask	50 μM	5 day	Slightly decreased	Phenolic	3.2	[34]
Echinacea purpurea	HR	Flask/Bioreactor	400 μM	7 day	Not significant	Triterpenoid	4.5	[78]
Polygonum multiflorum	AR	Bioreactor	50 μM	7 day	Slightly decreased	Phenolic	2.1	[62]
Panax ginseng	AR	Bioreactor	100 μM	10 day	Decreased	Ginsenoside	11-fold in Rb group	[66]
Panax ginseng and Panax quinquefolium	AR	Bioreactor	200 μM	9 day	Not significant	Saponin	4	[12,13]
Eleutherococcus koreanum	AR	Bioreactor	50 μM	7 day	Not significant	Eleutherosides B and E, and chlorogenic acid	1.1, 1.4 and 1.2	[79]
Olopanax elatus	AR	Bioreactor	200 μM	8 day	Decreased	Polysaccharide, phenolic	1.83 and 1.49	[80]
Panax ginseng	CS	Bioreactor	200 μM	8 day	Decreased	Ginsenoside	1.6–3.7	[81]
Rhodiola sachalinensis	CS	Bioreactor	125 μM	7 day	Decreased	Salidroside	5	[82]
Eleutherococcus senticosus	SE	Bioreactor	200 μM	7 day	Decreased	Eleutheroside	7.3	[83]

[z] *CS* cell suspension culture, *AR* adventitious root culture, *HR* hairy root culture, *SE* somatic embryos culture.

4.2. Application of MJ for Enhancement of Secondary Metabolites in Bioreactor Cultures

Inhibition of root growth and decreased biomass accumulation are the major challenges with MJ treated cell and organ suspension cultures, this might be due to enhanced accumulation of ROS, which adversely affects root biomass. Airlift bioreactors appear to be ideal for plant cell and organ cultures by efficiently controlling the culture environment, foam generation, shear stress, and oxygen supply [78,83,84]. They are suitable for the cultivation of cell, hairy root, adventitious root and embryo suspension cultures of various medicinal plants [32,78,84,85], and are also applied in MJ treatment to reduce the inhibition of root biomass by two 2-stage cultures. The first step involves the optimization of root growth in bioreactors without elicitor and then concentration and time of elicitor application will have to be standardized in the second step. A numerous study has been reported in using balloon-type bubble bioreactor (BTBB) for both biomass production and enhancement of secondary metabolite treated with MJ. After 6 weeks of culture, the addition of 50 μM MJ for 1 week was found to be the optimal concentration for eleutheroside B and E, and chlorogenic acid production in adventitious root culture of *Eleutherococcus koreanum* [80]. Thanh et al. [80] used 200 μM MJ on day-15 during culture period in cell suspension culture of *Panax ginseng* showed the highest ginsenoside yield after 8 days of treatment. The ginsenoside, Rb group and Rg group content increased 2.9, 3.7, and 1.6-fold after treatment, respectively. In the study of the somatic embryo of *Eleutherococcus senticosus*, the total eleutheroside was increased 7.3-fold after treated with 200 μM MJ [83]. The maximum content of salidroside (4.74 mg·g^{-1} DW) was observed at 125 μM MJ, which was 5-fold higher than the control in *Rhodiola sachalinensis* callus culture [81]. Jiang et al. [80] reported that when added 200 μM MJ to culture medium after 30 days of culture, maximum productivity of bioactive compound was found 8 days after treatment in the adventitious culture of *Olopanax elatus*.

The effect of MJ on biomass and secondary metabolite production in adventitious root culture of *Polygonum multiflorum* and *Echinacea purpurea* is presented in Table 2 and Figure 7. The treatment with MJ inhibited root growth in both medicinal plants after 7 days of treatment. The biomass was decreased by 5.8% (in DW) of *P. multiflorum* and 22.97% (in DW) of *E. purpurea*. The roots also turn to brown with such treatments (Figure 7). However, MJ at 50 μM enhanced the phenolic content approximately 2-fold in *P. multiflorum* compared to non-treated, and also reached to higher the total phenolic compounds when compared to field-grown plants (Table 2). Whereas, total caffeic acid derivatives in *E. purpurea* was also increased 1.16-fold compared to the control with the treatment of 200 μM MJ. Moreover, total productivity was also increased in both plants with 1.92-fold and 1.12-fold compared with non-treated roots.

Table 2. Effect of MJ on the accumulation of biomass and secondary metabolites in *Polygonum multiflorum* and *Echinacea purpurea* adventitious root culture system.

Systems	FW (g·L^{-1})	DW (g·L^{-1})	Dry Matter (%)	Growth Ratio [z]	Total Bioactive Compounds (mg·g^{-1} DW) [y]	Total Productivity (mg·L^{-1}) [x]
Polygonum multiflorum [w]						
Control	93.95 ± 3.23	10.61 ± 0.32	11.30 ± 0.11	20.23 ± 0.64	11.20 ± 0.17	118.82 ± 1.78
MJ50	90.09 ± 1.51	9.99 ± 0.23	11.10 ± 0.44	18.98 ± 0.46	22.83 ± 0.30	228.08 ± 3.00
5-year-old root					24.78 ± 0.82	
Echinacea purpurea [v]						
Control	83.81 ± 2.12	8.10 ± 0.72	9.63 ± 0.64	15.19 ± 1.45	25.22 ± 1.27	204.16 ± 10.24
MJ200	67.26 ± 1.72	6.24 ± 0.40	9.27 ± 0.53	11.48 ± 0.80	38.32 ± 0.34	239.20 ± 4.44

Data present mean ± SE. [v] Adventitious root of *E. purpurea* culture in 20-L bioreactor in 3 weeks, treated with 200 μM MJ for 7 days before harvest. [w] Adventitious root of *P. multiflorum* culture in 20-L bioreactor in 3 weeks, treated with 100 μM MJ for 7 days before harvest. [x] Total productivity (mg·L^{-1} medium) = Total bioactive compound content (mg·g^{-1} DW) x DW (mg·L^{-1}). [y] Total bioactive compounds (mg·g^{-1} DW) = Total phenolic compounds in *P. multiflorum*, and total caffeic acid derivatives in *E. purpurea*. [z] Growth ratio = (DW—Initial DW)/ Initial DW.

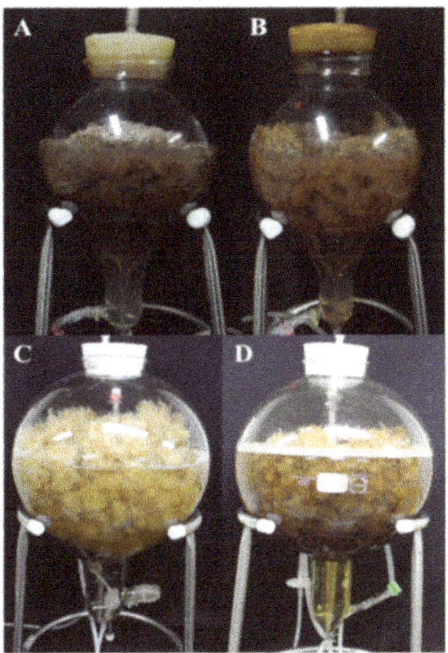

Figure 7. Effect of MJ on biomass production in *Polygonum multiflorum* and *Echinacea purpurea* adventitious root culture system. (**A,B**) control and 100 μM MJ in *P. multiflorum*, (**C,D**) control and 100 μM MJ in *E. purpurea*, respectively.

In addition, the treatment of MJ for improving bioactive compounds is also successfully applied in pilot-scale bioreactors up to 10,000 L [3,84–86]. A pilot-scale (500-L) bioreactor cultures were established in adventitious root culture of *Echinacea angustifolia* [87]. In this study, the authors achieved that although root biomass and growth ratio slightly decreased in the 500-L BTBB compared to the 5- and 20-L BTBBs, the highest concentrations of total phenolics (60.41 mg·g^{-1} DW), flavonoids (16.45 mg·g^{-1} DW), and total caffeic acid derivatives (33.44 mg·g^{-1} DW) were observed in the 500-L BTBB when added 100 μM MJ in the culture after 4 weeks. Especially, the accumulation of echinacoside (the major bioactive compound) in MJ-treated adventitious roots grown in the 500-L bioreactor was the highest (12.3 mg·g^{-1} DW), which is approximately three-fold higher than the non-MJ-treated roots cultured in 5- and 20-L bioreactors [80]. Baque et al. [88] achieved the highest accumulation of secondary metabolites in *Morinda citriflora* adventitious root cultures and cultures which were treated with 100 μM MJ after 4 weeks of culture. They reported higher accumulation of anthraquinones (205.75 mg·g^{-1} DW), phenolics (90.26 mg·g^{-1} DW) and flavonoids (93.34 mg·g^{-1} DW) in pilot-scale (500-L) bioreactor cultures when compared to small-scale bioreactor (3-L) cultures (approximately 1.8-fold, 1.3-fold, and 1.55-fold, respectively). In summary, two-stage culture systems should be addressed within in vitro culture system. First, a stage cultivation of cell suspension, adventitious roots and hairy roots for biomass production, followed by addition of MJ (at potential concentration and exposure time) in the second stage for enhancing metabolite production without decreasing root biomass. The application of scale-up bioreactor cultures is promising in the production of biomass and bioactive compounds and such materials/products will provide material for the pharmaceutical and cosmetic industry.

4.3. Effect of MJ on Gene Expression and Secondary Metabolite Accumulation

The variation of the expression of genes encoding key enzymes in a biosynthetic pathway directly influences the accumulation of the corresponding secondary metabolites [72,89,90]. The expression levels of the genes responsible for PAL, hydroxycinnamate coenzyme A ligase (4CL), cinnamic acid 4-hydroxylase (C4H), tyrosine aminotransferase (TAT), 4-Hydroxyphenylpyruvate reductase (HPPR), and rosmarinic acid synthase (RAS) were enhanced after 6-days of MJ treatment [72]. These enzymes were involved in the phenolic acid biosynthetic pathways which resulted in 197-fold increment phenolics in *Salvia miltiorrhiza* hairy roots [66]. Rahimi et al. [5] also demonstrated the enhanced regulation of genes involved in MEP (methylerythritol phosphate) and MVA (mevalonate) pathways with the treatment of MJ, in *Panax ginseng* [5]. Genes involved in andrographolide (diterpene) synthesis were influenced by MJ treatment in *Andrographis paniculata* [91]. The MEP pathway genes namely DXS (1-deoxy-D-xylulose 5-phosphate synthase), DXR (1-deoxy-D-xylulose 5-phosphate reductoisomerase), HDS (4-hydroxy-3-methylbut2-enyl-diphosphate synthase) and ISPH (1-Hydroxy-2-methyl-2-(E)-butenyl 4-diphosphate reductase) were up-regulated under MJ stimulation. Similarly, genes involved in the MVA pathway specially HMGS (3-hydroxy-3-methylglutaryl-Co-A synthase) and HMGR (3-hydroxy-3-methylglutaryl-CoA reductase) were also enhanced (Figure 8) by MJ treatment. A number of studies demonstrated that MJ treatment affected on MEP and MVA pathways and secondary metabolites biosynthesis [5,73,92,93].

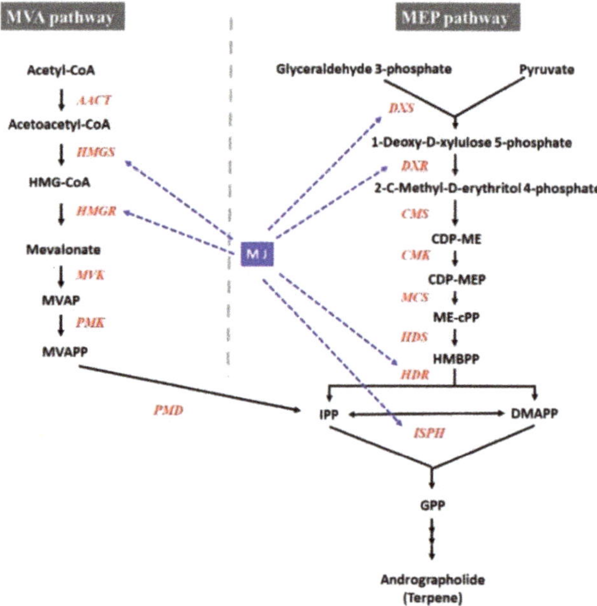

Figure 8. Andrographolide biosynthetic gene activation by MJ elicitation in *Andrographis paniculata* [91]. Dashed blue arrows show the relationship between MJ signalling and andrographolide biosynthetic genes. HMG-CoA, 3-Hydroxy-3-methylglutaryl CoA. MVA, mevalonic acid. MVAP, mevalonic acid 5-phosphate. MVAPP, mevalonic acid 5-diphosphate. IPP, isopentenyl diphosphate. DMAPP, dimethylallyl diphosphate. MEP, 2-C-methyl-D-erythritol 4-phosphate. CDP-ME, 4-diphosphocytidyl-2-C-methyl-D-erythritol. CDP-MEP, 4-diphosphocytidyl2-C-methyl-D-erythritol 2-phosphate. ME-cPP, 2-C-methyl-D-erythritol 2,4-cyclodiphosphate. HMBPP, 1-hydroxy-2-methyl-2-(E)-butenyl 4-diphosphate. Enzymes of the MVA pathway: HMGS, HMG-CoA synthase. HMGR, HMG-CoA reductase. MVK, MVA kinase. PMK, MVAP kinase. PMD, MVAPP decarboxylase. Enzymes of the MEP pathway: DXS, DOXP synthase. DXR, DOXP reductoisomerase. CMS, CDP-ME synthase. CMK, CDP-ME kinase. MCS, ME-2,4cPP synthase. HDS, HMBPP synthase. HDR, HMBPP reductase, ISPH, 1-Hydroxy-2-methyl-2-(E)-butenyl 4-diphosphate reductase.

Author Contributions: T.-T.H. contributed to writing of manuscript, H.N.M. revised for important intellectual content. S.-Y.P. made substantial contributions to interpreted data, conception and design this study. All authors have read and agree to the published version of the manuscript.

Funding: This work was funded by a grant from the Next Generation BioGreen21 Program (Project No. PJ013689), Rural Development Administration, Republic of Korea.

Acknowledgments: Authors were supported by Brain Korea (BK) 21 Plus Program through the National Research Foundation (NRF) of Korea.

Conflicts of Interest: The authors declare no conflict of interest.

Abbreviations

4CL	Hydroxycinnamate coenzyme A ligase
AOC	Allene oxide cyclase
AOS	Allene oxide synthase
APX	Ascorbate peroxidase
ASA	Ascorbate
AsA-GSH	Ascorbateglutahione
C4H	Cinnamic acid 4-hydroxylase
CAT	Catalase
DHAR	Dehydroascorbate reductase
DXS	1-Deoxy-d-xylulose 5-phosphate synthase
ET	Ethylene
FT-IR	Fourier-transform infrared spectroscopy
GGPPS	Geranylgeranyl diphosphate synthase
G-POD	Guaiacol peroxidase
GR	Glutathione reductase
GSH	Glutathione
H_2O_2	Hydrogen peroxide
HDR	1-Hydroxy-2-methyl-2-(E)-butenyl-4-diphosphate reductase
HPLC	High-performance liquid chromatography
HPPL	4-Hydroxyphenylpyruvate reductase
ISPH	1-Hydroxy-2-methyl-2-(E)-butenyl 4-diphosphate reductase
JA	Jasmonic acid
JAs	Jasmonate
JA-Ile	(+)-7-iso-jasmonoyl-L-isoleucine
LOX	Lipoxygenase
MJ	Methyl jasmonate
MEP	Methylerythritol phosphate
MDHAR	Monodehydroascorbate reductase
MVA	Mevalonate
NO	Nitric oxide
PAL	Phenylalanine ammonia-lyase
RAS	Rosmarinic acid synthase
RNS	Reactive nitrogen species
ROS	Reactive oxygen species
SA	Salicylic acid
SOD	Superoxide dismutase
TAT	Tyrosine aminotransferase

References

1. Ferrari, F.; Pasqua, G.; Monacelli, B. Xanthones from calli of *Hypericum perforatum* subsp. *perforatum*. *Nat. Prod. Res.* **2005**, *19*, 171–176. [CrossRef] [PubMed]

2. Brugger, G.A.; Lamotte, O.; Vandelle, E.; Bourque, S.; Lecourieux, D.; Poinssot, B.; Wendehenne, D.; Pugin, A. Early signalling events induced by elicitors of plant defences. *Mol. Plant Microbe Interact.* **2006**, *19*, 711–724. [CrossRef] [PubMed]

3. Paek, K.Y.; Murthy, H.N.; Hahn, E.J.; Zhong, J.J. Large scale culture of ginseng adventitious roots for production of ginsenosides. *Adv. Biochem. Eng. Biotechnol.* **2009**, *113*, 151–176.

4. Murthy, H.N.; Lee, E.J.; Paek, K.Y. Production of secondary metabolites from cell and organ cultures: Strategies and approaches for biomass improvement and metabolite accumulation. *Plant Cell Tissue Organ Cult.* **2014**, *118*, 1–16. [CrossRef]

5. Rahimi, S.; Kim, Y.J.; Yang, D.C. Production of ginseng saponins: Elicitation strategy and signal transductions. *Appl. Microbiol. Biotechnol.* **2015**, *99*, 6987–6996. [CrossRef] [PubMed]

6. Foyer, C.H.; Harbinson, J. Oxygen metabolism and the regulation of photosynthetic electron transport. In *Causes of Photooxidative Stresses and Amelioration of Defense Systems in Plants*; Foyer, C.H., Mullineaux, P., Eds.; CRC Press: Boca Raton, FL, USA, 1994; pp. 1–42.

7. Blokhina, O.; Fagerstedt, K.V. Reactive oxygen species and nitric oxide in plant mitochondria: Origin and redundant regulatory systems. *Physiol. Plant.* **2010**, *138*, 447–462. [CrossRef]

8. Mittler, R. Oxidative stress, antioxidants and stress tolerance. *Trends Plant Sci.* **2002**, *7*, 405–410. [CrossRef]

9. Hu, W.H.; Song, X.S.; Shi, K.; Xia, X.J.; Zhou, Y.H.; Yu, J.Q. Changes in electron transport, superoxide dismutase and ascorbate peroxidase isoenzymes in chloroplasts and mitochondria of cucumber leaves as influenced by chilling. *Photosynthetica* **2008**, *46*, 581–588. [CrossRef]

10. Sharma, P.; Jha, A.B.; Dubey, R.S.; Pessarakli, M. Reactive oxygen species, oxidative damage, and antioxidative defense mechanism in plants under stressful conditions. *J. Bot.* **2012**, *2012*, 1–26. [CrossRef]

11. Desikan, R.; Mackerness, S.A.H.; Hancock, S.J.T.; Neill, S.J. Regulation of the *Arabidopsis* transcriptome by oxidative stress. *Plant Physiol.* **2001**, *127*, 9–172. [CrossRef]

12. Ali, M.B.; Yu, K.W.; Hahn, E.J.; Paek, K.Y. Differential responses of anti-oxidants enzymes, lipoxygenase activity, ascorbate content and the production of saponins in tissue cultured root of mountain *Panax ginseng* C.A. Mayer and *Panax quinquefolium* L. in bioreactor subjected to methyl jasmonate stress. *Plant Sci.* **2005**, *169*, 83–192.

13. Ali, M.B.; Yu, K.W.; Hahn, E.J.; Paek, K.Y. Methyl jasmonate and salicylic acid elicitation induces ginsenosides accumulation, enzymatic and non-enzymatic antioxidant in suspension culture *Panax ginseng* roots in bioreactors. *Plant Cell Rep.* **2006**, *25*, 613–620. [CrossRef] [PubMed]

14. Ali, M.B.; Hahn, E.J.; Paek, K.Y. Methyl jasmonate and salicylic acid induced oxidative stress and accumulation of phenolics in *Panax ginseng* bioreactor root suspension culture. *Molecules* **2007**, *12*, 607–621. [CrossRef] [PubMed]

15. Ho, T.T. Strategies and Approaches for Improvement of Biomass and Bioactive Compounds in Adventitious Root and Hairy Root Cultures in *Polygonum multiflorum*. Ph.D. Thesis, Chungbuk National University, Cheongju, Korea, 2018.

16. Ho, T.T.; Lee, K.J.; Lee, J.D.; Bhushan, S.; Paek, K.Y.; Park, S.Y. Adventitious root culture of *Polygonum multiflorum* for phenolic compounds and its pilot-scale production in 500 L-tank. *Plant Cell Tissue Organ Cult.* **2017**, *130*, 167–181. [CrossRef]

17. Ho, T.T.; Murthy, H.N.; Dalawai, D.; Bhat, M.A.; Paek, K.Y.; Park, S.Y. Attributes of *Polygonum multiflorum* to transfigure red biotechnology. *App. Microbiol. Biotechnol.* **2019**, *103*, 3317–3326. [CrossRef] [PubMed]

18. Demole, E.; Lederer, E.; Mercier, D. Isolement et détermination de la structure du jasmonate de méthyle, constituant odorant caractéristique de lessence de jasmin. *Helv. Chim. Acta* **1962**, *45*, 675–685. [CrossRef]

19. Aldridge, D.C.; Galt, S.; Giles, D.; Turner, W.B. Metabolites of *Lasiodiplodia theobromae*. *J. Chem. Soc. C* **1971**, 1623–1627. [CrossRef]

20. Hyun, Y.; Choi, S.; Hwang, H.J.; Yu, J.; Nam, S.J.; Ko, J.; Park, J.Y.; Seo, Y.S.; Kim, E.Y.; Ryu, S.B.; et al. Cooperation and functional diversification of two closely related galactolipase genes for jasmonate biosynthesis. *Dev. Cell* **2008**, *14*, 183–192. [CrossRef]

21. Santino, A.; Taurino, M.; De Domenico, S.; Bonsegna, S.; Pastor, V.; Flors, V. Jasmonate signalling in plant development and defence response to multiple abiotic stresses. *Plant Cell Tissue Organ Cult.* **2013**, *32*, 1085–1098.

22. Ziegler, J.; Stenzel, I.; Hause, B.; Mauche, H.; Hamberg, M.; Grimm, R.; Ganal, M.; Wasternack, C. Molecular cloning of allene oxide cyclase. The enzyme establishing the stereochemistry of octadecanoids and jasmonates. *J. Biol. Chem.* **2000**, *275*, 19132–19138. [CrossRef]

23. Wasternack, C.; Hause, B. Jasmonates and octadecanoids: Signal in plant stress responses and development. *Prog. Nucl. Res. Mol. Biol.* **2002**, *72*, 165–221.

24. Yan, J.; Zhang, C.; Gu, M.; Bai, Z.; Zhang, W.; Qi, T.; Cheng, Z.; Peng, W.; Luo, H.; Nan, F.; et al. The *Arabidopsis* CORONATINE INSENSITIVE1 Protein is a jasmonate receptor. *Plant Cell* **2009**, *21*, 2220–2236. [CrossRef] [PubMed]

25. Ruan, J.; Zhou, Y.; Zhou, M.; Yan, J.; Khurshid, M.; Weng, W.; Cheng, J.; Zhang, K. Jasmonic acid signalling pathways in plants. *Int. J. Mol. Sci.* **2019**, *20*, 2479. [CrossRef] [PubMed]

26. Creelman, R.A.; Mullet, J.E. Jasmonic acid distribution and action in plants: Regulation during development and response to biotic and abiotic stress. *Proc. Natl. Acad. Sci. USA* **1995**, *92*, 4114–4119. [CrossRef]

27. Bertini, L.; Palazzi, L.; Proietti, S.; Pollastri, S.; Arrigoni, G.; de Laureto, P.; Carusco, C. Proteomic analysis of MeJa-induced defense responses in rice against wounding. *Int. J. Mol. Sci.* **2019**, *2020*, 2525. [CrossRef]

28. Bertini, L.; Proietti, S.; Focaracci, F.; Sabatin, B.; Carusco, C. Epigenetic control of dense genes following MeJA-induced priming in rice (*O. sativa*). *J. Plant Physiol.* **2018**, *228*, 166–177.

29. Giri, C.C.; Zaheer, M. Chemical elicitors versus secondary metabolite production in vitro using plant cell, tissue and organ cultures: Recent trends and a sky eye view appraisal. *Plant Cell Tissue Organ Cult.* **2016**, *126*, 1–18. [CrossRef]

30. Baenas, N.; Garcia-Viguera, C.; Moreno, D.A. Elicitation: A tool for enriching the bioactive composition of foods. *Molecules* **2014**, *19*, 13541–13563. [CrossRef]

31. Zhao, J.; Davis, L.C.; Verpoorte, R. Elicitor signal transduction leading to production of plant secondary metabolites. *Biotechnol. Adv.* **2005**, *23*, 283–333. [CrossRef]

32. Shohael, A.M.; Murthy, H.N.; Lee, H.L.; Hahn, E.J.; Paek, K.Y. Increased eleutheroside production in *Eleutherococcus sessiliflorus* embryogenic suspension cultures with methyl jasmonate treatment. *Biochem. Eng. J.* **2008**, *38*, 270–273. [CrossRef]

33. Durango, D.; Pulgarin, N.; Echeverri, F.; Escobar, G.; Quiñones, W. Effect of salicylic acid and structurally related compounds in the accumulation of phytoalexins in cotyledons of common bean (*Phaseolus vulgaris* L.) cultivars. *Molecules* **2013**, *18*, 10609–10628. [CrossRef] [PubMed]

34. Ho, T.T.; Lee, J.D.; Ahn, M.S.; Kim, S.W.; Park, S.Y. Enhanced production of phenolic compounds in hairy root cultures of *Polygonum multiflorum* and its metabolite discrimination using HPLC and FT-IR methods. *Appl. Microbiol. Biotechnol.* **2018**, *102*, 9563–9575. [CrossRef] [PubMed]

35. Vasconsuelo, A.; Boland, R. Molecular aspects of the early stages of elicitation of secondary metabolites in plants. *Plant Sci.* **2007**, *172*, 861–875. [CrossRef]

36. Mir, M.Y.; Hamid, S.; Kamili, A.N.; Hassan, Q.P. Sneak peek of *Hypericum perforatum* L.: Phytochemistry, phytochemical efficacy and biotechnological interventions. *J. Plant Biochem. Biotechnol.* **2019**, *28*, 357–375. [CrossRef]

37. Scandalios, L.G. Oxygen stress and superoxide dismutase. *Plant Physiol.* **1993**, *101*, 7–12. [CrossRef] [PubMed]

38. Yu, D.Q.; Cen, C.; Yang, M.L. Studies on the salicylic acid induced lipid peroxidation and defence gene expression in tobacco cell culture. *Acta Bot. Sin.* **1999**, *41*, 977–982.

39. Chong, T.M.; Abdullah, M.A.; Fadzillah, N.M.; Lai, O.M.; Lajis, N.H. Jasmonic acid elicitation of anthraquinones with some associated. *Enzyme Microb. Technol.* **2005**, *36*, 469–477. [CrossRef]

40. Creelman, R.A.; Mullet, J.E. Biosynthesis and action of jasmonates in plants. *Annu. Rev. Plant. Physiol. Plant Mol. Biol.* **1997**, *48*, 355–381. [CrossRef]

41. Asada, K. The water-water cycle in chloroplasts: Scavenging of active oxygens and dissipation of excess photons. *Annu. Rev. Plant Physiol. Plant Mol. Biol.* **1999**, *50*, 601–639. [CrossRef]

42. Seo, Y. Antioxidant activity of the chemical constituents from the flower buds of *Magnolia denudata*. *Biotechnol. Bioprocess Eng.* **2010**, *15*, 400–406. [CrossRef]

43. Parra-Lobato, M.G.; Fernandez-Garcia, N.; Olmos, E.; Alvarez-Tinaut, M.G.; Gomez-Jiménez, M.C. Methyl jasmonate-induced antioxidant defence in root apoplast from sunflower seedlings. *Environ. Exp. Bot.* **2009**, *66*, 9–17. [CrossRef]

44. Wasternack, C.; Parthier, B. Jasmonate signaled plant gene expression. *Trends Plant Sci.* **1997**, *2*, 302–307. [CrossRef]

45. Yamauchi, Y.; Furutera, A.; Seki, K.; Toyoda, Y.; Tanaka, K.; Sugimoto, Y. Malondialdehyde generated from peroxidized linolenic acid causes protein modification in heat-stressed plants. *Plant Physiol. Biochem.* **2008**, *8–9*, 786–793. [CrossRef] [PubMed]

46. Welinder, K.G. Superfamily of plant, fungal and bacterial peroxidases. *Curr. Opin. Struct. Biol.* **1992**, *2*, 388–393. [CrossRef]

47. Farooq, M.A.; Gill, R.A.; Islam, F.; Ali, B.; Liu, H.; Xu, J.; He, S.; Zhou, W. Methyl jasmonate regulates antioxidant defence and suppresses arsenic uptake in *Brassica napus* L. *Front. Plant Sci.* **2016**, *7*, 468. [CrossRef]

48. Wasternack, C. Jasmonates: An update on biosynthesis, signal transduction and action in plant stress response, growth and development. *Ann. Bot.* **2007**, *100*, 681–697. [CrossRef]

49. Wasternack, C.; Hause, B. Jasmonates: Biosynthesis, perception, signal transduction and action in plant stress response, growth and development. An update to 2007 review in Annals of Botany. *Ann. Bot.* **2013**, *111*, 1021–1058. [CrossRef]

50. Rahimi, S.; Devi, B.S.R.; Khorolragcha, A.; Kim, Y.J.; Kim, J.H.; Jung, S.K.; Yang, D.C. Effect of salicylic acid and yeast extract on the accumulation of jasmonic acid and sesquiterpenoids in *Panax ginseng* adventitious roots. *Russ. J. Plant Physiol.* **2014**, *61*, 811–817. [CrossRef]

51. Kim, O.T.; Bang, K.H.; Kim, Y.C.; Hyun, D.Y.; Kim, M.Y.; Cha, S.W. Upregulation of ginsenoside and gene expression related to triterpene biosynthesis in ginseng hairy root cultures elicited by methyl jasmonate. *Plant Cell Tissue Organ Cult.* **2009**, *98*, 25–33. [CrossRef]

52. Xie, D.X.; Feys, B.F.; James, S.; Nieto-Rostro, M.; Turner, J.G. COL1, an Arabidopsis gene required for jasmonae-regulated defense and fertility. *Science* **1998**, *280*, 1091–1094. [CrossRef]

53. Staswick, P.E.; Tiryaki, I. The oxylipin signal jasmonic acid is activated by an enzyme that conjugates it it to isoleucine in *Arabidopsis*. *Plan Cell* **2004**, *16*, 2117–2127. [CrossRef] [PubMed]

54. Lorenzo, O.; Chico, J.M.; Sanchez-Serrano, J.J.; Solano, R. JASMONATE INSENSTIVE1 encodes a MYC transcription factor essential to discriminate between different jasmonate-regulated defense responses in *Arabidopsis*. *Plant Cell* **2004**, *16*, 1938–1950. [CrossRef] [PubMed]

55. Yang, J.; Duan, G.; Li, C.; Liu, L.; Han, G.; Zhang, Y.; Wang, C. The crosstalk between jasmonic acid and other plant hormone sginalling highlight the involvement of jasmonic acid as a core component in plant reponses to biotic and abiotic stresses. *Front. Plant Sci.* **2019**, *10*, 1349. [CrossRef] [PubMed]

56. Hu, F.X.; Zhong, J.J. Jasmonic acid mediates gene transcription of ginsenoside biosynthesis in cell cultures of *Panax notoginseng* treated with chemically synthesized 2-hydroxyethyl jasmonate. *Process Biochem.* **2008**, *43*, 113–118. [CrossRef]

57. Xu, X.; Hu, X.; Neill, S.J.; Fang, J.; Cai, W. Fungal elicitor induces singlet oxygen generation, ethylene release and saponin synthesis in cultured cells of *Panax ginseng* CA Meyer. *Plant Cell Physiol.* **2005**, *46*, 947–954. [CrossRef]

58. Tewari, R.K.; Paek, K.Y. Salicylic acid-induced nitric oxide and ROS generation stimulate ginsenoside accumulation in *Panax ginseng* roots. *J. Plant Growth Regul.* **2011**, *30*, 396–404. [CrossRef]

59. Jeong, G.T.; Park, D.H. Enhancement of growth and secondary metabolite biosynthesis: Effect of elicitor derived from plant and insects. *Biotechnol. Bioprocess Eng.* **2005**, *10*, 73–77. [CrossRef]

60. Yin, Z.; Shangguan, X.; Chen, J.; Zhao, Q.; Li, D. Growth and triterpenic acid accumulation of *Cyclocarya paliurus* cell suspension cultures. *Biotechnol. Bioprocess Eng.* **2013**, *18*, 606–614. [CrossRef]

61. Gorelick, J.; Bernstein, N. Elicitation: An underutilized tool in the development of medicinal plants as a source of therapeutic secondary metabolites. *Adv. Agron.* **2014**, *124*, 201–230.

62. Ho, T.T.; Lee, J.D.; Jeong, C.S.; Paek, K.Y.; Park, S.Y. Improvement of biosynthesis and accumulation of bioactive compounds by elicitation in adventitious root cultures of *Polygonum multiflorum*. *Appl. Microbiol. Biotechnol.* **2018**, *102*, 199–209. [CrossRef]

63. Belhadj, A.; Teleg, N.; Saigne, C.; Cluzet, S.; Barrieu, F.; Hamdi, S.; Merillon, J.M. Effect of methyl jasmonate in combination with carbohydrates on gene expression of PR proteins, stilbene and anthocyanin accumulation in grapevine cell cultures. *Plant Physiol. Biochem.* **2008**, *46*, 493–499. [CrossRef]

64. Zhou, M.L.; Zhu, X.M.; Shao, J.R.; Tang, J.X.; Wu, Y.M. Production and metabolic engineering of bioactive substances in plant hairy root culture. *Appl. Microbiol. Biotechnol.* **2011**, *90*, 1229–1239. [CrossRef] [PubMed]

65. Han, R.B.; Yuan, Y.J. Oxidative burst in suspension culture of *Taxus cuspidata* induced by a laminar shear stress in short-term. *Biotechnol. Prog.* **2004**, *20*, 507–513. [CrossRef]

66. Kim, Y.S.; Hahn, E.J.; Murthy, H.N.; Paek, K.Y. Adventitious root growth and ginsenoside accumulation in *Panax ginseng* cultures as affected by methyl jasmonate. *Biotechnol. Lett.* **2004**, *26*, 1619–1622. [CrossRef]

67. Cai, J.; Ma, Y.; Hu, P.; Zhang, Y.; Chen, J.; Li, X. Elicitation of furanocoumarins in *Changium smyrnioides* suspension cells. *Plant Cell Tissue Organ Cult.* **2017**, *130*, 1–12. [CrossRef]

68. Sampedro, M.A.S.; Tárrago, J.F.; Corchete, P. Yeast extract and methyl jasmonate-induced silymarin production in cell cultures of *Silybum marianum* (L.) Gaertn. *J. Biotechnol.* **2005**, *119*, 60–69. [CrossRef] [PubMed]

69. Baek, S.E.; Ho, T.T.; Lee, H.S.; Jung, G.Y.; Kim, Y.E.; Jeong, C.S.; Park, S.Y. Enhanced biosynthesis of triterpenoids in *Centella asiatica* hairy root culture by precursor feeding and elicitation. *Plant Biotechnol. Rep.* **2019**. [CrossRef]

70. Shilpha, J.; Satish, L.; Kavikkuil, M.; Largia, M.J.V.; Ramesh, M. Methyl jasmonate elicits the solasodine production and antioxidant activity in hairy root cultures of *Solanum trilobatum* L. *Ind. Crop. Prod.* **2015**, *71*, 54–64. [CrossRef]

71. Sivanandhan, G.; Dev, K.G.; Jeyaraj, M.; Rajesh, M.; Arjunan, A.; Muthuselvam, M.; Manickavasagam, M.; Ganapathi, A. Increased production of withanolide A, withanone and withaferin A in hairy root cultures of *Withania somnifera* (L.) Dunal elicited with methyl jasmonate and salicylic acid. *Plant Cell Tissue Organ Cult.* **2013**, *114*, 121–129. [CrossRef]

72. Xing, B.; Yang, D.; Liu, L.; Han, R.; Sun, Y.; Liang, Z. Phenolic acid production is more effectively enhanced than tanshinone production by methyl jasmonate in *Salvia miltiorrhiza* hairy roots. *Plant Cell Tissue Organ Cult.* **2018**, *134*, 119–129. [CrossRef]

73. Wang, W.; Zhao, Z.J.; Xu, Y.; Qian, X.; Zhong, J.J. Efficient induction of ginsenoside biosynthesis and alteration of ginsenoside heterogeneity in cell cultures of *Panax notoginseng* by using chemically synthesized 2-hydroxyethyl jasmonate. *Appl. Microbiol. Biotechnol.* **2006**, *70*, 298–307. [CrossRef]

74. Komaraiah, P.; Kavi-Kishor, P.B.; Carlsson, M.; Magnusson, K.E.; Mandenius, C.F. Enhancement of anthraquinone accumulation in *Morinda citrifolia* suspension cultures. *Plant Sci.* **2005**, *168*, 1337–1344. [CrossRef]

75. Krzyzanowska, J.; Czubacka, A.; Pecio, L.; Przybys, M.; Doroszewska, T.; Stochmal, A.; Oleszek, W. The effects of jasmonic acid and methyl jasmonate on rosmarinic acid production in *Mentha x piperita* cell suspension cultures. *Plant Cell Tissue Organ Cult.* **2012**, *108*, 73–81. [CrossRef]

76. Kang, S.M.; Jung, H.Y.; Kang, Y.M.; Yun, D.J.; Bahk, J.D.; Yang, J.K. Effects of methyl jasmonate and salicylic acid on the production of tropane alkaloids and the expression of PMT and H6H in adventitious root cultures of *Scopolia praviflora. Plant Sci.* **2004**, *166*, 745–751. [CrossRef]

77. Zaker, A.; Sykora, C.; Gössnitzer, F.; Abrishamchi, P.; Asili, J.; Mousavi, S.H.; Wawrosch, C. Effects of some elicitors on tanshinone production in adventitious root cultures of *Perovskia abrotanoides* Karel. *Ind. Crop. Prod.* **2015**, *67*, 97–102. [CrossRef]

78. Lee, E.J.; Park, S.Y.; Paek, K.Y. Enhancement strategies of bioactive compound production in adventitious root cultures of *Eleutherococcus koreanum* Nakai subjected to methyl jasmonate and salicylic acid elicitation through airlift bioreactors. *Plant Cell Tissue Organ Cult.* **2015**, *120*, 1–10. [CrossRef]

79. Jiang, X.L.; Piao, X.C.; Gao, R.; Jin, M.Y.; Jiang, J.; Jin, X.H.; Lian, M.L. Improvement of bioactive compound accumulation in adventitious root cultures of an endangered plant species, *Oplopanax elatus. Acta Physiol. Plant.* **2017**, *39*, 226. [CrossRef]

80. Thanh, N.T.; Murthy, H.N.; Yu, K.W.; Hahn, E.J.; Paek, K.Y. Methyl jasmonate elicitation enhanced synthesis of ginsenoside by cell suspension cultures of *Panax ginseng* in 5-l balloon type bubble bioreactor. *Appl. Micorbiol. Biotechnol.* **2005**, *67*, 197–201. [CrossRef] [PubMed]

81. Li, Y.; Shao, C.H.; Park, S.Y.; Piao, X.C.; Lian, M.L. Production of salidroside and polysaccharides in *Rhodiola sachalinensis* using airlift bioreactor systems. *Acta Physiol. Plant.* **2014**, *36*, 2975–2983. [CrossRef]

82. Shohael, A.M.; Murthy, H.N.; Lee, H.L.; Hahn, E.J.; Paek, K.Y. Methyl jasmonate induced overproduction of eleutherosides in somatic embryos of *Eleutherococcus senticosus* cultures in bioreactors. *Electron. J. Biotechnol.* **2007**, *10*, 633–637. [CrossRef]

83. Jeong, J.A.; Wu, C.H.; Murthy, H.N.; Hahn, E.J.; Paek, K.Y. Application of an airlift bioreactor system for the production of adventitious root biomass and caffeic acid derivatives of *Echinacea purpurea. Biotechnol. Bioprocess Eng.* **2009**, *14*, 91–98. [CrossRef]

84. Baque, M.A.; Moh, S.H.; Lee, E.J.; Zhong, J.J.; Paek, K.Y. Production of biomass and useful compounds from adventitious roots of high-value added medicinal plants using bioreactor. *Biotechnol. Adv.* **2012**, *30*, 1255–1267. [CrossRef] [PubMed]
85. Murthy, H.N.; Dandin, V.S.; Paek, K.Y. Tools for biotechnological production of useful phytochemicals from adventitious root cultures. *Phytochem. Rev.* **2016**, *15*, 129–145. [CrossRef]
86. Wang, G.R.; Qi, N.M. Influence of mist intervals and aeration rate on growth and second metabolite production of *Pseudostellaria heterophylla* adventitious roots in a siphon-mist bioreactor. *Bioprocess Eng.* **2010**, *15*, 1059–1098. [CrossRef]
87. Cui, H.Y.; Baque, M.A.; Lee, E.J.; Paek, K.Y. Scale-up of adventitious root cultures of *Echinacea angustifolia* in a pilot-scale bioreactor for the production of biomass and caffeic acid derivatives. *Plant. Biotechnol. Rep.* **2013**, *7*, 297–308. [CrossRef]
88. Baque, M.A.; Murthy, H.N.; Paek, K.Y. Adventitious root culture of *Morinda citrifolia* in bioreactors for production of bioactive compounds. In *Production of Biomass and Bioactive Compounds Using Bioreactor Technology*, 1st ed.; Paek, K.Y., Murthy, H.N., Zhong, J.J., Eds.; Springer: Dordrecht, The Netherlands, 2014; pp. 185–222.
89. Yu, H.; Liu, X.; Gao, S.; Han, X.; Cheng, A.; Lou, H. Molecular cloning and functional characterization of a phenylalanine ammonialyase from liverwort *Plagiochasma appendiculatum*. *Plant Cell Tissue Organ Cult.* **2014**, *117*, 265–277. [CrossRef]
90. Mueller, M.J.; Brodschelm, W.; Spannagl, E.; Zenk, M.H. Signaling in the elicitation process is mediated through the octadecanoid pathway leading to jasmonic acid. *Proc. Natl. Acad. Sci. USA* **1993**, *90*, 7490–7494. [CrossRef]
91. Sinha, R.K.; Sharma, S.N.; Verma, S.S.; Zha, J. Effects of lovastin, fosmidomycin and methyl jasmonate on andrographolide biosynthesis in the *Andrographis paniculata*. *Acta Physiol. Plant.* **2008**, *40*, 165. [CrossRef]
92. Yan, Y.; Stolz, S.; Chetelat, A.; Reymond, P.; Pagni, M.; Dubugnon, L.; Farmer, E.E. A downstream mediator in the growth repression limb of the jasmonate pathway. *Plant Cell* **2007**, *19*, 2470–2483. [CrossRef]
93. Murthy, H.N.; Georgiev, M.I.; Kim, Y.S.; Jeong, C.S.; Kim, S.J.; Park, S.Y.; Paek, K.Y. Ginsenosides: Prospective for sustainable biotechnological production. *Appl. Microbiol. Biotechnol.* **2014**, *98*, 6243–6254. [CrossRef]

© 2020 by the authors. Licensee MDPI, Basel, Switzerland. This article is an open access article distributed under the terms and conditions of the Creative Commons Attribution (CC BY) license (http://creativecommons.org/licenses/by/4.0/).

International Journal of
Molecular Sciences

Article

Influence of Chloroplast Defects on Formation of Jasmonic Acid and Characteristic Aroma Compounds in Tea (*Camellia sinensis*) Leaves Exposed to Postharvest Stresses

Jianlong Li [1,†], Lanting Zeng [2,3,†], Yinyin Liao [2,3], Dachuan Gu [2,3], Jinchi Tang [1,*] and Ziyin Yang [2,3,*]

[1] Tea Research Institute, Guangdong Academy of Agricultural Sciences & Guangdong Provincial Key Laboratory of Tea Plant Resources Innovation and Utilization, Dafeng Road 6, Tianhe District, Guangzhou 510640, China; skylong.41@163.com

[2] Guangdong Provincial Key Laboratory of Applied Botany & Key Laboratory of South China Agricultural Plant Molecular Analysis and Genetic Improvement, South China Botanical Garden, Chinese Academy of Sciences, Xingke Road 723, Tianhe District, Guangzhou 510650, China; zenglanting@scbg.ac.cn (L.Z.); honey_yyliao@scbg.ac.cn (Y.L.); gdcawang@126.com (D.G.)

[3] College of Advanced Agricultural Sciences, University of Chinese Academy of Sciences, No. 19A Yuquan Road, Beijing 100049, China

* Correspondence: tangjinchi@126.com (J.T.); zyyang@scbg.ac.cn (Z.Y.); Tel.: +86-20-8516-1049 (J.T.); +86-20-3807-2989 (Z.Y.)

† These authors contributed equally to this work.

Received: 23 January 2019; Accepted: 22 February 2019; Published: 27 February 2019

Abstract: Characteristic aroma formation in tea (*Camellia sinensis*) leaves during the oolong tea manufacturing process might result from the defense responses of tea leaves against these various stresses, which involves upregulation of the upstream signal phytohormones related to leaf chloroplasts, such as jasmonic acid (JA). Whether chloroplast changes affect the formation of JA and characteristic aroma compounds in tea leaves exposed to stresses is unknown. In tea germplasms, albino-induced yellow tea leaves have defects in chloroplast ultrastructure and composition. Herein, we have compared the differential responses of phytohormone and characteristic aroma compound formation in normal green and albino-induced yellow tea leaves exposed to continuous wounding stress, which is the main stress in oolong tea manufacture. In contrast to single wounding stress (from picking, as a control), continuous wounding stress can upregulate the expression of *CsMYC2*, a key transcription factor of JA signaling, and activate the synthesis of JA and characteristic aroma compounds in both normal tea leaves (normal chloroplasts) and albino tea leaves (chloroplast defects). Chloroplast defects had no significant effect on the expression levels of *CsMYC2* and JA synthesis-related genes in response to continuous wounding stress, but reduced the increase in JA content in response to continuous wounding stress. Furthermore, chloroplast defects reduced the increase in volatile fatty acid derivatives, including jasmine lactone and green leaf volatile contents, in response to continuous wounding stress. Overall, the formation of metabolites derived from fatty acids, such as JA, jasmine lactone, and green leaf volatiles in tea leaves, in response to continuous wounding stress, was affected by chloroplast defects. This information will improve understanding of the relationship of the stress responses of JA and aroma compound formation with chloroplast changes in tea.

Keywords: albino; aroma; *Camellia sinensis*; chloroplast; jasmonic acid; light-sensitive; stress; tea; volatile

---

1. Introduction

Plants generally synthesize large amounts of volatile compounds in response to environmental stresses. These volatile metabolites contribute to plant defense against environmental stresses [1,2] and can also be regarded as important quality components in crops [3]. Utilizing the stress response to improve the natural quality components of horticultural crops has recently attracted increasing attention. Tea (*Camellia sinensis*) plants are famous horticultural and representative plants in China, with their leaves used to make the second most popular beverage, globally, after water. Aroma is an important factor affecting the character and quality of tea [4]. Aroma formation in tea leaves can result from the defense responses of tea leaves against various stresses during the preharvest (tea growth) and postharvest (tea manufacture) processes [3,4]. During tea growth, attack by insects, such as tea green leafhoppers [5,6], and light conditions, such as dark, blue-light, and red-light treatments [7,8], can significantly increase endogenous aroma compounds or produce new aroma compounds in tea leaves. During the tea manufacturing process, especially that of oolong tea, tea leaves are exposed to various stresses, including plucking (wounding), solar withering (drought, heat, and UV radiation), indoor withering (drought), and turnover (continuous wounding) [5,9]. Aroma formation in tea leaves during the oolong tea manufacturing process might result from the defense responses of tea leaves against these various stresses [3]. For example, levels of three characteristic aroma compounds, including indole, jasmine lactone, and (*E*)-nerolidol, were significantly enhanced during the turnover stage of oolong tea manufacture owing to continuous wounding stress [9–13]. As these aroma compounds are biosynthesized through different metabolic pathways, it has been proposed that common upstream signals regulate the multiple biosynthetic pathways of tea aromas under these stresses [3]. Phytohormones, especially jasmonic acid (JA), have been reported to regulate the biosynthesis of volatiles and act as important upstream signal chemicals [14]. Several studies have shown that JA is involved in the formation of tea aroma compounds [15,16]. In some plants, the main upstream synthetic pathways involved in JA formation (linoleic acid (LA) is released from chloroplast lipids, then converted to 12-oxo-phytodienoic acid by lipoxygenases (LOX), allene oxide synthase (AOS), and allene oxide cyclase (AOC) as catalysts) have been reported to mostly occur in the chloroplast [17–20]. In our previous study, CsAOS2 isolated from *Camellia sinensis* was found to have a role in JA synthesis, and was located in the chloroplast membrane [21]. This suggested that chloroplasts might be related to JA synthesis in tea. This inspired our interest in whether chloroplast defects affect JA synthesis and, consequently, aroma formation in tea leaves.

Generally, tea cultivars naturally grown as green leaves are usually processed into tea products. However, tea mutants whose young shoots are grown in white or yellow color in some conditions have recently attracted a significant increasing attention from the researchers and manufactures in the related areas, because they are supposed as potential raw materials to be processed into "high-quality" tea products [22]. Light-sensitive and low-temperature-sensitive types are mainly albino teas, and their new grown shoots exhibit white or yellow color [23,24]. In light-sensitive mutants, strong light illumination is a determinant factor, while in low-temperature-sensitive types, low temperature is a prerequisite [25–27]. Former studies showed that a significant deficiency of chlorophylls mainly result in white or yellow shoots in these albino tea mutants [23,24,28]. In some recent studies, the samples used to investigate the underlying mechanism of albino tea leaves are under different genetic background, geography, climate, amongst other factors. Therefore, to avoid the interference from these factors, *C. sinensis* cv. Yinghong No. 9 (an original cultivar grown with green leaves), and its light-sensitive mutant (albino-induced yellow leaves in strong light illumination) were collected in pairs under the same growth conditions as used in our previous experiments [22,28]. This light-sensitive mutant shared the same genetic background of the original cultivar. Under no-stress conditions, the freshly picked albino-induced yellow tea leaves had lower contents of free tea aroma compounds than the freshly picked normal green tea leaves [22]. Furthermore, the albino-induced yellow tea leaves had defects in the chloroplast ultrastructure and composition [28]. This allowed us to use albino-induced yellow tea leaves as a model for studying the influence of chloroplast defects on the formation of characteristic aroma compounds in tea leaves exposed to continuous wounding stress, which is a major stress in the oolong tea manufacturing process. In this study, we compared the endogenous free tea aroma compound contents of yellow leaves and green leaves either under single wounding stress (from picking, as a control) or continuous wounding stress (from tea manufacture). Furthermore, expression levels of key synthetic genes involved in the formation of characteristic aroma compounds in both tea leaves under continuous wounding stress were investigated. Finally, the tea aroma formation-related upstream signals in both tea leaves under continuous wounding stress were studied.

2. Results

2.1. Effect of Chloroplast Defects on Characteristic Aroma Compound Contents in Response to Continuous Wounding Stress

We first investigated whether chloroplast defects affected the characteristic aroma compound contents in response to continuous wounding stress. The mass chromatogram of identified aroma compounds is shown in Figure S1. When tea leaves are picked, they are exposed to a single wounding stress as one main stress. Therefore, picked tea leaves left to stand for a certain time were the single wounding stress treatment group used as a control. Stress-induced aroma compound formation mainly occurs during the turnover stage of oolong tea manufacturing process, which involves continuous wounding stress. A shaking machine was employed to simulate continuous wounding stress from the turnover stage. Tea leaves subjected to this processing were regarded as the continuous wounding stress treatment group. In contrast to single wounding stress, continuous wounding stress upregulated the contents of (*E*)-nerolidol, jasmine lactone, and green leaf volatiles (including (*Z*)-3-hexenol, 1-hexenal, and 2-hexenal), and indole, and reduced the linalool content (Figure 1). Chloroplast defects reduced the accumulation of volatile fatty acid derivatives, including jasmine lactone and green leaf volatiles (Figure 1B), but had no significant effect on (*E*)-nerolidol and indole accumulation in response to continuous wounding stress (Figure 1A,C).

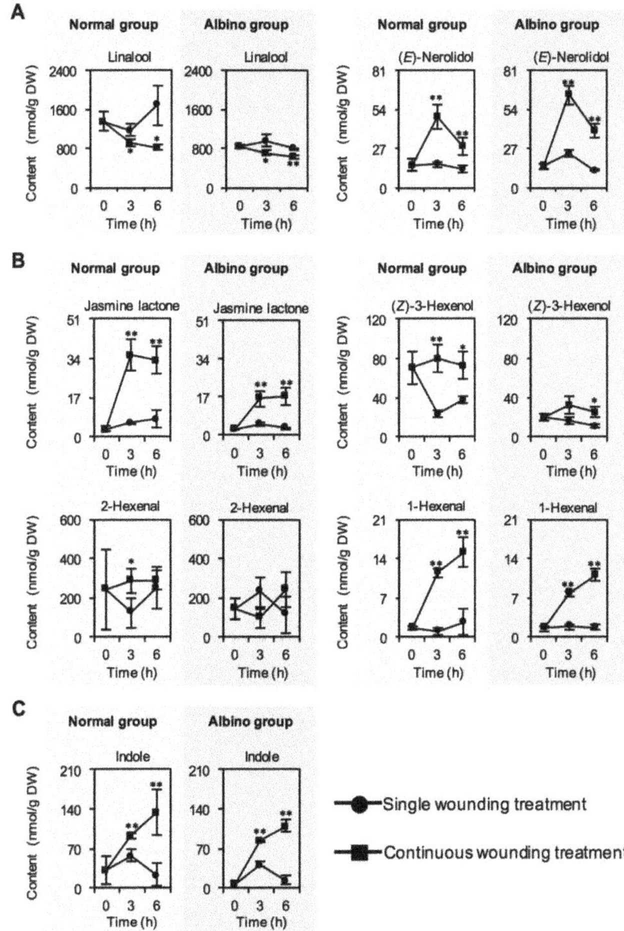

Figure 1. Changes in contents of characteristic aroma compound biosynthetic genes of normal tea leaves and albino tea leaves exposed to single wounding treatment and continuous wounding treatment, respectively. Data are expressed as mean ± SD ($n = 3$). * $p \leq 0.05$; **, $p \leq 0.01$, comparison between single wounding treatment and continuous wounding treatment at the same treatment time. (**A**) Volatile terpenes. (**B**) Volatile fatty acid derivatives. (**C**) Indole.

2.2. Effect of Chloroplast Defects on Expression Levels of Characteristic Genes for Aroma Compound Biosynthesis in Response to Continuous Wounding Stress

We next investigated whether chloroplast defects affected the expression levels of characteristic genes for aroma compound biosynthesis in response to continuous wounding stress. In contrast to single wounding stress, continuous wounding stress upregulated the expression levels of key characteristic genes for aroma compound biosynthesis, including *CsNES1* and *CsNES2* (responsible for (*E*)-nerolidol synthesis), *CsLIS* (responsible for linalool synthesis), *CsLOX1*, *CsLOX2*, and *CsHPL* (responsible for jasmine lactone and green leaf volatile syntheses), and *CsTSB2* (responsible for indole synthesis). No significant difference was observed in the continuous wounding response patterns of most characteristic genes for aroma compound biosynthesis between normal tea leaves and albino tea leaves, while *CsHPL* expression was upregulated only in normal tea leaves exposed to continuous wounding stress (Figure 2B). This suggested that chloroplast defects affected *CsHPL* expression, which might lead to less accumulation of green leaf volatiles in albino tea leaves exposed to continuous wounding stress (Figure 1).

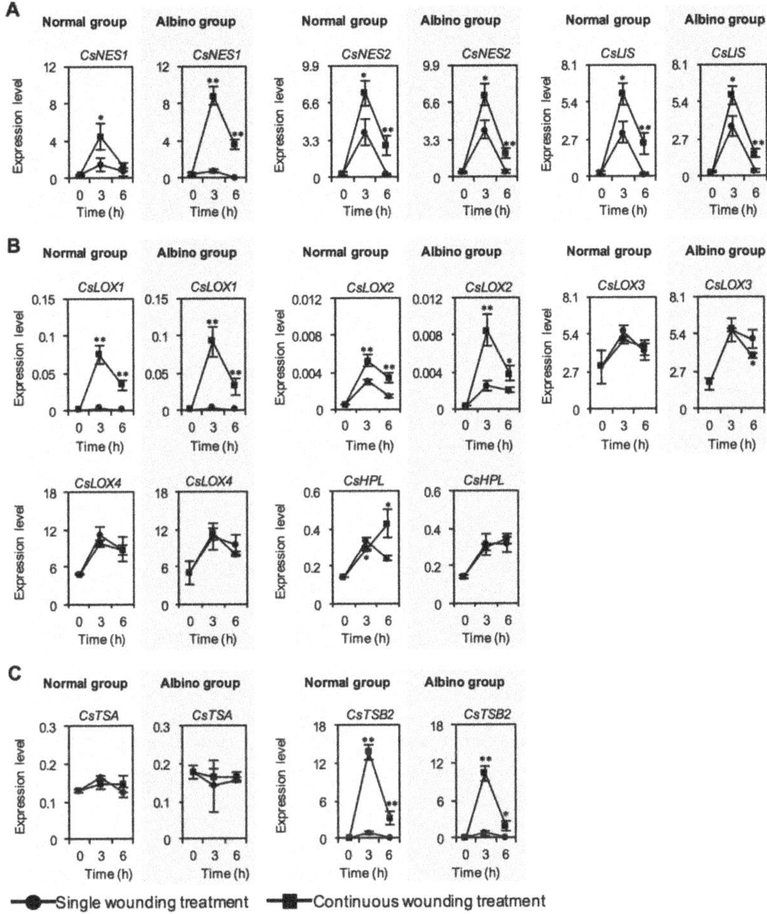

Figure 2. Changes in expression level of characteristic aroma compound biosynthetic genes of normal tea leaves and albino tea leaves exposed to single wounding treatment and continuous wounding treatment, respectively. Data are expressed as mean ± SD (*n* =3). * $p \leq 0.05$; ** $p \leq 0.01$, comparison between single wounding treatment and continuous wounding treatment at the same treatment time. *NES, (E)-nerolidol synthase; LIS, linalool synthase; LOX, lipoxygenase; HPL, hydroperoxide lyase; TSA, tryptophan synthase α-subunit; TSB, tryptophan synthase β-subunit.* (**A**) Genes involved in formations of volatile terpenes. (**B**) Genes involved in formations of volatile fatty acid derivatives. (**C**) Genes involved in formation of indole.

2.3. Effect of Chloroplast Defects on Phytohormone Contents in Response to Continuous Wounding Stress

We also investigated whether chloroplast defects affected phytohormone contents in response to continuous wounding stress. In contrast to single wounding stress, continuous wounding stress significantly increased the JA content, but had no significant effect on abscisic acid (ABA) and salicylic acid (SA) contents both in normal tea leaves and albino tea leaves (Figure 3). Furthermore, chloroplast defects reduced the increase in JA content in response to continuous wounding stress, although there was no significant difference between the continuous wounding response patterns of phytohormone contents in normal tea leaves and albino tea leaves (Figure 3).

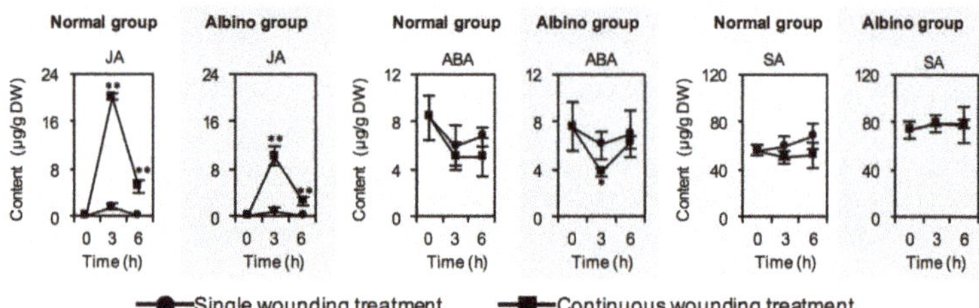

Figure 3. Changes in contents of phytohormones of normal tea leaves and albino tea leaves exposed to single wounding treatment and continuous wounding treatment, respectively. Data are expressed as mean $\pm$ SD (n =3). * $p \leq 0.05$; ** $p \leq 0.01$, comparison between single wounding treatment and continuous wounding treatment at the same treatment time. JA, jasmonic acid; ABA, abscisic acid; SA, salicylic acid.

2.4. Effect of Chloroplast Defects on Expression Levels of JA Synthesis-Related Genes in Response to Continuous Wounding Stress

As the JA content was significantly increased by continuous wounding stress both in normal tea leaves and albino tea leaves, we investigated whether chloroplast defects affected the expression levels of JA synthesis-related genes in response to continuous wounding stress. In contrast to single wounding stress, continuous wounding stress upregulated expression levels of most genes involved in JA synthesis, both in normal tea leaves and albino tea leaves, with no significant difference between the continuous wounding response patterns in normal tea leaves and albino tea leaves (Figure 4). This suggested that chloroplast defects did not significantly affect expression levels of JA synthesis-related genes in response to continuous wounding stress.

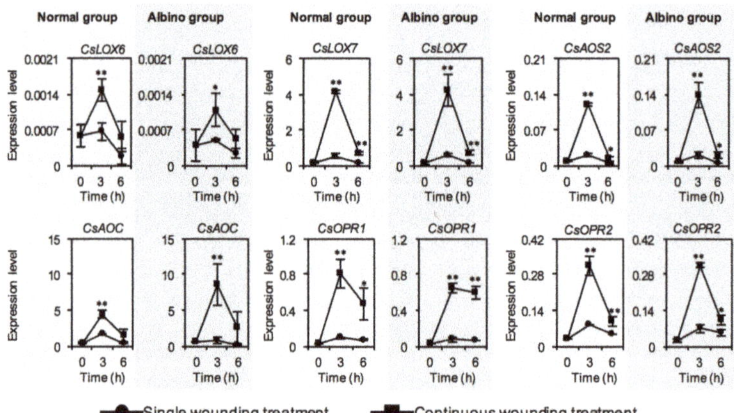

Figure 4. Changes in expression level of jasmonic acid (JA) biosynthetic genes of normal tea leaves and albino tea leaves exposed to single wounding treatment and continuous wounding treatment, respectively. Data are expressed as mean $\pm$ SD (n =3). * $p \leq 0.05$; ** $p \leq 0.01$, comparison between single wounding treatment and continuous wounding treatment at the same treatment time. *LOX, lipoxygenase; AOS, allene oxide synthase; AOC, allene oxide cyclase; OPR, 12-oxo-phytodienoic acid reductase.*

2.5. Effect of Chloroplast Defects on Expression Level of CsMYC2, a Key Transcription Factor of JA Signaling, in Response to Continuous Wounding Stress

We next investigated whether chloroplast defects affected the expression levels of *CsMYC2*, a key transcription factor of JA signaling, in response to continuous wounding stress. In contrast to single wounding stress, continuous wounding stress upregulated expression levels of *CsMYC2*, both in normal tea leaves and albino tea leaves, and no significant difference was observed between the continuous wounding response patterns in normal tea leaves and albino tea leaves (Figure 5). This suggested that chloroplast defects did not significantly affect the expression levels of *CsMYC2* in response to continuous wounding stress.

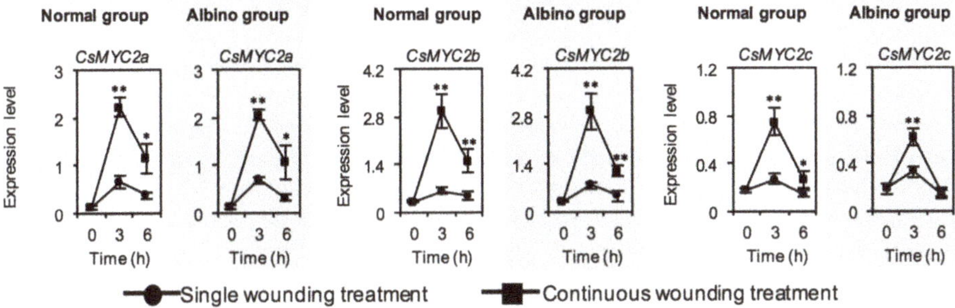

Figure 5. Changes in expression level of *CsMYC2*, a key transcription factor of jasmonic acid (JA), of normal tea leaves and albino tea leaves exposed to single wounding treatment and continuous wounding treatment, respectively. Data are expressed as mean ± SD (n =3). * $p \leq 0.05$; ** $p \leq 0.01$, comparison between single wounding treatment and continuous wounding treatment at the same treatment time.

3. Discussion

Recently, using stress responses to improve tea aromas has attracted increased attention [3]. This approach is not only related to the tea growth process, but also explains aroma formation during the tea manufacturing process. Among the six types of tea, aroma formation during the oolong tea manufacturing process is a representative example of stress-induced tea aroma, because oolong tea manufacture involves the most stresses. Among stresses applied in the oolong tea manufacturing process, continuous wounding in the turnover stage significantly affects the high accumulation of tea aroma compounds, which contributes to the floral odor of oolong tea [3,9–11,13]. Continuous wounding can induce the high accumulation of characteristic aroma compounds, such as indole, jasmine lactone, and (*E*)-nerolidol. Furthermore, these compounds are produced by the activation of key aroma synthetic genes in response to continuous wounding stress [10,11,13]. These observations were made using normal green tea leaves with normal chloroplasts. In this study, normal tea leaves with normal chloroplasts showed similar tea aroma formation results in response to continuous wounding stress (Figure 1), although a different cultivar (cv. Yinghong No. 9) was used. This suggested that the relationship between continuous wounding stress and tea aroma formation occurs widely in different tea cultivars. In the present study, chloroplast was partly defective in the light-sensitive mutant (albino-induced yellow) of cv. Yinghong No. 9 original cultivar (green) [28]. Microscopy observations indicated that the albino tea leaves exhibited a significant reduction in number of chloroplasts, and some chloroplasts showed grana thylakoid structures that were damaged or developing (called etioplasts) [28]. This mutant was used to investigate whether chloroplast defects affected tea aroma compound formation in response to continuous wounding stress. The results showed that chloroplast defects did not significantly affect the formation of indole and (*E*)-nerolidol (Figures 1 and 2). Furthermore, the linalool content was not increased by continuous wounding stress both in normal tea leaves and albino tea leaves (Figure 1A), although expression of its synthetic gene,

CsLIS, was significantly upregulated by continuous wounding stress both in normal tea leaves and albino tea leaves (Figure 2A). The change in linalool formation under continuous wounding stress showed a similar trend to that during the turnover stage of the oolong tea manufacturing process [9,12]. However, it is yet to be determined whether these changes are related to linalool glycosidation, metabolism, or emission. Notably, chloroplast defects reduced the accumulation of volatile fatty acid derivatives, including jasmine lactone and green leaf volatiles, in response to continuous wounding stress (Figure 1B). In general, the final synthetic steps of these volatiles occur in the cytosol [29], but their formations were affected by chloroplast defects in tea leaves. Therefore, it would be interesting to investigate whether CsHPL located in the cytosol is affected by chloroplast defects and, consequently, affects green leaf volatile formation.

In the investigations of *C. sinensis*, a mature genetic transformation system has not been firmly established. Therefore, most genes related to the formation of tea aroma compounds have not been functionally characterized, in vivo, in tea plants over the past decade [3]. These genes were usually expressed in *Escherichia coli*, yeast, or an insect cell system. In addition, some of these genes were further functionally characterized in transient overexpression model plant systems [3]. Many genes involved in the final biosynthetic step of several important aroma compounds, including (*S*)-linalool, (*E*)-nerolidol, and indole, have been functionally characterized, and the related contents have been summarized in our previous review [3]. *CsTPSs* are responsible for the formation of (*S*)-linalool and (*E*)-nerolidol, and *CsTSA* and *CsTSBs* play roles in indole synthesis in tea. *CsNES1* and *CsNES2*, identified in the present study (Figure 2A), have been proposed to be responsible for (*E*)-nerolidol formation based on their subcellular localization in the cytosol and function validation in *Escherichia coli* and transient overexpression model plant systems [11,30]. *CsLIS*, also found in the present study (Figure 2A), was proposed to be responsible for (*S*)-linalool formation based on its subcellular localization in plastids and functional validation in *Escherichia coli* and transient overexpression model plant systems [7,30]. Our previous study showed that the protein mixture of CsTSA (tryptophan synthase α-subunit) and CsTSB2 (tryptophan synthase β-subunit) catalyzes indole formation in vitro [10], suggesting that CsTSA and CsTSB2 might be a protein complex in *C. sinensis* that is similar to the $\alpha_2\beta_2$ tetramer in bacteria [31]. In the present study, only *CsTSB2* was significantly upregulated by continuous wounding stress both in normal tea leaves and albino tea leaves, while *CsTSA* was not significantly affected (Figure 2C), suggesting that *CsTSB2* was more sensitive to wounding stress compared with *CsTSA*. Furthermore, a few enzymes and genes involved in the upstream pathways responsible for tea aroma compound formation, such as CsLOX1 for jasmine lactone formation [13], have also been functionally characterized. Based on correlation analysis of *CsLOXs* gene expression and the stresses and tissue distributions of expression of different *CsLOX* genes, *CsLOX1*, *CsLOX2*, *CsLOX3*, and *CsLOX4*, found in the present study (Figure 2B), were proposed to be closely involved in the syntheses of green leaf volatiles [32]. *CsHPL*, found in the present study (Figure 2B), has also been proposed to be involved in green leaf volatile synthesis based on function validation in *E. coli* [33], because functional identification of *CsHPL* showed that recombinant CsHPL can catalyze 13-hydroperoxy-9(*Z*), 11(*E*), 15(*Z*)-octadecatrienoic acid into 3-(*Z*)-hexenal, which is a key precursor of 3-(*Z*)-hexenol, 2-(*Z*)-hexenal, and 2-(*Z*)-hexenol [34]. In the present study, the levels of key genes involved in tea aroma formation were investigated to compare the tea aroma biosynthesis abilities of albino-induced yellow leaves and normal green tea leaves exposed to wounding stress (Figure 2). As the functions of most genes involved in tea aroma formation investigated in this study have been validated previously, the differences in metabolic flux between albino-induced yellow leaves and normal green tea leaves were more reliable.

Phytohormones are key upstream signals, especially in regulating the formation of plant volatiles in response to environmental stresses. Among phytohormones, JA has mostly been reported to be related to plant volatiles. In contrast to phytohormone formation in preharvest tea leaves exposed to stresses, little is known about phytohormone formation under postharvest stresses. This study found that continuous wounding stress from postharvest tea processing had no significant effect on the ABA and SA contents, but increased the JA content both in normal tea leaves and albino tea leaves (Figure 3), which was attributed to activation of the expression of most JA synthetic genes (Figure 4). Furthermore, chloroplast defects reduced the increase in JA content in response to continuous wounding stress (Figure 3). Current knowledge concerning phytohormone and metabolite biosynthesis in tea plants is mostly based on findings reported for other plant species. Based on correlation analysis of expression of different *CsLOX* genes and stresses, and tissue distributions of expression of different *CsLOX* genes, *CsLOX1* (Figure 2), *CsLOX6*, and *CsLOX7* (Figure 4) were proposed to be closely involved in JA synthesis [32]. Furthermore, using a transient expression system in *Nicotiana benthamiana* plants, *CsAOS2*, a gene involved in JA synthesis, was validated to have a function in JA synthesis and located in the chloroplast membrane [21]. MYC2 is the core transcription factor of JA signaling. Three *Arabidopsis MYC2* homologue genes were found in the *C. sinensis* genome database (Figure 5). Among these three MYC2s, CsMYC2a was clustered together with the reported functional MYC2 in *Arabidopsis* [35]. In the study, the expression levels of *CsMYC2* were not significantly affected by chloroplast defection in the albino leaves. Besides *CsMYC2* and JA synthetic genes, other factors—such as precursor metabolites—involved in JA synthesis may affect JA synthesis in the albino leaves under wounding stress, since other metabolites that are shared with the partly fatty acid-derived pathways of JA, such as jasmine lactone and green leaf volatiles, had similar effects resulting from chloroplast defects (Figure 1B). As authentic standards of many precursors of JA synthesis are unavailable, we did not confirm this hypothesis in the present study. In future work, we will try to obtain authentic standards of the key precursors of JA synthesis and confirm this hypothesis. In the present study, although the precursor of ABA is also synthesized in plastid, ABA biosynthesis was not significantly influenced by chloroplast defects (Figure 3). Wounding stress was the main stress in the treatments of the present study. Compared with ABA synthesis, JA synthesis was generally more sensitive to the wounding stress. It remains to be determined whether ABA biosynthesis is influenced by chloroplast defection under other stresses such as drought stress.

4. Materials and Methods

4.1. Plant Materials and Treatments

One bud and three leaves of *C. sinensis* cv. Yinghong No. 9 and its yellow mutant (a light-sensitive variant) were plucked and used in the present study. These tea samples were picked at the Tea Research Institute, Guangdong Academy of Agricultural Sciences (Yingde, Guangdong, China), in November 2018.

The plucked tea leaves were shaken using a shaking table at $23 \pm 1\ °C$ and 60% humidity, and collected after continuous shaking for 0, 3, and 6 h (continuous wounding treatment). The tea leaves, without continuous shaking, stored under the same conditions for 0, 3, and 6 h were used as controls (single wounding treatment). After treatment, the samples were frozen immediately with liquid nitrogen and stored at –80 °C for further study.

4.2. Extraction and Analysis of Aroma Compounds in Tea Leaves

According to our previous studies [12,13], direct organic solvent extraction was applied to investigate content changes in endogenous aroma compounds of finely powered tea leaves [12,13]. Dichloromethane (1.8 mL) was used to extract aroma compounds from 300 mg (fresh weight) of tea samples, and 5 nmol of ethyl decanoate was added to the organic solvent as an internal standard. The mixture was treated in a shaker at room temperature, and the extract was collected after overnight extraction. Anhydrous sodium sulfate was used to dry the extract, and nitrogen was applied to condense the extract into a 200 µL volume. The extract (1 µL) was then subjected to gas chromatography–mass spectrometry (GC–MS) analysis conforming on a GC–MS QP2010 SE (Shimadzu Corporation, Kyoto, Japan) equipped with GCMS Solution software (Version 2.72, Shimadzu Corporation, Japan). The sample was injected in splitless mode for 1 min under 230 °C of GC port. A SUPELCOWAX 10 column (30 m × 0.25 mm × 0.25 µm, Supelco Inc., Bellefonte, PA, USA) was used to separate the aroma compounds, with helium as a carrier gas in the flow rate of 1 mL/min. Initially, the temperature of GC oven was kept at 60 °C for 3 min, then ramped to 240 °C at 4 °C/min, and held for another 20 min at 240 °C. A full scan mode was applied, and the mass spectrometry ranged from *m/z* 40 to *m/z* 200. The authentic standards were used to make the identification of aroma compounds, and quantitative analyses of compounds were constructed according to the calibration curves by plotting the concentration against the peak area of the authentic standard [22]. The dry weight of the sample was calculated based on the weight of fresh leaves and dried leaves obtained after drying.

4.3. Transcript Expression Analysis of the Related Genes

Quick RNA Isolation Kit (Huayueyang Biotechnology Co., Ltd., Beijing, China) was used to isolate total RNA from tea samples. The obtained RNA was purified after the removal of genomic DNA (gDNA) and reversely transcribed into cDNA using PrimeScript RT Reagent Kit with gDNA Eraser (Takara Bio Inc., Kyoto, Japan). Quantitative real time PCR (qRT-PCR) were applied to analyze the transcript expression level of gene [10,12,13]. The reaction system (20 μL) contained 10 μL iTaq^{TM} Universal SYBR$^{®}$ Green Supermix (Bio-Rad, Hercules, CA, USA), 0.4 μL of each specific primer (10 μM), 2 μL cDNA (diluted into 20-fold), and 7.2 μL ddH$_2$O. The qRT-PCR analysis was performed on a Roche LightCycle 480 (Roche Applied Science, Mannheim, Germany). One cycle of 95 °C for 60 s, and 40 cycles of 95 °C for 15 s and 60 °C for 30 s were used as PCR conditions. At the end of each PCR reaction, to verify the specificity of PCR product, a melt curve was carried out. Calculation of the relative expression level of genes was according to the $2^{-\Delta\Delta Ct}$ method, and based on the normalization to mRNA level of the reference gene. Identification and evaluation of reliable reference genes for qRT-PCR analysis in tea plants showed that encoding elongation factor1 (*CsEF1*) was the most stable reference gene in diurnal expression series [36]. In the study, the samples were collected during 6 h, and *CsEF1* was used as a reference gene. The primers used for qRT–PCR analysis are provided in Table S1.

4.4. Analysis of Phytohormone Contents in Tea Leaves

Finely powdered sample (300 mg, fresh weight) was extracted with 3 mL ethyl acetate by vortexing for 30 s followed by ultrasonic extraction in ice-cold water for 20 min. [^{2}H$_5$]JA, [^{2}H$_4$]SA, and [^{2}H$_6$]ABA were added to the mixture as internal standards. After centrifuging at 10,000× *g* for 5 min at 4 °C, 2.9 mL supernatants were collected and then dried under a stream of nitrogen. The residue was re-dissolved in 200 μL methanol. The supernatants were filtered through a 0.22 μm membrane, and subjected to an ultra performance liquid chromatography/quadrupole time-of-flight mass spectrometry (UPLC–QTOF-MS) (Acquity UPLC I-Class/ Xevo®G2-XS QTOF, Waters Corporation, MA, USA). Each sample (5 μL) was injected onto a Waters ACQUITY UPLC HSS T3 C18 column (2.1 mm × 100 mm, 1.8 μm). Solvent A was Milli-Q water with 0.1% (*v/v*) formic acid. Solvent B was acetonitrile with 0.1% (v/v) formic acid. The solvent gradient was started at 20% B, then linearly increased to 35% within 10 min, and later increased to 95% B in 0.1 min and kept for 3 min. In that moment, it suddenly dropped to 20% in 0.1 min and was maintained for 3 min. The flow rate was 0.4 mL/min. The column temperature was 30 °C. The electrospray ionization operated on negative mode. The MS conditions were capillary voltage: 1.5 kV; source temperature: 100 °C; desolvation temperature: 300 °C; cone gas flow: 50 L/h; and desolvation gas flow: 600 L/h. The quantitative analyses of phytohormones were based on calibration curves, which were constructed by plotting the concentration of each phytohormone against the peak area of the authentic standard.

4.5. Statistical Analysis

Statistical analysis was performed using SPSS software, version 18.0 (SPSS Inc., Chicago, IL, USA). Two-tailed student's *t* test was used to determine the differences between the single wounding and continuous wounding treatment.

5. Conclusions

In the present study, continuous wounding stress, which is a major stress in the postharvest tea manufacturing process, can upregulate expression of *CsMYC2*, a key transcription factor of JA signaling, and activate the synthesis of JA and characteristic aroma compounds (including (*E*)-nerolidol, indole, jasmine lactone, and green leaf volatiles) either in normal tea leaves (normal chloroplasts) or albino tea leaves (chloroplast defects). Furthermore, chloroplast defects did not significantly affect the expression levels of JA synthesis-related genes and *CsMYC2* in response to continuous wounding stress, but reduced the increase in JA content in response to continuous wounding stress. Furthermore, chloroplast defects reduced the increase in volatile fatty acid derivatives, including jasmine lactone and green leaf volatile contents, in response to continuous wounding stress. Overall, the formation of metabolites derived from fatty acids, such as JA, jasmine lactone, and green leaf volatiles in tea leaves, in response to continuous wounding stress, were affected by chloroplast defects (Figure 6). Although fresh albino tea leaves contained relatively low contents of aroma compounds compared with fresh normal tea leaves [22], the stress responses of aroma compounds occurred regardless of chloroplast defects. The information presented here will improve understanding of the relationship between stress responses of phytohormones and aroma compounds, and chloroplast changes. Furthermore, these results provide essential information for the future utilization of stress responses to improve the weak aroma quality of albino tea leaves.

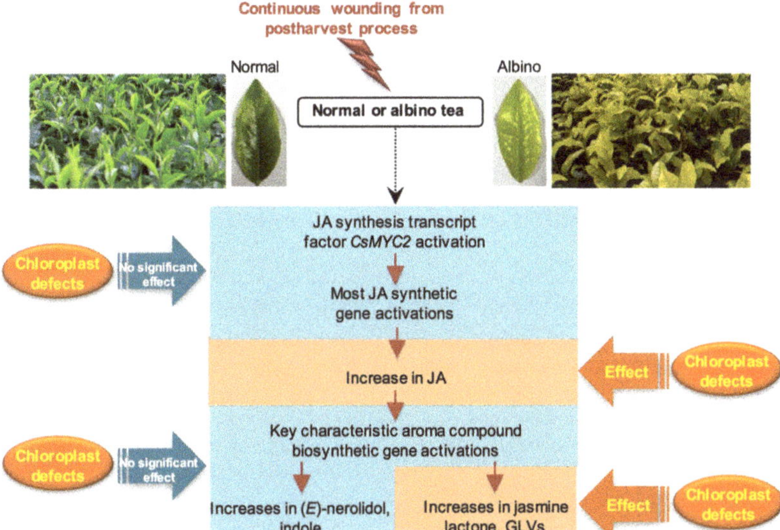

Figure 6. Formations of jasmonic acid (JA) and characteristic aroma compounds in normal tea leaves and albino tea leaves exposed to continuous wounding stress. GLVs, green leaf volatiles.

Supplementary Materials: Supplementary materials can be found at http://www.mdpi.com/1422-0067/20/5/1044/s1.

Author Contributions: Z.Y. and J.T. conceived and designed the experiments.; J.L., L.Z., and Y.L. conducted the experiments; Z.Y., L.Z., and J.L. analyzed the results and wrote the manuscript; D.G. gave instructive suggestions on the manuscript revision; all authors read and approved the final manuscript.

Funding: This study was supported by the financial support from the National Natural Science Foundation of China (31600559), the China Postdoctoral Science Foundation (2018M640837), the Foundation of Science and Technology Program of Guangzhou (201804010097), the Guangdong Natural Science Foundation for Distinguished Young Scholar (2016A030306039), the Guangdong Innovation Team of Modern Agricultural Industry Technology

System (2018LM1092), and the Academic Team Construction Projects of Guangdong Academy of Agricultural Sciences (201627TD).

Conflicts of Interest: The authors declare no conflict of interest.

Abbreviations

ABA	abscisic acid
AOC	allene oxide cyclase
AOS	allene oxide synthase
EF1	encoding elongation factor 1
GC–MS	gas chromatography–mass spectrometry
HPL	hydroperoxide lyase
JA	jasmonic acid
LIS	linalool synthase
LOX	lipoxygenase
NES	(E)-nerolidol synthase
OPR	12-oxo-phytodienoic acid reductase
SA	salicylic acid
TSA	tryptophan synthase α-subunit
TSB	tryptophan synthase β-subunit
UPLC–QTOF-MS	ultra performance liquid chromatography/quadrupole time-of-flight mass spectrometry

References

1. Pichersky, E.; Gershenzon, J. The formation and function of plant volatiles: Perfumes for pollinator attraction and defense. *Curr. Opin. Plant Biol.* **2002**, *5*, 237–243. [CrossRef]
2. Dong, F.; Fu, X.M.; Watanabe, N.; Su, X.G.; Yang, Z.Y. Recent advances in the emission and functions of plant vegetative volatiles. *Molecules* **2016**, *21*, 124. [CrossRef] [PubMed]
3. Zeng, L.T.; Watanabe, N.; Yang, Z.Y. Understanding the biosyntheses and stress response mechanisms of aroma compounds in tea (*Camellia sinensis*) to safely and effectively improve tea aroma. *Crit. Rev. Food Sci.* **2018**. [CrossRef] [PubMed]
4. Yang, Z.Y.; Baldermann, S.; Watanabe, N. Recent studies of the aroma compounds in tea. *Food Res. Int.* **2013**, *53*, 585–599. [CrossRef]
5. Cho, J.Y.; Mizutani, M.; Shimizu, B.; Kinoshita, T.; Ogura, M.; Tokoro, K.; Lin, M.L.; Sakata, K. Chemical profiling and gene expression profiling during the manufacturing process of Taiwan oolong tea "Oriental Beauty". *Biosci. Biotechnol. Biochem.* **2007**, *71*, 1476–1486. [CrossRef] [PubMed]
6. Mei, X.; Liu, X.Y.; Zhou, Y.; Wang, X.Q.; Zeng, L.T.; Fu, X.M.; Li, J.L.; Tang, J.C.; Dong, F.; Yang, Z.Y. Formation and emission of linalool in tea (*Camellia sinensis*) leaves infested by tea green leafhopper (*Empoasca (Matsumurasca) onukii* Matsuda. *Food Chem.* **2017**, *237*, 356–363. [CrossRef] [PubMed]
7. Fu, X.M.; Chen, Y.Y.; Mei, X.; Katsuno, T.; Kobayashi, E.; Dong, F.; Watanabe, N.; Yang, Z.Y. Regulation of formation of aroma compounds of tea (*Camellia sinensis*) leaves by single light wavelength. *Sci. Rep.* **2015**, *5*, 16858. [CrossRef] [PubMed]
8. Yang, Z.Y.; Kobayashi, E.; Katsuno, T.; Asanuma, T.; Fujimori, T.; Ishikawa, T.; Tomomura, M.; Mochizuki, K.; Watase, T.; Nakamura, Y.; et al. Characterisation of volatile and non-volatile metabolites in etiolated leaves of tea (*Camellia sinensis*) plants in the dark. *Food Chem.* **2012**, *135*, 2268–2276. [CrossRef] [PubMed]
9. Gui, J.D.; Fu, X.M.; Zhou, Y.; Katsuno, T.; Mei, X.; Deng, R.F.; Xu, X.L.; Zhang, L.Y.; Dong, F.; Watanabe, N.; et al. Does enzymatic hydrolysis of glycosidically bound aroma compounds really contribute to the formation of aroma compounds during the oolong tea manufacturing process? *J. Agric. Food Chem.* **2015**, *63*, 6905–6914. [CrossRef] [PubMed]
10. Zeng, L.T.; Zhou, Y.; Gui, J.D.; Fu, X.M.; Mei, X.; Zhen, Y.P.; Ye, T.X.; Du, B.; Dong, F.; Watanabe, N.; et al. Formation of volatile tea constituent indole during the oolong tea manufacturing process. *J. Agric. Food Chem.* **2016**, *64*, 5011–5019. [CrossRef] [PubMed]
11. Zhou, Y.; Zeng, L.T.; Liu, X.Y.; Gui, J.D.; Mei, X.; Fu, X.M.; Dong, F.; Tang, J.C.; Zhang, L.Y.; Yang, Z.Y. Formation of (E)-nerolidol in tea (*Camellia sinensis*) leaves exposed to multiple stresses during tea manufacturing. *Food Chem.* **2017**, *231*, 78–86. [CrossRef] [PubMed]

12. Zeng, L.T.; Zhou, Y.; Fu, X.M.; Mei, X.; Cheng, S.H.; Gui, J.D.; Dong, F.; Tang, J.C.; Ma, S.Z.; Yang, Z.Y. Does oolong tea (*Camellia sinensis*) made from a combination of leaf and stem smell more aromatic than leaf-only tea? Contribution of the stem to oolong tea aroma. *Food Chem.* **2017**, *237*, 488–498. [CrossRef] [PubMed]

13. Zeng, L.T.; Zhou, Y.; Fu, X.M.; Liao, Y.Y.; Yuan, Y.F.; Jia, Y.X.; Dong, F.; Yang, Z.Y. Biosynthesis of jasmine lactone in tea (*Camellia sinensis*) leaves and its formation in response to multiple stresses. *J. Agric. Food Chem.* **2018**, *66*, 3899–3909. [CrossRef] [PubMed]

14. Dudareva, N.; Klempien, A.; Muhlemann, J.K.; Kaplan, I. Biosynthesis, function and metabolic engineering of plant volatile organic compounds. *New Phytol.* **2013**, *198*, 16–32. [CrossRef] [PubMed]

15. Dong, F.; Yang, Z.Y.; Baldermann, S.; Sato, Y.; Asai, T.; Watanabe, N. Herbivore-induced volatiles from tea (*Camellia sinensis*) plants and their involvement in intraplant communication and changes in endogenous nonvolatile metabolites. *J. Agric. Food Chem.* **2011**, *59*, 13131–13135. [CrossRef] [PubMed]

16. Zeng, L.T.; Liao, Y.Y.; Li, J.L.; Zhou, Y.; Tang, J.C.; Dong, F.; Yang, Z.Y. α-Farnesene and ocimene induce metabolite changes by volatile signaling in neighboring tea (*Camellia sinensis*) plants. *Plant Sci.* **2017**, *264*, 29–36. [CrossRef] [PubMed]

17. Wasternack, C.; Hause, B. Jasmonates: Biosynthesis, perception, signal transduction and action in plant stress response, growth and development. An update to the 2007 review in Annals of Botany. *Ann. Bot.* **2013**, *111*, 1021–1058. [CrossRef] [PubMed]

18. Song, W.C.; Brash, A.R. Purification of an allene oxide synthase and identification of the enzyme as a cytochrome P-450. *Science* **1991**, *253*, 781–784. [CrossRef] [PubMed]

19. Song, W.C.; Funk, C.D.; Brash, A.R. Molecular cloning of an allene oxide synthase: A cytochrome P-450 specialized for metabolism of fatty acid hydroperoxides. *Proc. Natl. Acad. Sci. USA* **1993**, *90*, 8519–8523. [CrossRef] [PubMed]

20. Turner, J.G.; Ellis, C.; Devoto, A. The jasmonate signal pathway. *Plant Cell* **2002**, *14*, S153–S164. [CrossRef] [PubMed]

21. Peng, Q.Y.; Zhou, Y.; Liao, Y.Y.; Zeng, L.T.; Xu, X.L.; Jia, Y.X.; Dong, F.; Li, J.L.; Tang, J.C.; Yang, Z.Y. Functional characterization of an allene oxide synthase involved in biosynthesis of jasmonic acid and its influence on metabolite profiles and ethylene formation in tea (*Camellia sinensis*) flowers. *Int. J. Mol. Sci.* **2018**, *19*, 2440. [CrossRef] [PubMed]

22. Dong, F.; Zeng, L.T.; Yu, Z.M.; Li, J.L.; Tang, J.C.; Su, X.G.; Yang, Z.Y. Differential accumulation of aroma compounds in normal green and albino-induced yellow tea (*Camellia sinensis*) leaves. *Molecules* **2018**, *23*, 2677. [CrossRef] [PubMed]

23. Wang, K.R.; Li, M.; Zhang, L.J.; Liang, Y.R. Studies on classification of albino tea resources. *J. Tea* **2015**, *41*, 126–129. (In Chinese)

24. Wang, W.; Guo, Y.L. Development and application of albino tea varieties. *J. Food Safety Quality* **2017**, *8*, 3104–3110. (In Chinese)

25. Wang, K.K.; Li, N.N.; Du, Y.Y.; Liang, Y.R. Effect of sunlight shielding on leaf structure and amino acids concentration of light sensitive albino tea plan. *Afr. J. Biotechnol.* **2013**, *12*, 5535–5539.

26. Song, L.; Ma, Q.; Zou, Z.; Sun, K.; Yao, Y.; Tao, J.; Kaleri, N.A.; Li, X. Molecular Link between Leaf coloration and gene expression of flavonoid and carotenoid biosynthesis in *Camellia sinensis* cultivar 'Huangjinya'. *Front. Plant Sci.* **2017**, *8*, 803. [CrossRef] [PubMed]

27. Liu, G.F.; Han, Z.X.; Feng, L.; Gao, L.P.; Gao, M.J.; Gruber, M.Y.; Zhang, Z.L.; Xia, T.; Wan, X.C.; Wei, S. Metabolic flux redirection and transcriptomic reprogramming in the albino tea cultivar 'Yu-Jin-Xiang' with an emphasis on catechin production. *Sci. Rep.* **2017**, *7*, 45062. [CrossRef] [PubMed]

28. Cheng, S.H.; Fu, X.M.; Liao, Y.Y.; Xu, X.L.; Zeng, L.T.; Tang, J.C.; Li, J.L.; Lai, J.H.; Yang, Z.Y. Differential accumulation of specialized metabolite L-theanine in green and albino-induced yellow tea (*Camellia sinensis*) leaves. *Food Chem.* **2019**, *276*, 93–100. [CrossRef] [PubMed]

29. Pichersky, E.; Noel, J.P.; Dudareva, N. Biosynthesis of plant volatiles: nature's diversity and ingenuity. *Science* **2006**, *331*, 808–811. [CrossRef] [PubMed]

30. Liu, G.F.; Liu, J.J.; He, Z.R.; Wang, F.M.; Yang, H.; Yan, Y.F.; Gao, M.J.; Gruber, M.Y.; Wan, X.C.; Wei, S. Implementation of CsLIS/NES in linalool biosynthesis involves transcript splicing regulation in *Camellia sinensis*. *Plant Cell Environ.* **2018**, *41*, 176–186. [CrossRef] [PubMed]

31. Rhee, S.; Parris, K.D.; Ahmed, S.A.; Miles, E.W.; Davies, D.R. Exchange of K$^+$ or Cs$^+$ for Na$^+$ induces local and long-range changes in the three-dimensional structure of the tryptophan synthase $\alpha_2\beta_2$ complex. *Biochemistry* **1996**, *35*, 4211–4221. [CrossRef] [PubMed]

32. Zhu, J.Y.; Wang, X.W.; Guo, L.X.; Xu, Q.S.; Zhao, S.Q.; Li, F.D.; Yan, X.M.; Liu, S.R.; Wei, C.L. Characterization and alternative splicing profiles of lipoxygenase gene family in tea plant (*Camellia sinensis*). *Plant Cell Physiol.* **2018**, *59*, 1765–1781. [CrossRef] [PubMed]

33. Deng, W.W.; Wu, Y.L.; Li, Y.Y.; Tan, Z.; Wei, C.L. Molecular cloning and characterization of hydroperoxide lyase gene in the leaves of tea plant (*Camellia sinensis*). *J. Agric. Food Chem.* **2016**, *64*, 1770–1776. [CrossRef] [PubMed]

34. Negre-Zakharov, F.; Long, M.C.; Dudareva, N. Floral scents and fruit aromas inspired by nature. In *Plant-Derived Natural Products*, 1st ed.; Osbourn, A.E., Lanzotti, V., Eds.; Springer: New York, NY, USA, 2009; pp. 405–431.

35. Dombrecht, B.; Xue, G.P.; Sprague, S.J.; Kirkegaard, J.A.; Ross, J.J.; Reid, J.B.; Fitt, G.P.; Sewelam, N.; Schenk, P.M.; Manners, J.M.; et al. MYC2 differentially modulates diverse jasmonate-dependent function in *Arabidopsis*. *Plant Cell* **2007**, *19*, 2225–2245. [CrossRef] [PubMed]

36. Hao, X.; Horvath, D.P.; Chao, W.S.; Yang, Y.; Wang, X.; Xiao, B. Identification and evaluation of reliable reference genes for quantitative real-time PCR analysis in tea plant (*Camellia sinensis* (L.) O. Kuntze). *Int. J. Mol. Sci.* **2014**, *15*, 22155–22172. [CrossRef] [PubMed]

 © 2019 by the authors. Licensee MDPI, Basel, Switzerland. This article is an open access article distributed under the terms and conditions of the Creative Commons Attribution (CC BY) license (http://creativecommons.org/licenses/by/4.0/).

International Journal of
Molecular Sciences

MDPI

Article

PgMYB2, a MeJA-Responsive Transcription Factor, Positively Regulates the Dammarenediol Synthase Gene Expression in *Panax Ginseng*

Tuo Liu [1,†], Tiao Luo [1,2,†], Xiangqian Guo [1], Xian Zou [1], Donghua Zhou [2], Sadia Afrin [1], Gui Li [1], Yue Zhang [1], Ru Zhang [3] and Zhiyong Luo [1,*]

[1] Department of Biochemistry and Molecular Biology, School of Life Sciences, Central South University, Changsha 410008, China; lt1994@csu.edu.cn (T.L.); tiaooul96@163.com (T.L.); gxq199x@163.com (X.G.); zx13618463547@163.com (X.Z.); nilabotdu@yahoo.com (S.A.); ligui20061029@126.com (G.L.); zhang1045242781@126.com (Y.Z.)

[2] School of Stomatology of Changsha Medical University, Changsha 410006, China; csyxyzdh@163.com

[3] College of Chemistry and Chemical Engineering, Hunan Institute of Engineering, Xiangtan 411104, China; zhangru2002@126.com

* Correspondence: luozhiyong@csu.edu.cn; Tel.: +86-731-8480-5449

† These authors contributed equally to this work.

Received: 13 March 2019; Accepted: 23 April 2019; Published: 6 May 2019

Abstract: The MYB transcription factor family members have been reported to play different roles in plant growth regulation, defense response, and secondary metabolism. However, MYB gene expression has not been reported in *Panax ginseng*. In this study, we isolated a gene from ginseng adventitious root, PgMYB2, which encodes an R2R3-MYB protein. Subcellular localization revealed that PgMYB2 protein was exclusively detected in the nucleus of *Allium cepa* epidermis. The highest expression level of PgMYB2 was found in ginseng root and it was significantly induced by plant hormones methyl jasmonate (MeJA). Furthermore, the binding interaction between PgMYB2 protein and the promoter of dammarenediol synthase (DDS) was found in the yeast strain Y1H Gold. Moreover, the electrophoretic mobility shift assay (EMSA) identified the binding site of the interaction and the results of transiently overexpressing PgMYB2 in plants also illustrated that it may positively regulate the expression of PgDDS. Based on the key role of PgDDS gene in ginsenoside synthesis, it is reasonable to believe that this report will be helpful for the future studies on the MYB family in *P. ginseng* and ultimately improving the ginsenoside production through genetic and metabolic engineering.

Keywords: *Panax ginseng*; gene expression; ginsenoside; methyl jasmonate; MYB transcription factor; dammarenediol synthase

---

1. Introduction

Panax ginseng C. A. Meyer is a kind of Araliaceae plant, which has been regarded as an important conventional Chinese herb for a long time [1]. *P. ginseng* contains many active substances. The most important substance is thought to be triterpene saponin, also known as ginsenoside. Medicinally, the root is considered most valuable in providing the pharmacologically active ginsenoside and is widely used for the treatment of tumor, obesity, angiocardiopathy, diabetes mellitus and senile diseases [2]. However, natural ginseng plant contains a very low amount of ginsenoside. Metabolic engineering applications will be an attractive strategy to improve ginsenoside production in ginseng.

Based on previous reports and work [3–5], the biosynthetic process of dammarane-type ginsenoside can be summarized as the following three steps (Figure 1). Step 1 is the biochemical synthesis of isoprene pyrophosphate (IPP); Step 2 is the synthesis of 2,3-oxidosqualene and Step 3 involves the cyclization, hydroxylation, and glycosidation of 2,3-oxidosqualene. There are many rate-limiting enzymes in the ginsenoside biosynthesis pathway, such as farnesyl diphosphate synthase (FPS), squalene epoxidase (SE), squalene synthase (SS), dammarenediol synthase (DDS), cytochrome P450 (CYP450) and glycosyltransferase (UGT). Among them, the first key enzyme is dammarenediol synthase [6], which catalyzes the cyclization of 2,3-oxidosqualene into dammarenediol-II [7]. It is the most important biosynthetic branching of ginsenoside synthesis. Although there have some reports on the involvement of DDS in ginsenoside biosynthesis [6,8–10], there is no specific report on regulating DDS expression to improve dammarane-type ginsenoside. The underlying mechanism of regulation is unclear.

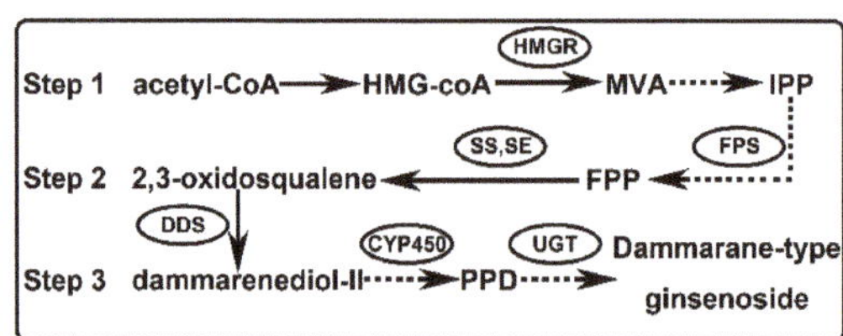

Figure 1. Biosynthetic pathway map of dammarane-type ginsenoside in *P. ginseng*. FPP: farnesyl pyrophosphate; PPD: protopanaxadiol.

In plants, many metabolic pathways are regulated at the transcriptional level, and the expression pattern of related genes is often influenced by plant growth, environment and phytohormones [11]. Several transcription factors involved in the regulation of genes related metabolic pathway have been reported, such as AP2/ERF, bHLH, MYB, WRKY and so on [11,12]. Among them, the MYB protein family is the most widely distributed and functional transcription factor. The MYB domain is usually composed of one to four incomplete repeats, each containing approximately 52 amino acid residues [13]. The first identified MYB gene was the v-MYB gene of avian myeloblastosis virus (AMV) [14]. Pazares et al. were the first to clone plant MYB gene in *Maize* [15]. MYB proteins can be divided into four categories according to the number of incomplete repeat sequences in the structural domain: 1R-MYB usually has only one repeat; R2R3-MYB has two repeats; while 3R-MYB and 4R-MYB have three and four repeats, respectively. In land plants, R2R3-MYB is the most abundant among these groups and is widely involved in various aspects of plant physiological metabolism [16]. In the past few years, many MYB genes have been shown to be involved in the response of different plant hormones such as ABA, SA, and MeJA, which related to plant defense and secondary metabolism [17–19]. In *Nicotiana tabacum*, NtMYBJS1 regulates the production of phenylpropanoid in a MeJA-dependent manner [20], whereas NtMYB1 and NtMYB2 are regulated by SA and are involved in plant defense [21]. In *Arabidopsis thaliana*, numerous MYB genes have been studied for their involvement in response to external stress and plant hormones [22]. Although there has been so much progress about the MYB genes in model plant species, almost no MYB genes studies have been reported in *P. ginseng*.

In our previous work, we obtained 71,095 raw data of transcriptome using next-generation sequencing (NGS) technology from adventitious root treated with MeJA [5]. After comparison using different databases, 163 unigenes were identified to be annotated as putative MYB genes. Based on the number of repeat sequences in the MYB protein domain, 30 unigenes with the R2R3 domain were obtained. Based on the expression level of these 30 genes under the induction of MeJA (Figure S1) and our previous screening experiment (Figure S2), we ultimately chose PgMYB2 (Unigene21198) as the research object.

In this study, we used bioinformatics methods to predict the protein structure and physicochemical properties of PgMYB2 (Figures S3 and S4). In order to find the target gene of PgMYB2, we conducted the yeast one-hybrid assay. Moreover, we used RT-PCR and qRT-PCR to detect the expression patterns of the PgMYB2 and the candidate target gene PgDDS under different conditions treated by MeJA. The role of PgMYB2 in ginsenoside biosynthesis has been elucidated. This research is the first to analyze the function of hormone-responsive MYB gene, which may be helpful for the study of secondary metabolites in *P. ginseng*.

2. Results

2.1. Characterization of PgMYB2 and Bioinformatics Analysis

Sequence analysis displayed that PgMYB2 comprises 1401 nucleotides with an 843bp ORF. The gene encoded 280 amino acids, and the predicted molecular weight of the protein was 31.71kDa with an isoelectric point of 9.27. Amino acid alignment confirmed that PgMYB2 is a member of the R2R3-MYB subfamily (Figure 2A). The three-dimensional structure of PgMYB2 was constructed by the Swiss-Model software, using the Trichomonas vaginalis MYB3 DNA binding domain as the template (SMTL ID: 3zqc.3, 37.72% sequence identity) (Figure 2B) [23].

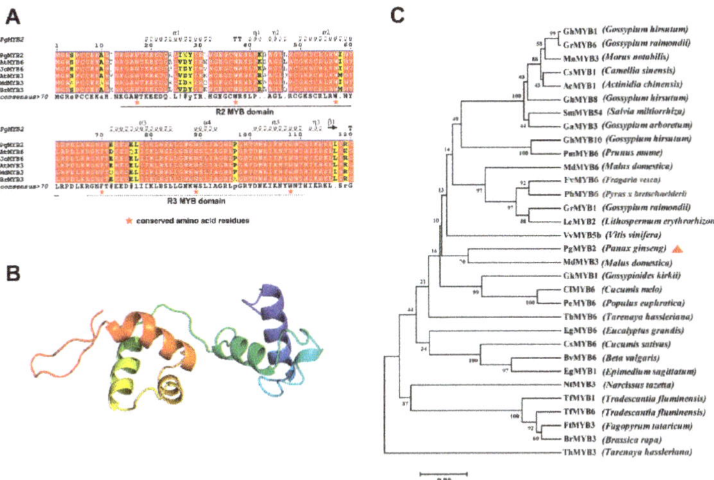

Figure 2. Bioinformatics analysis of PgMYB2 amino acids. (**A**) Amino acid sequence alignment between PgMYB2 and other plant MYB proteins. Red stars indicate conserved tryptophan (W) and phenylalanine (F) residues. (**B**) The three-dimensional structure diagram of the PgMYB2, constructed by Swiss-Model software. (**C**) The phylogenetic tree of PgMYB2 with 31 other plant MYB proteins was built using the MEGA 7.0. The procedure performed 1000 repetitions under the Neighbor-Joining method. Full names of the respective plant species are mentioned in brackets. All the accession number of amino acid sequences are listed below: PgMYB2 (API61854.1); AtMYB6 (XP_002872444.1); JcMYB6 (XP_012075785.1); AtMYB3 (NP_564176.2); MdMYB3 (AEX08668); BrMYB3 (XP_009115618.1); GhMYB1 (AAA33067.1); GrMYB6 (XP_011096483.1); MnMYB3 (XP_010104477.1); CsMYB1 (AEI83425.1); AcMYB1 (AHB17741.1); GhMYB8 (ABR01221.1); SmMYB54 (AGN52078.1); GaMYB3 (KHG11058.1); GhMYB10 (ABR01222.1); PmMYB6 (XP_008219033.1); MdMYB6 (XP_008378762.1); FvMYB6 (XP_004299892.1); PbMYB6 (XP_009362465.1); GrMYB1 (AAN28271.1); LeMYB2 (AIS39993.1); VvMYB5b (AAX51291.3); GkMYB1 (AAN28273.1); ClMYB6 (XP_008459665.1); PeMYB6 (XP_011001250.1); ThMYB6(XP_010530161.1); EgMYB6 (XP_010061981.1); CsMYB6 (XP_004141649.1); BvMYB6 (XP_010680433.1); EgMYB1 (AFH03053.1); NtMYB3 (AIU39031.1); TfMYB1 (AAS19475.1); TfMYB6 (AAS19480.1); FtMYB3 (AEC32977.1); BrMYB3 (XP_013706502.1); ThMYB3 (XP_010535219.1).

2.2. Homology Analysis of PgMYB2 Protein

To further identify the characteristics of PgMYB2, the phylogenetic tree was constructed. The analysis involved PgMYB2 with 31 MYB amino acid sequences from other species (Figure 2C). Interestingly, a MYB protein similar to PgMYB2 seems to be involved in plant secondary metabolism. The result indicated that PgMYB2 has high homology to MdMYB3, which is involved in transcriptional regulation of the flavonoid synthesis pathway and regulates the accumulation of anthocyanin [24]. Therefore, we can speculate that PgMYB2 protein might participate in secondary metabolism in the certain tissues of *P. ginseng*.

2.3. Subcellular Localization of PgMYB2

The PgMYB2 subcellular localization was observed by the co-expression of GFP-PgMYB2 under the control of the 35S promoter. The GFP-PgMYB2 could be expressed in onion epidermal cells through the Agrobacterium EHA105 infection. The onion epidermal cells infected by EHA105 with the empty pCAMBIA1302 vector were also observed as the control group. The experiment revealed that the GFP-PgMYB2 recombinant protein was specifically located in the nucleus, whereas empty control distributed evenly throughout the whole onion cell (Figure 3). So we can infer that PgMYB2 is a protein located in the nucleus from these results.

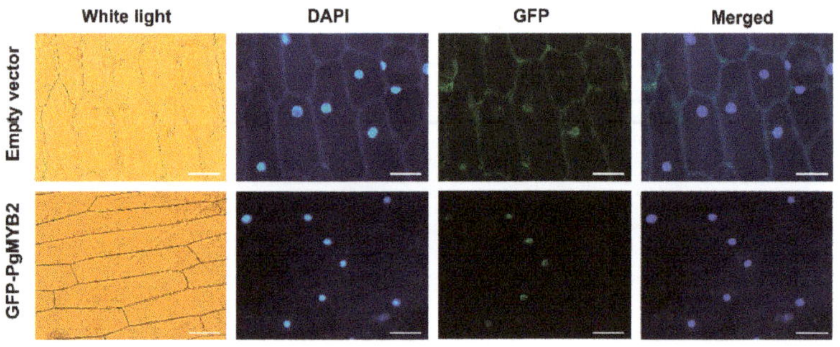

Figure 3. The subcellular localization of PgMYB2 protein in onion epidermal cells. Scale bar = 100 μm.

2.4. Expression Analysis of PgMYB2 in Different Tissue of P. ginseng

We used RT-PCR and qRT-PCR to analyze the expression differences of PgMYB2 in different ginseng tissues. The result revealed that the expression of PgMYB2 was vastly detected in roots (4.66-fold) and lateral roots (3.53-fold) compared to leaves (normalized as 1-fold). Likewise, just a slight increase of PgMYB2 was detected in stems (1.66-fold) and seeds (1.19-fold). The qRT-PCR results were consistent with RT-PCR (Figure 4A,B). These data indicated that the expression pattern of PgMYB2 in tissues of ginseng is significantly different.

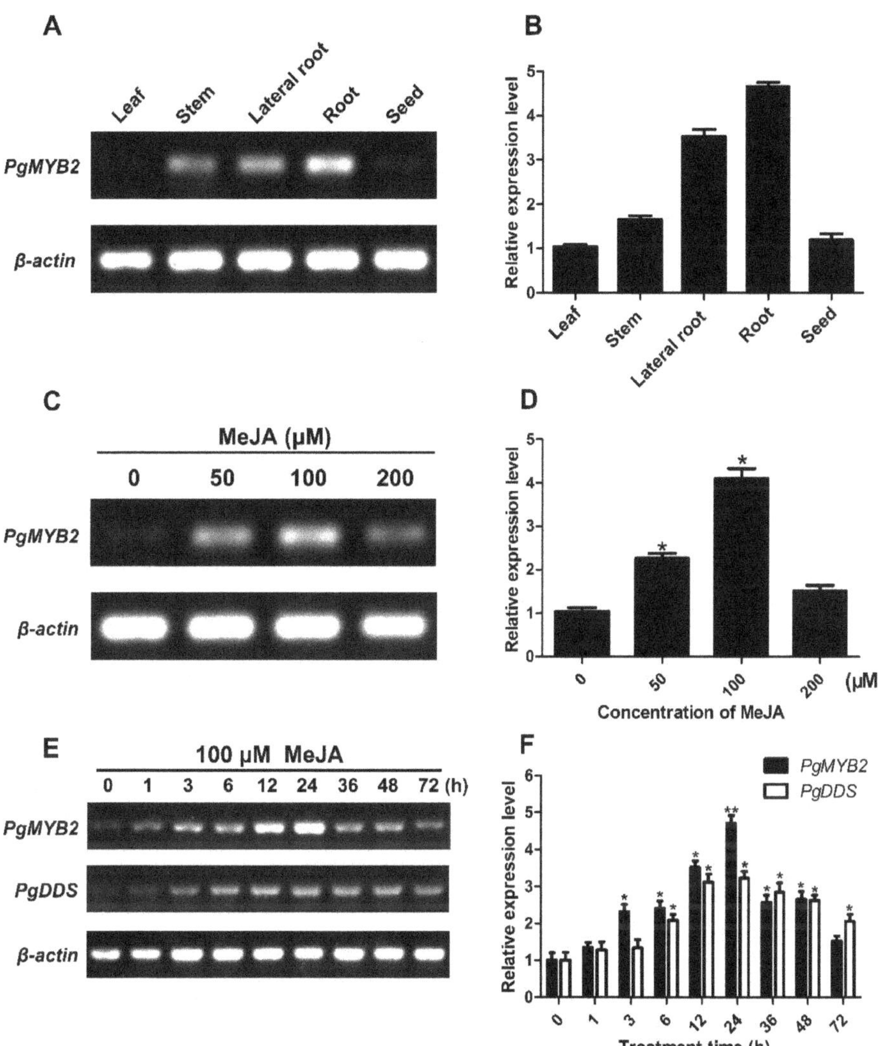

Figure 4. Expression analysis of PgMYB2 and PgDDS. RT-PCR (**A**) and qRT-PCR (**B**) analyzed the expression of PgMYB2 in different tissues. The relative expression level was calculated according to the expression of the corresponding gene in the leaves. RT-PCR (**C**) and qRT-PCR (**D**) analyzed the expression levels of PgMYB2 under different concentration of MeJA for 12 h. The relative expression level was calculated according to the expression of PgMYB2 at 0 μM MeJA. RT-PCR (**E**) and qRT-PCR (**F**) analyzed the expression levels of PgMYB2 and PgDDS under the treatment of 100 μM MeJA at different time points. The relative expression level was calculated according to the expression of PgMYB2 and PgDDS at 0 h. All expression levels were normalized according to the β-actin expression level. The standard deviations from three independent repeated trials were indicated by the error bars. Asterisks indicated a significant difference by *t*-test, * $p < 0.05$, ** $p < 0.01$.

2.5. Expression Analysis of PgMYB2 and PgDDS under MeJA Treatments

As a plant hormone, MeJA is involved in the synthesis of many secondary metabolites [25]. After the ginseng hairy roots induced by MeJA in a concentration-dependent manner for 12 h, the expression levels of PgMYB2 were measured by RT-PCR and qRT-PCR. The result indicated that the PgMYB2 expression increased immediately and reached the maximum at 100 µM MeJA treatment, and then declined obviously at 200 µM MeJA (Figure 4C,D). Therefore, we chose 100 µM as the optimal MeJA concentration to detect the expression of PgMYB2 and PgDDS.

To investigate the expression level of PgMYB2 and PgDDS at different time points (0, 1, 3, 6, 12, 24, 36, 48 and 72 h), 100 µM MeJA was supplemented to the liquid culture medium of the hairy root. The RT-PCR analysis revealed that both PgMYB2 and PgDDS slightly expressed without MeJA treatment, but the expression level significantly increased to the highest point after 24 h treatment of MeJA, and then gradually decreased (Figure 4E). Furthermore, the qRT-PCR analysis showed an obvious increase of PgMYB2 at 6–12 h of MeJA treatment and the relative expression level reached the highest point accounting for approximately a 4.71-fold increase at 24 h compared to 0 h (Figure 4F). Interestingly, the same tendency was shown on the relative expression level of PgDDS. These results suggested that there may be some relationships between the expression of PgMYB2 and PgDDS.

2.6. DNA Binding Activity of PgMYB2

Based on the fact that transcription factors can bind to specific sequences on target gene promoters [26], the promoter of the PgDDS gene which aliased as DDSpro was cloned by our laboratory. To explore whether PgMYB2 can bind the DDSpro, we conducted the yeast one-hybrid assay (Figure 5). The DDSpro was cloned into the pAbAi bait vector and the recombinant plasmids were transformed into Y1H Gold competent cells. Background expression test of Aureobasidin A (AbA) resistance showed that 200 ng/mL AbA could almost inhibit the basal expression of the pAbAi-DDSpro bait yeast strain without the prey protein. PgMYB2 was cloned into the pGADT7 prey vector to construct the pGADT7-PgMYB2 recombinant vector. Then the pGADT7-PgMYB2 and the empty pGADT7 vector were respectively transformed into the Y1H Gold which contained the recombinant bait vector pAbAi-DDSpro. All yeast cells transformed twice were cultured on SD/−Ura/−Leu selective medium. As predicted, only the yeast cells with pGADT7-PgMYB2 were able to grow on the selective medium containing 200 ng/mL AbA, suggesting that PgMYB2 can bind to DDSpro and activated transcription in the yeast system.

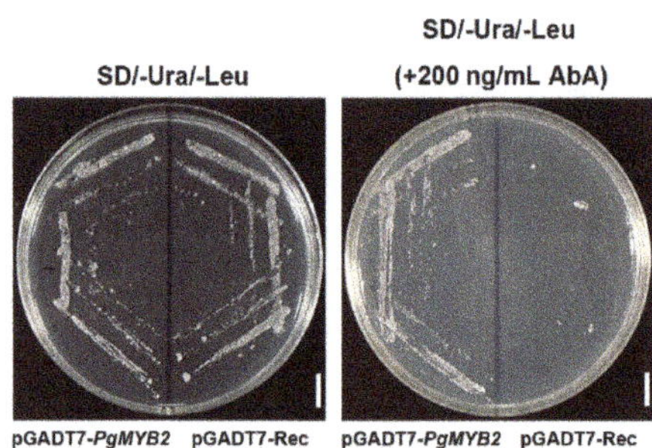

Figure 5. The interaction of PgMYB2 and DDSpro in Y1H Gold. As shown in the left half of the plate, only the yeast cells with pGADT7-PgMYB2 were able to grow on the SD/−Ura/−Leu selective medium added with 200 ng/mL AbA. Scale bar = 1 cm.

Based on the result of yeast one-hybrid assay, we found the DDSpro contains two MYB binding sites (MBS and MBSII) [27,28]. The MBS (TAACTG) positioned in the promoter area between −841 and −836 bp, while the MBSII (AAAATTTAGTTA) located in the section between −406 to −395 bp (Figure 6A). The results of EMSA revealed that PgMYB2 protein could bind to the MBSII in DDSpro (Figure 6B).

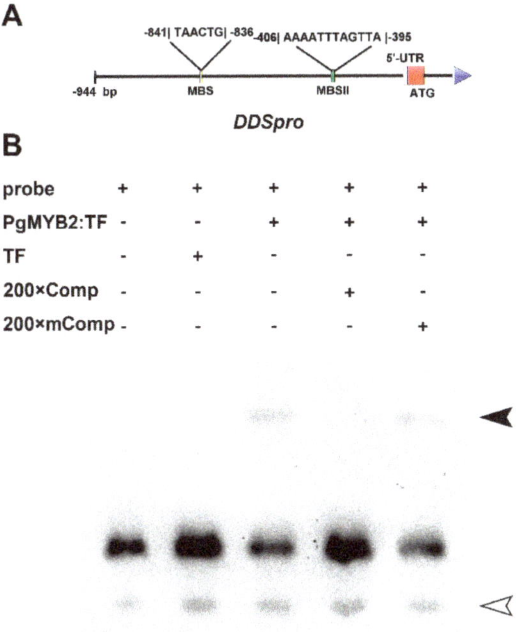

Figure 6. Binding assay of PgMYB2 to the MBSII. (**A**) The diagram shows the relative location of the MBS and MBSII in the region of DDSpro. (**B**) The fusion protein of PgMYB2 binds to the MBSII of DDSpro. The reaction system from left to right: biotin-labeled probe (containing MBSII site); labeled probe and trigger factor as negative control (TF, a kind of chaperonin); labeled probe and PgMYB2: TF protein; labeled probe, PgMYB2: TF protein and 200× Comp (200 times unlabeled competitive probe); labeled probe, PgMYB2: TF protein and 200× mComp (200 times unlabeled competitive mutant probe, the MBSII site was mutated). The protein-probe complexes were indicated with a solid arrow and the free probes were indicated with a hollow arrow.

2.7. PgMYB2 Activates the Expression of PgDDS in A. thaliana Protoplasts

To further explore whether the PgMYB2 combine with the DDSpro in plants and reveal the specific role played by PgMYB2 in this process, we conducted the dual-luciferase reporter assay. In this assay, the DDSpro was cloned into the vector pGreenII-0800-LUC as the reporter, and the PgMYB2 gene was cloned into the vector pEGAD-MYC as the effector. Both recombinant plasmids were co-transformed into the protoplasts of *A. thaliana* according to the different molar ratio. In addition, we also constructed the DDSpro genes with MBSII site knockout. Meanwhile, the empty vector pEGAD-MYC was used as control (Figure 7A). The results displayed that a more significant increase of relative LUC/REN ratio was induced with the increase of pEGAD-MYC-PgMYB2 concentration (Figure 7B). Compared with the control group (pGreenII-0800-LUC-DDSpro+pEGAD-MYC vector), the relative activity of group C (pGreenII-0800-LUC-DDSpro and pEGAD-MYC-PgMYB2 at a molar ratio of 1:1) increased about 14.2-fold. However, the relative activity level decreased to 2.7-fold with the absence of MBSII in DDSpro. These results further demonstrated that PgMYB2 could bind to the DDSpro and strongly activated the PgDDS expression in *A. thaliana* protoplasts.

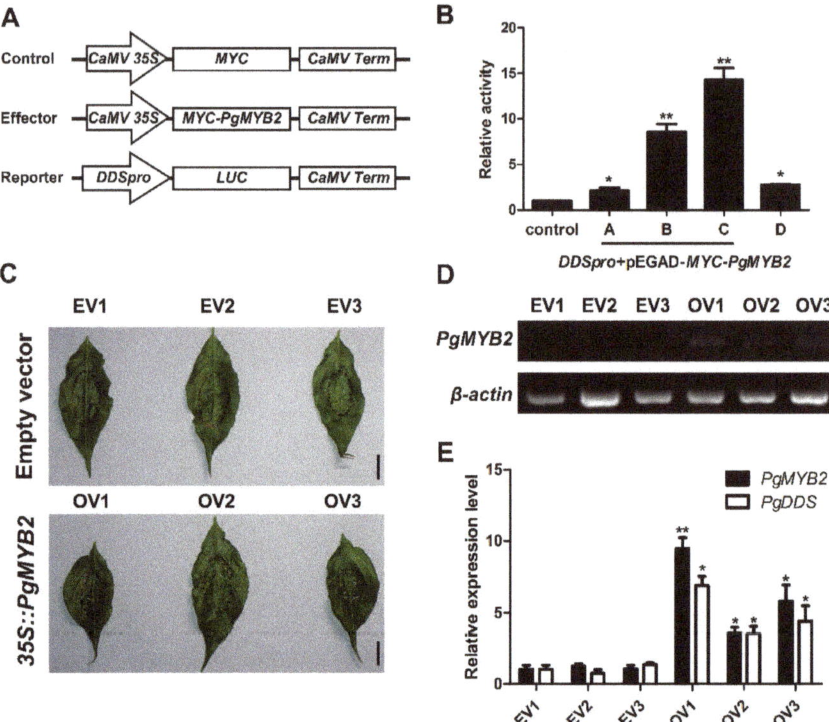

Figure 7. Transient expression assays of PgMYB2. (**A**) Sketch map of reporter and effector constructs used for transient expression system. The empty vector carrying CaMV 35S::MYC was constructed as a control. The CaMV Term box indicates the terminator. (**B**) The relative activity of PgMYB2 in the transient expression assay. Group A–C represent for the different molar ratio (4:1, 2:1, 1:1) of reporter and effector. Group D represents the absence of MBSII in DDSpro. The relative activity of the control group was set as 1, and the relative activity of other groups was calculated according to the LUC/REN ratio. (**C**) The phenotype of ginseng leaves after 2 d of Agrobacterium infection. EV1-3 represent the leaves injected with empty vector, while OV1-3 were injected with the pCAMBIA1302-PgMYB2 vector. Scale bar = 1 cm. (**D**) RT-PCR was performed to confirm the overexpression of PgMYB2 in ginseng leaves after 2 d of Agrobacterium injection. (**E**) The expression levels of PgMYB2 and PgDDS in ginseng leaves were quantitatively analyzed by qRT-PCR. The related genes expression levels of the EV1 were set as control, and the expression levels of other groups were calculated according to this group. All expression levels were normalized according to the expression level of β-actin. The diagram represents the average values and error bars represent the SDs from three replicate experiments. Asterisks indicated a significant difference between the EV1 and experimental groups, * $p < 0.05$, ** $p < 0.01$.

2.8. Transient Expression of PgMYB2 in Ginseng Leaves Promote the Expression of PgDDS

In order to explore the function of PgMYB2 in native plants, Agrobacterium strain EHA105 harboring pCAMBIA1302-PgMYB2 plasmid (35S::PgMYB2) was injected into the lower epidermis of ginseng leaves. Meanwhile, EHA105 cells harboring empty pCAMBIA1302 vector were also injected into leaves as the control. The RT-PCR analysis indicated that PgMYB2 expressed in the leaves (Figure 7D). Furthermore, the relative expression level of PgDDS in the experimental group was higher than those groups with lower PgMYB2 expression level (Figure 7E). The results revealed that transient expression of PgMYB2 in ginseng leaves may promote the expression of PgDDS.

3. Discussion

MeJA is a plant-specific signaling regulator which regulates many physiological and developmental processes, including resisting pests and synthesizing a series of secondary metabolites [29]. Increasing evidence showed that the syntheses of many secondary metabolites were increased under the induction of MeJA [30]. In *N. tabacum*, there was an increased accumulation of hydroxylcinnamoly-polyamine conjugates and the NtMYBJS1 gene was reported to have interactions with several phenylpropanoid synthesis genes in a MeJA-dependent manner. [20]. In *Pinus taeda*, PtMYB14 has a similar expression pattern to isoprenoid biosynthesis genes under MeJA treatments and may have a close relationship to sesquiterpene production [31]. In *P. ginseng*, the accumulation of ginsenoside increased with the upregulation of the key enzyme genes such as PgDDS after MeJA induction [32–34]. However, the transcription factors regulating PgDDS have not been reported. In this study, we found that PgMYB2 directly binds to the promoter region of PgDDS and has a parallel expression pattern with PgDDS under the induction of MeJA. This evidence suggested that similar regulatory mechanisms may exist in different plants. However, the activation mechanism of transcription factors is complicated, PgMYB2 may be stimulated directly by MeJA or regulated by other transcription factors [35]. Therefore, expression regulation of PgMYB2 provides a flexible network for the study of transcription factors in *P. ginseng*.

Among the transcription factors, the MYB is one of the largest family in plant [36], it plays an important role in regulating specific plant physiological processes and plant secondary metabolism [13, 37]. Recently a report revealed that some SmMYBs were potential positive regulators of terpenoid biosynthesis in *S. miltiorrhiza* [38]. It was also reported that MYB107 positively regulates suberin biosynthesis by binding to the promoter of suberin biosynthetic genes in *A. thaliana* [39]. Based on previous transcriptome data, we found two MYB binding sites (MBS and MBSII) in the DDSpro [27,28]. However, the experiments showed that PgMYB2 could bind to the MBSII binding site rather than MBS binding site in DDSpro (Figure S5). Therefore, it is reasonable to believe that PgMYB2 could also be involved in the ginsenoside biosynthesis by binding to the DDSpro in *P. ginseng*. However, besides the MYB binding site, many other regulatory elements are also contained in the upstream region of DDSpro, and the expression of PgDDS may also be regulated by other transcription factors. Based on the above, how these transcription factors take part in the regulation of PgDDS will be an interesting subject to explore.

Triterpenoid saponins are important compounds in plant secondary metabolism and act as the defender against fungi and microbes. The pharmacological activity of triterpenoid saponins has been confirmed by many clinical trials [40–42]. Saponins such as ginsenoside are synthesized through the mevalonate pathway, with acetyl-CoA as the precursor of the reaction (Figure 1). However, the yield of ginsenoside has been limited due to the long growth cycle of *P. ginseng*. Therefore, it is particularly important to further elucidate the biosynthetic pathway and regulatory mechanism of ginsenoside. It has been reported that overexpressing the VvMYB5b gene led to an increase of terpenoid metabolism in tomato [43]. Moreover, in *C. roseus*, overexpression of CrBPF1 (MYB like protein) resulted in increased expression levels for some genes in the terpenoid indole alkaloid (TIA) biosynthesis [44]. In our study, we also showed that overexpression of PgMYB2 in *P. ginseng* leaves would lead to increased expression of PgDDS (Figure 7E). Based on the fact that the expression of PgDDS increases the yield of ginsenoside [9,45], it is reasonable to speculate that PgMYB2 could also lead to an increase of ginsenoside by promoting the expression of PgDDS. Based on the above experimental results, we present an updated model for the ginsenoside pathway in *P. ginseng* (Figure 8). The proposed model clearly illustrates the mechanism of how PgMYB2 regulates the expression of PgDDS, so we assumed that PgMYB2 could improve ginsenoside production through this pathway. It also provides a new strategy to engineer the ginseng plant for more efficient ginsenoside production.

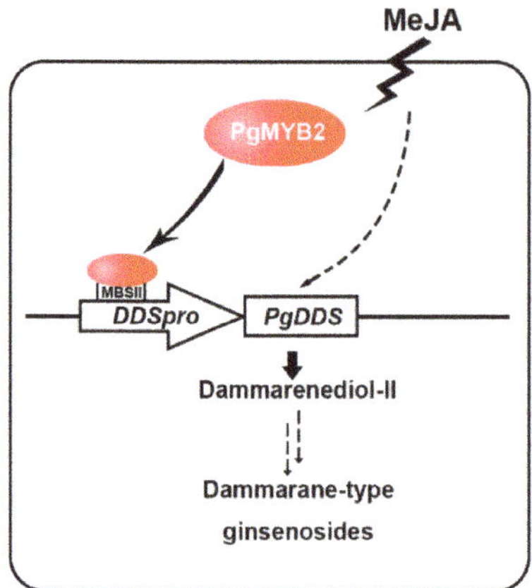

Figure 8. A proposed model describing the function of PgMYB2 in the ginsenoside pathway.

However, our work still has some limitations. Although it has been preliminarily confirmed that PgMYB2 could positively regulate PgDDS, whether it can achieve the expected effect after overexpression of PgMYB2 remains to be further studied. We are still trying to develop the stable overexpressed PgMYB2 hairy root lines, then check whether this transcription factor participates in the accumulation of secondary metabolites.

4. Materials and Methods

4.1. Plant Materials and Culture Environment

Ginseng adventitious roots and other tissues were induced and subcultured in our laboratory. MeJA-induced and absolute ethyl alcohol-treated (as control) adventitious roots were grown in 1/2 MS liquid medium under a stable environment of 25 °C with 24 h dark treatment. We selected well-growing ginseng tissue for subsequent experiments.

4.2. Total RNA Extraction and First Strand cDNA Synthesis

Total RNA from all plant tissues were extracted with the E.Z.N.A.® Plant RNA kit (Omega Biotek, Guangzhou, China). For each sample, about 100 mg adventitious roots powder was digested in 75 µL RNase-free DNase I (Solarbio, Beijing, China) to eliminate the effects of genomic DNA. After that, RNA concentration and purity were determined by a spectrophotometer (Thermo Fisher Scientific, Waltham, MA, USA). The first strand of cDNA was synthesized using the Maxima H Minus First Strand cDNA Synthesis Kit (Thermo Fisher Scientific, Waltham, USA) and stored at −20 °C for later use.

4.3. Bioinformatics Analysis and Prediction of PgMYB2

The ORF lookup was carried out in the ORF finder (https://www.ncbi.nlm.nih.gov/orffinder/). The DNA conserved binding domain was found by the NCBI conserved domains finder (https://www.ncbi.nlm.nih.gov/Structure/cdd/wrpsb.cgi) [46]. The theoretical PI, MW, and other physiochemical characteristics were predicted by the ProtParam tool (https://web.expasy.org/protparam/) [47]. The phylogenetic tree was constructed through the MEGA 7.0 software [48] and multiple sequence alignment was performed by the DNAMAN software. The TMHMM Server v. 2.0 (http://www.cbs.dtu.dk/services/TMHMM/) [49], the NPS@ server (https://npsa-prabi.ibcp.fr/) [50], and the Swiss-Model tool (https://swissmodel.expasy.org/) [51] were used to predict the transmembrane domain, the protein secondary structure, and the three-dimensional structure of PgMYB2 protein, respectively.

The comparison of PgMYB2 with other homologous proteins was performed on the ESPript 3.0 (http://espript.ibcp.fr/ESPript/ESPript/) [52].

4.4. Subcellular Localization of PgMYB2

Complete PgMYB2 (GenBank: KU096984.1) sequence was cloned into an overexpression vector pCAMBIA1302 containing the GFP gene. EHA105 cells carrying the recombinant plasmid pCAMBIA1302-PgMYB2 were injected into the inner epidermis of onion (*Allium cepa*). The onion inner epidermis cells infected with EHA105 carrying empty pCAMBIA1302 plasmid were operated in the same way as a negative control. After the dark culture process occurred for 36 h at 28 °C, the result of subcellular localization was observed by the fluorescence microscope (Nikon Eclipse 80i, Tokyo, Japan).

4.5. Hormone Treatments

To explore the responses of PgMYB2 and PgDDS (GenBank: AB265170.1) to MeJA treatments, 4 weeks cultured adventitious roots were transferred in 1/2 MS liquid media containing 50, 100 and 200 μM MeJA, then kept at 25 °C for shaking under 110 rpm. About 200 mg of adventitious roots were taken from each group at the corresponding time points from 0 to 72 h and stored in liquid nitrogen for subsequent RNA extraction. Equivalent volumes of ethanol were used for the controls and maintained at the same conditions.

4.6. The Expression Analysis of Related Genes by RT-PCR and qRT-PCR

The expression level of related genes in ginseng hairy roots and different tissues were quantified by RT-PCR and qRT-PCR. The RT-PCR procedure was set as follows: 98 °C for 15 s; 98 °C, 10 s; 58 °C, 5 s; 72 °C, 10 s; 30 cycles; and 72 °C for 5 min. The reaction mixture system was prepared according to the reagent instructions of PrimeSTAR® Max DNA Polymerase (Takara, Dalian, China). Amplified fragments from each sample were analyzed by 1% agarose gel electrophoresis.

The qRT-PCR experiment was performed by the three steps method on qTOWER 2.2 (Analytik Jena, Jena, Germany). The amplification program was set as follows: 95 °C for 5 min; 95 °C, 10 s, 58 °C, 20 s; 72 °C, 20 s; 40 cycles. The melting curve analysis was set as follows: 95 °C for 15 s; 60 °C for 60 s and 95 °C for 15 s. The reaction mixture system was prepared according to the reagent instructions of UltraSYBR Mixture (CWBIO, Beijing, China). The relative expression level of related genes was computed based on the $2^{-\Delta\Delta Ct}$ method [53]. All expression levels were normalized according to the expression level of β-actin (GenBank: AY907207). Three sets of parallel replicates were set up in all experiments.

4.7. Analysis of Transcriptional Activity in Yeast of PgMYB2

In order to obtain the promoter fragment of PgDDS, we used plant genomic DNA kit (TIANGEN, Beijing, China) to extract genomic DNA of *P. ginseng* The full-length fragment of DDSpro (GenBank: GU323921.1) was amplified by the gene-specific primers (Table S1). The reaction mixture system was prepared according to the FastPfu DNA Polymerase reagent instructions (TransGen Biotech, Beijing, China). The amplification procedure was set as follows: 95 °C for 5 min; 95 °C, 30 s; 58 °C, 30 s; 72 °C, 60 s; 35 cycles; and 72 °C for 10 min. Then the amplified products were purified and inserted into pAbAi (Takara, Dalian, China) as the bait vector using the XhoI and SmaI restriction endonuclease. Meanwhile, the PgMYB2 sequence was cloned into pGADT7 (Takara, Dalian, China) as the prey vector. The pABAi-DDSpro recombinant plasmid was transformed into the Y1H Gold yeast strain. After 3 days of culture at 28 °C, the transformed yeast cells were selected on the SD/−Ura selection plate added with 200 ng/mL AbA. Then the pGADT7-PgMYB2 recombinant plasmid was transformed into positive colonies. After the second transformation, the yeast solution was uniformly coated on the selected medium (SD/−Ura/−Leu) containing 200 ng/mL AbA. The yeast cells carried with the empty vector (pGADT7-Rec) were cultured as a negative control.

4.8. Expression of Fusion PgMYB2 Protein and Purification

The PgMYB2 protein fused with a trigger factor (TF) and 6× histidine (His) tag was obtained by prokaryotic expression with the pCold/TF vector. Isopropyl β-D-thiogalactoside (IPTG) at 1 mM was used to induce the strain bacteria E. coli BL21 that carried the pCold/TF-PgMYB2 recombinant plasmid and the empty vector (as control) for 24 h at 16 °C. The PgMYB2 recombinant protein with His-tag was purified using Ni-NTA agarose (QIAGEN, Frankfurt, Germany) following the operating instructions and subsequently dialyzed at 8000× *g* for 20 min using the Amicon® Ultra-4 centrifugal filter (Millipore, Massachusetts, USA). The protein was stored at −80 °C with the 20 mM Tris-HCl buffer, pH = 7.2.

4.9. Electrophoretic Mobility Shift Assay

The EMSA was performed with the LightShift® Chemiluminescent EMSA Kit (Thermo Fisher Scientific, Waltham, USA) following the manufacturer's protocol. Two pairs of DNA oligonucleotide probes were synthesized, including the corresponding sequence of MBSII and mutant MBSII (Table S1). The annealed probe reacted with the purified protein at room temperature for 20 min. Then the complexes were separated by native polyacrylamide gel electrophoresis.

4.10. Transient Expression Analysis of PgMYB2

We used the Dual-Luciferase® Reporter Assay System (Promega, Madison, USA) to detect the transient expression of PgMYB2 in the protoplasts of *A. thaliana*. In the assay of the dual-luciferase reporter gene, the pGreenII 0800-DDSpro-LUC reporter and the pEGAD-MYC-PgMYB2 effector were co-transfected into Arabidopsis protoplasts using PEG-Ca. The knockout of MBSII was performed with the Fast Mutagenesis Kit V2 (Vazyme, Nanjing, China). The luminescent signals were detected by EnSpire® Multilabel Reader (PerkinElmer, Waltham, USA).

In the experiment of transiently overexpressing PgMYB2 in ginseng leaves, we used the same EHA105 strain in the subcellular localization. The EHA105 cells were cultured in LB medium (containing 0.01 M MES and 40 μM acetosyringone) and grew overnight on a shaker at 250 rpm. The bacteria pellets were collected by centrifugation, resuspended to OD600 = 1.0 with a solution of 10 mM $MgCl_2$ and 0.2 mM acetylclogenone, and kept it standing more than 3 h. The EHA105 suspension was injected into the lower epidermis of ginseng leaves in the flourishing period (about 2 months) by a sterile syringe. The injected ginseng plants were grown at 25 °C for two days, then the infiltrated leaves were cut off for subsequent RNA extraction. Three sets of parallel replicates were set up in all experiments.

Supplementary Materials: Supplementary materials can be found at http://www.mdpi.com/1422-0067/20/9/2219/s1. Figure S1: Heatmap of 30 R2R3-MYB unigenes; Figure S2: The screening assay of 3 unigenes in Y1H Gold; Figure S3: The transmembrane domain of PgMYB2; Figure S4: The MYB DNA binding sites of PgMYB2; Figure S5. Binding assay of PgMYB2 to MBS site; Table S1: Specific primers used in the study.

Author Contributions: T.L. (Tuo Liu) and X.G. designed and conducted experiments, edited figures, and wrote the original draft; T.L. (Tiao Luo) and X.Z. assisted in performing experiments and collating data; S.A. helped to organize the data; G.L., Y.Z. and D.Z. helped in methodology and software analysis. R.Z. gave valuable suggestions and modified the draft, and Z.L. supervised the whole process and reviewed the final version of the manuscript.

Funding: The work was funded by National Natural Science Foundation of China (Grant No.81673544, 81874332 and 81071821), Hunan Province Undergraduate Research-based Learning and Innovative Experimental Program (XiangJiaoTong 2016 No.283-774), Independent Exploration and Innovation Project for Postgraduates of Central South University (Grant No.2018zzts395), and Central South University Innovation and Entrepreneurship Project by Teacher and Students (Grant No.2017gczd028).

Acknowledgments: We sincerely thank Feng Yu (Hunan University) and Ying Ruan (Hunan Agricultural University) for providing the protoplast of *A. thaliana*. We would also like to thank Jianbin Yan and Daoxin Xie (Tsinghua University) for their guidance on the yeast one-hybrid assay.

Conflicts of Interest: The authors declare no conflict of interest.

References

1. Yun, T.K. Brief introduction of *Panax ginseng* C.A. Meyer. *J. Korean Med. Sci.* **2001**, *16*, S3–S5. [CrossRef]
2. Chu, S.F.; Zhang, J.T. New achievements in ginseng research and its future prospects. *Chin. J. Integr. Med.* **2009**, *15*, 403–408. [CrossRef]
3. Kim, Y.-S.; Han, J.-Y.; Lim, S.; Choi, Y.-E. Ginseng metabolic engineering: Regulation of genes related to ginsenoside biosynthesis. *J. Med. Plants Res.* **2010**, *3*, 1270–1276.
4. Augustin, J.M.; Kuzina, V.; Andersen, S.B.; Bak, S. Molecular activities, biosynthesis and evolution of triterpenoid saponins. *Phytochemistry* **2011**, *72*, 435–457. [CrossRef]
5. Cao, H.; Nuruzzaman, M.; Xiu, H.; Huang, J.; Wu, K.; Chen, X.; Li, J.; Wang, L.; Jeong, J.H.; Park, S.J. Transcriptome Analysis of Methyl Jasmonate-Elicited *Panax ginseng* Adventitious Roots to Discover Putative Ginsenoside Biosynthesis and Transport Genes. *Int. J. Mol. Sci.* **2015**, *16*, 3035–3057. [CrossRef]
6. Tansakul, P.; Shibuya, M.; Kushiro, T.; Ebizuka, Y. Dammarenediol-II synthase, the first dedicated enzyme for ginsenoside biosynthesis, in *Panax ginseng*. *Febs Lett.* **2006**, *580*, 5143–5149. [CrossRef]
7. Kushiro, T.; Ohno, Y.; Shibuya, M.; Ebizuka, Y. In Vitro Conversion of 2,3-Oxidosqualene into Dammarenediol by *Panax ginseng* Microsomes. *Biol. Pharm. Bull.* **1997**, *20*, 292. [CrossRef]
8. Kim, Y.J.; Zhang, D.; Yang, D.C. Biosynthesis and biotechnological production of ginsenosides. *Biotechnol. Adv.* **2015**, *33*, 717–735. [CrossRef]
9. Han, J.Y.; Yong, S.K.; Yang, D.C.; Jung, Y.R.; Yong, E.C. Expression and RNA interference-induced silencing of the dammarenediol synthase gene in *Panax ginseng*. *Plant Cell Physiol.* **2006**, *47*, 1653. [CrossRef]
10. Liang, Y.; Zhao, S.; Zhang, X. Antisense Suppression of Cycloartenol Synthase Results in Elevated Ginsenoside Levels in *Panax ginseng* Hairy Roots. *Plant Mol. Biol. Report.* **2009**, *27*, 298–304. [CrossRef]
11. Zhou, M.; Memelink, J. Jasmonate-responsive transcription factors regulating plant secondary metabolism. *Biotechnol. Adv.* **2016**, *34*, S0734975016300118. [CrossRef]
12. Ambawat, S.; Sharma, P.; Yadav, N.R.; Yadav, R.C. MYB transcription factor genes as regulators for plant responses: An overview. *Physiol. Mol. Biol. Plants* **2013**, *19*, 307–321. [CrossRef]
13. Christian, D.; Ralf, S.; Erich, G.; Bernd, W.; Cathie, M.; Loïc, L. MYB transcription factors in Arabidopsis. *Trends Plant Sci.* **2010**, *15*, 573–581. [CrossRef]
14. Klempnauer, K.H.; Gonda, T.J.; Bishop, J.M. Nucleotide sequence of the retroviral leukemia gene v-myb and its cellular progenitor c-myb: The architecture of a transduced oncogene. *Cell* **1982**, *31*, 453–463. [CrossRef]
15. Pazares, J.; Ghosal, D.; Wienand, U.; Peterson, P.A.; Saedler, H. The regulatory c1 locus of Zea mays encodes a protein with homology to myb proto-oncogene products and with structural similarities to transcriptional activators. *Embo J.* **1987**, *6*, 3553–3558. [CrossRef]
16. Du, H.; Wang, Y.B.; Xie, Y.; Liang, Z.; Jiang, S.J.; Zhang, S.S.; Huang, Y.B.; Tang, Y.X. Genome-Wide Identification and Evolutionary and Expression Analyses of MYB-Related Genes in Land Plants. *DNA Res.* **2013**, *20*, 437–448. [CrossRef]
17. Reyes, J.L.; Chua, N.H. ABA induction of miR159 controls transcript levels of two MYB factors during Arabidopsis seed germination. *Plant J.* **2010**, *49*, 592–606. [CrossRef]
18. Zhang, W.; Xu, F.; Cheng, S.; Liao, Y. Characterization and functional analysis of a MYB gene (GbMYBFL) related to flavonoid accumulation in Ginkgo biloba. *Genes Genom.* **2018**, *40*, 49. [CrossRef]
19. Gajjeraman, P.; Doddananjappa Theertha, P. Functional characterization of sugarcane MYB transcription factor gene promoter (PScMYBAS1) in response to abiotic stresses and hormones. *Plant Cell Rep.* **2012**, *31*, 661–669.
20. Gális, I.; Simek, P.; Narisawa, T.; Sasaki, M.; Horiguchi, T.; Fukuda, H.; Matsuoka, K. A novel R2R3 MYB transcription factor NtMYBJS1 is a methyl jasmonate-dependent regulator of phenylpropanoid-conjugate biosynthesis in tobacco. *Plant J.* **2010**, *46*, 573–592. [CrossRef]
21. Sugimoto, K.; Takeda, S.; Hirochika, H. MYB-Related Transcription Factor NtMYB2 Induced by Wounding and Elicitors is a Regulator of the Tobacco Retrotransposon Tto1 and Defense-Related Genes. *Plant Cell* **2000**, *12*, 2511–2527. [CrossRef]
22. Chen, Y.; Yang, X.; He, K.; Liu, M.; Li, J.; Gao, Z.; Lin, Z.; Zhang, Y.; Wang, X.; Qiu, X. The MYB Transcription Factor Superfamily of Arabidopsis: Expression Analysis and Phylogenetic Comparison with the Rice MYB Family. *Plant Mol. Biol.* **2006**, *60*, 107–124.

23. Wei, S.Y.; Lou, Y.C.; Jia-Yin, T.; Meng-Ru, H.; Chun-Chi, C.; Rajasekaran, M.; Hong-Ming, H.; Tai, J.H.; Chwan-Deng, H.; Chen, C. Structure of theTrichomonas vaginalisMyb3 DNA-binding domain bound to a promoter sequence reveals a unique C-terminal β-hairpin conformation. *Nucleic Acids Res.* **2012**, *40*, 449–460. [CrossRef]

24. Vimolmangkang, S.; Han, Y.; Wei, G.; Korban, S.S. An apple MYB transcription factor, MdMYB3, is involved in regulation of anthocyanin biosynthesis and flower development. *BMC Plant Biol.* **2013**, *13*, 176. [CrossRef]

25. Cheong, J.J.; Yang, D.C. Methyl jasmonate as a vital substance in plants. *Trends Genet.* **2003**, *19*, 409–413. [CrossRef]

26. Yousaf, N.; Gould, D. Demonstrating Interactions of Transcription Factors with DNA by Electrophoretic Mobility Shift Assay. *Methods Mol. Biol.* **2017**, 11–21.

27. Lescot, M.; Déhais, P.; Thijs, G.; Marchal, K.; Moreau, Y.; Peer, Y.V.D.; Rouz, P.; Rombauts, S. PlantCARE, a database of plant cis-acting regulatory elements and a portal to tools for in silico analysis of promoter sequences. *Nucleic Acids Res.* **2002**, *30*, 325–327. [CrossRef]

28. Solano, R.; Nieto, C.; Avila, J.; Canas, L.; Diaz, I.; Paz-Ares, J. Dual DNA binding specificity of a petal epidermis-specific MYB transcription factor (MYB.Ph3) from Petunia hybrida. *EMBO J.* **1995**, *14*, 1773–1784. [CrossRef]

29. Farmer, E.E. Plant biology: Jasmonate perception machines. *Nature* **2007**, *448*, 659–660. [CrossRef]

30. Choi, D.W.; Jung, J.; Ha, Y.I.; Park, H.W.; In, D.S.; Chung, H.J.; Liu, J.R. Analysis of transcripts in methyl jasmonate-treated ginseng hairy roots to identify genes involved in the biosynthesis of ginsenosides and other secondary metabolites. *Plant Cell Rep.* **2005**, *23*, 557–566. [CrossRef] [PubMed]

31. Bedon, F.; Bomal, C.; Caron, S.; Levasseur, C.; Boyle, B.; Mansfield, S.D.; Schmidt, A.; Gershenzon, J.; Grimapettenati, J.; Séguin, A. Subgroup 4 R2R3-MYBs in conifer trees: Gene family expansion and contribution to the isoprenoid- and flavonoid-oriented responses. *J. Exp. Bot.* **2010**, *61*, 3847–3864. [CrossRef]

32. Lee, M.H.; Jeong, J.H.; Seo, J.W.; Shin, C.G.; Kim, Y.S.; In, J.G.; Yang, D.C.; Yi, J.S.; Choi, Y.E. Enhanced triterpene and phytosterol biosynthesis in *Panax ginseng* overexpressing squalene synthase gene. *Plant Cell Physiol.* **2004**, *45*, 976–984. [CrossRef]

33. Oktae, K.; Bang, K.H.; Youngchang, K.; Dongyun, H.; Minyoung, K.; Seonwoo, C. Upregulation of ginsenoside and gene expression related to triterpene biosynthesis in ginseng hairy root cultures elicited by methyl jasmonate. *Plant Cell Tissue Organ Cult.* **2009**, *98*, 25–33.

34. Kim, Y.S.; Hahn, E.J.; Murthy, H.N.; Paek, K.Y. Adventitious root growth and ginsenoside accumulation in *Panax ginseng* cultures as affected by methyl jasmonate. *Biotechnol. Lett.* **2004**, *26*, 1619–1622. [CrossRef]

35. Zhao, J.; Davis, L.C.; Verpoorte, R. Elicitor signal transduction leading to production of plant secondary metabolites. *Biotechnol. Adv.* **2005**, *23*, 283–333. [CrossRef]

36. Martin, C.; Paz-Ares, J. MYB transcription factors in plants. *Trends Genet.* **1997**, *13*, 67–73. [CrossRef]

37. Stracke, R.; Werber, M.; Weisshaar, B. The R2R3-MYB gene family in *Arabidopsis thaliana*. *Curr. Opin. Plant Biol.* **2001**, *4*, 447–456. [CrossRef]

38. Li, C.; Lu, S. Genome-wide characterization and comparative analysis of R2R3-MYB transcription factors shows the complexity of MYB-associated regulatory networks in Salvia miltiorrhiza. *BMC Genom.* **2014**, *15*, 277. [CrossRef]

39. Gou, M.; Hou, G.; Yang, H.; Zhang, X.; Cai, Y.; Kai, G.; Liu, C.J. The MYB107 Transcription Factor Positively Regulates Suberin Biosynthesis. *Plant Physiol.* **2017**, *173*, 1045. [CrossRef]

40. Wu, Q.; Wang, Y.; Guo, M. Triterpenoid Saponins from the Seeds of Celosia argentea and Their Anti-inflammatory and Antitumor Activities. *Chem. Pharm. Bull.* **2011**, *59*, 666–671. [CrossRef]

41. Osbourn, A.; Goss, R.J.M.; Field, R.A. The saponins—polar isoprenoids with important and diverse biological activities. *Nat. Prod. Rep.* **2011**, *28*, 1261–1268. [PubMed]

42. Thang, N.V.; Thu, V.K.; Nhiem, N.X.; Dung, D.T.; Quang, T.H.; Tai, B.H.; Anh, H.L.T.; Yen, P.H.; Ngan, N.T.T.; Hoang, N.H. Oleanane-type Saponins from Glochidion hirsutum and Their Cytotoxic Activities. *Chem. Biodivers.* **2017**, *14*, e1600445. [CrossRef] [PubMed]

43. Mahjoub, A.; Hernould, M.; Joubès, J.; Decendit, A.; Mars, M.; Barrieu, F.; Hamdi, S.; Delrot, S. Overexpression of a grapevine R2R3-MYB factor in tomato affects vegetative development, flower morphology and flavonoid and terpenoid metabolism. *Plant Physiol. Biochem.* **2009**, *47*, 551–561. [CrossRef] [PubMed]

44. Zhou, M.L.; Hou, H.L.; Zhu, X.M.; Shao, J.R.; Wu, Y.M.; Tang, Y.X. Molecular regulation of terpenoid indole alkaloids pathway in the medicinal plant, Catharanthus roseus. *J. Med. Plant Res.* **2011**, *425*, 2760–2772.

45. Hu, W.; Liu, N.; Tian, Y.; Zhang, L. Molecular cloning, expression, purification, and functional characterization of dammarenediol synthase from *Panax ginseng*. *Biomed Res. Int.* **2012**, *2013*, 285740. [PubMed]

46. Marchler-Bauer, A.; Bo, Y.; Han, L.; He, J.; Lanczycki, C.J.; Lu, S.; Chitsaz, F.; Derbyshire, M.K.; Geer, R.C.; Gonzales, N.R.; et al. CDD/SPARCLE: Functional classification of proteins via subfamily domain architectures. *Nucleic Acids Res.* **2017**, *45*, D200–D203. [CrossRef]

47. Wilkins, M.R.; Gasteiger, E.; Bairoch, A.; Sanchez, J.C.; Williams, K.L.; Appel, R.D.; Hochstrasser, D.F. Protein Identification and Analysis Tools in the ExPASy Server. *Methods Mol. Biol.* **1999**, *112*, 531.

48. Kumar, S.; Stecher, G.; Tamura, K. MEGA7: Molecular Evolutionary Genetics Analysis version 7.0 for bigger datasets. *Mol. Biol. Evol.* **2016**, *33*, 1870. [CrossRef]

49. Krogh, A.; Larsson, B.; Von, H.G.; Sonnhammer, E.L. Predicting transmembrane protein topology with a hidden Markov model: Application to complete genomes. *J. Mol. Biol.* **2001**, *305*, 567–580. [CrossRef]

50. Combet, C.; Blanchet, C.; Geourjon, C.; Deléage, G. NPS@: Network Protein Sequence Analysis. *Trends Biochem. Sci.* **2000**, *25*, 147–150. [CrossRef]

51. Waterhouse, A.; Bertoni, M.; Bienert, S.; Studer, G.; Tauriello, G.; Gumienny, R.; Heer, F.T.; de Beer, T.A.P.; Rempfer, C.; Bordoli, L.; et al. SWISS-MODEL: Homology modelling of protein structures and complexes. *Nucleic Acids Res.* **2018**, *46*, W296–W303. [CrossRef]

52. Robert, X.; Gouet, P. Deciphering key features in protein structures with the new ENDscript server. *Nucleic Acids Res.* **2014**, *42*, W320–W324. [CrossRef]

53. Livak, K.J.; Schmittgen, T.D. Analysis of relative gene expression data using real-time quantitative PCR and the 2(-Delta Delta C(T)) Method. *Methods* **2001**, *25*, 402–408. [CrossRef] [PubMed]

 © 2019 by the authors. Licensee MDPI, Basel, Switzerland. This article is an open access article distributed under the terms and conditions of the Creative Commons Attribution (CC BY) license (http://creativecommons.org/licenses/by/4.0/).

International Journal of
Molecular Sciences

Article

Proteomic Analysis of MeJa-Induced Defense Responses in Rice against Wounding

Laura Bertini [1,†], Luana Palazzi [2,†], Silvia Proietti [1], Susanna Pollastri [3], Giorgio Arrigoni [4,5], Patrizia Polverino de Laureto [2,*] and Carla Caruso [1,*]

[1] Department of Ecological and Biological Sciences, University of Tuscia, 01100 Viterbo, Italy; lbertini@unitus.it (L.B.); s.proietti@unitus.it (S.P.)

[2] Department of Pharmaceutical and Pharmacological Sciences, University of Padova, 35131 Padova, Italy; luana.palazzi@unipd.it

[3] Institute for Sustainable Plant Protection, National Research Council of Italy, Sesto Fiorentino, 50019 Florence, Italy; susanna.pollastri@ipsp.cnr.it

[4] Department of Biomedical Sciences, University of Padova, 35131 Padova, Italy; giorgio.arrigoni@unipd.it

[5] Proteomics Center of Padova University and Azienda Ospedaliera di Padova, 35131 Padova, Italy

* Correspondence: patrizia.polverinodelaureto@unipd.it (P.P.d.L.); caruso@unitus.it (C.C.); Tel.: +39-049-8276157 (P.P.d.L.); Tel.: +39-0761-357330 (C.C.)

† These authors contributed equally to this work.

Received: 16 April 2019; Accepted: 20 May 2019; Published: 22 May 2019

Abstract: The role of jasmonates in defense priming has been widely recognized. Priming is a physiological process by which a plant exposed to low doses of biotic or abiotic elicitors activates faster and/or stronger defense responses when subsequently challenged by a stress. In this work, we investigated the impact of MeJA-induced defense responses to mechanical wounding in rice (*Oryza sativa*). The proteome reprogramming of plants treated with MeJA, wounding or MeJA+wounding has been in-depth analyzed by using a combination of high throughput profiling techniques and bioinformatics tools. Gene Ontology analysis identified protein classes as defense/immunity proteins, hydrolases and oxidoreductases differentially enriched by the three treatments, although with different amplitude. Remarkably, proteins involved in photosynthesis or oxidative stress were significantly affected upon wounding in MeJA-primed plants. Although these identified proteins had been previously shown to play a role in defense responses, our study revealed that they are specifically associated with MeJA-priming. Additionally, we also showed that at the phenotypic level MeJA protects plants from oxidative stress and photosynthetic damage induced by wounding. Taken together, our results add novel insight into the molecular actors and physiological mechanisms orchestrated by MeJA in enhancing rice plants defenses after wounding.

Keywords: MeJA; priming; rice; proteomics; ROS; chlorophyll fluorescence imaging

---

1. Introduction

Plants are exposed to a variety of external factors that unfavorably affect their growth and development, and are generally classified into biotic (microbial pathogens and insect herbivores) and abiotic (extreme temperature, water logging, drought, high salinity or toxic compounds, etc.) stresses. Adaptation to these environmental stresses is essential for survival, growth and reproduction [1]. Among the defense strategies that plants have evolved, some are constitutive whereas other are induced in response to stimuli, thus being more specific [2]. It is widely recognized that the identification of elicitors triggers the activation of peculiar subsets of defense responses [3]. Furthermore, plants are able to recognize non-self molecules or signals from their own damaged cells and consequently to activate an efficient immune response against the stress they encounter [4–6]. It has been shown that phytohormones such as salicylic acid (SA), jasmonic acid (JA), ethylene (ET), abscisic acid (ABA), cytokinin, brassinosteroids and auxin are the main players in coordinating signaling networks involved in the adaptive response of plants to its (a)biotic environment [6,7]. These signal-transduction pathways in turn activate large suites of genes, including those coding for transcription factors, enzymes involved in the production of plant toxins, plant volatiles and reactive oxygen species (ROS) [8]. Generally, SA induces defense responses against biotrophic pathogens, whereas JA and ET are important hormonal regulators of induced reaction against necrotrophic pathogens [6]. Moreover, it has been shown that either JA or ABA induce plant defenses against herbivorous insects, and that both JA and its methyl ester (MeJA) are key components of a wound signal transduction cascade in plants [6,9]. Indeed, application of exogenous JA induces the expression of genes, such as phenylalanine ammonia lyase and proteinase inhibitors, known to be responsive to wounding [10]. Furthermore, using defective tomato mutants in both JA biosynthesis and perception in grafting experiments, it has been further demonstrated that JA or one of its derivatives may also act as a long-distance transmissible wound signal [10–13].

Recent evidences show that plants can be primed for more rapid and robust activation of defense response to biotic or abiotic stresses [14,15]. Defense priming is considered to be an adaptive, low-cost defensive strategy since defense responses are not, or only slightly and transiently, activated by a given priming agent. Conversely, defense responses are activated in a faster, stronger, and/or more persistent manner following the perception of a later challenging signal [15]. Effectively, primed plants possess molecular mechanisms that allow them to memorize previous priming events and generate memory imprints during the establishment of priming [16–19].

The primed state can be induced by a pre-exposition of the plants to low doses of natural or synthetic (a) biotic stress inducers, among which are chemical compounds (hormones, pipecolic acid, hexanoic acid, volatile organic compounds), pathogens, insect herbivores, beneficial microorganisms or environmental cues [3,20–23]. To date, induction of priming by chemicals has been observed in many plant species, such as parsley, tobacco, *Arabidopsis thaliana* as well as in many others monocots and dicot species [24,25]. The mediation of hormones in the primed responses is mainly restricted to SA, ET, JA and ABA [16,26,27]. Among them, JA was studied in relation to resistance induction, demonstrating that Arabidopsis plants primed with JA showed protection and reduction of infection symptoms by *P. cucumerina* and *A. brassicicola* [28]. The role of JA as a priming hormone was also studied in rice following *Rhizoctonia solani* infection [29]. Moreover, Methyl Jasmonate (MeJA)-induced priming was studied in the herbaceous monocotyledon *Calla lily*, infected with the necrotrophic bacterium *Pectobacterium carotovorum*, highlighting decreased necrosis in infected plant tissues [30]. Unraveling the molecular basis of priming has recently received increasing attention [17,21,31,32]. Depending on the nature of the priming agent and the stressor, priming can involve diverse mechanisms. Priming could be related to the accumulation of key cellular proteins in their inactive state, which could be readily activated following exposure to biotic or abiotic stress speeding up the signal amplification cascade [33]. Another hypothesis on the molecular mechanism of priming suggested that epigenetic mechanisms get ready the defense genes in a permissive modified state facilitating quicker and more potent responses to subsequent attacks [34]. Additionally, several studies have shown the relevance of epigenetic mechanisms underlying priming phenomenon [17,18,35,36].

Rice (*Oryza sativa* L.) is an important food crop worldwide. In rice, priming was mainly used to improve seeds performance in terms of higher rate of germination and seedlings vigor under suboptimal environmental conditions. In the so-called "seed priming", controlled hydration of seeds is used to break dormancy, speed germination and improve germination under stress conditions [37]. Recently, "seed priming" was also exploited to enhance the tolerance against various abiotic stresses including drought, submergence, salinity, chilling, and heavy metals in various plant species [38–40].

Recent developments in "omics" disciplines have opened up new perspectives to achieve a comprehensive understanding of biological processes related to stress responses in plants. In the post-genomic era, the enormous amounts of high throughput -omics data along with robust bioinformatics and data mining tools can potentially provide a global view on physiological processes triggered by stresses and also support the identification of novel signaling nodes in the plant defense signaling. Indeed, advances in transcriptomic, metabolomic and proteomic technologies allowed highlighting new hallmarks of biotic and abiotic stress responses in several plant species [41–45]. In particular, proteomics could be crucial to understanding physiological processes that are not accounted at genomic level. The few proteomic studies published so far on the priming role during environmental stresses identified key protein targets and signaling pathways, which are involved in mitigating negative effects of stress factors [19]. Recently, proteomics has been exploited to characterize the response of monocots to MeJA [46–48]. In particular, a proteomic analysis has suggested a role for MeJA in enhancing fungal disease resistance in rice [49]. Remarkably, to date there are not many studies highlighting the role of MeJA in protecting plants from wounding in rice. Mechanical wounding, which is induced by biotic (e.g., herbivore attack and pathogens infection) and abiotic (e.g., raining, wind, touching, and hailing) factors in plants, exists widely in nature [50]. Wounding stress is pretty deleterious since it can open the way to the invasion by microbial pathogens, providing nutrients to pathogens and facilitating their entry into the tissue and subsequent infection [51].

In this work, we investigated the impact of MeJA-induced priming on the efficacy of the rice plant response to mechanical wounding. The proteome profiling of MeJA-primed plants has been in-depth analyzed by using a combination of high throughput profiling techniques and bioinformatics tools. Moreover, Gene Ontology (GO) analysis has been carried out to obtain more comprehensive insight into the biological processes affected by MeJA, wounding and MeJA + wounding treatments. Moreover, we showed that low doses of MeJA prime plants for augmented level of a subset of proteins, upon wounding. Interestingly, while some of them are defense-related, others are involved in oxidative stress responses and photosynthesis. Finally, phenotypic analysis performed on primed and not primed rice plants strengthened the role of MeJA in protecting plants against potential oxidative stresses and photosynthetic alterations due to mechanical stress. To the best of our knowledge, this is the first study performed by shot-gun proteomics-based approach to investigate the role of MeJA as priming agent against wounding in rice.

2. Results and Discussion

Jasmonates (JAs) are plant-specific signaling molecules that steer a broad set of physiological as well as defense processes. Pathogen attack and wounding caused by herbivores induce the biosynthesis of JAs, activating defense responses both locally and systemically [52]. To shed some light on the effect caused by MeJA-induced priming on defenses against herbivorous, we investigated the whole proteome changes of rice plants subjected to mechanical wounding following or not priming treatment by comparative proteomic analysis. An overview of the experimental workflow is shown in Figure 1.

To induce priming state, 21-day-old rice plants (3–4 leaves stage) were sprayed with 10 μM MeJA solution, a dose not able to induce direct defense response, as previously demonstrated [53]. Twenty-four hours after spraying, both mock and primed plants were wounded. Since plant response to MeJA is quite fast, the time gap elapsed between hormone treatment and wounding is reasonably enough to allow the establishment of the priming effect [53,54]. Leaf samples were harvested at 48 hours-post-wounding (hpw) and proteins were extracted from mock (M), wounded (W), primed with MeJA (P) and wounded after priming treatment (P + W) rice plants (Figure 1A). Comparative quantitative proteomic analysis (Figure 1B) was performed analyzing the proteome of the W, P and P + W rice leaves with respect to the plants grown under physiological conditions (M), allowing the identification of differentially expressed proteins (DEPs) in the treated samples. In the M sample, a total of 1417 proteins was identified, while in P, W, and P + W samples, the proteins detected were 1448, 1430 and 1447, respectively. Three biological replicates were performed and the numbers reported above encompass only the proteins overlapping between the three replicates. The differences in their levels were evaluated by label-free quantification approach, chosen to avoid excessive manipulation of samples and artifacts, using the MaxQuant software (Table S1).

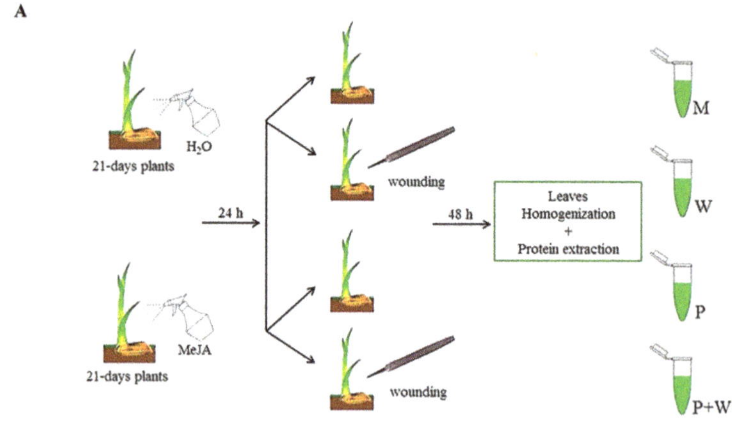

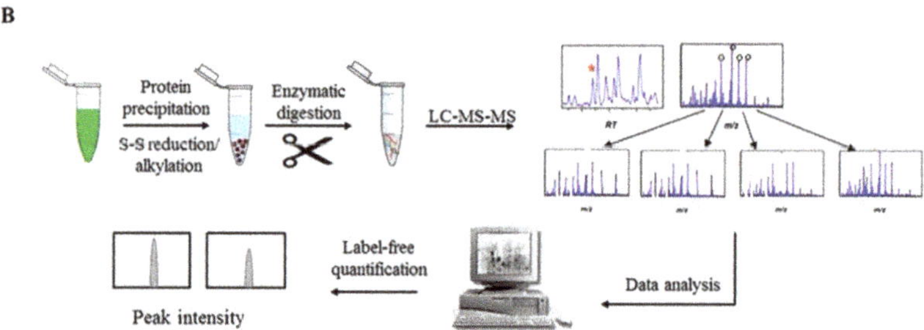

Figure 1. Experimental workflow. (**A**) Plant treatments. Mock (M), leaves after wounding (W), leaves primed with MeJA (P), and leaves primed with MeJA followed by wounding (P + W). (**B**) Protein treatment and proteomic analysis.

2.1. MeJA Treatment Modulates Broad Spectrum Biological Processes

The list of Differentially Expressed Proteins (DEPs) after MeJA treatment (P samples) compared to the mock is reported in Table 1.

Actually, listed DEPs are only those identified by at least three unique peptides, showing a level greater than two-fold or lower than two-fold (log$_2$ fold change >1 or <−1, respectively) compared to the mock, and with a p-value ≤ 0.05. The analysis disclosed 32 proteins that underwent significant quantitative variations in plants treated by MeJA. Among these, 21 proteins showed a log$_2$ fold change value greater than 1, indicating over-expression, whereas 11 proteins were found to be under-expressed with log$_2$ fold change value less than -1, as compared to the mock.

Gene Ontology analysis was performed by the Protein Annotation Through Evolutionary Relationship (PANTHER) software to classify DEPs into two major categories: biological processes and protein classes (Figure 2A,B, respectively).

Table 1. Differentially expressed proteins (DEPs) after MeJA treatment (P), compared to mock (M).

UniProt Code	MSU ID Code	Protein Name	Log$_2$ Fold-Change	*p*-Value
Q10D65	LOC_Os03g52860	Lipoxygenase Linoleate 9S-lipoxygenase 2	6.66	0.018
Q53LW0	LOC_Os11g20160	O-methyltransferase	4.59	0.013
Q01HV9	no code	Arginine decarboxylase	3.67	0.005
Q0JR25	LOC_Os01g03360	Bowman–Birk type bran trypsin inhibitor	3.25	0.014
B7E4J4	LOC_Os05g31750	Os05g0382600	2.98	0.02
Q5WMX0	LOC_Os05g15770	DIP3	2.58	0.002
Q5Z678	LOC_Os06g47620	IAA-amino acid hydrolase ILR1-like 6	2.41	0.037
Q8LMW8	LOC_Os10g11500	Os10g0191300 protein (putative PRB1-2)	2.26	0.022
Q5U1I3	no code	Peroxidase	2.24	0.01
Q8S3P3	LOC_Os04g56430	DUF26-like protein	2.23	0.018
Q69JF3	LOC_Os09g36700	Os9g0538000	2.14	0.016
Q7XAD8	no code	Os07g0126400 protein (putative Prb1)	2.07	0.024
Q5ZCA9	no code	Bowman–Birk type bran trypsin inhibitor (Fragment)	2.0	0.008
Q6YZZ7	LOC_Os08g08970	Germin-like protein 8-3	2.0	0.015
Q75T45	LOC_Os12g36830	Os12g0555000 (root PR10)	1.77	0.014
Q8L6H4	LOC_Os03g32314	Allene oxide cyclase, chloroplastic	1.72	0.05
Q6YXT5	LOC_Os08g02230	Os08g0114300 protein	1.58	0.02
Q33E23	LOC_Os04g45970	Glutamate dehydrogenase 2, mitochondrial	1.33	0.024
Q40707	LOC_Os12g36880	PBZ1	1.26	0.01
B9F4F6	no code	Citrate synthase	1.1	0.046
Q9ATR3	no code	Glucanase	1.07	0.0012
Q0JG75	LOC_Os01g71190	Photosystem II reaction center Psb28 protein	−1.09	0.002
C5MRM9	no code	PsbA (Fragment)	−1.09	0.019
A2YVX9	no code	Putative uncharacterized protein (Germin-like protein 8-14)	−1.09	0.044
Q5QLS1	LOC_Os01g47780	Arabinogalactan protein-like	−1.1	0.047
B7EKW3	LOC_Os07g26690	Aquaporin	−1.18	0.03
Q5Z5A8	LOC_Os06g51330	Photosystem II stability/assembly factor HCF136 chloroplastic	−1.32	0.045
Q6ERW9	LOC_Os09g23540	Probable cinnamyl alcohol dehydrogenase 8B	−1.36	0.041
Q7F8S5	LOC_Os02g09940	Peroxiredoxin-2E-2, chloroplastic	−1.64	0.029
B9FY06	LOC_Os07g38300	Ribosome-recycling factor, chloroplastic	−1.89	0.014
H2KW47	LOC_Os11g13890	Chlorophyll A-B binding protein, chloroplastic	−1.94	0.014
J3RG68	no code	Photosystem I iron-sulfur center	−2.06	0.013

GO analysis highlighted biological processes affected by MeJA treatment, largely represented by metabolic processes (43%) as well as cellular processes (29%) (Figure 2A). Moreover, our analysis revealed some proteins involved in developmental processes (14%). Indeed, it is well known from the literature that JAs are involved in the regulation of many developmental processes, including male fertility, fruit ripening, and root growth [52]. Within this group of proteins, we found a "probable cinnamyl alcohol dehydrogenase 8B" (OsCAD8B) (UniProt code Q6ERW9, Table 1), involved in lignin biosynthesis catalyzing the final step of the production of lignin monomers [55]. Interestingly, it has been reported that CAD genes are also stress-responsive [56]. The rice genome contains 12 different CAD genes distributed at nine different loci and expression patterns have been reported only for few of them; moreover, it has been hypothesized that the rice CAD genes could share similar expression profiles with orthologs in other plant species [55]. The OsCAD8B closest related gene is LpCAD1, characterized in *Lolium perenne*, which was found was found to be wound induced within six hours, but its level dropped down between 24–48 h [57]. According to this, in our experimental condition,

OsCAD8B was found to be slightly down-regulated 48 hours after treatment with low doses of MeJA, but its induction at earlier time point after treatment cannot be ruled out. Our *in silico* functional characterization highlighted also that 14% of the input proteins belongs to the "Response to stimulus" group (Figure 2A). A representative protein of this category is germin-like protein 8-3 (UniProt code Q6YZZ7, Table 1) that resulted over-expressed in our experimental conditions and it is known to play a role in broad-spectrum disease resistance [58]. Members of the *Oryza sativa* 12 germin-like protein (OsGLP) gene cluster are located on chromosome 8 in the major-effect quantitative trait loci (QTL) for fungal blast resistance. In particular, proteins belonging to the OsGLP family were shown to contribute to disease resistance as silencing of several genes confers susceptibility to two distinct fungal pathogens, *Magnaporthe oryzae* and *Rhizoctonia solani*, the sheath blight pathogen [58]. In general, germins and germin-like proteins (GLPs) constitute a large plant gene family and they were first identified searching for germination-specific proteins in wheat (*Triticum aestivum*) [58,59]. They are present as glycoproteins often retained in the extracellular matrix by ionic bonds. Most are very stable oligomers [60,61]. They are structurally related to members of the cupin superfamily, that includes isomerases, sugar- or auxin-binding proteins, cyclases, dioxygenases, and monomeric or dimeric globulin seed storage proteins, such as phaseolin [62,63]. Germins and GLPs are known to play a wide variety of roles as enzymes, structural proteins, or receptors [60]. As Enzymes, germins have oxalate oxidase activity [64,65] and some GLPs have superoxide dismutase (SOD) activity [60], highlighting a role in defense responses since both of them can produce hydrogen peroxide (H_2O_2) in plants [66]. Some studies have demonstrated that germins and GLPs modulate plant responses to abiotic and biotic stresses [61,62]. Moreover, according to our findings, they are responsive to MeJA [67].

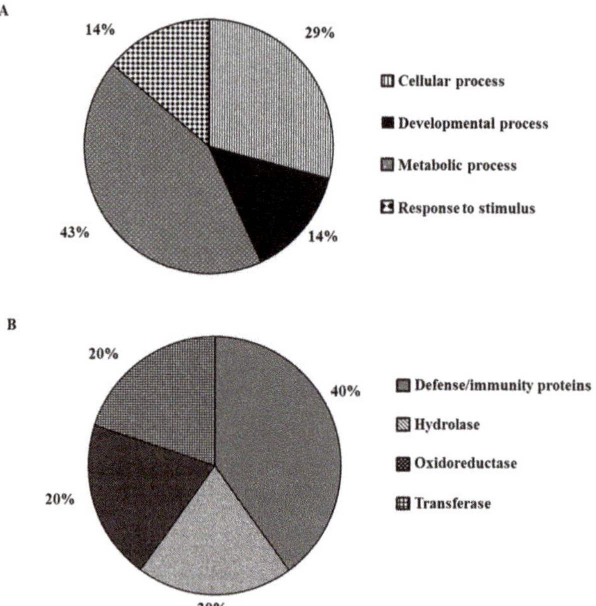

Figure 2. Functional classification of the 32 DEPs after MeJA treatment, using Protein Annotation Through Evolutionary Relationship (PANTHER) gene ontology (GO) analysis. The proteins were classified into (**A**) biological processes and (**B**) protein classes.

As shown in Figure 2B, DEPs are mostly "Defense/immunity proteins" (40%), whereas other protein classes are represented by Hydrolase, Oxidoreductase and Transferase (20% each). Among defense/immunity proteins are putative Pathogenesis-Related (PR) protein PRB1-2 (UniProt code

Q8LMW8, Table 1) and Prb1 (UniProt code Q7XAD8, Table 1) belonging to the PR1 family, both over-expressed in our experimental conditions. PR1 family is a dominant protein group induced by pathogens and is commonly used as a marker for SA-related systemic acquired resistance (SAR) [68]. In rice it has been shown that Prb1 proteins were induced in roots of seedlings after salt stress or JA treatment, as well as in JA-treated stems or leaves [69–72]. Within the "Hydrolase" group a drought-induced gene, DIP3, encoding a chitinase III protein (UniProt code Q5WMX0, Table 1) was found over-expressed in our experimental condition. Chitinases (EC 3.2.1.14) catalyze the hydrolytic cleavage of the β-1,4-glycosidic bond in N-acetyl-glucosamine biopolymers largely found in chitin [73,74]. One of the physiological roles of plant chitinases is the protection against fungal pathogens by degrading chitin [75]. Remarkably, some chitinases do not show any antifungal activities [76]. Chitinases also respond to abiotic stress, and are involved in developmental processes or growth [74,77]. Notably, it has been demonstrated that treatment by JA induces the accumulation of chitinases in rice [78], according with our results. The protein class "Transferase" includes the protein o-methyltrasferase (UniProt code Q53LW0, Table 1), putatively involved in serotonin and melatonin biosynthesis, which was found strongly over-expressed in our experimental system. Melatonin (*N*-acetyl-5-methoxytryptamine) has been characterized as an important bioactive molecule that is not only an animal hormone, but also plays a role in plant growth and development [79]. Although significant advances in elucidating the physiological roles and biochemical pathways of melatonin in animals have been achieved, studies on melatonin in plants are at their infancy, but advancing rapidly [80]. Very recently, it has been reported that its functions in plants include also the ability to reduce susceptibility to diseases [81]. Our results corroborate the role of MeJA in rewiring a broad spectrum of biological processes in rice plants, even at low doses.

2.2. Wounding Induces Proteome Changes on Immunity-Related Proteins and Enzymes

The list of DEPs after wounding (W samples) compared to the mock is reported in Table 2.

The analysis disclosed 11 proteins subjected to significant quantitative variations in wounded plants. Among these, 10 proteins showed a $\log_2$ fold change value greater than 1, indicating over-expression, whereas only 1 protein showed a $\log_2$ fold change <-1, as compared to the mock. Analysis was performed by PANTHER software and DEPs were classified into the category of protein classes. As shown in Figure 3, DEPs can be grouped in the following functional groups: "Defense/immunity proteins", "Hydrolase", "Isomerase", "Lyase" and "Oxidoreductase" (20% each).

Table 2. Differentially expressed proteins (DEPs) after wounding, compared to mock.

UniProt Code	MSU ID Code	Protein Name	Log$_2$ Fold-Change	*p*-Value
Q306J3	LOC_Os12g14440	Dirigent protein	5.11	0.001
Q10D65	LOC_Os03g52860	Linoleate 9S-lipoxygenase 2	4.17	0.019
Q75T45	LOC_Os12g36830	Os12g0555000 (root PR10)	3.52	0.001
Q945E9	LOC_Os03g18850	JIOsPR10	2.38	0.011
Q40707	LOC_Os12g36880	PBZ1	2.04	0.035
Q5WMX0	LOC_Os05g15770	DIP3	1.63	0.001
Q8S3P3	LOC_Os04g56430	DUF26-like protein	1.63	0.012
Q5ZCA9	no code	Bowman–Birk type bran trypsin inhibitor (Fragment)	1.32	0.007
Q0JR25	LOC_Os01g03360	Bowman–Birk type bran trypsin inhibitor	1.2	0.025
Q7XAD8	no code	Os07g0126400 protein (putative Prb1)	1.14	0.036
Q9FTN5	LOC_Os01g01660	Os01g0106400 (putative isoflavone)	−1.56	0.044

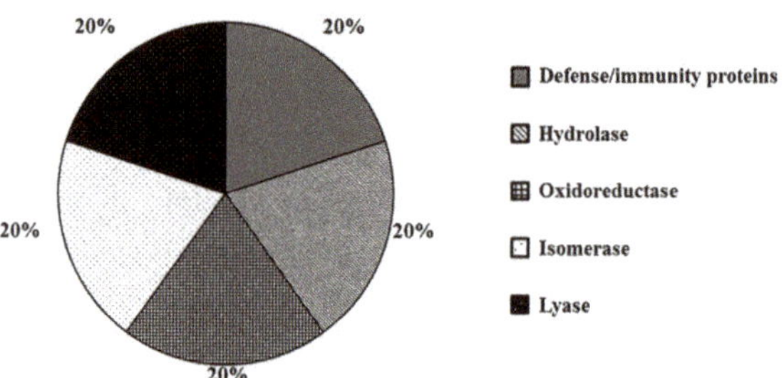

Figure 3. Functional classification of the 11 DEPs after wounding, using PANTHER gene ontology (GO) analysis. The proteins were classified into protein classes.

The only protein significantly repressed by wounding is categorized into "Oxidoreductase" and it is represented by a putative isoflavone reductase (UniProt code Q9FTN5, Table 2). Isoflavone reductases are enzymes involved in the biosynthesis of isoflavonoid phytoalexins in plants. They play essential roles in response to several biotic and abiotic stresses and are restricted to the plant kingdom. Isoflavonoid phytoalexins are small anti-microbial compounds produced by plants upon pathogen attack, exposure to elicitor molecules, or other biotic and abiotic stresses [82]. In rice, an isoflavone reductase-like, OsIRL, was found to be regulated by phytohormones either positively through JA or negatively through SA and ABA [82]. Moreover, when produced in combination with JA, upon wounding or herbivory, ABA acts synergistically on the expression of the MYC branch of the JA response pathway, while it antagonizes the ERF branch, induced by JA and ET [6,83]. A role of ABA in defense against insects has been suggested also in Arabidopsis [84]. Moreover, ABA has been demonstrated to be involved in gene regulation in response to wounding; in fact, endogenous ABA levels rise in plants after mechanical damage, both locally and systemically [69]. In light of this and according to previous findings [82], we may speculate that in our experimental conditions the isoflavone reductase-like under study could be down-regulated after wounding due to the wound-induced increase of ABA.

The GO category "Defense/immunity proteins" includes the protein Prb1 (UniProt code Q7XAD8, Table 2). We previously found that Prb1 is over-expressed also by MeJA (Table 1) and we discussed about its role in plant defense. Accordingly, it has been demonstrated that PR1 gene family is over-expressed after wounding in rice [85]. Moreover, the Arabidopsis Prb1 ortholog (At2g14580) is also involved in response to wounding (source: TAIR). The Protein class "Hydrolase" includes DIP3, a chitinase III protein (UniProt code Q5WMX0, Table 2). In our hands we found DIP3 up-regulated also by MeJA treatment (Table 1) and this is not surprising since it is well known that chitinases are induced by different abiotic stresses such as salt, cold, osmosis and heavy metals. For instance, in Arabidopsis, chitinase activity is induced by heat shock, UV light, and wounding [74]. Our study suggests that proteome reprogramming induced by wounding involve a broad variety of proteins functionally related with immunity processes and mainly aiming to boost defense responses.

2.3. Combined MeJA Treatment and Wounding Affect the Level of Proteins Related to Defense Processes

The list of DEPs after MeJA treatment followed by wounding (P + W sample), as compared to the mock is reported in Table 3.

Table 3. Differentially expressed proteins (DEPs) after MeJA treatment followed by wounding, compared to mock.

UniProt Code.	MSU ID Code	Protein Name	Log$_2$ Fold-Change	p-Value
Q10D65	LOC_Os03g52860	Linoleate 9S-lipoxygenase 2	5.3	0.015
Q5U1I3	no code	Peroxidase	5.1	0.039
Q306J3	LOC_Os12g14440	Dirigent protein	4.3	0.005
Q8S3P3	LOC_Os04g56430	DUF26-like protein	3.2	0.01
Q01HV9	no code	Arginine decarboxylase	3.1	0.033
Q5WMX0	LOC_Os05g15770	DIP3	3.0	0.01
Q9ATR3	no code	Glucanase	2.9	0.043
B7E4J4	LOC_Os05g31750	Os05g0382600	2.7	0.016
Q10N98	LOC_Os03g16950	33 kDa secretory protein, putative expressed	2.7	0.042
Q75T45	LOC_Os12g36830	Os12g0555000 (root PR10)	2.7	0.018
Q69JX7	LOC_Os09g36680	Drought-induced S-like ribonuclease	2.7	0.017
Q7XAD8	no code	Os07g0126400 protein (putative Prb1)	2.4	0.029
Q8LMW8	LOC_Os10g11500	Os10g0191300 protein (putative PRB1-2)	2.2	0.001
Q0JR25	LOC_Os01g03360	Bowman–Birk type bran trypsin inhibitor	2.2	0.037
Q69JF3	LOC_Os09g36700	Os09g0538000	2.1	0.008
Q6YXT5	LOC_Os08g02230	Os08g0114300 protein	1.9	0.008
Q5ZCA9	no code	Bowman–Birk type bran trypsin inhibitor (Fragment)	1.8	0.005
Q8L6H4	LOC_Os03g32314	Allene oxide cyclase, chloplastic	1.7	0.031
Q5Z7J2	LOC_Os06g35520	Peroxidase	1.7	0.033
Q40707	no code	PBZ1	1.7	0.037
Q33E23	LOC_Os04g45970	Glutamate dehydrogenase 2, mitochondrial	1.6	0.013
Q6ZI95	LOC_Os08g41880	Purple acid phosphatase	1.6	0.036
Q0D3V1	no code	Os07g0664300 protein	1.6	0.046
B9F4F6	no code	Citrate synthase	1.6	0.024
S4U072	LOC_Os04g39150	OSJNBb0048E02.12 protein	−1.0	0.023
Q10A54	LOC_Os10g05069	Alpha-mannosidase	−2.3	0.047

The analysis disclosed 26 proteins that underwent significant quantitative variations with respect to mock. Among these, 24 proteins showed a log$_2$ fold change value greater than 1, indicating over-expression, whereas 2 proteins were found to be under-expressed (log$_2$ fold change value less than −1), as compared to the mock. Results of Gene Ontology analysis performed by PANTHER is shown in Figure 4.

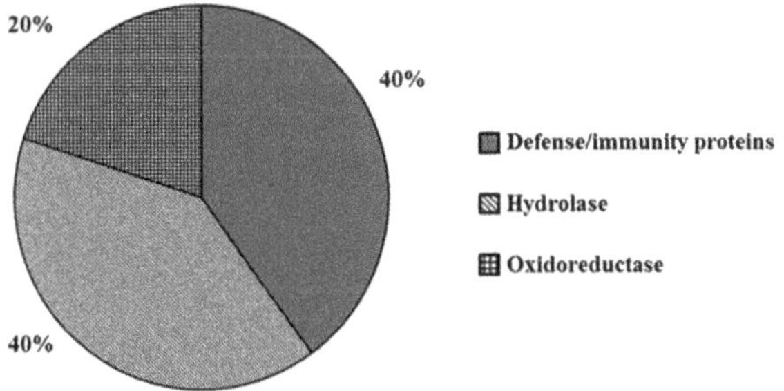

Figure 4. Functional classification of the 26 DEPs after MeJA + W, using PANTHER gene ontology (GO) analysis. The proteins were classified into protein classes.

DEPs are classified into "Defense/immunity proteins" (40%), "Hydrolase" (20%) and "Oxidoreductase" (40%). The protein class "Defense/immunity proteins" includes the Pathogenesis-related proteins PRB1-2 (UniProt code Q8LMW8, Table 3) and Prb1 (UniProt code Q7XAD8, Table 3). As described above, both are significantly over-expressed by MeJA (Table 1), but only Prb1 was found significantly over-expressed also by wounding (Table 2).

Analogously, within the "Hydrolase" group, we disclosed DIP3 (UniProt code Q5WMX0, Table 3) also found over-expressed by MeJA (Table 1) and wounding (Table 2), implying that the double treatment is not essential for its induction. Within the "Hydrolase" group another representative protein is a purple acid phosphatase encoded by *NPP1* (UniProt code Q6ZI95, Table 3). Plant acid phosphatases are involved in phosphate acquisition and utilization and their synthesis is affected by developmental as well as environmental cues. Phosphate starvation induces *de novo* synthesis of extra- and intra-cellular acid phosphatases, that might be one of the strategies plants have evolved to cope with phosphate-limiting conditions [86]. Purple acid phosphatases have mainly been studied in Arabidopsis, especially for their response to phosphorus starvation. Induction at both mRNA or protein level in roots and in leaves under phosphate deficiency suggests that they may function in scavenging phosphate from the soil as well as recycling it within the plant [87]. Interestingly, purple acid phosphatases share similar transcriptional regulation features to *Arabidopsis Vegetative Storage Protein2* (*AtVSP2*) gene. Basically, *AtVSP2* is a gene induced by wounding, MeJA and insect feeding. Moreover, the defense function of AtVSP2 is correlated with its acid phosphatase activity [88]. In addition, the Arabidopsis ortholog of the purple acid phosphatase encoded by *NPP1*, *AtPAP27* (At5g50400), shared 65% of amino acid sequence identity with rice NPP1. By querying Genevestigator V3 [89] we found a strong induction of *AtPAP27* following MeJA treatment or wounding in different plant developmental stages, suggesting a similar role of rice NPP1 in plant defense mediated by MeJA and wound signaling.

Within the "Oxidoreductase" group, Os07g0664300 protein was found as differentially expressed (UniProt code Q0D3V1, Table 3). This protein belongs to the Short-chain Dehydrogenases/Reductases (SDR) family. SDR comprises a broad family of NAD(P)H-dependent oxidoreductases represented in plant kingdom. Functions of SDRs include many aspects of primary (chlorophyll biosynthesis, lipid synthesis, or degradation) and secondary (steroids, terpenoids, phenolics and alkaloids) metabolism. In analogy with animal SDRs, it may be rational to assume that several SDRs play a major role regarding hormone metabolism, including ABA biosynthesis [90]. Our results corroborate the evidence that MeJA and wounding signaling could overlap in inducing proteins with key roles in defense responses.

2.4. Priming-Regulated Proteins Correlate with Defense Processes

Priming mechanisms include the accumulation of proteins in an inactive form that are rapidly modulated upon exposure to stress, resulting in a more efficient and robust defense mechanism [34]. Our ultimate goal was to highlight proteins specifically affected by the priming treatment, i.e., all proteins that after MeJA treatment and subsequent wounding (P + W) showed a alevel greater than two fold ($\log_2$ fold change >1) compared to both wounding (W) and MeJA (P) single treatment and having a *p*-value ≤ 0.05. This comparison was performed in order to exclude the contribution of single treatments to the protein level occurred in the double treatment and to characterize molecules regulating plant priming as well as the potential interplay between them at proteome level. These proteins are listed in Table 4.

Table 4. Priming-regulated proteins. Log2 fold-change after MeJA+wounding, compared to wounding (P + W/W), log$_2$ fold-change after MeJA+wounding, compared to MeJA (P + W/P) and corresponding p-values are shown.

UniProt Code	MSU ID Code	Protein Name	Log$_2$ Fold-Change (P + W/W)	p-Value	Log$_2$ Fold-Change (P + W/P)	p-Value
Q7F2G3	LOC_Os01g45274	Carbonic anhydrase, chloroplast precursor, putative, expressed	1.48	0.006	1.33	0.002
Q943K1	LOC_Os01g64960	Chlorophyll A-B binding protein, putative, expressed	1.22	9.29×10^{-5}	1.21	0.001
Q84NW1	LOC_Os07g32880	ATP synthase gamma chain, putative, expressed	1.13	0.015	1.32	0.027
Q9SDJ2	LOC_Os01g17170	Magnesium-protoporphyrin IX monomethyl ester cyclase, chloroplast precursor, putative, expressed	1.54	0.003	1.42	0.005
Q10S82	LOC_Os03g03910	Catalase domain containing protein	1.33	0.005	1.48	0.007
Q7XSU8	LOC_Os04g59190	Peroxidase precursor, putative, expressed	1.21	0.049	1.14	0.028

The protein carbonic anhydrase (UniProt code Q7F2G3) encoded by *Os01g0639900* gene, belongs to the large family of Carbonic Anhydrases (CAs). CAs are zinc metalloenzymes that catalyze the interconversion of CO_2 and bicarbonate. CAs are ubiquitous in nature and they play essential roles in all photosynthetic organisms [91]. In plants, CAs are involved in various physiological processes such as photosynthesis, stomatal movement, development, amino acid biosynthesis, metabolism of nitrogen-fixing root nodules and lipid biosynthesis [91]. CAs are also involved in biotic and abiotic stress responses in both monocots and dicots [91]. In particular, many of them have been reported as involved in response against various pathogens and pests [92–94]. Moreover, there are evidences of CAs involvement in plant response to MeJA. Recombinant inbred lines of Arabidopsis resistant to the herbivore insect *Plutella xylostella* showed a limited oxidative stress, due to a 2-fold increase in abundance of AtbCA1 and AtbCA4 proteins [94]. Moreover, a proteomic study demonstrated that CA1 and CA2 from Arabidopsis are strongly up-regulated by MeJA [41].

The importance in restraining oxidative stress induced by (a)biotic cues is emphasized by the presence of ROS scavengers among the priming-regulated proteins disclosed in this study. Biotic and abiotic stresses can induce an oxidative burst, which is followed by rapid changes in hydrogen peroxide (H_2O_2) levels, leading to a variety of physiological responses in plants. Catalases (CATs) and peroxidases (Prxs) are heme enzymes that are able to detoxify H_2O_2, protecting cells from its toxic effects. In our study, a catalase encoded by *Os03g03910* (UniPROT code Q10S82) and a peroxidase encoded by *Os04g0688300* (UniProt code Q7XSU8) were found involved in priming phenomenon. Our peroxidase belongs to class III peroxidases which are glycoproteins located in vacuoles and cell walls [95]. They are part of a large multigenic family with 138 members in rice and 73 members in Arabidopsis [96]. Prxs belong to the PR9 family [97] and are involved in a broad spectrum of physiological processes, probably due to the high number of enzymatic isoforms (isoenzymes) and to the versatility of their enzyme-catalyzed reactions [95]. Indeed, plant Prxs are involved in lignin and suberin formation, cross-linking of cell wall components, auxin metabolism, phytoalexin synthesis and metabolism of ROS [95]. Prxs ability to catalyze the synthesis of bioactive plant products enables them to exert a role in plant defense. For example, Prxs are induced in host plant tissues by pathogen infection and are expressed to limit cellular spreading of the infection through the formation of structural barriers [95]. The stress-induced expression of Prx is conferred by the nature of the 5′ flanking regions of the genes that contain many different potential stress-responsive cis-elements [96]. According with our results, it has been widely reported that JA, MeJA and beneficial microbes with priming effects positively regulate *prx* gene expression [23,53,98].

Catalases (CATs) are major antioxidant enzymes primarily located in peroxisomes that detoxify hydrogen peroxide, produced from various metabolic reactions and environmental stresses, into oxygen and water [99]. Studies indicates that catalases play an important role in plant defense, aging, and senescence [100]. Furthermore, CATs are involved in the resistance of plant cell wall and they also act as a signal for the induction of defense genes playing a crucial role in maintaining active the defensive responses [101]. In rice, three classes of CATs have been identified as CatA, CatB, CatC, which are involved in environmental stress response, root growth, and photorespiration, respectively [102]. Interestingly, CATs are also involved in resistance against insects. It has been demonstrated that aphid resistance in tobacco plants infested with *Bemisia tabaci* nymphs is associated with enhanced antioxidant activities in which CAT may play a dominant role. Moreover, a proteomic study highlighted that CAT2 and CAT3 from Arabidopsis were strongly up-regulated by MeJA [41]. It has been shown that MeJA mediates intra- and inter-plant communications and modulates plant defense responses, including antioxidant systems [103]. In our systems, up-regulation of proteins involved in ROS scavenging corroborate the evidence that priming-induced plant resistance can be triggered by activation of redox-sensitive genes, as previously found [103].

Among the protein significantly involved in the priming phenomenon there is also a "Putative ATP synthase gamma chain 1, chloroplast (H(+)-transporting two-sector ATPase/F(1)-ATPase/ATPC1)" (UniProt code Q84NW1) encoded by *Os07g0513000* gene, belonging to the ATP synthesis-coupled proton transport. ATP synthase is a greatly conserved enzyme catalyzing the synthesis of ATP from ADP and phosphate through a flux of protons over an electrochemical gradient. Interestingly, proteolytic fragments of chloroplastic ATP synthase have been found to mediate plant perception of herbivory through the induction of volatile, phenylpropanoids, protease inhibitors and hormones, including MeJA [104].

By using the freely available STRING program, we unraveled the interaction pattern of proteins involved in the priming phenomenon (Figure 5).

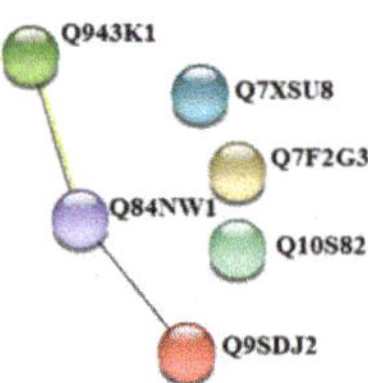

Figure 5. Interaction map of priming-regulated proteins. Network was built by using STRING 10.5 software, at 0.4 confidence level. Prediction was performed on proteins listed in Table 4.

Analyzing the STRING output, we found out that ATP synthase (UniPROT code: Q84NW1) interacts with Magnesium-protoporphyrin IX monomethyl ester [oxidative] cyclase (UniPROT code: Q9SDJ2), encoded by *ZIP1* gene and with a Putative photosystem II subunit PsbS (UniPROT code: Q943K1), encoded by *Os01g0869800* gene. Mg-protoporphyrin IX monomethyl ester cyclase is involved in the chlorophyll biosynthetic pathway [105]. The PsbS protein is a key component in the regulation of non-photochemical quenching (NPQ) in the photosynthesis of higher plant. The PsbS subunit of photosystem II (PSII) plays a crucial role in pH- and xanthophyll-dependent nonphotochemical quenching of excess absorbed light energy, thus contributing to the defense mechanism against photo-oxidative damage [106]. Taken together, our results highlighted that the MeJA priming brings about up-regulation of proteins involved in ROS scavenging and photosynthesis. This suggests an inter-pathway crosstalk between ROS, phytohormone signaling and photosynthesis that allows plants to efficiently respond to stress inputs, as previously reported [107]. It is worthwhile mentioning that the increased photosynthesis rates suggest a boost of primary metabolism probably due to the need of energy and carbon skeletons necessary for the synthesis of secondary metabolites. In general, it has been suggested that alterations in primary metabolism allow the plant to tolerate herbivores while minimizing impact on fitness traits. Therefore, alterations in the levels of key primary metabolites might themselves have the potential for a defensive mode of action.

2.5. MeJA Protects Plants from Effects of Wounding-Triggered H_2O_2 Production

It is widely recognized that biotic and abiotic stresses induce ROS production in plant cells. Under adverse conditions ROS may play two very different roles: activation of signaling leading to defense responses or exacerbating damage. In fact, ROS can have a deleterious effect on cells since they can modify biomolecules such as nucleic acids, proteins, and lipids leading to cell damage and death [108]. At basal level, hydrogen peroxide (H_2O_2), the simplest peroxide recognized as ROS, plays important roles in several developmental and physiological processes When H_2O_2 accumulates in response to biotic and abiotic stresses, it is responsible of several phenomena, including stomatal closure and cell death [109]. Our aim was to highlight the presence of H_2O_2 in wounded rice leaves and to verify if MeJA can affect wounding-dependent H_2O_2 production exerting a priming role. To this end, three-week-old rice leaves from M, P, W and P + W plants were incubated with 2′,7′-dichlorofluorescein diacetate (2′,7′-DCFH$_2$-DA) to detect the presence of H_2O_2. This compound exerts its function into the cytoplasm where is deacetylated by intracellular esterase and subsequently oxidized by H_2O_2 producing the green fluorescent dye dichlorofluorescein (2′,7′-DCF). A negative control is represented by leaves incubated with buffer only. Fluorescence was detected with a confocal microscope. The green fluorescence of the probe was revealed using a 488 filter whereas the auto-fluorescence of the chlorophyll was detected using a 563 nm filter. The experiment was performed three times independently and representative results are shown in Figure 6.

The negative control showed the red fluorescence due to the chlorophyll (Figure 6A). In panel B, wounded leaves exhibited very strong green fluorescence due to 2′,7′-DCF highlighting the presence of high levels of H_2O_2 localized at the level of stomata and vascular tissue. Noteworthy, in P + W green fluorescence was not observed anymore, suggesting a beneficial effect of the hormone in protecting plants from oxidative damage. It has been shown that H_2O_2 participates as a pivotal signal messenger in response to wounding in several species, including rice [110]. It is conceivable that ABA could represents one of the stimuli triggering H_2O_2 production during wounding [111]. The accumulation of H_2O_2 detected in stomata could depend on ABA-induced stomata closure, a phenomenon that exert a protection of leaves against further damage by subsequent threats [108,110]. Additionally, the presence of this compound in the vascular tissue could be due to its nature since it could migrate from the synthesis site to the neighboring vascular tissues or leaves, exerting a defense role against biotic agents [50]. Unfortunately, H_2O_2 levels are usually high under stress conditions so its effect as strong toxic oxidant agent could lead to cell damage or cell death since the ROS scavenging is compromised [112]. These latter considerations strengthen the role of MeJA in ameliorating plant cell life during adverse conditions.

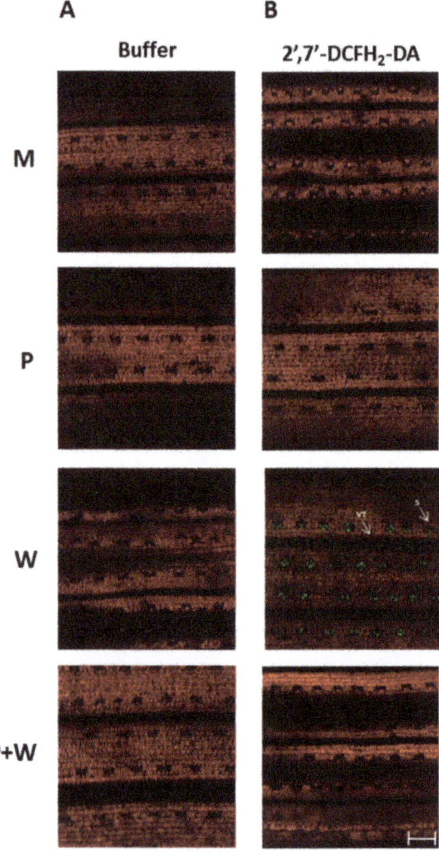

Figure 6. Detection of H_2O_2 in rice leaves using buffer (negative technical control) (**A**) or $2',7'$ DCFH$_2$-DA (**B**). Detection was performed in rice leaves sprayed with a mock solution (M), with 10 μM MeJA solution (P), wounded (W), or sprayed with 10 μM MeJA and wounded (P + W) and harvested at 48 h-post-wounding. Fluorescence was observed under an LSM 710 confocal microscope with Planneofluoar 40/1.30 objective. Two laser excitations lines were used (i.e., 488 for probe detection and 563 nm for chlorophyll autofluorescence). S: stomata. VT: vascular tissue. Bar corresponds to 200 μm.

2.6. MeJA Protects Plants from Photosynthetic Damage

Chlorophyll fluorescence is a non-destructive method used to study plant photosynthetic performance in response to biotic and abiotic stresses [113] In this study, we used the chlorophyll fluorescence imaging tool in order to evaluate the effect of MeJA in wounded and not wounded leaves (Figure 7).

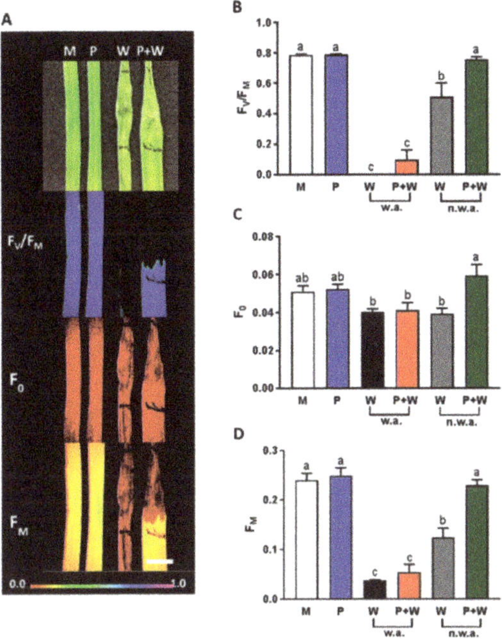

Figure 7. Measurement of chlorophyll fluorescence. Representative images of rice leaves mock (M); primed with MeJA (P); wounded (W); primed and wounded (P + W) in brightfield, maximum quantum efficiency of Photosystem II (F_V/F_M), minimal (F_0) and maximal fluorescence (F_M) (panel **A**). The color bar below shows the range of fluorescence values. Scale bar: 5 mm. Means ($n = 5$) ± SE of F_V/F_M (**B**), F0 (**C**) and FM (**D**) values are shown. In all bar panels, white bars represent M, blue bars P, black bars W (w.a. = wounded area)), red bars P + W (w.a.), grey bars W (n.w.a = near wounded area), green bars P + W (n.w.a.). A one-way ANOVA followed by Tukey's test was performed to define statistical significance ($p < 0.05$) of differences among means. Data not sharing the same letters are statistically significantly different.

To this purpose, three-week-old rice leaves from M, P, W and P + W plants were imaged 48 hours after wounding. We measured the maximum quantum yield of photosynthesis, F_V/F_M, as a plant stress indicator whose decline refers to a compromised photosynthetic performance [114]. The maximum quantum yield of photosynthesis in M and P leaves was around a value of 0.8, indicating healthy leaves (Figure 7B). P + W leaves, compared to W, showed a reduced damaged area (Figure 7A) and a higher maximum quantum yield of photosynthesis (Figure 7B). In Figure 7A, the healthy part of the leaves fluoresced whilst the damaged part appeared dark. In the wounded area (w.a.) F_V/F_M dropped to zero due to very low F_M values comparable to F_0 (Figure 7B–D). In the regions near the wounded area (n.w.a.), P + W showed a F_V/F_M similar to M and P whereas in W it was significantly decreased (Figure 7B).

Other studies already showed a reduced chlorophyll fluorescence in wounded leaves [115] after herbivorous insect attack [116] and fungal infection [117,118]. The decrease in F_V/F_M, in our case, is due to low F_M values. F_M reduction can be related to photoinactivation of photosystem II (PSII) reaction centers or changes in PSII fraction due to modifications in thylakoid membrane structure and organization [119,120].

Our data suggests that MeJA treatment protects PSII reaction centers and maintain structural integrity of chloroplast, as already reported in salt stress conditions [121,122]. Therefore, we can put forward the hypothesis that low doses of MeJA configure the priming condition and that the hormone can exert this action inducing proteins that are able to reduce the damaged area and to protect the photosynthetic system.

3. Materials and Methods

3.1. Plant Material and Treatments

Rice seeds (*Oryza sativa* spp. *Japonica* cv. Carnaroli), supplied by Ente Nazionale Risi (Milano, Italy), were surface sterilized using 10% (*v/v*) H_2O_2 solution for 10 min. Seeds were washed with 70% (*v/v*) EtOH solution for 5 min, and soaked in water overnight. After incubation at 37 °C for 2 days on sterile water-imbibed filter paper, coleoptiles were transferred into alveolar trays and grown in hydroponic culture in Yoshida nutrient solution, in a growth chamber under the following conditions: 28°C, 14 h light / 23°C, 10 h dark, with 60% ± 5% relative humidity.

Three-week-old rice plants (3–4 leaves stage) were sprayed with 10 µM MeJA (Sigma; St. Louis, MO, USA) and 1% (*v/v*) TWEEN 20 solution to induce priming. Mock plants were sprayed with sterile water and 1% (*v/v*) Tween 20 solution only. Each plant was sprayed, making sure that droplets were uniformly distributed. Both mock and primed plants were wounded 24 h after MeJA treatment squeezing the leaves at the base and in the middle with a clamp and scraping the epidermal layer with carborundum in three different areas equally spaced over the length of the leaf. Leaf samples were harvested 48 hours-post-wounding (hpw) and homogenized by grinding with a pestle under continuous addition of liquid nitrogen.

3.2. Protein Sample Preparation

Grinded leaves were suspended in a lysis buffer containing 10% TCA in acetone and 10 mM DTT, left for 2 h at −20°C and then centrifuged at 13500 rpm for 14 min at 4°C. Pellet was washed in acetone, containing 10 mM DTT, 2 mM EDTA and 1 mM PMSF and centrifuged again under the same conditions. The obtained pellet was dried in Speed Vac Concentrator (Savant, ThermoFisher Scientific, Waltham, MA, USA).

Samples were solubilized in 100 mM Tris-HCl, pH 8.5, containing 8 M urea and 7.5 mM DTT, and sonicated by using 2 min-cycles (6 times) at 40 KHz and 4°C. Samples were subsequently centrifuged at 45,000 rpm for 10 min at 4°C. Protein quantification was conducted by BCA assay (Thermo Scientific, Rockford, IL, USA) in triplicate. Disulfide bridges reduction was performed by 10 mM DTT for 45 min at 30°C. Alkylation was obtained by 50 mM 2-iodoacetamide for 20 min, under dark. Protein digestion was performed by treating the diluted samples with two proteases. LysC digestion (LysC Mass spectrometry grade, WAKO, Neuss, Germany) was carried out by using an enzyme to protein ratio of 1:100, with an incubation of 4 h at 37 °C. The resulting digestion mixture was treated with Trypsin (Promega, Fitchburg, WI, USA) by using an enzyme to protein ratio of 1:50, incubating the samples overnight at 37 °C. Reactions were stopped by adding TFA to a final concentration of 0.5% and the mixture was desalted by RP-HPLC with a Zorbax column C18 eluted with a methanol gradient from 2 to 40% in 8 min, at a flow rate of 0.6 mL/min. Eluates were dried in Speed Vac Concentrator (Savant).

3.3. Proteomic Profiling and Data Analysis

LC-MS/MS analyses were conducted with a LTQ-Orbitrap XL mass spectrometer (ThermoFisher Scientific, Waltham, MA, USA) coupled online with a nano-HPLC Ultimate 3000 (Dionex–ThermoFisher Scientific, Waltham, MA, USA) using a 10 cm pico-frit capillary column (75 µm Internal diameter (I.D.), 15 µm tip, New Objective) packed in-house with C18 material (Aeris Peptide 3.6 um XB-C18, Phenomenex, Torrance, CA, USA). Peptides were eluted with a linear gradient from 3 to 50% of acetonitrile containing 0.1% formic acid in 160 min at a flow rate of 250 nL/min. The capillary voltage was set at 1.2 kV and the source temperature at 200°C. For the analysis a full scan at 60,000 resolution on the Orbitrap was followed by MS-MS fragmentation scans on the four most intense ions acquired with collision-induced dissociation (CID) fragmentation in the linear trap (data-dependent acquisition

- DDA). For each analysis, about 1 µg of protein extract was used. Protein identification and quantification was performed by the software MaxQuant [123]. For each analysis, three biological replicates were analyzed. The database used for protein identification was the *Oryza sativa* section of the Uniprot database (version 20150805). Enzyme specificity was set to trypsin with 2 missed cleavages. The mass tolerance window was fixed to 20 ppm for parent mass and to 0.5 Da for fragment ions. Carbamidomethylation of cysteine residues was set as fixed modification and methionine oxidation as variable modification. Proteins were filtered with a false discovery rate (FDR) $\leq$ 0.01. Data from different samples were compared using a T-test with a level of significance of 95% ($p \leq 0.05$). For Gene Ontology analysis, data were analyzed by the PANTHER version 11.0 [124]. The PANTHER classification system combines gene function, ontology, pathways and statistical analysis tools allowing to analyze large-scale data from sequencing, proteomics or gene expression data. PANTHER is based on 82 complete genomes data organized in gene families and subfamilies. Genes are classified according to their function, with families and subfamilies annotated with ontology terms (Gene Ontology (GO) and PANTHER protein classes). STRING analysis (Search Tool for the Retrieval of Interacting Genes/Proteins, http://string.embl.de/) has been carried out using STRING-10.5 server to predict the protein-protein interaction of priming targets [125]. STRING database employs a mixture of prediction approaches and a combination of experimental data (neighborhood, gene fusion, co-expression, experiments, databases, text mining, co-occurrence). Network was completed at 0.4 confidence level.

3.4. ROS Detection in Rice Leaves

ROS detection was performed as previously described [126]. Briefly, H_2O_2 production was revealed by the specific probe 2′,7′-dichlorofluorescein diacetate (DCFH$_2$-DA; Sigma Aldrich, St. Louis, MO, USA), which is rapidly oxidized to highly fluorescent dichlorofluorescein (DCF) in the presence of H_2O_2. Three-week-old rice plants were treated with MeJA or with sterile water (mock), as described above. Both mock and primed plants were wounded 24 h after MeJA treatment. For each treatment, two leaves from five plants were collected at 48 hpw Half number of leaves was incubated in a solution containing 20 mM DCFH$_2$-DA in 10 mM Tris-HCl (pH 7.4) for 45 min under dark. The remaining half leaves was incubated in 10 mM Tris-HCl (pH 7.4) only, under the same conditions (negative technical control). After staining, leaves were washed three times in fresh buffer for 10 min and mounted on slides A LSM 710 confocal microscope (Carl Zeiss Microscopy GmbH, Jena, Germany) with Planneofluoar 40/1.30 objective, was used to detect the fluorescence. Two laser excitations lines were used (i.e., 488 nm for probe detection and 563 nm for chlorophyll auto-fluorescence). Data were managed using Image J software 1.46r (http://rsbweb.nih.gov/ij/) (LOCI, University of Wisconsin, Wisconsin, UW, USA). The experiment was performed three times independently.

3.5. Chlorophyll Fluorescence

Plants were dark-adapted for 30 min before chlorophyll fluorescence measurements. The minimal fluorescence (F_0), maximal fluorescence (F_M) and maximum quantum efficiency of Photosystem II ($F_V/F_M = F_M - F_0/F_M$) were measured in single leaves using an Imaging Pam M-series fluorimeter (Heinz Walz GmbH, Effeltrich, Germany) [116]. ImagingWin software (Heinz Walz GmbH, Effeltrich, Germany) allowed to select regions of interest in wounded leaves and to refer measurements to wounded (w.a.) and near wounded areas (n.w.a.). Data are shown as means ± standard errors (SEs). The normality of data distribution was tested using the Shapiro–Wilk Normality Test. Significant differences ($p < 0.05$) were analyzed using a one-way analysis of variance (ANOVA) followed by Tukey post-hoc test. SigmaPlot was used for the analysis (Systat Software Inc., San Jose, CA, USA).

4. Conclusions

Priming encompasses accumulation of latent signaling components that are quickly activated when plants are exposed to a stress. Therefore, it is interesting to exploit comparative proteomic analysis in plants treated with chemical priming agents before they encounter stress conditions.

Our results strengthen the awareness that LC-MS/MS-based proteomic approach is an exceptional analytical tool for a better understanding of plant defense molecular mechanisms and of the proteome reprogramming modulated by different treatments. Using this approach, we highlighted proteins involved in ROS signaling and photosynthesis that could cooperate in regulating priming-dependent defense responses. Some of the currently identified proteins had previously been shown to play a role in defense responses; however, our study revealed a role of MeJA-priming in protecting rice plants from mechanical damages. In the future, it would be interesting to further investigate the exact role of these proteins in priming phenomenon.

Supplementary Materials: Supplementary materials can be found at http://www.mdpi.com/1422-0067/20/10/2525/s1. Table S1. List of proteins identified by MaxQuant analysis, using the database Uniprot, taxonomy *Oryza sativa* (version 20150805). Enzyme specificity was set to trypsin with 2 missed cleavages. The mass tolerance window was set to 20 ppm for parent mass and to 0.5 Da for fragment ions. Carbamidomethylation of cysteine residues was set as fixed modification and methionine oxidation as variable modification. Content of columns: A, identified proteins by Entry name; B, identified proteins and other relevant information for the identification; C-F, unique peptides media relative to the three replicated of samples M, W, P, P + W; G-I, label-free intensities media relative to three replicates, standard deviation and percentage of coefficient variation of sample M; J-L, label-free intensities media relative to three replicates, standard deviation and percentage of coefficient variation of sample W; M-O, label-free intensities media relative to three replicates, standard deviation and percentage of coefficient variation of sample P; P-R, label-free intensities media relative to three replicates, standard deviation and percentage of coefficient variation of sample P + W.

Author Contributions: Conceptualization, L.B., C.C. and P.P.d.L.; Methodology, L.B., S.P., L.P., S.P. and G.A.; Validation, L.B., S.P. and L.P.; Formal Analysis, L.B., S.P., G.A., L.P. and S.P.; Investigation, L.B., S.P. and L.P.; Resources, C.C. and P.P.d.L.; Data Curation, S.P., L.P. and L.B.; Writing—Original Draft Preparation, S.P; Writing—Review and Editing, S.P., L.B., C.C. and P.P.d.L.; Visualization, S.P., L.B., C.C. and P.P.d.L.; Supervision, C.C. and P.P.D.L.; Project Administration, C.C.; Funding Acquisition, C.C.

Funding: This research was founded by Ministero dell'Istruzione, Università e Ricerca Scientifica (MIUR), project MIUR_FIRB of CNR-IGV, Consiglio Nazionale delle Ricerche-Istituto di Genetica Vegetale, Portici, Italy (grant number RBNE01KZE7) and by MIUR-PNRA (Programma Nazionale Ricerche in Antartide) (grant number PNRA16_00068).

Conflicts of Interest: The authors declare no conflict of interest.

Abbreviations

MeJA	Methyl-jasmonate
M	Mock
P	MeJA- treated plants
W	Wounded plants
P + W	Wounded plants after MeJA treatment

References

1. Rasmann, S.; De Vos, M.; Jander, G. Ecological role of transgenerational resistance against biotic threats. *Plant Signal. Behav.* **2012**, *7*, 447–449. [CrossRef] [PubMed]
2. Frost, C.J.; Mescher, M.C.; Carlson, J.E.; De Moraes, C.M. Plant defense priming against herbivores: Getting ready for a different battle. *Plant Physiol.* **2008**, *146*, 818–824. [CrossRef] [PubMed]
3. Aranega-Bou, P.; De la, O.; Leyva, M.; Finiti, I.; García-Agustín, P.; González-Bosch, C. Priming of plant resistance by natural compounds. Hexanoic acid as a model. *Front. Plant Sci.* **2014**, *5*, 488. [CrossRef]
4. Howe, G.A.; Jander, G. Plant immunity to insect herbivores. *Annu. Rev. Plant Biol.* **2008**, *59*, 41–66. [CrossRef]
5. Jones, J.D.G.; Dangl, J.L. The plant immune system. *Nature* **2006**, *444*, 323–329. [CrossRef] [PubMed]
6. Pieterse, C.M.J.; Van der Does, D.; Zamioudis, C.; Leon-Reyes, A.; Van Wees, S.C.M. Hormonal modulation of plant immunity. *Annu. Rev. Cell Dev. Biol.* **2012**, *28*, 489–521. [CrossRef]
7. Robert-Seilaniantz, A.; Grant, M.; Jones, J.D.G. Hormone crosstalk in plant disease and defense:more than just JASMONATE-SALICYLATE antagonism. *Annu. Rev. Phytopathol.* **2011**, *49*, 317–343. [CrossRef]
8. Baxter, A.; Mittler, R.; Suzuki, N. ROS as key players in plant stress signaling. *J. Exp. Bot.* **2013**, *65*, 1229–1240. [CrossRef] [PubMed]

9. Vos, I.A.; Pieterse, C.M.J.; Van Wees, S.C.M. Costs and benefits of hormone-regulated plant defences. *Plant Pathol.* **2013**, *62*, 43–55.

10. Howe, G.A.; Lightner, J.; Browse, J.; Ryan, C.A. An octadecanoid pathway mutant (JL5) of tomato is compromised in signaling for defense against insect attack. *Plant Cell* **1996**, *8*, 2067–2077. [CrossRef] [PubMed]

11. McConn, M.; Creelman, R.A.; Bell, E.; Mullet, J.E.; Browse, J. Jasmonate is essential for insect defense in Arabidopsis. *Proc. Natl. Acad. Sci. USA* **1997**, *94*, 5473–5477. [CrossRef] [PubMed]

12. Li, L.; Li, C.; Lee, G.I.; Howe, G.A. Distinct roles for jasmonate synthesis and action in the systemic wound response of tomato. *Proc. Natl. Acad. Sci. USA* **2002**, *99*, 6416–6421. [CrossRef] [PubMed]

13. Lee, G.I.; Howe, G.A. The tomato mutant *spr1* is defective in systemin perception and the production of a systemic wound signal for defense gene expression. *Plant J.* **2003**, *33*, 567–576. [CrossRef]

14. Conrath, U. Systemic Acquired Resistance. *Plant Signal. Behav.* **2006**, *1*, 179–184. [CrossRef] [PubMed]

15. Martinez-Medina, A.; Flors, V.; Heil, M.; Mauch-Mani, B.; Pieterse, C.M.J.; Pozo, M.J.; Ton, J.; van Dam, N.M.; Conrath, U. Recognizing plant defense priming. *Trends Plant Sci.* **2016**, *21*, 818–822. [CrossRef]

16. Conrath, U. Molecular aspects of defence priming. *Trends Plant Sci.* **2011**, *16*, 524–531. [CrossRef]

17. Jaskiewicz, M.; Conrath, U.; Peterhänsel, C. Chromatin modification acts as a memory for systemic acquired resistance in the plant stress response. *EMBO Rep.* **2011**, *12*, 50–55. [CrossRef]

18. Slaughter, A.; Daniel, X.; Flors, V.; Luna, E.; Hohn, B.; Mauch-Mani, B. Descendants of primed Arabidopsis plants exhibit resistance to biotic stress. *Plant Physiol.* **2012**, *158*, 835–843. [CrossRef]

19. Tanou, G.; Fotopoulos, V.; Molassiotis, A. Priming against environmental challenges and proteomics in plants: Update and agricultural perspectives. *Front. Plant Sci.* **2012**, *3*, 216. [CrossRef]

20. Conrath, U.; Beckers, G.J.M.; Langenbach, C.J.G.; Jaskiewicz, M.R. Priming for enhanced defense. *Annu. Rev. Phytopathol.* **2015**, *53*, 97–119. [CrossRef]

21. Bäurle, I. Plant Heat Adaptation: Priming in response to heat stress. *F1000Research* **2016**, *5*. [CrossRef] [PubMed]

22. Švecová, E.; Proietti, S.; Caruso, C.; Colla, G.; Crinò, P. Antifungal activity of Vitex agnus-castus extract against Pythium ultimum in tomato. *Crop Prot.* **2013**, *43*, 223–230. [CrossRef]

23. De Palma, M.; D'Agostino, N.; Proietti, S.; Bertini, L.; Lorito, M.; Ruocco, M.; Caruso, C.; Chiusano, M.L.; Tucci, M. Suppression Subtractive Hybridization analysis provides new insights into the tomato (*Solanum lycopersicum* L.) response to the plant probiotic microorganism *Trichoderma longibrachiatum* MK1. *J. Plant Physiol.* **2016**, *190*, 79–94. [CrossRef]

24. Beckers, G.J.M.; Conrath, U. Priming for stress resistance: From the lab to the field. *Curr. Opin. Plant Biol.* **2007**, *10*, 425–431. [CrossRef]

25. Beckers, G.J.; Jaskiewicz, M.; Liu, Y.; Underwood, W.R.; He, S.Y.; Zhang, S.; Conrath, U. Mitogen-activated protein kinases 3 and 6 are required for full priming of stress responses in *Arabidopsis thaliana*. *Plant Cell* **2009**, *21*, 944–953. [CrossRef] [PubMed]

26. Pastor, V.; Luna, E.; Ton, J.; Cerezo García, M.; García Agustín, P.; Flors, V. Fine tuning of reactive oxygen species homeostasis regulates primed immune responses in Arabidopsis. *Mol. Plant Microbe Interact.* **2013**, *26*, 1334–1344. [CrossRef]

27. Gamir, J.; Sánchez-Bel, P.; Flors, V. Molecular and physiological stages of priming: How plants prepare for environmental challenges. *Plant Cell Rep.* **2014**, *33*, 1935–1949. [CrossRef]

28. Ton, J.; Mauch-Mani, B. Amino-butyric acid-induced resistance against necrotrophic pathogens is based on ABA-dependent priming for callose. *Plant J.* **2004**, *38*, 119–130. [CrossRef] [PubMed]

29. Taheri, P.; Tarighi, S. Riboflavin induces resistance in rice against *Rhizoctonia solani* via jasmonate-mediated priming of phenylpropanoid pathway. *J. Plant Physiol.* **2010**, *167*, 201–208. [CrossRef]

30. Luzzatto, T.; Golan, A.; Yishay, M.; Bilkis, I.; Ben-Ari, J.; Yedidia, I. Priming of antimicrobial phenolics during induced resistance response towards *Pectobacterium carotovorum* in the ornamental monocot calla lily. *J. Agric. Food Chem.* **2007**, *55*, 10315–10322. [CrossRef]

31. Ding, Y.; Fromm, M.; Avramova, Z. Multiple exposures to drought 'train' transcriptional responses in *Arabidopsis. Nat. Commun.* **2012**, *3*, 740. [CrossRef]

32. Sani, E.; Herzyk, P.; Perrella, G.; Colot, V.; Amtmann, A. Hyperosmotic priming of *Arabidopsis* seedlings establishes a long-term somatic memory accompanied by specific changes of the epigenome. *Genome Biol.* **2013**, *14*, R59. [CrossRef]

33. Conrath, U.; Beckers, G.J.M.; Flors, V.; García-Agustín, P.; Jakab, G.; Mauch, F.; Newman, M.A.; Pieterse, C.M.; Poinssot, B.; Pozo, M.J.; et al. Priming: Getting ready for battle. *Mol. Plant Microbe Interact.* **2006**, *19*, 1062–1071. [CrossRef]

34. Bruce, T.J.A.; Matthes, M.C.; Napier, J.A.; Pickett, J.A. Stressful 'memories' of plants: Evidence and possible mechanisms. *Plant Sci.* **2007**, *173*, 603–608. [CrossRef]

35. Luna, E.; Bruce, T.J.A.; Roberts, M.R.; Flors, V.; Ton, J. Next-generation systemic acquired resistance. *Plant Physiol.* **2012**, *158*, 317–327. [CrossRef]

36. Kathiria, P.; Sidler, C.; Golubov, A.; Kalischuk, M.; Kawchuk, L.M.; Kovalchuk, I. Tobacco mosaic virus infection results in an increase in recombination frequency and resistance to viral, bacterial, and fungal pathogens in the progeny of infected tobacco plants. *Plant Physiol.* **2010**, *153*, 1859–1870. [CrossRef]

37. Liu, Y.; Bino, R.J.; Van der Burg, W.J.; Groot, S.P.C.; Hilhorst, H.W.M. Effects of osmotic priming on dormancy and storability of tomato (*Lycopersicon esculentum* Mill.) seeds. *Seed Sci. Res.* **1996**, *6*, 49–55. [CrossRef]

38. Jisha, K.C.; Vijayakumari, K.; Puthur, J.T. Seed priming for abiotic stress tolerance: An overview. *Acta Physiol. Plant* **2013**, *35*, 1381–1396. [CrossRef]

39. Paparella, S.; Arau, J.S.S.; Rossi, G.; Wijayasinghe, M.; Carbonera, D.; Balestrazzi, A. Seed priming: State of the art and new perspectives. *Plant Cell Rep.* **2015**, *34*, 1281–1293. [CrossRef]

40. Hussain, S.; Yin, H.; Peng, S.; Khan, F.A.; Khan, F.; Sameeullah, M.; Hussain, H.A.; Huang, J.; Cui, K.; Nie, L. Comparative transcriptional profiling of primed and non-primed rice seedlings under submergence stress. *Front. Plant Sci.* **2016**, *7*, 1125. [CrossRef]

41. Proietti, S.; Bertini, L.; Timperio, A.M.; Zolla, L.; Caporale, C.; Caruso, C. Crosstalk between salicylic acid and jasmonate in Arabidopsis investigated by an integrated proteomic and transcriptomic approach. *Mol. Biosyst.* **2013**, *9*, 1169–1187. [CrossRef]

42. Kushalappa, A.C.; Gunnaiah, R. Metabolo-proteomics to discover plant biotic stress resistance genes. *Trends Plant Sci.* **2013**, *18*, 522–531. [CrossRef]

43. Obata, T.; Fernie, A.R. The use of metabolomics to dissect plant responses to abiotic stresses. *Cell Mol. Life Sci.* **2012**, *69*, 3225–3243. [CrossRef]

44. Gupta, B.; Saha, J.; Sengupta, A.; Gupta, K. Plant Abiotic Stress: 'Omics' Approach. *J. Plant Biochem. Physiol.* **2013**, *1*, e108. [CrossRef]

45. Cramer, G.R.; Urano, K.; Delrot, S.; Pezzotti, M.; Shinozaki, K. Effects of abiotic stress on plants: A systems biology perspective. *BMC Plant Biol.* **2011**, *11*, 163. [CrossRef]

46. Li, G.; Wu, Y.; Liu, G.; Xiao, X.; Wang, P.; Gao, T.; Xu, M.; Han, Q.; Wang, Y.; Guo, T.; Kang, G. Large-scale Proteomics Combined with Transgenic Experiments Demonstrates an Important Role of Jasmonic Acid in Potassium Deficiency Response in Wheat and Rice. *Mol. Cell. Proteom.* **2017**, *16*, 1889–1905. [CrossRef]

47. Zhang, Y.T.; Zhang, Y.L.; Chen, S.X.; Yin, G.H.; Yang, Z.Z.; Lee, S.; Liu, C.G.; Zhao, D.D.; Ma, Y.K.; Song, F.Q.; et al. Proteomics of methyl jasmonate induced defense response in maize leaves against Asian corn borer. *BMC Genom.* **2015**, *16*, 224. [CrossRef]

48. Dhakarey, R.; Raorane, M.L.; Treumann, A.; Peethambaran, P.K.; Schendel, R.R.; Sahi, V.P.; Hause, B.; Bunzel, M.; Henry, A.; Kohli, A.; et al. Physiological and Proteomic Analysis of the Rice Mutant cpm2 Suggests a Negative Regulatory Role of Jasmonic Acid in Drought Tolerance. *Front Plant Sci.* **2017**, *8*, 1903. [CrossRef]

49. Li, Y.; Nie, Y.; Zhang, Z.; Ye, Z.; Zou, X.; Zhang, L.; Wang, Z. Comparative proteomic analysis of methyl jasmonate-induced defense responses in different rice cultivars. *Proteomics* **2014**, *14*, 1088–1101. [CrossRef]

50. Si, T.; Wang, X.; Wu, L.; Zhao, C.; Zhang, L.; Huang, M.; Cai, J.; Zhou, Q.; Dai, T.; Zhu, J.K.; et al. Nitric Oxide and Hydrogen Peroxide Mediate Wounding-Induced Freezing Tolerance through Modifications in Photosystem and Antioxidant System in Wheat. *Front. Plant Sci.* **2017**, *8*, 1284. [CrossRef]

51. Savatin, D.V.; Gramegna, G.; Modesti, V.; Cervone, F. Wounding in the plant tissue: The defense of a dangerous passage. *Front Plant Sci.* **2014**, *5*, 470. [CrossRef] [PubMed]

52. Pauwels, L.; Morreel, K.; De Witte, E.; Lammertyn, F.; Van Montagu, M.; Boerjan, W.; Inzé, D.; Goossens, A. Mapping methyl jasmonate-mediated transcriptional reprogramming of metabolism and cell cycle progression in cultured Arabidopsis cells. *Proc. Natl. Acad. Sci. USA* **2008**, *105*, 1380–1385. [CrossRef] [PubMed]

53. Bertini, L.; Proietti, S.; Focaracci, F.; Sabatini, B.; Caruso, C. Epigenetic control of defense genes following MeJA-induced priming in rice (O. sativa). *J. Plant Physiol.* **2018**, *228*, 166–177.

54. Hickman, R.J.; Van Verk, M.C.; Van Dijken, A.J.H.; Pereira Mendes, M.; Vroegop-Vos, I.A.; Caarls, L.; Steenbergen, M.; Van der Nagel, I.; Wesselink, G.J.; Jironkin, A.; et al. Architecture and dynamics of the jasmonic acid gene regulatory network. *Plant Cell* **2017**, *29*, 2086–2105. [CrossRef]

55. Tobias, C.M.; Chow, E.K. Structure of the cinnamyl-alcohol dehydrogenase gene family in rice and promoter activity of a member associated with lignification. *Planta* **2005**, *220*, 678–688. [CrossRef] [PubMed]

56. Mitchell, H.J.; Hall, J.L.; Barber, M.S. Elicitor-induced cinnamyl alcohol dehydrogenase activity in lignifying wheat (*Triticum aestivum* L.) leaves. *Plant Physiol.* **1994**, *104*, 551–556. [CrossRef]

57. Lynch, D.; Lidgett, A.; McInnes, R.; Huxley, H.; Jones, E.; Mahoney, N.; Spangenberg, G. Isolation and characterisation of three cinnamyl alcohol dehydrogenase homologue cDNAs from perennial ryegrass (*Lolium perenne* L.). *J. Plant Physiol.* **2002**, *159*, 653–660.

58. Davidson, R.M.; Manosalva, P.M.; Snelling, J.; Bruce, M.; Leung, H.; Leach, J.E. Rice germin-like proteins: Allelic diversity and relationships to early stress responses. *Rice* **2010**, *3*, 43–55. [CrossRef]

59. Grzelczak, Z.F.; Lane, B.G. Signal resistance of a soluble protein to enzymic proteolysis: An unorthodox approach to the isolation and purification of germin, a rare growth-related protein. *Can. J. Biochem. Cell Biol.* **1984**, *62*, 1351–1353. [CrossRef]

60. Bernier, F.; Berna, A. Germins and germin-like proteins: Plant do-all proteins. But what do they do exactly? *Plant Physiol. Biochem.* **2001**, *39*, 545–554. [CrossRef]

61. Lane, B.G. Oxalate, germins, and higher-plant pathogens. *IUBMB Life* **2002**, *53*, 67–75. [CrossRef]

62. Dunwell, J.M.; Khuri, S.; Gane, P.J. Microbial relatives of the seed storage proteins of higher plants: Conservation of structure and diversification of function during evolution of the cupin superfamily. *Microbiol. Mol. Biol. Rev.* **2000**, *64*, 153–179. [CrossRef]

63. Druka, A.; Kudrna, D.; Kannangara, C.G.; Wettstein, D.V.; Kleinhofs, A. Physical and genetic mapping of barley (*Hordeum vulgare*) germin-like cDNAs. *Proc. Natl. Acad. Sci. USA* **2002**, *99*, 850–855. [CrossRef]

64. Lane, B.G.; Dunwell, J.M.; Ray, J.A.; Schmitt, M.R.; Cuming, A.C. Germin, a protein marker of early plant development, is an oxalate oxidase. *J. Biol. Chem.* **1993**, *268*, 12239–12242.

65. Lane, B.G. Oxalate oxidases and differentiating surface structure in wheat: Germins. *Biochem. J.* **2000**, *349*, 309–321. [CrossRef]

66. Lamb, C.; Dixon, R.A. The oxidative burst in plant disease resistance. *Annu. Rev. Plant Physiol. Plant Mol. Biol.* **1997**, *48*, 251–275. [CrossRef] [PubMed]

67. Lou, Y.; Baldwin, I.T. Silencing of a Germin-Like gene in *Nicotiana attenuate* improves performance of native herbivores. *Plant Physiol.* **2006**, *140*, 1126–1136. [CrossRef] [PubMed]

68. Van Loon, L.C.; Bakker, P.A.H.M.; Pieterse, C.M.J. Systemic resistance induced by rhizosphere bacteria. *Annu. Rev. Phytopathol.* **1998**, *36*, 453–483. [CrossRef]

69. Moons, A.; Prinsen, E.; Bauw, G.; Van Montagu, M. Antagonistic effects of abscisic acid and jasmonates on salt stress-inducible transcripts in rice roots. *Plant Cell* **1997**, *9*, 2243–2259. [CrossRef]

70. Rakwal, R.; Komatsu, S. Role of jasmonate in the rice (*Oryza sativa* L.) self-defense mechanism using proteome analysis. *Electrophoresis* **2000**, *21*, 2492–2500. [CrossRef]

71. Miché, L.; Battistoni, F.; Gemmer, S.; Belghazi, M.; Reinhold-Hurek, B. Upregulation of jasmonate-inducible defense proteins and differential colonization of roots of *Oryza sativa* cultivars with the endophyte *Azoarcus* sp. *Mol. Plant Microb. Interact.* **2006**, *19*, 502–511. [CrossRef]

72. Xie, X.Z.; Xue, Y.J.; Zhou, J.J.; Zhang, B.; Chang, H.; Takano, M. Phytochromes regulate SA and JA signaling pathways in rice and are required for developmentally controlled resistance to *Magnaporthe grisea*. *Mol. Plant* **2011**, *4*, 688–696. [CrossRef]

73. Cohen-Kupiec, R.; Chet, I. The molecular biology of chitin digestion. *Curr. Opin. Biotechnol.* **1998**, *9*, 270–277. [CrossRef]

74. Guo, X.L.; Bai, L.R.; Su, C.Q.; Shi, L.R.; Wang, D.W. Molecular cloning and expression of drought-induced protein 3 (DIP3) encoding a class III chitinase in upland rice. *Genet. Mol. Res.* **2013**, *12*, 6860–6870. [CrossRef]

75. Schlumbaum, A.; Mauch, F.; Vögeli, U.; Boller, T. Plant chitinases are potent inhibitors of fungal growth. *Nature* **1986**, *12*, 6860–6870. [CrossRef]

76. Taira, T.; Toma, N.; Ishihara, M. Purification, characterization, and antifungal activity of chitinases from pineapple (*Ananas comosus*) leaf. *Biosci. Biotechnol. Biochem.* **2005**, *69*, 189–196. [CrossRef]

77. Collinge, D.B.; Kragh, K.M.; Mikkelsen, J.D.; Nielsen, K.K.; Rasmussen, U.; Vad, K. Plant chitinases. *Plant J.* **1993**, *3*, 31–40. [CrossRef]

78. Schweizer, P.; Buchala, A.; Silverman, P.; Seskar, M.; Raskin, I.; Métraux, J.-P. Jasmonate-inducible genes are activated in rice by pathogen attack without a concomitant increase in endogenous jasmonic acid levels. *Plant Physiol.* **1997**, *114*, 79–88. [CrossRef]

79. Byeon, Y.; Choi, G.-H.; Lee, H.Y.; Back, K. Melatonin biosynthesis requires N-acetylserotonin methyltransferase activity of caffeic acid O-methyltransferase in rice. *J. Exp. Bot.* **2015**, *66*, 6917–6925. [CrossRef]

80. Tan, D.X. Preface: Melatonin and plants. *J. Exp. Bot.* **2015**, *66*, 625–626. [CrossRef]

81. Lee, H.Y.; Byeon, Y.; Back, K. Melatonin as a signal molecule triggering defense responses against pathogen attack in *Arabidopsis* and tobacco. *J. Pineal. Res.* **2014**, *57*, 262–268. [CrossRef]

82. Kim, S.; Cho, K.S.; Kim, S.G.; Kang, S.Y.; Kang, K.Y. A rice isoflavone reductase-like gene, OsIRL, is induced by rice blast fungal elicitor. *Mol. Cells* **2003**, *16*, 224–231.

83. Adie, B.A.T.; Perez-Perez, J.; Perez-Perez, M.M.; Godoy, M.; Sanchez-Serrano, J.J.; Schmelz, E.A.; Solano, R. ABA is an essential signal for plant resistance to pathogens affecting JA biosynthesis and the activation of defenses in *Arabidopsis*. *Plant Cell* **2007**, *19*, 1665–1681. [CrossRef] [PubMed]

84. Bodenhausen, N.; Reymond, P. Signaling pathways controlling induced resistance to insect herbivores in Arabidopsis. *Mol. Plant Microb. Interact.* **2007**, *20*, 1406–1420. [CrossRef]

85. Mitsuhara, I.; Iwai, T.; Seo, S.; Yanagawa, Y.; Kawahigasi, H.; Hirose, S.; Ohkawa, Y.; Ohashi, Y. Characteristic expression of twelve rice PR1 family genes in response to pathogen infection, wounding, and defense-related signal compounds (121/180). *Mol. Genet. Genomics* **2008**, *279*, 415–427. [CrossRef]

86. Duff, S.M.G.; Gautam, S.; Plaxton, W.C. The role of acid phosphatases in plant phosphorus metabolism. *Physiol. Plant* **1994**, *90*, 791–800. [CrossRef]

87. Del Pozo, J.C.; Allona, I.; Rubio, V.; Leyva, A.; de la Peña, A.; Aragoncillo, C.; Paz-Ares, J. A type 5 acid phosphatase gene from *Arabidopsis thaliana* is induced by phosphate starvation and by some other types of phosphate mobilising/oxidative stress conditions. *Plant J.* **1999**, *19*, 579–589. [CrossRef]

88. Liu, Y.; Ahn, J.-E.; Datta, S.; Salzman, R.A.; Moon, J.; Huyghues-Despointes, B.; Pittendrigh, B.; Murdock, L.L.; Koiwa, H.; Zhu-Salzman, K. Arabidopsis Vegetative Storage protein is an anti-insect acid phosphatase. *Plant Physiol.* **2005**, *139*, 1545–1556. [CrossRef]

89. Hruz, T.; Laule, O.; Szabo, G.; Wessendorp, F.; Bleuler, S.; Oertle, L.; Widmayer, P.; Gruissem, W.; Zimmermann, P. Genevestigator V3: A reference expression database for the meta-analysis of transcriptomes. *Adv. Bioinform.* **2008**, *2008*, 420747. [CrossRef]

90. Tonfack, L.B.; Moummou, H.; Latché, A.; Youmbi, E.; Benichou, M.; Pech, J.C.; Van Der Rest, B. The plant SDR superfamily: Involvement in primary and secondary metabolism. *Curr. Top. Plant Biol.* **2011**, *12*, 41–53.

91. Di Mario, R.J.; Clayton, H.; Mukherjee, A.; Ludwig, M.; Moroney, J.V. Plant Carbonic Anhydrases: Structures, Locations, Evolution, and Physiological Roles. *Mol. Plant* **2017**, *10*, 30–46. [CrossRef]

92. Slaymaker, D.H.; Navarre, D.A.; Clark, D.; del Pozo, O.; Martin, G.B.; Klessig, D.F. The tobacco salicylic acid-binding protein 3 (SABP3) is the chloroplast carbonic anhydrase, which exhibits antioxidant activity and plays a role in the hypersensitive defense response. *Proc. Natl. Acad. Sci. USA* **2002**, *99*, 11640–11645. [CrossRef]

93. Restrepo, S.; Myers, K.L.; del Pozo, O.; Martin, G.B.; Hart, A.L.; Buell, C.R.; Fry, W.E.; Smart, C.D. Gene profiling of a compatible interaction between *Phytophthora infestans* and *Solanum tuberosum* suggests a role for carbonic anhydrase. *Mol. Plant Microb. Interact.* **2005**, *18*, 913–922. [CrossRef]

94. Collins, R.M.; Afzal, M.; Ward, D.A.; Prescott, M.C.; Sait, S.M.; Rees, H.H.; Tomsett, A.B. Differential proteomic analysis of *Arabidopsis thaliana* genotypes exhibiting resistance or susceptibility to the insect herbivore, *Plutella xylostella*. *PLoS ONE* **2010**, *5*, e10103. [CrossRef] [PubMed]

95. Passardi, F.; Cosio, C.; Penel, C.; Dunand, C. Peroxidases have more functions than a Swiss army knife. *Plant Cell Rep.* **2005**, *24*, 255–265. [CrossRef]

96. Sasaki, K.; Yuichi, O.; Hiraga, S.; Gotoh, Y.; Seo, S.; Mitsuhara, I.; Ito, H.; Matsui, H.; Ohashi, Y. Characterization of two rice peroxidase promoters that respond to blast fungus-infection. *Mol. Genet. Genomics* **2007**, *278*, 709–722. [CrossRef]

97. Van Loon, L.C.; Rep, M.; Pieterse, C.M.J. Significance of inducible defense-related proteins in infected plants. *Annu. Rev. Phytopathol.* **2006**, *44*, 135–162. [CrossRef]

98. Almagro, L.; Gómez Ros, L.V.; Belchi-Navarro, S.; Bru, R.; Ros Barceló, A.; Pedreño, M.A. Class III peroxidases in plant defence reactions. *J. Exp. Bot.* **2009**, *60*, 377–390. [CrossRef]

99. Willekens, H.; Inze, D.; Van Montagu, M.; Van Camp, W. Catalases in plants. *Mol. Breed.* **1995**, *1*, 207–228. [CrossRef]

100. Yang, T.; Poovaiah, B.W. Hydrogen peroxide homeostasis: Activation of plant catalases by calcium/calmodulin. *Proc. Natl. Acad. Sci. USA* **2002**, *99*, 4097–4102. [CrossRef]

101. Zhao, H.; Sun, X.; Xue, M.; Zhang, X.; Li, Q. Antioxidant enzyme responses induced by whiteflies in tobacco plants in defense against aphids: Catalase may play a dominant role. *PLoS ONE* **2016**, *11*, e0165454. [CrossRef] [PubMed]

102. Joo, J.; Lee, Y.H.; Song, S.I. Rice CatA, CatB, and CatC are involved in environmental stress response, root growth, and photorespiration, respectively. *J. Plant Biol.* **2014**, *57*, 375–382. [CrossRef]

103. González-Bosch, C. Priming plant resistance by activation of redox-sensitive genes. *Free Radic. Biol. Med.* **2018**, *122*, 171–180. [CrossRef]

104. Schmelz, E.A.; Carroll, M.J.; LeClere, S.; Phipps, S.M.; Meredith, J.; Chourey, P.S.; Alborn, H.T.; Teal, P.E. Fragments of ATP synthase mediate plant perception of insect attack. *Proc. Natl. Acad. Sci. USA* **2006**, *103*, 8894–8899. [CrossRef] [PubMed]

105. Bollivar, D.W.; Beale, S.I. The Chlorophyll Biosynthetic Enzyme Mg-Protoporphyrin IX Monomethyl Ester (Oxidative) Cyclase (Characterization and Partial Purification from *Chlamydomonas reinhardtii* and *Synechocystis* sp. PCC 6803). *Plant Physiol.* **1996**, *112*, 105–114. [CrossRef]

106. Bergantino, E.; Segalla, A.; Brunetta, A.; Teardo, E.; Rigoni, F.; Giacometti, G.M. Light- and pH-dependent structural changes in the PsbS subunit of photosystem II. *Proc. Natl. Acad. Sci. USA* **2003**, *100*, 15265–15270. [CrossRef] [PubMed]

107. Fujita, M.; Fujita, Y.; Noutoshi, Y.; Takahashi, F.; Narusaka, Y.; Yamaguchi-Shinozaji, K. Cross talk between abiotic and biotic stress responses: A current view from the point of convergence in the stress signalling network. *Curr. Opin. Plant Biol.* **2006**, *9*, 436–442. [CrossRef]

108. Yi, H.; Liu, X.; Yi, M.; Chen, G. Dual Role of Hydrogen Peroxide in Arabidopsis Guard Cells in Response to Sulfur Dioxide. *Adv. Toxicol.* **2014**, *2014*, 9. [CrossRef]

109. Niu, L.; Liao, W. Hydrogen Peroxide Signaling in Plant Development and Abiotic Responses: Crosstalk with Nitric Oxide and Calcium. *Front. Plant Sci.* **2016**, *7*, 230. [CrossRef]

110. Orozco-Cardenas, M.; Ryan, C.A. Hydrogen peroxide is generated systemically in plant leaves by wounding and systemin via the octadecanoid pathway. *Proc. Natl. Acad. Sci. USA* **1999**, *96*, 6553–6557. [CrossRef]

111. Zhang, X.; Zhang, L.; Dong, F.; Gao, J.; Galbraith, D.W.; Song, C.-P. Hydrogen Peroxide Is Involved in Abscisic Acid-Induced Stomatal Closure in *Vicia faba*. *Plant Physiol.* **2001**, *126*, 1438–1448. [CrossRef]

112. Das, K.; Roychoudhury, A. Reactive oxygen species (ROS) and response of antioxidants as ROS-scavengers during environmental stress in plants. *Front. Environ. Sci.* **2014**, *2*, 53. [CrossRef]

113. Baker, N.R. Chlorophyll Fluorescence: A probe of photosynthesis in vivo. *Annu. Rev. Plant Biol.* **2008**, *59*, 89–113. [CrossRef]

114. Maxwell, K.; Johnson, G.N. Chlorophyll fluorescence-a practical guide. *J. Exp. Bot.* **2000**, *51*, 659–668. [CrossRef]

115. Quilliam, R.S.; Swarbrick, P.J.; Scholes, J.D.; Rolfe, S.A. Imaging photosynthesis in wounded leaves of Arabidopsis thaliana. *J. Exp. Bot.* **2006**, *57*, 55–69. [CrossRef]

116. Tang, J.Y.; Zielinski, R.E.; Zangerl, A.R.; Crofts, A.R.; May, R.; Berenbaum, M.R.; DeLucia, E.H. The differential effects of herbivory by first and fourth instars of *Trichoplusia ni* (Lepidoptera: Noctuidae) on photosynthesis in *Arabidopsis thaliana*. *J. Exp. Bot.* **2006**, *57*, 527–536. [CrossRef]

117. Scholes, J.D.; Rolfe, S.A. Chlorophyll fluorescence imaging as tool for understanding the impact of fungal diseases on plant performance; a phenomics perspective. *Funct. Plant Biol.* **2009**, *36*, 880–892. [CrossRef]

118. McElrone, A.J.; Hamilton, J.G.; Krafnick, J.; Aldea, M.; Knepp, R.G.; DeLuica, E.H. Combined effects of elevated CO2 and natural climatic variation on leaf spot diseases of redbud and sweetgum trees. *Environ. Pollut.* **2010**, *158*, 108–114. [CrossRef]

119. Adams, W.W., III; Demmig-Adams, B. Chlorophyll fluorescence as a tool to monitor plant response to the environment. In *Chlorophyll a Fluorescence: A Signature of Photosynthesis*; Papageorgiou, G.C., Govindjee, Eds.; Springer: Dordrecht, The Netherlands, 2004; pp. 583–604.

120. Melis, A. Photosystem II damage and repair cycle in chloroplasts: What modulates the rate of photodamage in vivo? *Trends Plant Sci.* **1999**, *4*, 130–135. [CrossRef]

121. Mittal, S.; Kumari, N.; Sharma, V. Differential response of salt stress on brassica juncea: Photosynthetic performance, pigment, proline, d1 and antioxidant enzymes. *Plant Physiol. Biochem.* **2012**, *54*, 17–26. [CrossRef]

122. Ji, B.; Li, Z.; Gu, W.; Li, J.; Xie, T.; Wei, S. Methyl jasmonate pretreatment promotes the growth and photosynthesis of maize seedlings under saline conditions by enhancing the antioxidant defense system. *Int. J. Agric. Biol.* **2018**, *20*, 1454–1462.

123. Cox, J.; Mann, M. MaxQuant enables high peptide identification rates, individualized p.p.b.-range mass accuracies and proteome-wide protein quantification. *Nat. Biotechnol.* **2008**, *26*, 1367–1372. [CrossRef]

124. Thomas, P.D.; Campbell, M.J.; Kejariwal, A.; Mi, H.; Karlak, B.; Daverman, R.; Diemer, K.; Muruganujan, A.; Narechania, A. PANTHER: A library of protein families and subfamilies indexed by function. *Genome Res.* **2003**, *13*, 2129–2141. [CrossRef] [PubMed]

125. Szklarczyk, D.; Morris, J.H.; Cook, H.; Kuhn, M.; Wyder, S.; Simonovic, M. The STRING database in 2017: Quality-controlled protein-protein association networks, made broadly accessible. *Nucleic Acids Res.* **2017**, *45*, D362–D368. [CrossRef] [PubMed]

126. Proietti, S.; Giangrande, C.; Amoresano, A.; Pucci, P.; Molinaro, A.; Bertini, L.; Caporale, C.; Caruso, C. *Xanthomonas campestris* lipooligosaccharides trigger innate immunity and oxidative burst in Arabidopsis. *Plant Physiol. Biochem.* **2014**, *85*, 51–62. [CrossRef] [PubMed]

 © 2019 by the authors. Licensee MDPI, Basel, Switzerland. This article is an open access article distributed under the terms and conditions of the Creative Commons Attribution (CC BY) license (http://creativecommons.org/licenses/by/4.0/).

International Journal of
Molecular Sciences

Article

Jasmonic Acid-Induced VQ-Motif-Containing Protein OsVQ13 Influences the OsWRKY45 Signaling Pathway and Grain Size by Associating with OsMPK6 in Rice

Yuya Uji, Keita Kashihara, Haruna Kiyama, Susumu Mochizuki, Kazuya Akimitsu and Kenji Gomi *

Plant Genome and Resource Research Center, Faculty of Agriculture, Kagawa University, Miki, Kagawa 761-0795, Japan; ijuayuy@gmail.com (Y.U.); 8743002z@stu.kagawa-u.ac.jp (K.K.); s18g621@stu.kagawa-u.ac.jp (H.K.); motti245@ag.kagawa-u.ac.jp (S.M.); kazuya@ag.kagawa-u.ac.jp (K.A.)
* Correspondence: gomiken@ag.kagawa-u.ac.jp; Tel.: +81-87-891-3111

Received: 23 May 2019; Accepted: 14 June 2019; Published: 14 June 2019

Abstract: Jasmonic acid (JA) is a plant hormone that plays an important role in the defense response and stable growth of rice. In this study, we investigated the role of the JA-responsive valine-glutamine (VQ)-motif-containing protein OsVQ13 in JA signaling in rice. OsVQ13 was primarily located in the nucleus and cytoplasm. The transgenic rice plants overexpressing *OsVQ13* exhibited a JA-hypersensitive phenotype and increased JA-induced resistance to *Xanthomonas oryzae* pv. *oryzae* (*Xoo*), which is the bacteria that causes rice bacterial blight, one of the most serious diseases in rice. Furthermore, we identified a mitogen-activated protein kinase, OsMPK6, as an OsVQ13-associating protein. The expression of genes regulated by OsWRKY45, an important WRKY-type transcription factor for *Xoo* resistance that is known to be regulated by OsMPK6, was upregulated in *OsVQ13*-overexpressing rice plants. The grain size of *OsVQ13*-overexpressing rice plants was also larger than that of the wild type. These results indicated that OsVQ13 positively regulated JA signaling by activating the OsMPK6–OsWRKY45 signaling pathway in rice.

Keywords: MAP kinase; jasmonate; rice bacterial blight; salicylic acid; grain development

---

1. Introduction

Rice (*Oryza sativa* L.) is a major crop in the world, and decreased crop yields caused by pathogen attacks is a serious problem in rice farming. Many studies have shown that rice plants have developed complex defense systems to protect themselves against various pathogens. Among the defense systems, two plant hormones, jasmonic acid (JA) and salicylic acid (SA), are important signaling compounds that help to regulate the expression of defense-related genes in rice [1]. Treatment with benzothiadiazole (BTH), an SA analog, upregulates many pathogenesis-related (PR) genes and defense-related transcription factors (TFs) in rice and enhances resistance to rice bacterial blight and rice blast caused by *Xanthomonas oryzae* pv. *oryzae* (*Xoo*) and *Pyricularia oryzae*, respectively, which are both hemibiotrophic pathogens causing two of the most serious rice diseases, rice bacterial blight and rice blast, respectively [2]. The WRKY-type TFs, identified as a TF family containing a conserved WRKY domain [3], have important roles in rice defense responses. As one of the WRKY-type TFs, OsWRKY45 plays a crucial role in the BTH-mediated defense response against *Xoo* and *P. oryzae* [2,4]. OsNPR1, a rice homologue to *Arabidopsis* NON-EXPRESSOR OF PATHOGENESIS-RELATED GENES1 (AtNPR1) [5], acts as a positive regulator of SA signaling and is involved in SA-mediated defense response in rice [6–8].

JA also plays an important role in the defense response against *Xoo*. Prior treatment with *Xoo*-derived cellulase (ClsA) and lipase/esterase (LipA) increases a rice plant's resistance to subsequent *Xoo* infection and upregulates the expression of JA-biosynthetic and JA-responsive genes [9]. The JA-upregulated rice jasmonate ZIM domain (JAZ) protein, OsJAZ8, interacts with the F-box protein CORONATINE INSENSITIVE 1 (COI1), which is the primary JA receptor, and acts as a repressor of the JA response, thus negatively regulating the expression of JA-responsive defense-related genes and resistance to *Xoo* [10]. OsWRKY45-2 is involved in the JA-mediated resistance to *Xoo* [11]. Activation of the Cysteine3Histidine (CCCH)-type zinc-finger DNA-binding protein has been reported to induce JA-mediated resistance to *Xoo* in rice [12]. The basic helix–loop–helix (bHLH)-type TF OsMYC2, which is the rice homologue of AtMYC2, positively regulates the JA-mediated defense response against *Xoo* in rice [13]. OsNINJA1, which is the rice homologue of *Arabidopsis* NOVEL INTERACTOR OF JAZ (AtNINJA) [14], acts as a negative regulator of the OsMYC2-mediated defense response against *Xoo* in rice [15]. JA-induced volatiles such as monoterpenes and sesquiterpenes act as antibacterial or signaling compounds in the defense response against *Xoo* [16–20]. Of these JA-induced monoterpenes, linalool functions as a signal molecule to induce the upregulation of defense-related genes in rice [17]. In addition, (*E*)-nerolidol and γ-terpinene exhibit antimicrobial activities against *Xoo* [19,20]: γ-terpinene induces antibacterial activity against *Xoo* by damaging the bacterial plasma membrane [19]. The JA-induced accumulation of some volatiles is regulated by OsJAZ8 [17,18]. These results suggest that the JA signaling pathway is necessary for inducing rice defense systems against *Xoo*.

Recent studies have reported that JA-responsive plant-specific valine-glutamine (VQ) (FxxxVQxLTG)-motif-containing proteins have been identified in many plant species [21–24]. In *Arabidopsis*, JASMONATE-ASSOCIATED VQ MOTIF GENE1 (JAV1, known as AtVQ22) acts as a negative regulator of JA-mediated plant defense, operating against necrotrophic pathogens and herbivorous insects [25]. MITOGEN-ACTIVATED PROTEIN KINASE4 SUBSTRATE1 (MKS1, also known as AtVQ21) is required for the activation of SA-dependent defense [26,27]. In addition, some VQ-motif-containing proteins have been shown to interact with numerous WRKY TFs [21]. AtVQ23 interacts with AtWRKY33, a key regulator of plant defense against necrotrophic pathogens [26,28,29]. These results suggest that VQ-motif-containing proteins act as modulators in JA- and SA-mediated plant defense. The rice genome contains 39 VQ-motif-containing protein family genes, some of which respond to pathogen attacks [22,30]. However, studies on VQ-motif-containing proteins in JA-mediated defense signaling in rice are lacking. We recently identified a few JA-responsive VQ-motif-containing genes in rice through a microarray analysis [10]. In the current study, we investigated the role of the JA-responsive rice VQ-motif-containing protein OsVQ13 in the JA-mediated defense response in rice. We also provide evidence regarding an OsVQ13-associating protein, which acts as a key regulator of OsWRKY45.

2. Results

2.1. Properties of OsVQ13

We carried out reverse transcription-quantitative PCR (RT-qPCR) analysis of *OsVQ13* to investigate its expression in response to JA treatment. The expression of *OsVQ13* reached its maximum level after 24 h of JA treatment (Figure 1A). To determine the subcellular localization of OsVQ13, we generated transgenic rice plants overexpressing the OsVQ13 green fluorescent protein (GFP) fusion protein (*OsVQ13GFP*-ox) and confirmed the expression of the transgene through RT-PCR (Figure 1B). The GFP signal in the root tissue of *OsVQ13GFP*-ox line 9 was observed by fluorescence microscopy. As shown in Figure 1C, the GFP fluorescent signals were detected in the nucleus and the cytoplasm, indicating that OsVQ13 was localized in these specific locations.

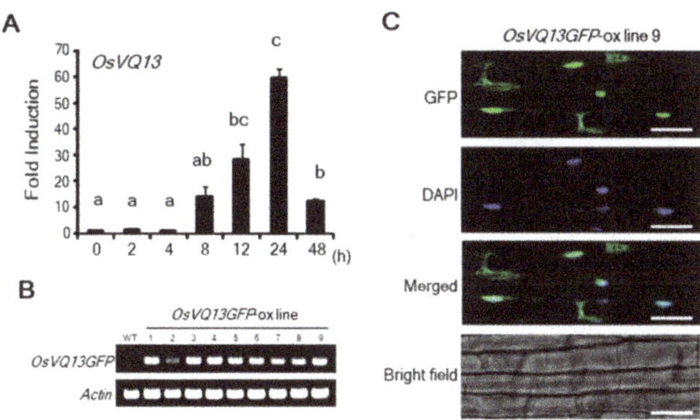

Figure 1. Jasmonic acid (JA)-induced expression and subcellular localization of OsVQ13. (**A**) Expression levels of *OsVQ13* in response to JA. Total RNA was extracted at the indicated time points after 100 µM of JA treatment. Values are means ± SE. Data were analyzed using Tukey's HSD test ($n = 4$ for each genotype). Bars with different letters are significantly different at $p < 0.05$. (**B**) Reverse transcription (RT)-PCR analysis of *OsVQ13GFP* and *actin* expression in wild-type (WT) and *OsVQ13GFP*-overexpressing rice plants (*OsVQ13GFP*-ox; lines 1–9). (**C**) Subcellular localization of OsVQ13. The green fluorescent protein (GFP) signal in the root tissue of *OsVQ13GFP*-ox (line 9) was visualized by fluorescence microscopy. Nuclear localization of OsVQ13GFP was confirmed by 4′,6-diamidino-2-phenylindole (DAPI) staining. Bars = 10 µm.

2.2. Phenotypes of OsVQ13-Overexpressing Rice Plants

We generated *OsVQ13*-overexpressing (*OsVQ13*-ox) rice plants (lines 2 and 8) and confirmed the expression of the transgene through RT-PCR (Figure 2A). To identify JA responses in these transgenic rice plants, we measured chlorophyll (Chl) contents after JA treatment, because it is known that JA promotes a Chl degradation in rice [13]. The Chl contents of *OsVQ13*-ox rice plants were significantly reduced at three days after JA treatment (Figure 2B).

To determine whether OsVQ13 is involved in JA-mediated resistance to *Xoo*, we performed a resistant test on these transgenic rice plants. The JA-treated or untreated rice plants were inoculated with a virulent *Xoo*, and the length of any blight lesions was measured 14 days after inoculation. The lengths of blight lesions in the *OsVQ13*-ox plants were significantly shorter than those in the wild-type (WT) plants without JA treatment (Figure 2C,D). Furthermore, JA-induced resistance was enhanced in the transgenic rice plants compared to the WT plants (Figure 2C,D).

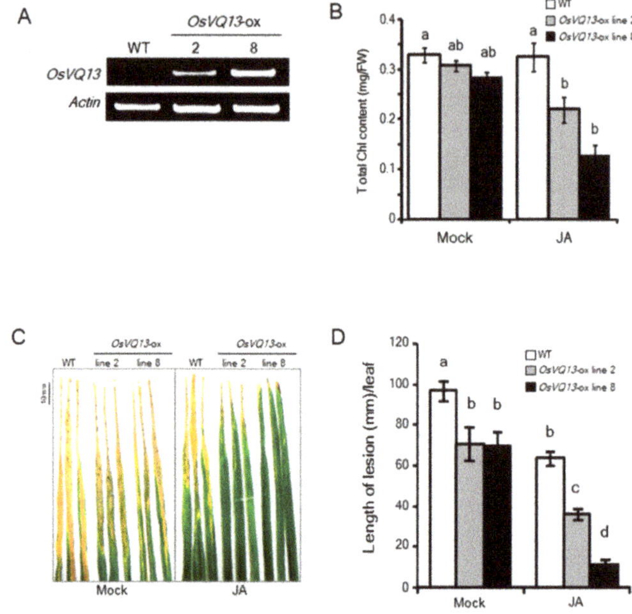

Figure 2. Phenotype of *OsVQ13*-overexpressing rice plants. (**A**) RT–PCR analysis of *OsVQ13* and *actin* expression in wild-type (WT) and *OsVQ13*-overexpressing rice plants (*OsVQ13*-ox; lines 2 and 8). (**B**) Total chlorophyll content in leaf blades after treatment with 100 μM of JA for 3 d in WT and *OsVQ13*-ox rice plants. Values are means ± SE. Data were analyzed using Tukey's HSD test (n = 4 for each genotype). Bars with different letters are significantly different at $p < 0.05$. (**C**) Disease symptoms of rice bacterial blight in WT and *OsVQ13*-ox with mock or 100 μM of JA pretreatment for 24 h. The fifth leaf blades were photographed 14 d after inoculation with *Xoo*. (**D**) The length of lesions on the fifth leaf blades at 14 d after inoculation with *Xoo* with pretreatment with 100 μM of JA for 24 h. Values are means ± SE. Data were analyzed using the Tukey–Kramer test (n = 12 for both WT mock and JA; n = 7 for line 2 mock; n = 12 for line 2 JA; n = 10 for line 8 mock; n = 12 for line 8 JA). Bars with different letters are significantly different at $p < 0.05$.

2.3. Identification of OsVQ13-Associating Proteins

Arabidopsis VQ proteins act as positive or negative regulators through interactions with various proteins in response to abiotic or biotic stresses [31]. To determine whether OsVQ13 associates with uncharacterized proteins in rice, we performed a co-immunoprecipitation assay on anti-GFP antibodies derived from *OsVQ13GFP*-ox rice plants. The *OsVQ13GFP*-ox rice plant exhibited a JA-hypersensitive phenotype similar to that of the *OsVQ13*-ox transgenic plant, indicating that the OsVQ13GFP protein was functional in the rice plant (Figure 3A). The *GFP*-ox transgenic rice plant (*GFP*-ox) was used as a negative control. After anti-GFP antibody precipitation, numerous proteins were detected in the *OsVQ13GFP*-ox rice plant sample, whereas only a small number of proteins were detected in the *GFP*-ox rice plant sample (Figure 3B,C). Experiments using *OsVQ13GFP*-overexpressing rice plants were repeated five times, and ultimately, the four proteins that were reproducibly detected at least twice in the immunoprecipitates were selected (Figure 3C). These selected proteins were analyzed using matrix-assisted laser desorption/ionization time-of-flight mass spectrometry (MALDI-TOF MS) and identified in a MASCOT database. As a result, one mitogen-activated protein kinase (MAPK), OsMPK6, and three TFs, namely OsNAC5, OsERF36, and OsMADS2, were identified (Table 1).

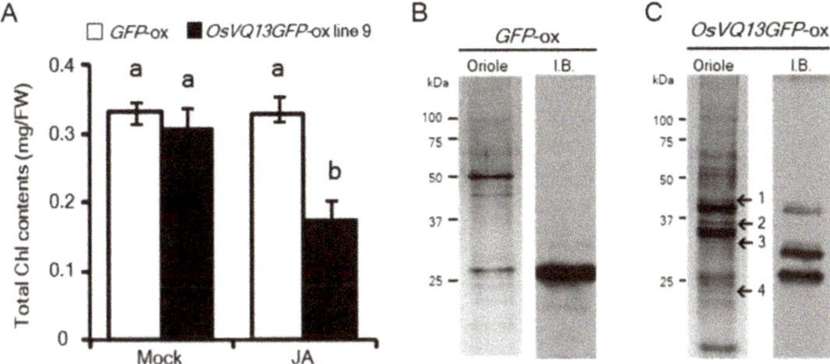

Figure 3. Identification of OsVQ13-associating proteins. (**A**) Confirmation of JA-hypersensitive phenotype in *OsVQ13GFP*-ox plant. Total chlorophyll content in leaf blades after treatment with 100 µM of JA for 3 d in WT and *OsVQ13GFP*-ox rice plants. Values are means ± SE. Data were analyzed using Tukey's HSD test (n = 4 for each genotype). Bars with different letters are significantly different at $p < 0.05$. (**B,C**) The proteins co-purified with GFP-Trap from *GFP*-overexpressing (*GFP*-ox: (**B**)) and *OsVQ13GFP*-overexpressing rice plants (*OsVQ13GFP*-ox: (**C**)) were separated through SDS-PAGE, and protein bands were visualized by Oriole straining (Oriole). GFP (**B**) and OsVQ13GFP (**C**) proteins were detected using anti-GFP antibody (I.B.). The numbers on the left indicate the position of the protein size markers in kDa. The putative molecular weight of each protein was as follows: GFP, 27 kDa; OsVQ13GFP, 41 kDa. The numbers on the right of the Oriole staining lane in (**C**) indicate protein bands excised for MALDI-TOF MS protein identification. (**C**) The gel is representative of five independent experiments.

Table 1. The list of proteins co-purified with OsVQ13.

No.	Locus No.	Protein Name	Predicted MW (kDa)	Occurrence/Total No. of Experiments
1	Os06g06090	Mitogen-activated protein kinase 6 (OsMPK6)	45	4/5
2	Os11g08210	NAC domain-containing protein 5 (OsNAC5)	35	3/5
3	Os10g41330	Ethylene response factor 36 (OsERF36)	29	2/5
4	Os01g66030	MADS-box transcription factor 2 (OsMADS2)	24	2/5

OsVQ13 co-IP was performed using *OsVQ13GFP*-overexpressing transgenic rice plants. Experiments were independently performed five times. Nos. 1–4 are as in Figure 3. MW: Molecular weight.

Among the OsVQ13-associating proteins, we focused on OsMPK6, which is known to be involved in the rice defense response [32,33]. We first found a direct interaction between OsVQ13 and OsMPK6 proteins using a yeast two-hybrid system (Figure 4A). We next performed Phos-tag® SDS-PAGE to reveal whether OsVQ13 in rice was phosphorylated. In SDS-PAGE using Phos-tag® acrylamide, phosphorylated proteins are trapped by the Phos-tag® sites, which leads to a delay in their migration and a resulting separation from unphosphorylated proteins. This makes it straightforward to identify the phosphorylated proteins from their observed positions on blots [34,35]. As a positive control, we used a commercially available α-casein, which contains both phosphorylated and unphosphorylated forms. When the α-casein was separated by Phos-tag® SDS-PAGE, a slow-migrating band was detected (Figure 4B). The band disappeared following treatment with protein phosphatase, indicating that the upper band was a highly phosphorylated form of α-casein, which could be distinguished from the unphosphorylated form through Phos-tag® SDS-PAGE (Figure 4B). In the case of the sample from the *OsVQ13GFP*-ox rice plant, however, there was no change in the banding pattern detected by anti-GFP antibodies between samples treated and not treated with phosphatase (Figure 4C).

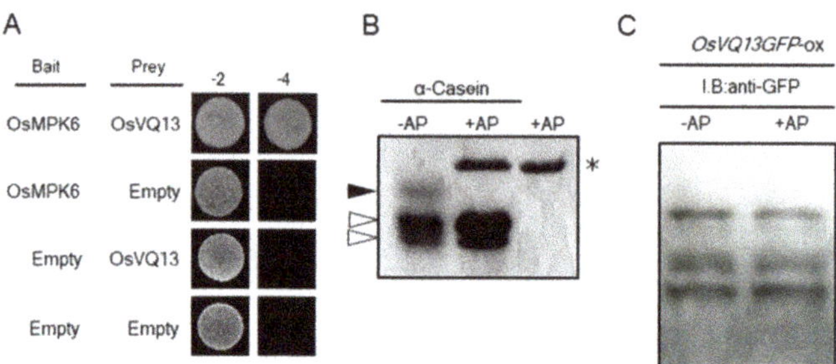

Figure 4. Phosphorylation assay of OsVQ13. (**A**) Interaction between OsVQ13 and OsMPK6 using yeast cells. The yeast AH109 strain was dropped on synthetic dropout (SD) glucose medium lacking Leu and Trp (−2) or on SD glucose medium lacking Ade, His, Leu, and Trp (−4). (**B**) Phosphorylation assay of the α-casein protein through Phos-tag® PAGE. Here, α-casein was a control protein treated with alkaline phosphatase (+AP) or untreated (-AP) (as indicated), separated through Phos-tag® SDS-PAGE. The black arrowhead indicates phosphorylated-α-casein, and the white arrowheads indicate dephosphorylated α-caseins. An asterisk indicates an AP protein. (**C**) Phosphorylation assay of the OsVQ13 protein through Phos-tag® PAGE. Purified OsVQ13GFP proteins were treated with AP (+AP) or were untreated (−AP). The treated proteins were separated through Phos-tag® SDS-PAGE and were detected using anti-GFP antibodies.

2.4. OsVQ13 Affected OsMPK6-Mediated Signaling Pathways in Rice

OsMPK6 is involved in the rice defense response through the phosphorylation of OsWRKY45, a TF that, in turn, plays a crucial role in defense responses in rice [2,4,36,37]. To investigate whether OsVQ13 affects the OsWRKY45-dependent pathway via its association with OsMPK6, we compared the expression levels in WT and *OsVQ13*-ox rice plants of the following OsWRKY45-responsive genes (identified by Nakayama et al. [4]): *OsWRKY62, cytochrome P450, OsOPR5, OsPrx126, OsPrx72, OsGSTU4, UDP-glucosyltransferase, Proteinase inhibitor I20,* and *beta-1,3-glucanase.* Genes *OsWRKY62, cytochrome P450, OsOPR5, OsPrx126, OsPrx72, OsGSTU4,* and *UDP-glucosyltransferase* tended to be upregulated in *OsVQ13*-ox rice plants compared to WT plants in the absence of JA. The expression levels of all genes tested were significantly higher in JA-treated *OsVQ13*-ox rice than in WT plants (Figure 5A). In contrast, the expression levels of endogenous *OsVQ13, OsMPK6,* and *OsWRKY45* in *OsVQ13*-ox plants with or without JA treatment were not significantly different from those in WT rice plants (Figure 5B).

It was also shown that the length, width, and weight of *OsVQ13*-ox grains were significantly larger than WT grains (Figure 6A–D). OsMPK6 has been shown to positively regulate grain size and weight in rice [38,39].

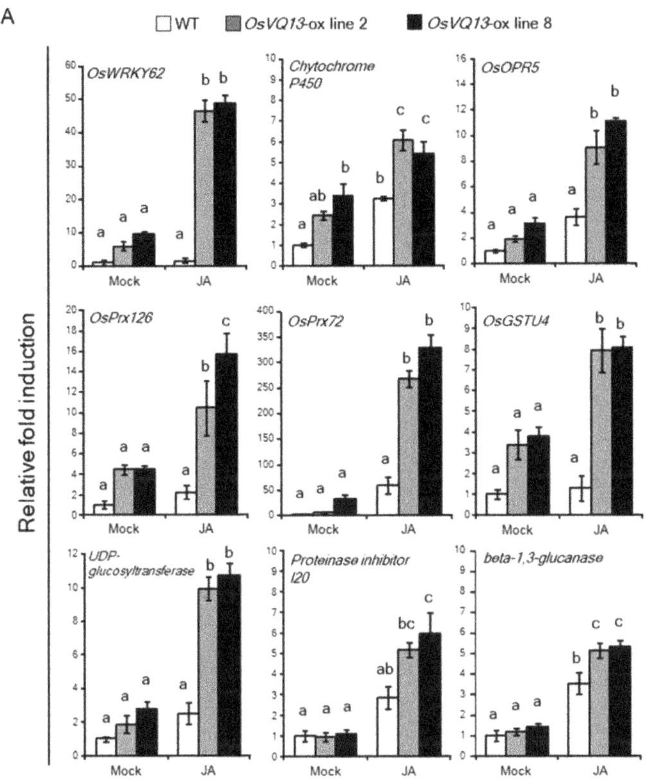

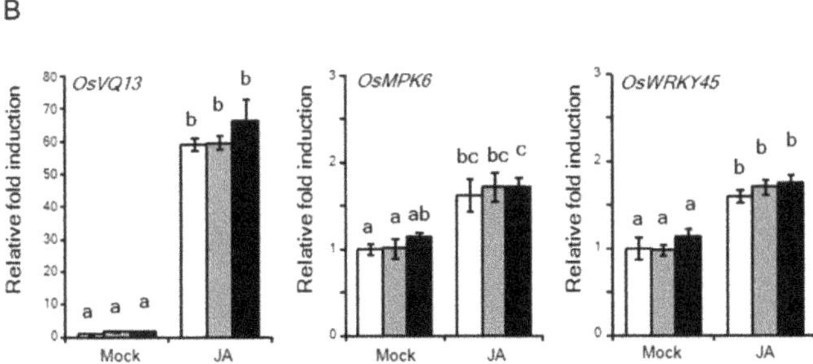

Figure 5. Expression of OsWRKY45-responsive defense-related genes in *OsVQ13*-overexpressing rice plants. (**A**) Expression levels of OsWRKY45-responsive genes after JA treatment. The OsWRKY45-responsive genes identified by Nakayama et al. [4] were used: *OsWRKY62, cytochrome P450, OsOPR5, OsPrx126, OsPrx72, OsGSTU4, UDP-glucosyltransferase, Proteinase inhibitor I20*, and *beta-1,3-glucanase*. (**B**) Expression levels of endogenous *OsVQ13, OsMPK6*, and *OsWRKY45* after JA treatment. (**A**,**B**) Values are means ± SE. Data were analyzed using Tukey's HSD test (*n* = 4 for each genotype). Bars with different letters are significantly different at *p* < 0.05.

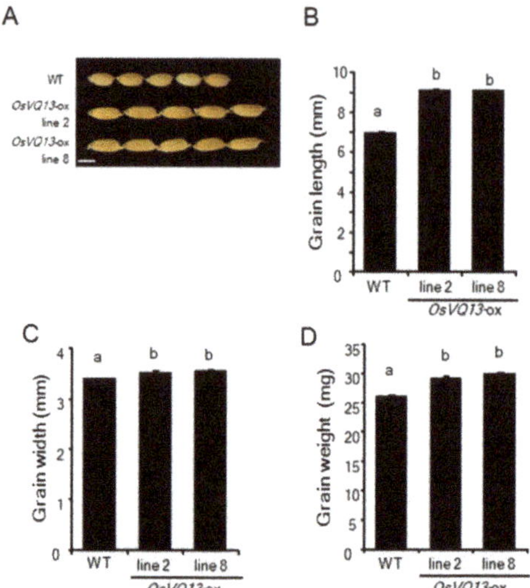

Figure 6. Grain size of *OsVQ13*-overexpressing rice plants. (**A**) Mature grains of the WT and *OsVQ13*-overexpresing rice plants (*OsVQ13*-ox: lines 2 and 8). Bar = 5 mm. The average grain length (**B**), width (**C**), and grain weight (**D**) of the WT and *OsVQ13*-ox plants. Values are means ± SE. Data were analyzed using Tukey's HSD test (*n* = 100 for each genotype). Bars with different letters are significantly different at *p* < 0.05.

3. Discussion

We previously demonstrated that the gene expression patterns in response to JA are broadly divided into two phases, the early (<12 h) and late (>12 h) phases, under our experimental conditions [13]. The expression level of *OsVQ13* reached a maximum at 24 h after JA treatment, indicating that OsVQ13 is a late JA-responsive gene in rice. It has been reported that the basal expression of *OsVQ13* in the resistant rice plants harboring a *Xoo*-resistance gene, *Xa3/Xa26*, was significantly higher than in corresponding susceptible rice plants [30]. When *OsVQ13* was overexpressed in rice in the current study, the transgenic plants exhibited a JA-hypersensitive phenotype and increased basal and JA-induced resistances against *Xoo*. The increased JA-induced resistance against *Xoo* in transgenic plants depended on the expression levels of the *OsVQ13* transgene. These results indicated that OsVQ13 acted as a positive regulator of JA signaling in rice.

In the present study, we revealed that OsVQ13 associated with OsMPK6 and positively regulated OsMPK6-mediated signaling pathways in rice. OsVQ13 possessed putative phosphorylation sites through MAPK. Furthermore, we suggested the possibility that OsVQ13 may not be a substrate of OsMPK6, although further analysis is needed to determine whether OsVQ13 is phosphorylated or not by OsMPK6 in rice. These results suggest that OsVQ13 acts as a regulatory protein toward OsMPK6, but not as a substrate for OsMPK6. Furthermore, expression patterns of *OsMPK6* and *OsWRKY45* were similar in WT and *OsVQ13*-ox rice plants, suggesting that OsVQ13 positively regulated the expression of OsWRKY45-responsive genes by activating OsMPK6 at the protein level but not at the transcriptional level. The activity of OsMPK6 is suppressed via interaction with the MAPK phosphatase OsMKP1, which dephosphorylates both serine/threonine and tyrosine residues of MAPKs in rice [40]. Thus, it is possible that OsVQ13 blocks the interaction between OsMPK6 and OsMKP1 to activate the OsMPK6–OsWRKY45 cascade in response to JA. Regulation of OsWRKY45 activity via the phosphorylation mediated by OsMPK6 has been analyzed in detail [37]. OsWRKY45 is phosphorylated at Thr266, Ser294, and Ser299 by OsMPK6. Phosphorylation of Ser294 and Ser299 is required for the full activation of OsWRKY45 in the defense response in rice [37]. Conversely, phosphorylation of

Thr266 negatively affects the defense response in rice [41]. In the current study, it was demonstrated that OsVQ13 positively regulated the OsWRKY45-dependent signaling pathway by associating with OsMPK6, although there was no direct evidence for the phosphorylation of OsWRKY45 by OsVQ13 in rice (Figure 7). Further studies are needed to test this hypothesis.

Both OsMPK6 and OsWRKY45 act as central regulators of SA signaling in rice [2,4,37]. In the current study, we demonstrated that the expression of some OsWRKY45-responsive genes was upregulated by JA, which also increased their upregulation in *OsVQ13*-ox rice plants. The expression of *OsMPK6* and *OsWRKY45* is upregulated through JA treatment [42,43]. Another important fact is that the expression of *OsVQ13* is also upregulated by BTH, an analog of SA [2]. These results suggest that OsVQ13 plays a critical role as an activator of the OsMPK6–OsWRKY45-dependent cascade involved in both JA- and SA-induced resistance against *Xoo*.

Generally, SA confers resistance against biotrophic and hemibiotrophic pathogens, whereas JA confers resistance against necrotrophic pathogens in plants. The relationship between JA and SA is antagonistic in many plants. As an example of an *Arabidopsis* VQ protein, plants exhibiting RNA interference (RNAi) of *JAV1* exhibited increased resistance against the necrotrophic pathogen *Botrytis cinerea* [25]. The *Arabidopsis mks1* mutant exhibited increased susceptibility to the biotrophic pathogen *Pseudomonas syringae* [27]. However, in rice, it is unclear whether this antagonistic cross-talk between JA- and SA-dependent defense signaling occurs. In addition to the current study, other studies have demonstrated that the JA-dependent signaling pathway has an important role in resistance against hemibiotrophic pathogens in rice. Accumulations of JA-isoleucine (JA-Ile), a bioactive form of JA that is induced through inoculation with *P. oryzae* [44], and the jasmonate-deficient rice mutants *cpm2* and *hebiba*, exhibited decreased resistance to an avirulent *P. oryzae* [45]. The *osjar1-2* mutant, which exhibits diminished JA-Ile accumulation in leaves, also showed decreased resistance to *P. oryzae* [46]. Expression of the microRNA *miR319b* was upregulated through inoculation with virulent *P. oryzae*. The upregulation of *miR319b* caused the suppression of its target gene, TEOSINTE BRANCHED1/CYCLOIDEA/PROLIFERATING CELL FACTOR1 (TCP)21 (*OsTCP21*). OsTCP21 acts as a positive TF regulating the JA-biosynthetic genes *lipoxygenase2* (*OsLOX2*) and *OsLOX5* [47]. On the other hand, *P. oryzae* converts JA into 12OH-JA by secreting a monooxygenase, Abm. The 12OH-JA suppresses the induction of JA-mediated defense responses [48]. It has been reported that 313 BTH-upregulated genes were identified by microarray analysis in rice [2]. When the expression levels of BTH-upregulated genes were compared to those of JA-responsive genes, as identified by microarray analysis in rice [10], more than half of the BTH-upregulated genes, including *OsVQ13*, were also upregulated by JA [42]. OsNPR1 is degraded by OsCUL3a to suppress OsNPR1-dependent cell death [8]. The expression of *OsNPR1* is also upregulated by JA [49]. Furthermore, an *oscul3a* mutant exhibited increased resistance to *P. oryzae* and *Xoo* by activating both the JA- and SA-signaling pathways [8]. Taken together, these results strongly indicate that JA and SA signaling can interact coordinately in an induced defense response and that OsVQ13 may be a key factor in the rice defense pathway induced by both JA and SA.

The overexpression of *OsVQ13* has resulted in larger grain sizes compared to WT. The overexpression of *OsMPK6* or a constitutively active version of *OsMPK6* (*CA-OsMPK6*) has also resulted in significantly larger grains [39]. Furthermore, an *oscoi1b* mutant has exhibited a significantly lower grain weight [50], suggesting that JA has a positive effect on the seed development process. However, there is not enough experimental evidence to discuss the function of OsVQ13 in JA-dependent seed development in rice (Figure 7). Further studies are needed.

Among the OsVQ13-associating proteins, the NAC-type TF OsNAC5 positively regulates some defense-related genes in rice, and the expression of *OsNAC5* responds to both JA and abscisic acid [51]. OsNAC5 is also required to express salt stress tolerance in rice [51]. *OsNAC5*-overexpressing rice plants have exhibited increased salt tolerance, whereas *OsNAC5*-RNAi rice plants have exhibited greater sensitivity than the WT to salt stress [51,52]. Recently, it has been reported that JA plays an important role in salt tolerance in many plant species, including rice [53]. RICE SALT SENSITIVE3 (RSS3) acts as a negative regulator of JA signaling by interacting with OsJAZ9 and OsJAZ11, and the *rss3* mutant exhibits the salt-hypersensitive phenotype [54]. OsNAC5 is also known to be negatively involved in leaf senescence in rice [55], whereas JA has a positive effect on leaf senescence in rice [50,56]. There is no information as to whether OsVQ13 is a senescence-associated gene in rice. Further analyses of OsVQ13-associating proteins such as OsNAC5 may provide new insights into the molecular mechanism of JA-dependent biotic and abiotic stress responses in rice.

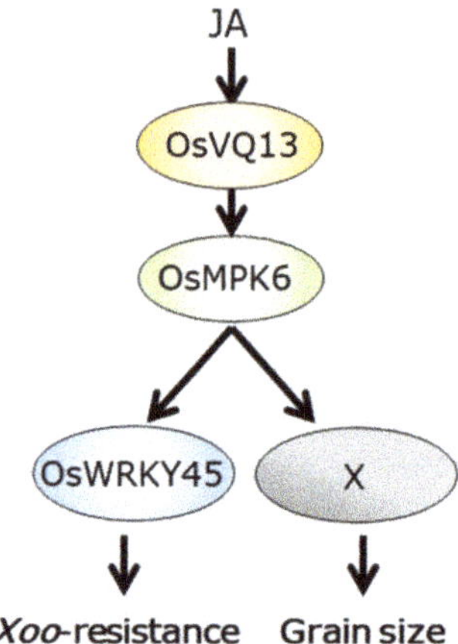

Figure 7. A model for OsVQ13/OsMPK6 complex-mediated signaling pathways in rice. OsVQ13 associates with OsMPK6 and acts as an activator of the OsWRKY45-dependent signaling pathway, contributing to resistance to *Xanthomonas oryzae* pv. *oryzae*. OsVQ13 also affects rice grain development by associating with OsMPK6.

4. Materials and Methods

4.1. Plant Materials, Jasmonate Treatment, and Bacterial Inoculation

Plant and bacteria growth conditions were as previously described, they are grown and cultured in our laboratory by ourselves [15]. The fully opened fifth leaf blades of rice plants were inoculated using the clipping inoculation technique [57]. The lengths of blight lesions on each leaf blade were measured for each leaf at 14 days post-inoculation.

To examine the effects of JA on rice growth and gene expression, rice plants were grown to the four-leaf stage in a growth chamber in Kimura-B liquid medium [58] at 25 °C (24 h light) and then incubated in the same medium supplemented with 100 μM of JA (Sigma-Aldrich, St. Louis, MO, USA). The fourth leaf blades were used for RT-qPCR analyses.

4.2. RT-qPCR

Total RNA was extracted from rice leaf blades from plants of different genotypes and treatments using Trizol (Thermo Fisher Scientific, Waltham, MA, USA) according to the manufacturer's instructions. RT-qPCR was performed as previously described [15,59]. The sequences of the gene-specific primers used in this study are presented in the Supplementary Table S1.

4.3. Construction of OsVQ13-Overexpressing Vectors and Rice Transformation

The *OsVQ13* cDNA (accession number: AK109228) was ligated into the pBI333-EN4 vector [60]. The pBI333-EN4-GFP binary vector was prepared by subcloning the GFP coding sequence into the pBI333-EN4 binary vector. The production of transgenic rice plants was performed as previously described [61,62]. Second- or third-generation plants were used for the experiments.

RT-PCR was performed using the OneStep RT-PCR kit (QIAGEN, Hilden, Germany) according to the manufacturer's instructions. The sequences used for RT-PCR were as follows: *OsVQ13F*, 5′-GACATGTTCGACTACGCGTC-3′ and t*NOSR*, 5′-GTATAATTGCGGGACTCTAATC-3′; *GFPR*, 5′-GCATGGCGGACTTGAAGA-3′; *actin*, forward, 5′-CCTGGAATCCATGAGACCAC-3′ and reverse, 5′-ACACCAACAATCCCAAACAGAG-3′.

4.4. Chlorophyll Content Measurement

Leaf blades treated with 100 μM of JA for four days were homogenized in 1 mL of 80% acetone and centrifuged at 15,000 rpm for 10 min. The specific chlorophyll content was determined as described by Arnon [63].

4.5. Subcellular Localization Assay

Seven-day-old rice seedlings were incubated at 25°C in CHU (N6) medium (Duchefa Biochemie, Haarlem, The Netherlands) with 50 mg/L of hygromycin. The GFP signal in the root tissue overexpressing *OsVQ13-GFP* was visualized by using a KEYENCE BIOREVO BZ-9000 fluorescence microscope (Keyence, Osaka, Japan). We used BZ filter GFP (excitation wavelength: 470/40 nm; absorption wavelength: 535/50 nm; dichroic mirror wavelength: 495 nm) and BZ filter DAPI (excitation wavelength: 360/40 nm; absorption wavelength: 460/50 nm; dichroic mirror wavelength: 400 nm). The root tissues were stained with 4′,6-diamidino-2-phenylindole (DAPI) solution (Dojindo, Kumamoto, Japan).

4.6. Identification of OsVQ13-Associating Proteins

4.6.1. In Vivo Purification of the OsVQ13GFP Protein

Rice leaf blades (2 g, fresh weight) were homogenized with liquid nitrogen, and the powder was resuspended in 30 mL of extraction buffer 1 (0.4 M sucrose, 10 mM Tris-HCl (pH 7.5), 5 mM β-mercaptoethanol (β-ME), 0.1 mM phenylmethylsulfonyl fluoride (PMSF), and 1% plant protease inhibitor cocktail (Sigma-Aldrich)). After filtration of the slurry through Miracloth (Merck, Darmstadt, Germany), the filtrate was centrifuged (2880× g, 20 min, 4 °C). The pellets containing nuclei were resuspended in 1 mL of extraction buffer 2 (0.25 M sucrose, 10 mM Tris-HCl (pH 7.5), 10 mM $MgCl_2$, 1% Triton X-100, 5 mM β-ME, 0.1 mM PMSF. and 1% plant protease inhibitor cocktail) and collected by centrifugation (12,000× g, 10 min, 4 °C). The pellets were resuspended in 300 μL of extraction buffer 3 (1.7 M sucrose, 10 mM Tris-HCl (pH 7.5), 0.15% Triton X-100, 2 mM $MgCl_2$, 5 mM β-ME, 0.1 mM PMSF, and 1% plant protease inhibitor cocktail), and the resuspended pellets were overlaid onto 300 μL of extraction buffer 3. The nuclear proteins were collected by centrifugation (16,000× g, 60 min, 4 °C) and suspended in nuclear lysis buffer (10 mM Tris-HCl (pH 7.5), 0.2 mM EDTA, 150 mM NaCl, 0.1% Triton X-100, 5 mM β-ME, and 1% plant protease inhibitor cocktail).

4.6.2. Co-Immunoprecipitation and Protein Gel Blot Analysis

The nuclear proteins were mixed with magnetic bead-conjugated GFP-Trap (Chromotek, Munich, Germany) and incubated overnight at 4 °C. The beads were then washed five times in 0.5 mL of wash buffer II (Chromotek, Munich, Germany). The binding proteins were eluted with sample buffer. The extracted proteins were separated by SDS-PAGE and transferred to a nitrocellulose membrane through semidry blotting. A protein gel blot analysis was performed as previously described [13]. We used anti-GFP (1:1000) and antirabbit IgG horseradish peroxidase-conjugated secondary antibodies (1:20,000).

4.6.3. Protein Identification Using TOF MS

The proteins that were purified by GFP-Trap were separated by 12% SDS-PAGE and stained with Oriole Fluorescent Gel Stain (Bio-Rad, Hercules, CA, USA) for 90 min. The stained protein bands were excised from the gels, washed twice with 100 mM of NH_4HCO_3 containing 30% acetonitrile (Wako,

Osaka, Japan), washed with 100% acetonitrile, and then dried in a vacuum concentrator. The gels were reductively alkylated with 10 mM of dithiothreitol (DTT) in 100 mM of NH_4HCO_3 for 60 min at 56 °C and with iodoacetamide in 100 mM of NH_4HCO_3 for 30 min at 25 °C. The gels were washed with 100 μL of 100-mM NH_4HCO_3 for 10 min and dehydrated through the addition of acetonitrile. The dried gels were treated with 2 μL of 0.5 μg/μL of trypsin (Promega, Madison, WI, USA) in 50 mM of NH_4HCO_3 and incubated at 37 °C for 12 h. Peptides were extracted with 50 mM of NH_4HCO_3 in 1% trifluoroacetic acid and 50% acetonitrile. Sample preparation for MALDI analysis was according to the method of Shevchenko et al. [64]. Peptides of proteins that associate with OsVQ13 were identified through MALDI-TOF MS analysis with a Voyager-DE STR (Thermo Fisher Scientific, Waltham, MA, USA). Experiments using *OsVQ13GFP*-overexpressing rice plants were repeated five times, and four out of the five stained gels are presented in the Supplementary Figure S1.

4.6.4. Phos-tag® SDS-PAGE

To determine whether OsVQ13 was phosphorylated, we used Phos-tag® SDS-PAGE (NARD Institute Ltd., Amagasaki, Japan). To achieve dephosphorylation, the beads were prepared in 50 μL of 1× alkaline phosphatase buffer (Thermo Fisher Scientific) and incubated with 2 units of fast alkaline phosphatase (Thermo Fisher Scientific) for 2 h. After incubation, the beads were then washed five times in 0.5 mL of extraction buffer, and the binding proteins were eluted with 2× SDS sample buffer.

4.7. Yeast Two-Hybrid System

An analysis of protein–protein interactions using a yeast two-hybrid system was performed as previously described [15]. Images of yeast cells were obtained at day 3.

4.8. Additional Information

The accession numbers for the genes discussed in this article are as follows: OsVQ13 (AK109228), OsMPK6 (AK111942), OsWRKY45 (AK066255), OsWRKY62 (AK067834), cytochrome P450 (AK072220), OsOPR5 (AK104193), OsPrx126 (AK061206), OsPrx72 (AK067416), OsGSTU4 (AK103453), UDP-glucosyltransferase (AK064395), proteinase inhibitor I20 (AK105387), and beta-1,3-glucanase (AK068247).

Supplementary Materials: Supplementary materials can be found at http://www.mdpi.com/1422-0067/20/12/2917/s1.

Author Contributions: K.G. designed the research project. Y.U., K.K., and H.K. performed the experiments. Y.U. and K.K. wrote the manuscript. S.M. and K.A. modified the manuscript. All of the authors reviewed and approved the manuscript.

Funding: This work was supported in part by the Japan Society for the Promotion of Science (JSPS) (Funding Program for Next-Generation World-Leading Researchers (No. GS022)).

Acknowledgments: We thank Y. Nishizawa (National Institute of Agrobiological Sciences, NIAS), H. Kaku (NIAS), and N. Tanaka (Kagawa University) for providing the binary vector, pE7133 vector, *Xoo* strain, and yeast strain, respectively. We also thank I. Kataoka (Kagawa University), M. Satoh (National Agricultural Research Center for the Kyushu Okinawa Region, NARO), and H. Kanno (NARO) for laying the foundation of a part of this study.

Conflicts of Interest: The authors declare no conflict of interest.

References

1. De Vleesschauwer, D.; Gheysen, G.; Höfte, M. Hormone defense networking in rice: Tales from a different world. *Trends Plant Sci.* **2013**, *18*, 555–565. [CrossRef]
2. Shimono, M.; Sugano, S.; Nakayama, A.; Jiang, C.J.; Ono, K.; Toki, S.; Takatsuji, H. Rice WRKY45 plays a crucial role in benzothiadiazole-inducible blast resistance. *Plant Cell* **2007**, *19*, 2064–2076. [CrossRef]
3. Bakshi, M.; Oelmüller, R. WRKY transcription factors: Jack of many trades in plants. *Plant Signal. Behav.* **2014**, *9*, e27700. [CrossRef]

4. Nakayama, A.; Fukushima, S.; Goto, S.; Matsushita, A.; Shimono, M.; Sugano, S.; Jiang, C.J.; Akagi, A.; Yamazaki, M.; Inoue, H.; et al. Genome-wide identification of WRKY45-regulated genes that mediate benzothiadiazole-induced defense responses in rice. *BMC Plant Biol.* **2013**, *13*, 150. [CrossRef]

5. Ding, Y.; Sun, T.; Ao, K.; Peng, Y.; Zhang, Y.; Li, X.; Zhang, Y. Opposite roles of salicylic acid receptors NPR1 and NPR3/NPR4 in transcriptional regulation of plant immunity. *Cell* **2018**, *173*, 1454–1467.e15. [CrossRef]

6. Chern, M.; Fitzgerald, H.A.; Canlas, P.E.; Navarre, D.A.; Ronald, P.C. Overexpression of a rice NPR1 homolog leads to constitutive activation of defense response and hypersensitivity to light. *Mol. Plant Microbe Interact.* **2005**, *18*, 511–520. [CrossRef]

7. Sugano, S.; Jiang, C.J.; Miyazawa, S.I.; Masumoto, C.; Yazawa, K.; Hayashi, N.; Shimono, M.; Nakayama, A.; Miyao, M.; Takatsuji, H. Role of OsNPR1 in rice defense program as revealed by genome-wide expression analysis. *Plant Mol. Biol.* **2010**, *74*, 549–562. [CrossRef]

8. Liu, Q.; Ning, Y.; Zhang, Y.; Yu, N.; Zhao, C.; Zhan, X.; Wu, W.; Chen, D.; Wei, X.; Wang, G.L.; et al. OsCUL3a negatively regulates cell death and immunity by degrading OsNPR1 in rice. *Plant Cell* **2017**, *29*, 345–359. [CrossRef]

9. Ranjan, A.; Vadassery, J.; Patel, H.K.; Pandey, A.; Palaparthi, R.; Mithöfer, A.; Sonti, R.V. Upregulation of jasmonate biosynthesis and jasmonate-responsive genes in rice leaves in response to a bacterial pathogen mimic. *Funct. Integr. Genom.* **2015**, *15*, 363–373. [CrossRef]

10. Yamada, S.; Kano, A.; Tamaoki, D.; Miyamoto, A.; Shishido, H.; Miyoshi, S.; Taniguchi, S.; Akimitsu, K.; Gomi, K. Involvement of OsJAZ8 in jasmonate-induced resistance to bacterial blight in rice. *Plant Cell Physiol.* **2012**, *53*, 2060–2072. [CrossRef]

11. Tao, Z.; Liu, H.; Qiu, D.; Zhou, Y.; Li, X.; Xu, C.; Wang, S. A pair of allelic WRKY genes play opposite roles in rice-bacteria interactions. *Plant Physiol.* **2009**, *151*, 936–948. [CrossRef]

12. Deng, H.; Liu, H.; Li, X.; Xiao, J.; Wang, S. A CCCH-type zinc finger nucleic acid-binding protein quantitatively confers resistance against rice bacterial blight disease. *Plant Physiol.* **2012**, *158*, 876–889. [CrossRef]

13. Uji, Y.; Taniguchi, S.; Tamaoki, D.; Shishido, H.; Akimitsu, K.; Gomi, K. Overexpression of *OsMYC2* results in the up-regulation of early JA-rresponsive genes and bacterial blight resistance in rice. *Plant Cell Physiol.* **2016**, *57*, 1814–1827. [CrossRef]

14. Pauwels, L.; Barbero, G.F.; Geerinck, J.; Tilleman, S.; Grunewald, W.; Pérez, A.C.; Chico, J.M.; Bossche, R.V.; Sewell, J.; Gil, E.; et al. NINJA connects the co-repressor TOPLESS to jasmonate signalling. *Nature* **2010**, *464*, 788–791. [CrossRef]

15. Kashihara, K.; Onohata, T.; Okamoto, Y.; Uji, Y.; Mochizuki, S.; Akimitsu, K.; Gomi, K. Overexpression of *OsNINJA1* negatively affects a part of OsMYC2-mediated abiotic and biotic responses in rice. *J. Plant Physiol.* **2019**, *232*, 180–187. [CrossRef]

16. Tanaka, K.; Taniguchi, S.; Tamaoki, D.; Yoshitomi, K.; Akimitsu, K.; Gomi, K. Multiple roles of plant volatiles in jasmonate-induced defense response in rice. *Plant Signal. Behav.* **2014**, *9*, e29247. [CrossRef]

17. Taniguchi, S.; Hosokawa-Shinonaga, Y.; Tamaoki, D.; Yamada, S.; Akimitsu, K.; Gomi, K. Jasmonate induction of the monoterpene linalool confers resistance to rice bacterial blight and its biosynthesis is regulated by JAZ protein in rice. *Plant Cell Environ.* **2014**, *37*, 451–461. [CrossRef]

18. Taniguchi, S.; Miyoshi, S.; Tamaoki, D.; Yamada, S.; Tanaka, K.; Uji, Y.; Tanaka, S.; Akimitsu, K.; Gomi, K. Isolation of jasmonate-induced sesquiterpene synthase of rice: Product of which has an antifungal activity against *Magnaporthe oryzae*. *J. Plant Physiol.* **2014**, *171*, 625–632. [CrossRef]

19. Yoshitomi, K.; Taniguchi, S.; Tanaka, K.; Uji, Y.; Akimitsu, K.; Gomi, K. Rice *terpene synthase 24* (*OsTPS24*) encodes a jasmonate-responsive monoterpene synthase that produces an antibacterial γ-terpinene against rice pathogen. *J. Plant Physiol.* **2016**, *191*, 120–126. [CrossRef]

20. Kiryu, M.; Hamanaka, M.; Yoshitomi, K.; Mochizuki, S.; Akimitsu, K.; Gomi, K. Rice *terpene synthase 18* (*OsTPS18*) encodes a sesquiterpene synthase that produces an antibacterial (*E*)-nerolidol against a bacterial pathogen of rice. *J. Gen. Plant Pathol.* **2018**, *84*, 221–229. [CrossRef]

21. Cheng, Y.; Zhou, Y.; Yang, Y.; Chi, Y.-J.; Zhou, J.; Chen, J.Y.; Wang, F.; Fan, B.; Shi, K.; Zhou, Y.H.; et al. Structural and functional analysis of VQ motif-containing proteins in Arabidopsis as interacting proteins of WRKY transcription factors. *Plant Physiol.* **2012**, *159*, 810–825. [CrossRef]

22. Kim, D.Y.; Kwon, S.I.; Choi, C.; Lee, H.; Ahn, I.; Park, S.R.; Bae, S.C.; Lee, S.C.; Hwang, D.J. Expression analysis of rice VQ genes in response to biotic and abiotic stresses. *Gene* **2013**, *529*, 208–214. [CrossRef]

23. Wang, X.; Zhang, H.; Sun, G.; Jin, Y.; Qiu, L. Identification of active VQ motif-containing genes and the expression patterns under low nitrogen treatment in soybean. *Gene* **2014**, *543*, 237–243. [CrossRef]

24. Song, W.; Zhao, H.; Zhang, X.; Lei, L.; Lai, J. Genome-wide identification of VQ motif-containing proteins and their expression profiles under abiotic stresses in maize. *Front. Plant Sci.* **2016**, *6*. [CrossRef]

25. Hu, P.; Zhou, W.; Cheng, Z.; Fan, M.; Wang, L.; Xie, D. *JAV1* controls jasmonate-regulated plant defense. *Mol. Cell* **2013**, *50*, 504–515. [CrossRef]

26. Andreasson, E.; Jenkins, T.; Brodersen, P.; Thorgrimsen, S.; Petersen, N.H.T.; Zhu, S.; Qiu, J.L.; Micheelsen, P.; Rocher, A.; Petersen, M.; et al. The MAP kinase substrate MKS1 is a regulator of plant defense responses. *EMBO J.* **2005**, *24*, 2579–2589. [CrossRef]

27. Petersen, K.; Qiu, J.L.; Lütje, J.; Fiil, B.K.; Hansen, S.; Mundy, J.; Petersen, M. *Arabidopsis* MKS1 is involved in basal immunity and requires an intact N-terminal domain for proper function. *PLoS ONE* **2010**, *5*, e14364. [CrossRef]

28. Qiu, J.L.; Fiil, B.K.; Petersen, K.; Nielsen, H.B.; Botanga, C.J.; Thorgrimsen, S.; Palma, K.; Suarez-Rodriguez, M.C.; Sandbech-Clausen, S.; Lichota, J.; et al. Arabidopsis MAP kinase 4 regulates gene expression through transcription factor release in the nucleus. *EMBO J.* **2008**, *27*, 2214–2221. [CrossRef]

29. Birkenbihl, R.P.; Diezel, C.; Somssich, I.E. Arabidopsis WRKY33 is a key transcriptional regulator of hormonal and metabolic responses toward *Botrytis cinerea* infection. *Plant Physiol.* **2012**, *159*, 266–285. [CrossRef]

30. Li, N.; Li, X.; Xiao, J.; Wang, S. Comprehensive analysis of VQ motif-containing gene expression in rice defense responses to three pathogens. *Plant Cell Rep.* **2014**, *33*, 1493–1505. [CrossRef]

31. Jing, Y.; Lin, R. The VQ motif-containing protein family of plant-specific transcriptional regulators. *Plant Physiol.* **2015**, *169*, 371–378. [CrossRef]

32. Kishi-Kaboshi, M.; Okada, K.; Kurimoto, L.; Murakami, S.; Umezawa, T.; Shibuya, N.; Yamane, H.; Miyao, A.; Takatsuji, H.; Takahashi, A.; et al. A rice fungal MAMP-responsive MAPK cascade regulates metabolic flow to antimicrobial metabolite synthesis. *Plant J.* **2010**, *63*, 599–612. [CrossRef]

33. Ma, H.; Chen, J.; Zhang, Z.; Ma, L.; Yang, Z.; Zhang, Q.; Li, X.; Xiao, J.; Wang, S. MAPK kinase 10.2 promotes disease resistance and drought tolerance by activating different MAPKs in rice. *Plant J.* **2017**, *92*, 557–570. [CrossRef]

34. Kinoshita, E.; Kinoshita-Kikuta, E.; Takiyama, K.; Koike, T. Phosphate-binding tag, a new tool to visualize phosphorylated proteins. *Mol. Cell Proteom.* **2006**, *5*, 749–757. [CrossRef]

35. Hosokawa, T.; Saito, T.; Asada, A.; Fukunaga, K.; Hisanaga, S. Quantitative measurement of in vivo phosphorylation states of Cdk5 activator p35 by Phos-tag SDS-PAGE. *Mol. Cell Proteom.* **2010**, *9*, 1133–1143. [CrossRef]

36. Ueno, Y.; Yoshida, R.; Kishi-Kaboshi, M.; Matsushita, A.; Jiang, C.J.; Goto, S.; Takahashi, A.; Hirochika, H.; Takatsuji, H. MAP kinases phosphorylate rice WRKY45. *Plant Signal. Behav.* **2013**, *8*, e24510. [CrossRef]

37. Ueno, Y.; Yoshida, R.; Kishi-Kaboshi, M.; Matsushita, A.; Jiang, C.J.; Goto, S.; Takahashi, A.; Hirochika, H.; Takatsuji, H. Abiotic Stresses Antagonize the Rice Defence Pathway through the Tyrosine-Dephosphorylation of OsMPK6. *PLoS Pathog.* **2015**, *11*, e1005231. [CrossRef]

38. Liu, S.; Hua, L.; Dong, S.; Chen, H.; Zhu, X.; Jiang, J.; Zhang, F.; Li, Y.; Fang, X.; Chen, F. OsMAPK6, a mitogen-activated protein kinase, influences rice grain size and biomass production. *Plant J.* **2015**, *84*, 672–681. [CrossRef]

39. Xu, R.; Duan, P.; Yu, H.; Zhou, Z.; Zhang, B.; Wang, R.; Li, J.; Zhang, G.; Zhuang, S.; Lyu, J.; et al. Control of grain size and weight by the OsMKKK10-OsMKK4-OsMAPK6 signaling pathway in rice. *Mol. Plant* **2018**, *11*, 860–873. [CrossRef]

40. Katou, S.; Kuroda, K.; Seo, S.; Yanagawa, Y.; Tsuge, T.; Yamazaki, M.; Miyao, A.; Hirochika, H.; Ohashi, Y. A calmodulin-binding mitogen-activated protein kinase phosphatase is induced by wounding and regulates the activities of stress-related mitogen-activated protein kinases in rice. *Plant Cell Physiol.* **2007**, *48*, 332–344. [CrossRef]

41. Ueno, Y.; Matsushita, A.; Inoue, H.; Yoshida, R.; Jiang, C.J.; Takatsuji, H. WRKY45 phosphorylation at threonine 266 acts negatively on WRKY45-dependent blast resistance in rice. *Plant Signal. Behav.* **2017**, *12*, e1356968. [CrossRef]

42. Tamaoki, D.; Seo, S.; Yamada, S.; Kano, A.; Miyamoto, A.; Shishido, H.; Miyoshi, S.; Taniguchi, S.; Akimitsu, K.; Gomi, K. Jasmonic acid and salicylic acid activate a common defense system in rice. *Plant Signal. Behav.* **2013**, *8*, e24260. [CrossRef]

43. Lee, M.O.; Cho, K.; Kim, S.H.; Jeong, S.H.; Kim, J.A.; Jung, Y.H.; Shim, J.; Shibato, J.; Rakwal, R.; Tamogami, S.; et al. Novel rice *OsSIPK* is a multiple stress responsive MAPK family member showing rhythmic expression at mRNA level. *Planta* **2008**, *227*, 981–990. [CrossRef]

44. Wakuta, S.; Suzuki, E.; Saburi, W.; Matsuura, H.; Nabeta, K.; Imai, R.; Matsui, H. OsJAR1 and OsJAR2 are jasmonyl-L-isoleucine synthases involved in wound- and pathogen-induced jasmonic acid signalling. *Biochem. Biophys. Res. Commun.* **2011**, *409*, 634–639. [CrossRef]

45. Riemann, M.; Haga, K.; Shimizu, T.; Okada, K.; Ando, S.; Mochizuki, S.; Nishizawa, Y.; Yamanouchi, U.; Nick, P.; Yano, M.; et al. Identification of rice *Allene Oxide Cyclase* mutants and the function of jasmonate for defence against *Magnaporthe oryzae*. *Plant J.* **2013**, *74*, 226–238. [CrossRef]

46. Shimizu, T.; Miyamoto, K.; Miyamoto, K.; Minami, E.; Nishizawa, Y.; Iino, M.; Nojiri, H.; Yamane, H.; Okada, K. OsJAR1 contributes mainly to biosynthesis of the stress-induced jasmonoyl-isoleucine involved in defense responses in rice. *Biosci. Biotechnol. Biochem.* **2013**, *77*, 1556–1564. [CrossRef]

47. Zhang, X.; Bao, Y.; Shan, D.; Wang, Z.; Song, X.; Wang, Z.; Wang, J.; He, L.; Wu, L.; Zhang, Z.; et al. *Magnaporthe oryzae* induces the expression of a microRNA to suppress the immune response in rice. *Plant Physiol.* **2018**, *177*, 352–368.

48. Patkar, R.N.; Benke, P.I.; Qu, Z.; Chen, Y.Y.C.; Yang, F.; Swarup, S.; Naqvi, N.I. A fungal monooxygenase-derived jasmonate attenuates host innate immunity. *Nat. Chem. Biol.* **2015**, *11*, 733–740. [CrossRef]

49. Yuan, Y.; Zhong, S.; Li, Q.; Zhu, Z.; Lou, Y.; Wang, L.; Wang, J.; Wang, M.; Li, Q.; Yang, D.; et al. Functional analysis of rice *NPR1*-like genes reveals that *OsNPR1/NH1* is the rice orthologue conferring disease resistance with enhanced herbivore susceptibility. *Plant Biotechnol. J.* **2007**, *5*, 313–324. [CrossRef]

50. Lee, S.H.; Sakuraba, Y.; Lee, T.; Kim, K.W.; An, G.; Lee, H.Y.; Paek, N.C. Mutation of *Oryza sativa CORONATINE INSENSITIVE 1b* (*OsCOI1b*) delays leaf senescence. *J. Integr. Plant Biol.* **2015**, *57*, 562–576. [CrossRef]

51. Takasaki, H.; Maruyama, K.; Kidokoro, S.; Ito, Y.; Fujita, Y.; Shinozaki, K.; Yamaguchi-Shinozaki, K.; Nakashima, K. The abiotic stress-responsive NAC-type transcription factor OsNAC5 regulates stress-inducible genes and stress tolerance in rice. *Mol. Genet. Genom.* **2010**, *284*, 173–183. [CrossRef]

52. Song, S.Y.; Chen, Y.; Chen, J.; Dai, X.Y.; Zhang, W.H. Physiological mechanisms underlying OsNAC5-dependent tolerance of rice plants to abiotic stress. *Planta* **2011**, *234*, 331–345. [CrossRef]

53. Kazan, K. Diverse roles of jasmonates and ethylene in abiotic stress tolerance. *Trends Plant Sci.* **2015**, *20*, 219–229. [CrossRef]

54. Toda, Y.; Tanaka, M.; Ogawa, D.; Kurata, K.; Kurotani, K.I.; Habu, Y.; Ando, T.; Sugimoto, K.; Mitsuda, N.; Katoh, E.; et al. RICE SALT SENSITIVE3 forms a ternary complex with JAZ and class-C bHLH factors and regulates jasmonate-induced gene expression and root cell elongation. *Plant Cell* **2013**, *25*, 1709–1725. [CrossRef]

55. Sakuraba, Y.; Piao, W.; Lim, J.H.; Han, S.H.; Kim, Y.S.; An, G.; Paek, N.C. Rice ONAC106 inhibits leaf senescence and increases salt tolerance and tiller angle. *Plant Cell Physiol.* **2015**, *56*, 2325–2339. [CrossRef]

56. Uji, Y.; Akimitsu, K.; Gomi, K. Identification of OsMYC2-regulated senescence-associated genes in rice. *Planta* **2017**, *245*, 1241–1246. [CrossRef]

57. Kauffman, H.E.; Reddy, A.P.K.; Hsieh, S.P.Y.; Merca, S.D. An improved technique for evaluating resistance of rice varieties to *Xanthomonas oryzae*. *Plant Dis. Rep.* **1973**, *57*, 537–541.

58. Sato, H.; Imiya, Y.; Ida, S.; Ichii, M. Characterization of four molybdenum cofactor mutants of rice, *Oryza sativa* L. *Plant Sci.* **1996**, *119*, 39–47. [CrossRef]

59. Gomi, K.; Satoh, M.; Ozawa, R.; Shinonaga, Y.; Sanada, S.; Sasaki, K.; Matsumura, M.; Ohashi, Y.; Kanno, H.; Akimitsu, K.; et al. Role of hydroperoxide lyase in white-backed planthopper (*Sogatella furcifera Horváth*)-induced resistance to bacterial blight in rice, *Oryza sativa* L. *Plant J.* **2010**, *61*, 46–57. [CrossRef]

60. Nishizawa, Y.; Nishio, Z.; Nakazono, K.; Soma, M.; Nakajima, E.; Ugaki, M.; Hibi, T. Enhanced resistance to blast (*Magnaporthe grisea*) in transgenic Japonica rice by constitutive expression of rice chitinase. *Theor. Appl. Genet.* **1999**, *99*, 383–390. [CrossRef]

61. Shen, W.J.; Forde, B.G. Efficient transformation of *Agrobacterium* spp. by high voltage electroporation. *Nucleic Acids Res.* **1989**, *17*, 8385.

62. Hiei, Y.; Ohta, S.; Komari, T.; Kumashiro, T. Efficient transformation of rice (*Oryza sativa* L.) mediated by *Agrobacterium* and sequence analysis of the boundaries of the T-DNA. *Plant J.* **1994**, *6*, 271–282. [CrossRef]

63. Arnon, D.I. Copper enzymes in isolated chloroplasts. Polyphenoloxidase in *Beta vulgaris*. *Plant Physiol.* **1949**, *24*, 1–15. [CrossRef]

64. Shevchenko, A.; Jensen, O.N.; Podtelejnikov, A.V.; Sagliocco, F.; Wilm, M.; Vorm, O.; Mortensen, P.; Shevchenko, A.; Boucherie, H.; Mann, M. Linking genome and proteome by mass spectrometry: Large-scale identification of yeast proteins from two dimensional gels. *Proc. Natl. Acad. Sci. USA* **1996**, *93*, 14440–14445. [CrossRef]

© 2019 by the authors. Licensee MDPI, Basel, Switzerland. This article is an open access article distributed under the terms and conditions of the Creative Commons Attribution (CC BY) license (http://creativecommons.org/licenses/by/4.0/).

International Journal of
Molecular Sciences

Communication

Identification of Jasmonic Acid Biosynthetic Genes in Sweet Cherry and Expression Analysis in Four Ancient Varieties from Tuscany

Roberto Berni [1,2], Giampiero Cai [1], Xuan Xu [3], Jean-Francois Hausman [3] and Gea Guerriero [3,*]

[1]   University of Siena, Department of Life Sciences, via P.A. Mattioli 4, 53100 Siena, Italy
[2]   Trees and timber institute-National research council of Italy (CNR-IVALSA), via Aurelia 49, 58022 Follonica (GR), Italy
[3]   Research and Innovation Department, Luxembourg Institute of Science and Technology, 5 avenue des Hauts-Fourneaux, L-4362 Esch/Alzette, Luxembourg
*   Correspondence: gea.guerriero@list.lu; Tel.: +352-275-888-5096

Received: 26 June 2019; Accepted: 18 July 2019; Published: 22 July 2019

Abstract: Sweet cherries are non-climacteric fruits whose early development is characterized by high levels of the phytohormone jasmonic acid (JA). Important parameters, such as firmness and susceptibility to cracking, can be affected by pre- and postharvest treatments of sweet cherries with JA. Despite the impact of JA on sweet cherry development and fruit characteristics, there are no studies (to the best of our knowledge) identifying the genes involved in the JA biosynthetic pathway in this species. We herein identify the sweet cherry members of the lipoxygenase family (*13-LOX*); allene oxide synthase, allene oxide cyclase and 12-oxo-phytodienoic acid reductase 3, as well as genes encoding the transcriptional master regulator MYC2. We analyze their expression pattern in four non-commercial Tuscan varieties ('Carlotta', 'Maggiola', 'Morellona', 'Crognola') having different levels of bioactives (namely phenolics). The highest differences are found in two genes encoding 13-LOX in the variety 'Maggiola' and one MYC2 isoform in 'Morellona'. No statistically-significant variations are instead present in the allene oxide synthase, allene oxide cyclase and 12-oxo-phytodienoic acid reductase 3. Our data pave the way to follow-up studies on the JA signaling pathway in these ancient varieties, for example in relation to development and post-harvest storage.

Keywords: *Prunus avium*; Tuscan varieties; jasmonic acid; lipoxygenase; bioinformatics; gene expression

---

1. Introduction

Prunus avium L. is an economically relevant tree of the *Rosaceae* family producing stone fruits. Its fruits are appreciated worldwide for their aroma, taste and richness in bioactive molecules. Sweet cherries are indeed rich in polyphenols and pentacyclic triterpenes [1,2]. They have a low content of calories. They are cholesterol-free and therefore valuable from a nutraceutical point of view. The ripening of sweet cherries is accompanied by an early increase in endogenous jasmonic acid (JA) levels [3]. This phytohormone is known to regulate different aspects of plant growth and development [4], as well as the response to exogenous stresses [5,6].

Given the physiological relevance of JA, many studies in the literature have specifically looked at the response of plants to this phytohormone in relation to flower and seed development [7], fruit ripening [8,9], senescence [10] and secondary metabolite production [11–13]. Concerning fruits, the majority of papers have dealt with the postharvest application of JA and its derivatives, which improved shelf life by protecting against pathogen attack and by enhancing the antioxidant content [14]. Studies have also investigated the role of JA in the ripening of climacteric and, to a lesser extent, in non-climacteric fruits: In the former (e.g., apples), jasmonates act with ethylene in regulating ripening [15], while in the latter (e.g., strawberries) the phytohormone affects cell walls, modifying enyzmes, anthocyanin accumulation and lignin biosynthetic genes [16]. Some studies are present in the literature concerning the response of sweet cherries to JA: More specifically, derivatives of this phytohormones were applied both pre- and postharvest to improve the content of bioactives and the firmness of the fruits [8,17,18]. The preharvest application of methyl jasmonate (MeJA) at a concentration of 10 mM improved fruit flesh firmness, slowed down color changes and delayed harvest by one week [8]. Additionally, MeJA (applied at a concentration of 0.4 mM) alone or in combination with 0.1 mM abscisic acid (ABA) significantly reduced the susceptibility to cracking after tests consisting of 6 h-water immersion [17]. Interestingly, this effect was observed over 2 consecutive years and on fruits harvested both at the stage of fruit set and color change. Postharvest treatment of sweet cherries with MeJA also increased the activities of a peroxidase and β-1,3-glucanase: Those findings suggest a possible protection against biotic stress, although little inhibitory activity was observed on *Monilinia fructicola* [18].

Despite the economic value of sweet cherries and the physiological relevance of JA and its derivatives, no study is available, to the best of our knowledge, on the identification of genes related to its biosynthesis and signaling *in P. avium*. The content of JA is known to be extremely low in sweet cherries at maturity [3]; therefore, we made the choice of focusing on JA-related genes in this study. Here we identify the sweet cherry members of the lipoxygenase family (*13-LOX*), *allene oxide synthase (AOS)*, *allene oxide cyclase (AOC)* and *12-oxo-phytodienoic acid reductase 3 (OPR3)*, as well as the transcription factor (TF) *MYC2*. We provide data concerning gene expression in non-commercial ancient fruits collected at maturity (60 days post anthesis, dpa) and belonging to the regional germplasm of Tuscany [19]. Since no similar study exists on *P. avium* (and even more so on these ancient fruits) and since maturity is the interesting stage for an exploitation, we decided to measure gene expression at commercial harvest. In particular, we focused our attention on 4 varieties which were previously described to contain varying levels of bioactives ('Carlotta' and 'Maggiola' that produce lower amounts of phenolics, 'Morellona' and 'Crognola' that are instead high producers) [1]. The goal is to pave the way to follow-up studies aiming at analyzing JA signaling in these ancient fruits; e.g., both during development and postharvest storage. A kinetic study could indeed provide additional insight into the relationship existing between JA biosynthesis and the content of phenolics in sweet cherries and ultimately unveil potential links between early development and the higher antioxidant capacities measured in some of the Tuscan varieties. A study on postharvest treatments with the phytohormone or its derivatives will evaluate the effects on the stability of the fruits produced by the varieties studied here. A high level of bioactives, together with an increase in postharvest stability are indeed important characteristics for a potential exploitation of these ancient fruits.

2. Results and Discussion

2.1. Identification of JA Biosynthetic Genes in Sweet Cherries

The BLAST analysis carried out to mine the *13-LOX* genes in the genome of sweet cherry [20] using the six known thale cress sequences [21] led to the identification of five *9-LOXs* and six members of the *13-LOX* family. This number falls in the same range as the total number of LOX found in *Carica papaya* and *Vitis vinifera*, where a total of 11 and 13 genes encoding LOXs were reported [22]. The maximum likelihood phylogenetic analysis showed that *P. avium*'s corresponding proteins nicely cluster into two separate branches, one for 9-LOXs (Pav_sc0000700.1_g070.1.mk/XP_021817420.1, Pav_sc0000700.1_g050.1.mk/XP_021817418.1, Pav_sc0000700.1_g080.1.mk/XP_021817422.1, Pav_sc0000861.1_g060.1.mk/XP_021820334.1, Pav_sc0000648.1_g280.1.mk/XP_021817417.1) and one for 13-LOXs (Figure 1).

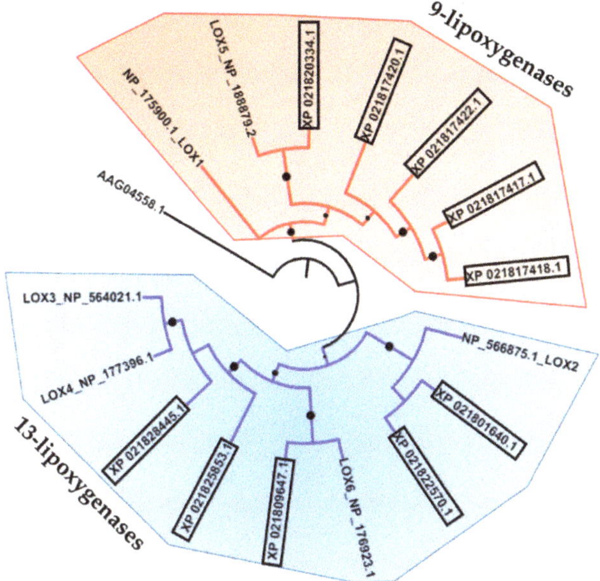

Figure 1. Maximum likelihood phylogenetic tree (number of bootstraps in ultrafast mode: 1000) of different LOX protein's full length sequences from thale cress and *P. avium* (boxed). Accession numbers from thale cress are NP_175900.1, NP_188879.2, NP_566875.1, NP_176923.1, NP_177396.1, NP_564021.1. All the other accessions in the tree are sequences from *P. avium* (see Table 1). Bootstrap values comprised of between 0.8 and 1 are displayed (the bigger the circle, the higher the bootstrap value). The tree was rooted with *Pseudomonas aeruginosa* lipoxygenase PAO1 (accession number AAG04558.1).

Table 1. Details of the identified JA biosynthetic genes (*13-LOX, AOS, AOC, OPR3, MYC2*) in *P. avium*, with proposed nomenclature, GDR (Genomic Database for *Rosaceae*) codes and NCBI accession numbers. Genes separated by/indicate paralogs.

Proposed Nomenclature	GDR Code	NCBI Accession (Best Hit)
PavLOX2	Pav_sc0001040.1_g410.1.mk	XP_021822553.1
PavLOX2.2	Pav_sc0001040.1_g480.1.mk	XP_021822570.1
PavLOX2.3	Pav_sc0005842.1_g020.1.mk/Pav_sc0001040.1_g230.1.br	XP_021801640.1/XP_021822570.1
PavLOX3	Pav_sc0001305.1_g640.1.mk	XP_021825853.1
PavLOX3.2	Pav_sc0001580.1_g270.1.mk	XP_021828445.1
PavLOX6	Pav_sc0000351.1_g320.1.mk	XP_021809647.1
PavAOS	Pav_sc0000890.1_g1300.1.mk/Pav_sc0004356.1_g050.1.mk	XP_021820947.1/XP_021800682.1
PavAOC	Pav_sc0000618.1_g380.1.mk	XP_021804158.1
PavAOC 2	Pav_sc0000567.1_g750.1.mk	XP_021814377.1
PavOPR3	Pav_sc0000129.1_g860.1.mk	XP_021804971.1
PavMYC2	Pav_sc0000107.1_g180.1.mk	XP_021803466.1
PavMYC2.2	Pav_sc0000652.1_g730.1.mk	XP_021816617.1
PavMYC2.3	Pav_sc0006499.1_g050.1.mk	XP_021802110.1

Based on the phylogenetic relationship with the thale cress 13-LOX, we could assign a nomenclature for the identified sweet cherry lipoxygenases (Table 1). The gene *PavLOX2.3* corresponds to two paralogs; however, it should be noted that Pav_sc0001040.1_g230.1.br is much shorter and may represent an incomplete sequence; indeed the best hit in NCBI corresponds to XP_021822570.1, which is 914 amino acids. The presence of a chloroplast signaling peptide was confirmed for all of the 13-LOX from *P. avium* (scores berween 0.509–0.565).

Additionally, we identified two *AOCs*, one *AOS* and one thale cress ortholog of *OPR3* (Table 1). Interestingly, three orthologs of the master regulator *MYC2* were retrieved from the genome of sweet cherry (Table 1): The gene named *MYC2.2* showed the highest percentage identity (obtained when using the blastx algorithm, and as input, the GDR nucleotide sequences indicated in Table 1) with *Arabidopsis* MYC2 (85.71%), while MYC2 and MYC2.3 had lower scores (37.88% and 52.51%, respectively).

The MYC2 phylogenetic tree shows that PavMYC2.3 branches together with *V. vinifera* MYC2 (Figure 2), while the other sweet cherry MYC2 proteins do not show obvious similarities with MYC2 from other plant species [23].

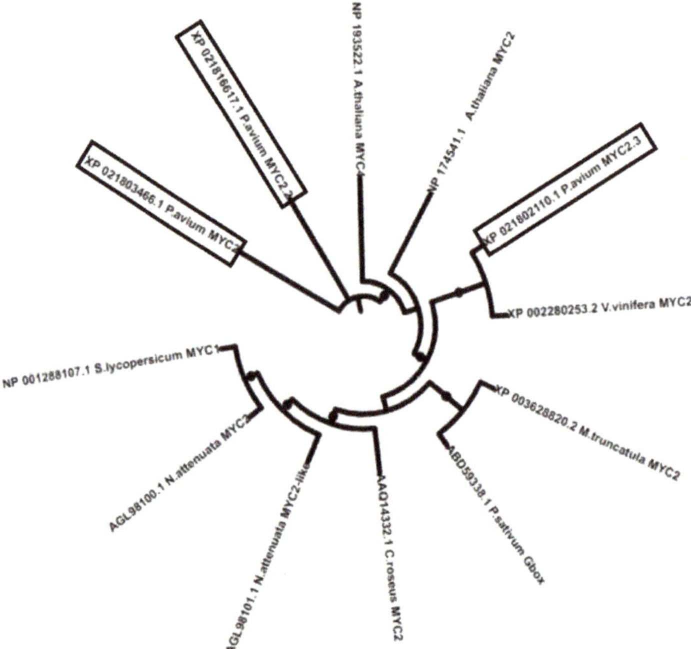

Figure 2. Unrooted maximum likelihood phylogenetic tree (number of bootstraps in ultrafast mode: 1000) of different MYC2 protein's full length sequences from *P. avium* (boxed) and other species taken from [23]. Bootstrap values comprised of between 0.8 and 1 are displayed (the bigger the circle, the higher the bootstrap value).

2.2. Total and Targeted Quantification of Phenolics in the Four Ancient Varieties

We previously characterized six ancient Tuscan varieties of *P. avium* sampled in 2017 from a nutraceutical point of view and ranked them according to their content in antioxidants, polyphenols, flavonoids and anthocyanins [1]. We here quantify the same parameters in a subset of four varieties known to show differences in the content of bioactives and sampled in the year 2018. We indeed wished to confirm the stability in the trend of the measured parameters over a different year of harvest. The results are shown in Table 2: Concerning the total antioxidants, the variety 'Crognola' ranks first, followed by 'Morellona', while 'Maggiola' is the lowest. The same ranking is observed for the polyphenols and flavonoids, while anthocyanins are highest in 'Morellona'. It should be noted that this variety is red fleshed, differently from all the others. Some differences are, however, observed with respect to the previously published data on fruits sampled in 2017 [1], where 'Carlotta' was the lowest producer of antioxidants, polyphenols and anthocyanins. In 2017, the highest amount of anthocyanins was observed for 'Crognola,' despite the absence of a red flesh.

Table 2. The table reports the values (± standard deviation, number of independent biological replicates = 3) obtained from the sweet cherry varieties sampled in 2018. The results show the total content of antioxidants expressed as mmol Fe^{2+} per 100 g fresh weight (FW), polyphenols as mg of GAE (gallic acid equivalents) per 100 g of FW, flavonoids as mg of QeE (quercetin equivalents) per 100 g of FW and anthocyanins as CyE (cyanidin-3-glucoside equivalents) per 100 g of FW. Different letters indicate statistically significant differences ($p < 0.05$) among groups at the one-way ANOVA with Tukey's post-hoc test.

Variety	Total Antioxidants (mmol Fe^{2+} mmol/100g FW)	Polyphenols (mg GAE/100g FW)	Flavonoids (mg QeE/100g FW)	Anthocyanins (mg CyE/100g FW)
'Carlotta'	1.73 ± 0.01 [b]	159.18 ± 0.41 [b]	49.33 ± 0.71 [b]	34.46 ± 0.81 [b]
'Morellona'	2.22 ± 0.02 [c]	313.11 ± 3.55 [c]	97.86 ± 1.56 [c]	65.27 ± 1.03 [d]
'Maggiola'	1.43 ± 0.02 [a]	137.70 ± 2.01 [a]	40.06 ± 1.23 [a]	32.41 ± 0.91 [a]
'Crognola'	3.07 ± 0.01 [d]	387.11 ± 1.29 [d]	102.05 ± 2.42 [d]	58.46 ± 1.11 [c]

The targeted quantification of phenolics on the cherries sampled in 2018 reflects the spectrophotometric analyses, with the only exception of *p*-coumaric acid, whose abundance is significantly higher in 'Maggiola' as compared to 'Carlotta' and even 'Morellona' (Table 3). Despite being fruits sampled in an experimental field, the standard devations are overall low: This is partly due to our choice of pooling several fruits per biological replicate and to the grinding of the whole tissues (exocarp, mesocarp and skin). Eventual differences in the content of bioactive molecules in the pericarp and skin are thus levelled across replicates. These results differ from the values reported in 2017, where the values were the lowest for 'Maggiola' [1]. The samples were harvested from trees grown in an experimental field and therefore exposed to the natural environment—the differences observed may have been due to the varying climatic conditions between 2017 and 2018. It will be interesting to carry out high-throughput studies involving proteomics and transcriptomics to understand the reasons for the recorded differences in the two years of harvest.

Table 3. The table reports the HPLC quantifications of specific bioactive molecules in the sweet cherry varieties sampled in 2018 expressed as µg per gram of fresh weight (FW). The values are indicated with the relative standard deviation (number of independent biological replicates = 3); different letters indicate statistically significant differences ($p < 0.05$) among groups at the one-way ANOVA with Tukey's post-hoc test.

Variety	Chlorogenic Acid (µg/g FW)	*p*-Coumaric Acid (µg/g FW)	(+)-Catechin (µg/g FW)	Rutin (µg/g FW)	Cyanidin-3-Glucoside (µg/g FW)
'Carlotta'	179.42 ± 0.83 [b]	29.02 ± 0.57 [a]	122.83 ± 4.96 [b]	32.54 ± 0.63 [b]	59.50 ± 1.02 [b]
'Morellona'	276.38 ± 0.98 [c]	51.17 ± 1.11 [b]	172.55 ± 1.10 [c]	39.05 ± 0.43 [b]	80.62 ± 0.77 [c]
'Maggiola'	99.23 ± 0.57 [a]	76.50 ± 0.88 [c]	46.74 ± 0.99 [a]	29.11 ± 0.72 [a]	31.26 ± 1.47 [a]
'Crognola'	312.67 ± 1.11 [d]	123.49 ± 0.64 [d]	219.44 ± 2.49 [d]	99.78 ± 0.55 [c]	149.77 ± 1.29 [d]

2.3. Gene Expression Analysis of JA Biosynthetic Genes in the Four Ancient Varieties

The qPCR analysis was performed on the same four varieties and on 8 targets whose expression was detectable in mature fruits, namely *LOX6*, *AOC2*, *AOS*, *LOX3*, *LOX3.2*, *MYC2.2*, *MYC2.3* and *OPR3*. We reasoned that these four varieties would be the best samples to highlight an eventual correlation between bioactive content and the expression of genes related to JA biosynthesis. As can be seen in Figure 3, the four ancient varieties differ in the expression of *LOX* genes and *MYC2.2*. More specifically, the variety 'Maggiola' shows the highest expression of *LOX3* and *LOX3.2*. *LOXs* code for non-haem iron-containing dioxygenases catalyzing the oxygenation of polyunsaturated fatty acids (PUFAs), such as α-linoleic or α-linolenic acid, and lead to the synthesis of molecules collectively known as oxylipins [24]. LOXs known as 13-LOXs catalyze the oxygenation of the C atom in position 13 in the PUFA α-linolenic acid which will then enter the JA biosynthetic pathway via the action of the enzymes AOS and AOC [25]. The higher expression of *LOX3* and *LOX3.2* in 'Maggiola,' coupled to the lack of any statistically significant differences in *AOC2* and *AOS* expressions in the Tuscan fruits, suggest that the identified genes partake in other biosynthetic branches (at least in the developmental stage herein analyzed). For example, the two genes may be involved in the LOX branch linked with hydroperoxide lyases (HPLs) which leads to the formation of volatiles responsible for the aroma of fruits. Plant HPLs have been classified into those acting on 13-hydroperoxylinoleic acid and 13-hydroperoxy-α-linolenic acid and those cleaving 9-hydroperoxy isomers of linoleic and α-linolenic acids [26]. In the literature, it was indeed shown that, at maturity, the content of C6 aldehydes increased in Chinese varieties of sweet cherry [27].

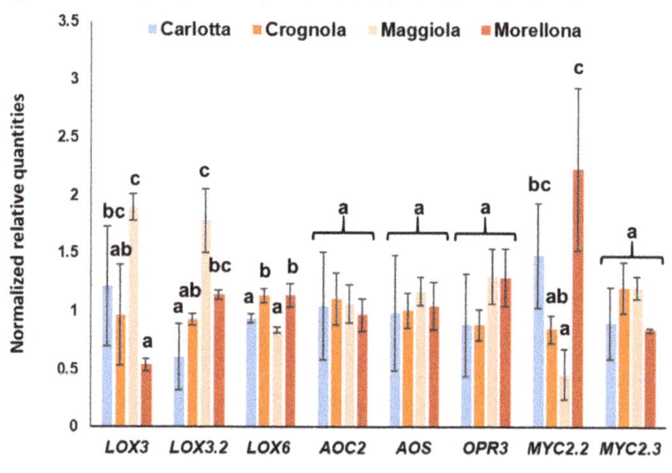

Figure 3. Relative expression of JA biosynthetic genes in the four Tuscan varieties. Error bars refer to the standard deviation (number of independent biological replicates = 4). Different letters indicate statistically significant differences among the groups of the one-way ANOVA with Tukey's post-hoc test.

LOX6 shows a different pattern (Figure 3): Even if slight, the higher expression in 'Crognola' and 'Morellona' is statistically significant and follows the same trend observed for the antioxidants, polyphenols, flavonoids and anthocyanins (Table 2). The Pearson correlation analysis showed a strong positive correlation between *LOX6* and total antioxidanty, polyphenols, flavonoids, and anthocyanins, as well as chlorogenic acid, catechin and cyanidin-3-glucoside (Supplementary File 1). Although JA is known to activate the late genes in the anthocyanin biosynthetic pathway via COI1 (Coronatine-Insensitive 1) in thale cress [28], a previous study comparing apple and sweet cherry fruits highlighted, in the latter, a lack of correlation between endogenous JA/MeJA content and anthocyanin accumulation at maturity [3]. Further experimental proofs are needed to understand the role of *P. avium* LOX6.

We also measured the expression of two genes encoding members of the master regulator MYC2: *MYC2.2* and *MYC2.3*, which show different trends, with the former expressed at the highest levels in 'Morellona' and the latter displaying no significant variations (Figure 3). MYC2 is known to be involved in flavonoid biosynthesis [29,30] and 'Morellona' is one of the highest producers; the gene is, however, expressed at lower levels in 'Crognola', despite its high levels of bioactives (Tables 2 and 3). RNA was extracted from samples comprising both the skin and mesocarp. The variety 'Crognola' is characterized by fruits displaying an intense red-colored skin but a yellow pulp: This may be due to a lower expression of *MYC2.2* with respect to 'Morellona'.

3. Materials and Methods

3.1. Fruit Harvesting

The 18-year-old sweet cherry trees were grown in an experimental field of the National Research Council (CNR) in Follonica (Tuscany, Italy). The coordinates of the experimental field were previously reported [31]. The fruits were harvested by picking them all over the tree canopy at an average height of 1.70–1.90 m from the ground. Sampling took place in 2018 (sampling day: 18 May 2018, sampling time: Between 9:00 and 10:00 a.m., daily maximum temperature: 24 °C, daily minimum temperature: 8 °C, humidity: 74%). The sampled fruits were immediately placed in liquid nitrogen, then brought to the laboratory for long term storage at −80 °C.

3.2. Chemical Assays, HPLC Analyses and Statistics

For the determination of the total antioxidants, polyphenols, flavonoids and anthocyanins, we followed the methods previously described [1]. The analyses were performed in 3 technical replicates and 3 biological replicates (each biological replicate consisted of a pool of 3 sweet cherry fruits). The quantification of targeted secondary metabolites was performed on an HPLC-DAD system (PerkinElmer Series 200 HPLC Systems, Diode Array Detector, Norwalk, CT, USA) and by using curves generated with standard molecules, as previously reported [1]. The values obtained by averaging 3 independent technical replicates were log-transformed, and a one-way ANOVA with a Tukey's post-hoc test was performed with IBM SPSS Statistics v19 (IBM SPSS, Chicago, IL, USA) to determine the statistically significant differences among groups.

3.3. RNA Extraction, Primer Design and Real-Time PCR Data Analysis

The RNA extraction method, together with quantification, assessment of integrities, reverse transcription and real-time PCR conditions were as previously reported [31]. The primers used for the reference genes have been previously published [2,31]; those for the JA biosynthetic genes are reported in Table 4. They were designed with Primer3Plus (http://www.bioinformatics.nl/cgi-bin/primer3plus/primer3plus.cgi, last accessed on: 20 May 2019) and the presence of self-dimers, hetero-dimers, hairpins was checked with OligoAnalyzer 3.1 (http://eu.idtdna.com/calc/analyzer, last accessed on: 20 May 2019).

Table 4. List of primers with details of the sequences, amplicon sizes and amplification efficiencies.

Name	Sequence (5′→3′)	R^2	Tm (°C)	Amplicon Size (bp)	Efficiency (%)
Pav_LOX3 Fwd Pav_LOX3 Rev	TCTTGACCTCATTGGGAACC ACCTGCTTGGATGGTGAATC	0.995	79.65	85	104.03
Pav_LOX3.2 Fwd Pav_LOX3.2 Rev	GCCATCAGTGAAGATTTGGTG CCTCCTTGATCTTGTTCCTCAC	0.995	81.39	106	96.7
Pav_LOX6 Fwd Pav_LOX6 Rev	CAGCATGGTGAAAGAGGTTC AAGCAAATCTATCCCCTTT	0.992	79.51	86	91.51
Pav_AOS Fwd Pav_AOS Rev	GGAGATGTTGTTCGGGTTTC CTACAAACTCCTCCGCT	0.995	78.98	74	104.42
Pav_AOC2 Fwd Pav_AOC2 Rev	CAATCTCTCGCATTCCTTCC AGTTCTTGGAGTTTGGGAA	0.996	80.32	71	92.44
Pav_OPR3 Fwd Pav_OPR3 Rev	CAAGTGGTGGAGCATTATCG AGTTTTGAGCCCCAGTCTTG	0.987	85.01	89	104.42
Pav_MYC2.2 Fwd Pav_MYC2.2 Rev	CCGCTCTGTTGTTCCAAATG TAGCCTCCAATTCCTCAACC	0.982	80.05	106	95.12
Pav_MYC2.3 Fwd Pav_MYC2.3 Rev	GGGTGAAGGGTTTTACAAGG GGACTTTTTTCCTGTACTCC	0.997	83.31	96	91.58

The gene expression values were determined with qBase^{PLUS} (version 3.2, Biogazelle, Ghent, Belgium) by using *actin7* and *PP2A* as reference genes. Statistics were performed with IBM SPSS Statistics v19. A one-way ANOVA with a Tukey's post-hoc test was carried out after having transformed the NRQs (Normalized Relative Quantities) in log2 values and after having confirmed the normal distribution with a quantile-quantile plot (Q-Q plot). Four independent biological replicates, each consisting of a pool of 3 fruits were used for the gene expression study.

3.4. Bioinformatics

Sweet cherry LOX, MYC, AOS, AOC, OPD3 were obtained by blasting the thale cress protein sequences in NCBI and the GDR (Genome Data for *Rosaceae*) database (https://www.rosaceae.org/blast/nucleotide/protein, last accessed on: 20 May 2019). The phylogenetic analysis was carried out by using the full-length LOX and MYC protein sequences identified in sweet cherry and thale cress, as well as other plant species. The pair-wise multiple alignment of LOX and MYC2 proteins from *P. avium*, *Arabidopsis thaliana* and the other plant species indicated in Figures 1 and 2 was performed with CLUSTAL-Ω (http://www.ebi.ac.uk/Tools/msa/clustalo/, last accessed on: 20 May 2019) [32], and the alignment was then used for the maximum likelihood phylogenetic tree using the online program W-IQ-TREE [33] (bootstraps in ultrafast mode: 1000; in auto mode, without selecting FreeRate Heterogeneity), available at http://iqtree.cibiv.univie.ac.at. The tree was visualized with iTOL (available at https://itol.embl.de/, last accessed on: 20 May 2019).

The chloroplast targeting sequence was verified with ChloroP [34] (available at http://www.cbs.dtu.dk/services/ChloroP/, last accessed on: 20 May 2019).

4. Conclusions

We have herein identified the JA biosynthetic genes in the economically relevant fruit tree *P. avium*, and measured their expression in four ancient varieties from Tuscany characterized by different levels of bioactives. The data suggest that *LOX3.2* in 'Maggiola' fruits at maturity is not linked to JA production, but to other biochemical branches; e.g., volatile production. *MYC2.2* is highest in 'Morellona,' a variety ranking first in terms of anthocyanins content. With this Communication, we wish to inspire future studies focusing on kinetics and postharvest treatment with JA (or its derivatives), to evaluate the relationship between bioactive contents and JA production, as well as to assess the stability of the Tuscan fruits during postharvest storage.

Supplementary Materials: The following are available online at http://www.mdpi.com/1422-0067/20/14/3569/s1.

Author Contributions: G.G. conceived the idea of writing the paper. R.B. performed the experiments and drafted the first version. R.B., G.C., X.X., J.-F.H., and G.G. contributed to the interpretation of the data and writing of the present paper.

Funding: This research received no external funding.

Acknowledgments: The authors are grateful to the Tuscany Region and the National Research Council (CNR-Italy) for support. R.B. acknowledges the Tuscany Region for financial support through the PhD fellowship "Pegaso."

Conflicts of Interest: The authors declare no conflict of interest.

References

1. Berni, R.; Romi, M.; Cantini, C.; Hausman, J.-F.; Guerriero, G.; Cai, G. Functional Molecules in Locally-Adapted Crops: The Case Study of Tomatoes, Onions, and Sweet Cherry Fruits From Tuscany in Italy. *Front. Plant Sci.* **2019**, *9*, 1983. [CrossRef]

2. Berni, R.; Hoque, M.Z.; Legay, S.; Cai, G.; Siddiqui, K.S.; Hausman, J.-F.; Andre, C.M.; Guerriero, G. Tuscan Varieties of Sweet Cherry Are Rich Sources of Ursolic and Oleanolic Acid: Protein Modeling Coupled to Targeted Gene Expression and Metabolite Analyses. *Molecules* **2019**, *24*, 1590. [CrossRef] [PubMed]

3. Kondo, S.; Tomiyama, A.; Seto, H. Changes of Endogenous Jasmonic Acid and Methyl Jasmonate in Apples and Sweet Cherries during Fruit Development. *J. Am. Soc. Hortic. Sci.* **2000**, *125*, 282–287. [CrossRef]

4. Huang, H.; Liu, B.; Liu, L.; Song, S. Jasmonate action in plant growth and development. *J. Exp. Bot.* **2017**, *68*, 1349–1359. [CrossRef] [PubMed]

5. Koo, A.J. Metabolism of the plant hormone jasmonate: A sentinel for tissue damage and master regulator of stress response. *Phytochem. Rev.* **2018**, *17*, 51–80. [CrossRef]

6. Ahmad, P.; Rasool, S.; Gul, A.; Sheikh, S.A.; Akram, N.A.; Ashraf, M.; Kazi, A.M.; Gucel, S. Jasmonates: Multifunctional Roles in Stress Tolerance. *Front. Plant Sci.* **2016**, *7*, 813. [CrossRef] [PubMed]

7. Wasternack, C.; Forner, S.; Strnad, M.; Hause, B. Jasmonates in flower and seed development. *Biochimie* **2013**, *95*, 79–85. [CrossRef] [PubMed]

8. Saracoglu, O.; Ozturk, B.; Yildiz, K.; Kucuker, E. Pre-harvest methyl jasmonate treatments delayed ripening and improved quality of sweet cherry fruits. *Sci. Hortic.* **2017**, *226*, 19–23. [CrossRef]

9. Asghari, M.; Hasanlooe, A.R. Methyl jasmonate effectively enhanced some defense enzymes activity and Total Antioxidant content in harvested "Sabrosa" strawberry fruit. *Food Sci. Nutr.* **2015**, *4*, 377–383. [CrossRef]

10. Kim, J.; Chang, C.; Tucker, M.L. To grow old: Regulatory role of ethylene and jasmonic acid in senescence. *Front. Plant Sci.* **2015**, *6*, 20. [CrossRef]

11. Mendoza, D.; Cuaspud, O.; Arias, J.P.; Ruiz, O.; Arias, M. Effect of salicylic acid and methyl jasmonate in the production of phenolic compounds in plant cell suspension cultures of *Thevetia peruviana*. *Biotechnol. Rep.* **2018**, *19*, e00273. [CrossRef] [PubMed]

12. Onofrio, C.D.; Cox, A.; Davies, C.; Boss, P.K. Induction of secondary metabolism in grape cell cultures by jasmonates. *Funct. Plant Biol.* **2009**, *36*, 323–338. [CrossRef]

13. Zaheer, M.; Reddy, V.D.; Giri, C.C. Enhanced daidzin production from jasmonic and acetyl salicylic acid elicited hairy root cultures of *Psoralea corylifolia* L. (Fabaceae). *Nat. Prod. Res.* **2016**, *30*, 1542–1547. [CrossRef] [PubMed]

14. Reyes-Díaz, M.; Lobos, T.; Cardemil, L.; Nunes-Nesi, A.; Retamales, J.; Jaakola, L.; Alberdi, M.; Ribera-Fonseca, A. Methyl Jasmonate: An Alternative for Improving the Quality and Health Properties of Fresh Fruits. *Molecules* **2016**, *21*, 567. [CrossRef] [PubMed]

15. Fan, X.; Mattheis, J.P.; Fellman, J.K. A role for jasmonates in climacteric fruit ripening. *Planta* **1998**, *204*, 444–449. [CrossRef]

16. Concha, C.M.; Figueroa, N.E.; Poblete, L.A.; Oñate, F.A.; Schwab, W.; Figueroa, C.R. Methyl jasmonate treatment induces changes in fruit ripening by modifying the expression of several ripening genes in Fragaria chiloensis fruit. *Plant Physiol. Biochem.* **2013**, *70*, 433–444. [CrossRef] [PubMed]

17. Balbontín, C.; Gutiérrez, C.; Wolff, M.; Figueroa, C.R.; Balbontín, C.; Gutiérrez, C.; Wolff, M.; Figueroa, C.R. Effect of abscisic acid and methyl jasmonate preharvest applications on fruit quality and cracking tolerance of sweet cherry. *Chil. J. Agric. Res.* **2018**, *78*, 438–446. [CrossRef]

18. Yao, H.; Tian, S. Effects of pre- and post-harvest application of salicylic acid or methyl jasmonate on inducing disease resistance of sweet cherry fruit in storage. *Postharvest Biol. Technol.* **2005**, *35*, 253–262. [CrossRef]

19. Berni, R.; Cantini, C.; Romi, M.; Hausman, J.-F.; Guerriero, G.; Cai, G. Agrobiotechnology Goes Wild: Ancient Local Varieties as Sources of Bioactives. *Int. J. Mol. Sci.* **2018**, *19*, 2248. [CrossRef]

20. Shirasawa, K.; Isuzugawa, K.; Ikenaga, M.; Saito, Y.; Yamamoto, T.; Hirakawa, H.; Isobe, S. The genome sequence of sweet cherry (*Prunus avium*) for use in genomics-assisted breeding. *DNA Res.* **2017**, *24*, 499–508. [CrossRef]

21. Umate, P. Genome-wide analysis of lipoxygenase gene family in *Arabidopsis* and rice. *Plant Signal. Behav.* **2011**, *6*, 335–338. [CrossRef] [PubMed]

22. Chen, Z.; Chen, D.; Chu, W.; Zhu, D.; Yan, H.; Xiang, Y. Retention and Molecular Evolution of Lipoxygenase Genes in Modern Rosid Plants. *Front. Genet.* **2016**, *7*, 176. [CrossRef] [PubMed]

23. Woldemariam, M.G.; Dinh, S.T.; Oh, Y.; Gaquerel, E.; Baldwin, I.T.; Galis, I. NaMYC2 transcription factor regulates a subset of plant defense responses in *Nicotiana attenuata*. *BMC Plant Biol.* **2013**, *13*, 73. [CrossRef] [PubMed]

24. Chauvin, A.; Lenglet, A.; Wolfender, J.-L.; Farmer, E.E. Paired Hierarchical Organization of 13-Lipoxygenases in *Arabidopsis*. *Plants* **2016**, *5*, 16. [CrossRef] [PubMed]

25. Wasternack, C.; Hause, B. Jasmonates: Biosynthesis, perception, signal transduction and action in plant stress response, growth and development. An update to the 2007 review in Annals of Botany. *Ann. Bot.* **2013**, *111*, 1021–1058. [CrossRef] [PubMed]

26. Noordermeer, M.A.; Veldink, G.A.; Vliegenthart, J.F. Fatty acid hydroperoxide lyase: A plant cytochrome p450 enzyme involved in wound healing and pest resistance. *Chembiochem* **2001**, *2*, 494–504. [CrossRef]

27. Zhang, X.; Jiang, Y.; Peng, F.; He, N.; Li, Y.; Zhao, D. Changes of Aroma Components in Hongdeng Sweet Cherry During Fruit Development. *Agric. Sci. China* **2007**, *6*, 1376–1382. [CrossRef]

28. Shan, X.; Zhang, Y.; Peng, W.; Wang, Z.; Xie, D. Molecular mechanism for jasmonate-induction of anthocyanin accumulation in *Arabidopsis*. *J. Exp. Bot.* **2009**, *60*, 3849–3860. [CrossRef]

29. Dombrecht, B.; Xue, G.P.; Sprague, S.J.; Kirkegaard, J.A.; Ross, J.J.; Reid, J.B.; Fitt, G.P.; Sewelam, N.; Schenk, P.M.; Manners, J.M.; et al. MYC2 Differentially Modulates Diverse Jasmonate-Dependent Functions in Arabidopsis. *Plant Cell* **2007**, *19*, 2225–2245. [CrossRef]

30. Kazan, K.; Manners, J.M. MYC2: The Master in Action. *Mol. Plant* **2013**, *6*, 686–703. [CrossRef]

31. Berni, R.; Piasecki, E.; Legay, S.; Hausman, J.-F.; Siddiqui, K.S.; Cai, G.; Guerriero, G. Identification of the laccase-like multicopper oxidase gene family of sweet cherry (*Prunus avium* L.) and expression analysis in six ancient Tuscan varieties. *Sci. Rep.* **2019**, *9*, 3557. [CrossRef] [PubMed]

32. McWilliam, H.; Li, W.; Uludag, M.; Squizzato, S.; Park, Y.M.; Buso, N.; Cowley, A.P.; Lopez, R. Analysis Tool Web Services from the EMBL-EBI. *Nucleic Acids Res.* **2013**, *41*, W597–W600. [CrossRef] [PubMed]

33. Trifinopoulos, J.; Nguyen, L.-T.; Von Haeseler, A.; Minh, B.Q. W-IQ-TREE: A fast online phylogenetic tool for maximum likelihood analysis. *Nucleic Acids Res.* **2016**, *44*, W232–W235. [CrossRef] [PubMed]

34. Emanuelsson, O.; Nielsen, H.; Von Heijne, G. ChloroP, a neural network-based method for predicting chloroplast transit peptides and their cleavage sites. *Protein Sci.* **1999**, *8*, 978–984. [CrossRef] [PubMed]

 © 2019 by the authors. Licensee MDPI, Basel, Switzerland. This article is an open access article distributed under the terms and conditions of the Creative Commons Attribution (CC BY) license (http://creativecommons.org/licenses/by/4.0/).

International Journal of
Molecular Sciences

Article

BrTCP7 Transcription Factor Is Associated with MeJA-Promoted Leaf Senescence by Activating the Expression of *BrOPR3* and *BrRCCR*

Yan-mei Xu [1], Xian-mei Xiao [1], Ze-xiang Zeng [1], Xiao-li Tan [1], Zong-li Liu [1], Jian-wen Chen [2,*], Xin-guo Su [3] and Jian-ye Chen [1,*]

[1]   State Key Laboratory for Conservation and Utilization of Subtropical Agro-bioresources/Guangdong Provincial Key Laboratory of Postharvest Science of Fruits and Vegetables/Engineering Research Center of Southern Horticultural Products Preservation, Ministry of Education, College of Horticulture, South China Agricultural University, Guangzhou 510642, China

[2]   Scientific Observing and Experimental Station of Crop Cultivation in South China, Ministry of Agriculture, College of Agriculture, South China Agricultural University, Guangzhou 510642, China

[3]   Department of Food Science, Guangdong Food and Drug Vocational College, Guangzhou 510520, China

*   Correspondence: chenjianwen@scau.edu.cn (J.-w.C.); chenjianye@scau.edu.cn (J.-y.C.); Tel.: +86-020-8528-5523 (J.-y.C.)

Received: 21 June 2019; Accepted: 13 August 2019; Published: 14 August 2019

Abstract: The plant hormone jasmonic acid (JA) has been recognized as an important promoter of leaf senescence in plants. However, upstream transcription factors (TFs) that control JA biosynthesis during JA-promoted leaf senescence remain unknown. In this study, we report the possible involvement of a TEOSINTE BRANCHED1/CYCLOIDEA/PCF (TCP) TF BrTCP7 in methyl jasmonate (MeJA)-promoted leaf senescence in Chinese flowering cabbage. Exogenous MeJA treatment reduced maximum quantum yield (Fv/Fm) and total chlorophyll content, accompanied by the increased expression of senescence marker and chlorophyll catabolic genes, and accelerated leaf senescence. To further understand the transcriptional regulation of MeJA-promoted leaf senescence, a class I member of TCP TFs BrTCP7 was examined. BrTCP7 is a nuclear protein and possesses trans-activation ability through subcellular localization and transcriptional activity assays. A higher level of *BrTCP7* transcript was detected in senescing leaves, and its expression was up-regulated by MeJA. The electrophoretic mobility shift assay and transient expression assay showed that BrTCP7 binds to the promoter regions of a JA biosynthetic gene *BrOPR3* encoding OPDA reductase3 (OPR3) and a chlorophyll catabolic gene *BrRCCR* encoding red chlorophyll catabolite reductase (RCCR), activating their transcriptions. Taken together, these findings reveal that BrTCP7 is associated with MeJA-promoted leaf senescence at least partly by activating JA biosynthesis and chlorophyll catabolism, thus expanding our knowledge of the transcriptional mechanism of JA-mediated leaf senescence.

Keywords: Chinese flowering cabbage; leaf senescence; JA; transcriptional activation

---

1. Introduction

As one of the major leafy vegetables, Chinese flowering cabbage (*Brassica rapa* ssp. *Parachinensis*) is a popular in the Asian diet due to its health-promoting compounds and valuable anticancer properties [1–3]. Generally, flowering shoots, stems, and younger leaves of this cabbage are harvested for eating. However, harvested cabbage leaves senesce rapidly during transportation and storage, wilt, and yellow, reducing product quality and commercial value [4–6]. Thus, there is a demand for the elucidation of the molecular mechanisms of leaf senescence of Chinese flowering cabbage. Such knowledge is important for developing practical solutions to maintain the quality and extend the shelf-life of this vegetable.

Previous genetic and molecular studies have demonstrated that leaf senescence is an active biological process that is tightly regulated by various intrinsic and external factors, such as developmental stage, natural plant hormones, and stresses [7–11]. Among the various phytohormones, jasmonic acid (JA) and its derivate methyl jasmonate (MeJA), have been reported to accelerate leaf senescence [12–14]. In a variety of plants such as Arabidopsis, rice, and maize, endogenous JA is found to be significantly accumulated during leaf senescence [15–17]. Exogenous JA/MeJA application can accelerate leaf yellowing and senescence and enhance the expression of several senescence-associated genes (*SAGs*) including chlorophyll catabolic genes (*CCGs*), as well as genes involved in the JA biosynthesis and signaling pathways [18].

Evidence suggests that transcription factors (TFs) are important regulatory proteins that control the onset and progression of JA-induced leaf senescence by altering the expression of *SAGs* [11,14]. For example, Arabidopsis MYC2, 3, and 4 function redundantly to activate JA-induced leaf senescence by directly activating the expression of *SAG29* and *CCGs* [18]. In contrast, bHLH subgroup IIId TFs bHLH03, 13, 14, and 1, target the *SAG29* promoter and repress its expression [19]. WRKY57 directly suppresses the transcription of *SEN4* and *SAG12*, acting as a negative regulator of JA-induced leaf senescence in Arabidopsis [20]. These findings highlight the complicated regulatory network of JA-promoted leaf senescence governed by a series of TFs, promoting the identification of other TFs associated with this process.

TEOSINTE BRANCHED1/CYCLOIDEA/PCF (TCP) proteins, with the conserved TCP domains, are plant-specific TFs that have been identified in many plant species including Arabidopsis [21], rice [22], maize [23], tomato [24], and Chinese cabbage [25]. TCP TFs have been well documented to participate in multiple biological processes such as cell proliferation and growth, and stress response [26]. Several TCP TFs have been shown to mediate hormone-induced changes in cell proliferation and act as modulators, or even as key players of hormone synthesis, transport, and signal transduction [26]. For instance, Arabidopsis TCP15 is required for the optimal balance between auxin levels and cytokinins responses in the developing carpel [27]. Rice OsTCP19 is strongly associated with abscisic acid (ABA)-mediated abiotic stress responses by binding and modulating the activity of *OsABI4* (ABSCISIC ACID INSENSITIVE4), and overexpression of OsTCP19 in Arabidopsis affects not only ABA but also auxin and JA signaling [28]. GhTCP19 from gladiolus was shown to be a positive regulator of the corm dormancy release process by repressing the expression of an ABA biosynthesis gene expression (*GhNCED*, 9-cis-epoxycarotenoid dioxygenase), as well as promoting cytokinins biosynthesis (*GhIPT*, isopentenyl transferase) and signal transduction (*GhARR*, ARABIDOPSIS RESPONSE REGULATORS) [29]. Genome-wide transcriptional analyses revealed that JA-related global responses are affected by gain or loss of TCP activity [30]. Breeze et al. conducted transcriptome analysis and suggested the role of TCPs during leaf senescence in Arabidopsis [31]. Three Arabidopsis TCP TFs, TCP20, TCP9, and TCP4, were further found to target the promoter of a JA biosynthetic gene *LOX2* (Lipoxygenase2), repressing and activating its transcription, thus acting antagonistically controlling JA-mediated leaf development and senescence [32]. The involvement of TCP members in JA-promoted leaf senescence as well as the associated regulatory mechanism remain to be elucidated.

Previously, we identified and characterized several TFs including BrWRKY65 [1], BrWRKY6 [2], BrERF72 [5], and BrNAC055 [6] involved in leaf senescence in Chinese flowering cabbage, of which BrERF72 was shown to enhance JA accumulation by inducing the expression of three JA biosynthetic genes (*BrLOX4*, *BrAOC3*, and *BrOPR3*) during MeJA-promoted leaf senescence [5]. Leaf senescence is a complex biological process mediated by numerous TFs, and the underpinning regulatory mechanisms vary from species to species and with growing conditions. The current study revealed that a TCP, TF BrTCP7, is positively associated with MeJA-promoted leaf senescence by the direct transcriptional activation of a JA biosynthetic gene *BrOPR3* and a *CCG BrRCCR*.

2. Results and Discussion

2.1. Leaf Senescence of Chinese Flowering Cabbage is Promoted by Exogenous MeJA Treatment

The phytohormone JA-induced leaf senescence has been observed in several plant species [14–19]. Our previous study showed that exogenous MeJA treatment promotes leaf senescence of Chinese flowering cabbage [5], which was further verified in the present work. As shown in Figure 1A, appearance and non-invasive chlorophyll fluorescence imaging with maximum quantum yield (*Fv/Fm*) default demonstrated that, in comparison to the control, MeJA-treated cabbage leaves exhibited a greater degree of yellowing on the fifth and seventh day of storage. As expected, the *Fv/Fm* value and total chlorophyll content were significantly lower in MeJA-treated leaves, which were 70.7% and 70.1% of the control leaves, respectively, on the fifth day of storage (Figure 1B). To investigate why MeJA promotes leaf senescence, we quantified the relative abundance of transcripts of two *SAGs* (*BrSAG12* and *BrSAG19*) and six *CCGs* (*BrPAO*, *BrPPH*, *BrNYC1*, *BrRCCR*, *BrSGR1*, and *BrSGR2*) after MeJA treatment. The transcript levels of all these genes increased more dynamically upon treatment with MeJA, with a 2.89-, 1.44-, 2.03-, 1.81-, 1.47-, 2.62-, 1.93-, and 1.83-fold induction for *BrNYC1*, *BrPPH*, *BrPAO*, *BrRCCR*, *BrSGR1*, *BrSGR2*, *BrSAG12*, and *BrSAG19*, respectively on the fifth day of storage compared with the control (Figure 1C).

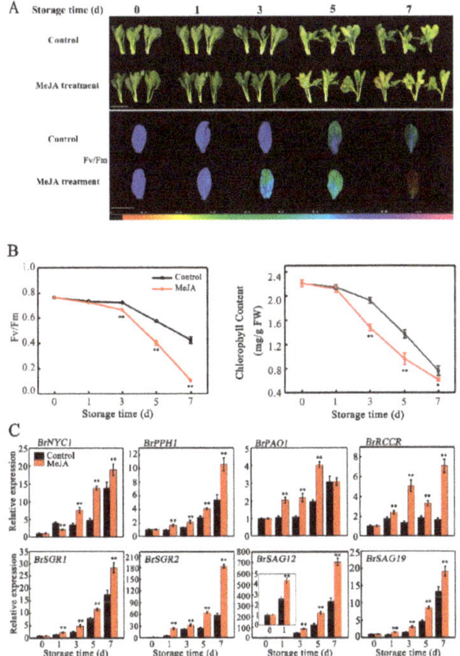

Figure 1. Methyl jasmonate (MeJA) treatment promotes leaf senescence of Chinese flowering cabbage. (**A**) Appearance and chlorophyll fluorescence imaging (Fv/Fm) of control and MeJA-treated cabbage leaves during senescence. The false color code depicted at the bottom of the image ranges from 0 (black) to 1.0 (purple). Bar = 10 cm. (**B**) Changes in Fv/Fm and total chlorophyll content in control and MeJA-treated cabbage leaves during senescence. (**C**) Relative expression of six *CCGs* (*BrNYC1*, *BrPPH1*, *BrPAO1*, *BrRCCR*, *BrSGR1* and *BrSGR2*) and two senescence-marker genes (*BrSAG12* and *BrSAG19*) in control and MeJA-treated cabbage leaves during senescence. Data presented in (**B**) and (**C**) are the mean ± SD of three biological replicates. Asterisks indicate a significant difference in MeJA-treated leaves compared with control leaves (Student's *t*-test: * *p* < 0.05 and ** *p* < 0.01).

2.2. BrTCP7 is A Member of Class I TCP

Transcriptional regulation mediated by TFs, such as MYCs, NACs, and WRKYs, plays an important role in JA-induced leaf senescence [14,33]. Numerous TCP TFs have been identified in plants; however, only few of them were reported to be involved in JA-mediated leaf development and senescence [26,32]. Thus, we focused on the identification of TCP TFs from our RNA-seq transcriptome database associated with Chinese flowering cabbage leaf senescence. Using a false discovery rate (FDR) of less than 0.05 and a fold change larger than 0.5 as thresholds, six genes annotated as TCP proteins were found to be up-regulated (unpublished data). Among these *TCPs*, the member (GenBank number, XM_009152613.2) that was most up-regulated during leaf senescence, was selected for further investigation.

The resulting amplified full-length gene was 753 bp, encoding a protein with 250 amino acid residues, with a calculated molecular weight of 26.97 kDa, and a *p*I value of 9.69. By blasting the NCBI database, we found that this gene shares high similarity (74.9%) with Arabidopsis AtTCP7, and thus was named BrTCP7. The most distinguished characteristic of TCP proteins is the presence of a ~59-amino-acid-long conserved region with a non-canonical basic helix-loop-helix (bHLH) structure (the TCP domain) in their N-terminal, which was proven to be responsible for nuclear targeting, DNA binding, and protein–protein interactions [25,34]. Multiple-alignment of BrTCP7 showed that, similar to other orthologous TCP proteins, BrTCP7 contains the conserved TCP domain (Figure 2A). On the basis of the TCP domain, the TCP members can be grouped into two distinct subfamilies: class I (PCF or TCP-P class) and class II (TCP-C class), among which class II can be further divided into subclades CIN and CYC/TB1 [26,35]. As shown in Figure 2B, the phylogenetic tree shows that BrTCP7 together with AtTCP7, AtTCP14, and SlTCP14 belong to class I.

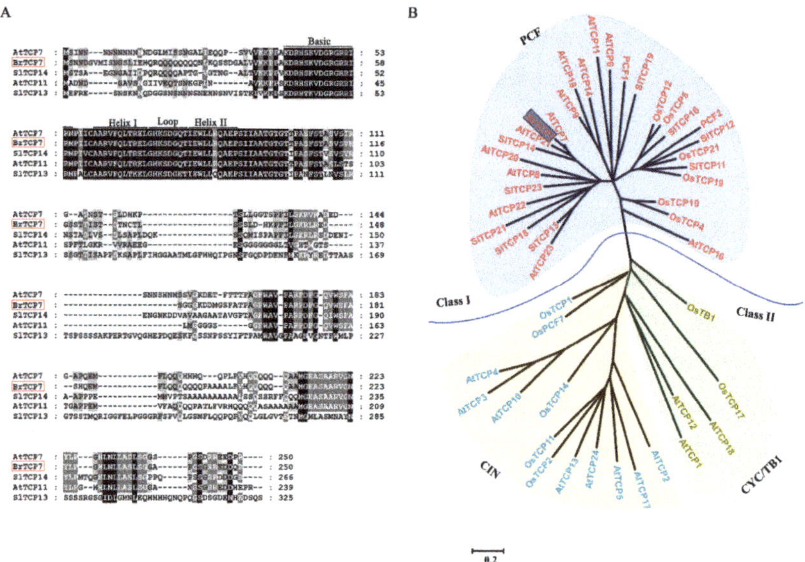

Figure 2. Multiple sequence alignment and phylogenetic analysis of BrTCP7. (**A**) Multiple alignment of BrTCP7 with other TCP members. The following proteins were used for analysis: AtTCP7 (NP_197719.1), AtTCP11 (NP_196450.1), SlTCP13 (NP_001233966.1), and SlTCP14 (NP_001293140.1). Identical and similar amino acids are shaded in black and grey, respectively. Single underlining indicates the conserved non-canonical basic helix–loop–helix structure. (**B**) Phylogenetic analysis of TCPs. BrTCP7 is boxed. The phylogenetic tree was constructed using the maximum likelihood inference method using the MEGA program (version 5.0, Institute of Molecular Evolutionary Genetics, The Pennsylvania State University, PA, USA).

2.3. Molecular Properties of BrTCP7

To evaluate whether *BrTCP7* is associated with MeJA-induced leaf senescence, we quantified its expression in Chinese flowering cabbage during leaf senescence. qRT-PCR analysis showed that, consistent with the RNA-seq data, the transcript level of *BrTCP7* increases during leaf senescence. More pronounced elevation of the *BrTCP7* transcript was observed in MeJA-treated leaves, which were ~0.85- and 0.51-fold higher than that of the control leaves on days 3 and 5 of storage, respectively (Figure 3A). Generally, TCP proteins are localized in nuclei [34,36,37]. Subcellular localization prediction also indicated that BrTCP7 is located in the nuclear region. To verify this prediction, *BrTCP7* was fused to green fluorescent protein (GFP) and transiently expressed in *Nicotiana benthamiana* leaves. Compared with the fluorescence of control 35S-GFP that is distributed throughout the cell, BrTCP7-GFP fusion protein, like the nuclear marker (NLS-mCherry), was exclusively detected in the nucleus (Figure 3B). To provide evidence for potential roles of BrTCP7 in transcriptional regulation, we assessed the transcriptional activity of BrTCP7 in vivo using the dual-luciferase reporter system. As shown in Figure 3C, similar with the activator control VP16, co-infiltration of *BrTCP7* with the reporter significantly increased the luciferase (LUC)/renilla luciferase (REN) ratio. These data reveal that BrTCP7 is a nuclear-localized transcriptional activator that is possibly associated with MeJA-induced leaf senescence of Chinese flowering cabbage.

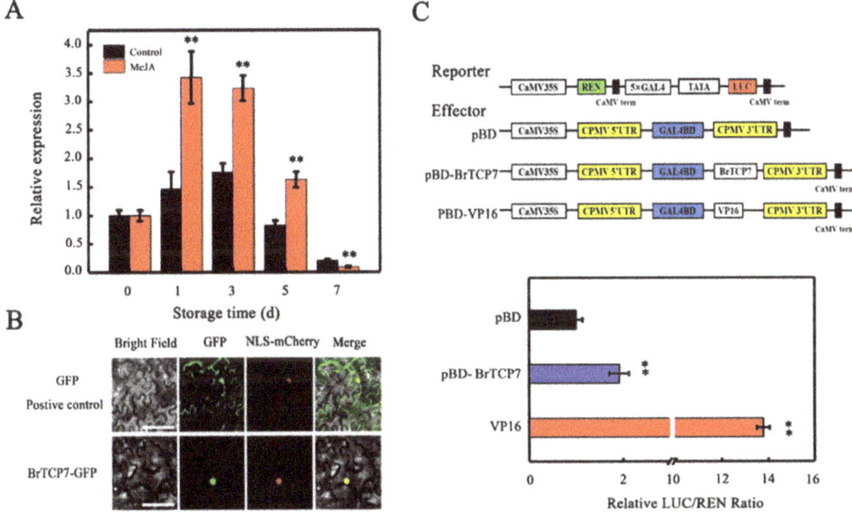

Figure 3. Molecular properties of BrTCP7. (**A**) Relative expression of BrTCP7 in control and MeJA-treated cabbage leaves during senescence. Each value represents the mean ± SD of three biological replicates. Asterisks indicate a significant difference in MeJA-treated leaves compared with control leaves (Student's *t*-test: * $p < 0.05$ and ** $p < 0.01$). (**B**) Subcellular localization of BrTCP7 in epidermal cells of *Nicotiana benthamiana* leaves. A plasmid harboring GFP or BrTCP7-GFP was transformed into *N. benthamiana* leaves by *Agrobacterium tumefaciens* strain EHA105. GFP signals were observed with a fluorescence microscope after two days of infiltration. Bars = 30 μm. (**C**) Trans-activation of BrTCP7 in *N. benthamiana* leaves. The trans-activation ability of BrTCP7 was indicated by the ratio of luciferase (LUC) to renilla luciferase (REN). The LUC/REN ratio of the empty pBD vector (negative control) was used as a calibrator (set to 1). pBD-VP16 was used as a positive control. Data are means ± SD of six independent biological replicates. Asterisks represent significant differences at the 0.01 level by Student's *t*-test compared to pBD.

2.4. BrTCP7 Directly Binds to the Promoter of BrOPR3 and BrRCCR

We then intended to reveal the regulatory mechanism of BrTCP7 involved in MeJA-induced leaf senescence of Chinese flowering cabbage. For this purpose, it is vital to identify the potential targets of BrTCP7. Molecular analyses showed that the predicted consensus binding site of class I TCP TFs is GGNCCCAC, especially the core sequence GCCCR [26,32,34]. After surveying the promoter sequences of three JA biosynthetic genes (*BrLOX4*, *BrAOC3*, and *BrOPR3*) as well as six *CCGs* (*BrPAO*, *BrPPH*, *BrNYC1*, *BrRCCR*, *BrSGR1*, and *BrSGR2*) that were previously reported to be associated with leaf senescence in Chinese flowering cabbage [2,5,38], the TCP binding site was found in *BrOPR3* and *BrRCCR* promoters (Text S1). To confirm the binding of BrTCP7 to *BrOPR3* and *BrRCCR* promoters, an in vitro electrophoretic mobility shift assay (EMSA) was performed. Results showed that purified GST-BrTCP7 protein (Figure 4A), not the GST protein, binds to the TCP binding elements presented in *BrOPR3* and *BrRCCR* promoters, causing a mobility shift band (Figure 4B). Adding excess amount (250-fold) of the cold probe (unlabeled wild-type DNA fragment) but not the mutant probe eliminated the binding complex (Figure 4B), indicating the specificity of the DNA-protein interaction. Expression analysis showed that *BrOPR3* is significantly upregulated by exogenous MeJA treatment (Figure S1).

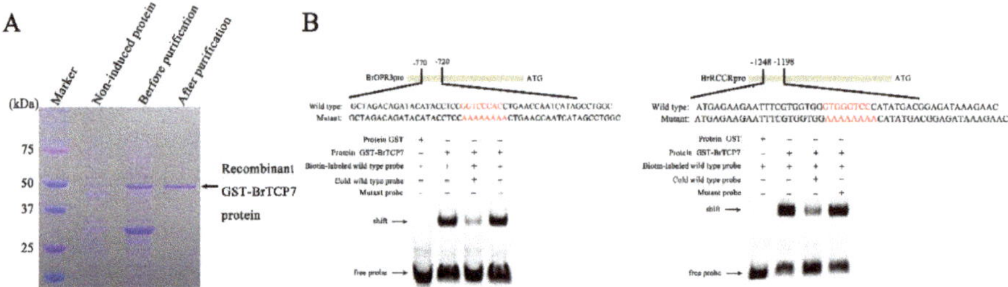

Figure 4. BrTCP7 directly binds to *BrOPR3* and *BrRCCR* promoters. (**A**) SDS-PAGE gel stained with Coomassie brilliant blue demonstrating affinity purification of the recombinant glutathione-S-transferases (GST)-tagged BrTCP7 protein used for EMSA. Recombinant GST-BrTCP7 protein is indicated by the arrow. (**B**) EMSA showing the binding of BrTCP7 to the TCP binding site of the *BrOPR3* or *BrRCCR* promoter. Purified GST-BrTCP7 protein was incubated with the biotin-labeled wild-type probe containing the TCP binding site, and the DNA–protein complexes were separated on native polyacrylamide gels. Sequences of both the wild-type and mutant probes are shown at the top of the image (wild-type and mutant TCP binding site are marked with red letters). The mutant probe was used to test binding specificity. Shifted bands, suggesting the formation of DNA–protein complexes, are indicated by arrows. − represents absence, + represents presence. Competition experiments were conducted by adding 250-fold molar excess of cold probe (unlabeled wild-type DNA fragment) or mutant probe.

2.5. BrTCP7 Activates the Transcription of BrOPR3 and BrRCCR

After deciphering *BrOPR3* and *BrRCCR* as the potential downstream targets of BrTCP7, and establishing BrTCP7 as a transcriptional activator, we hypothesized that BrTCP7 could act as a direct transcriptional activator of *BrOPR3* and *BrRCCR*. To investigate this hypothesis, *Nicotiana benthamiana* leaves were co-transformed with transient over-expression effector construct containing 35S-BrTCP7, and a dual-luciferase reporter construct carrying *BrOPR3* or *BrRCCR* promoter fused to LUC (Figure 5A). As illustrated in Figure 5B, expression of *BrTCP7* resulted in 3.07- and 2.11-fold increases in *LUC* expression driven by *BrOPR3* and *BrRCCR*, respectively, compared with the vector control, providing evidence for a positive trans-activation activity of BrTCP7. Collectively, BrTCP7 binds directly to the promoters of *BrOPR3* and *BrRCCR* and activates their transcriptions, rendering them direct targets of BrTCP7 transcriptional regulation. Further experiments, such as chromatin immunoprecipitation

(ChIP) assay or transient over-expression of *BrTCP7* in *Brassica* protoplasts, are needed to confirm the binding of BrTCP7 to *BrOPR3* and *BrRCCR* in vivo.

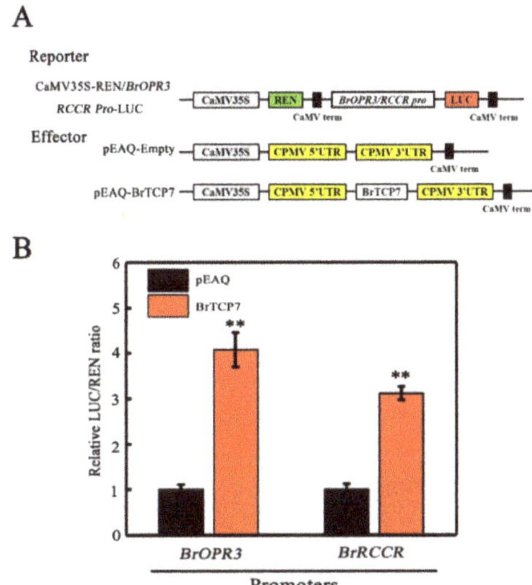

Figure 5. BrTCP7 enhances *BrOPR3* and *BrRCCR* transcriptions by transient transcription dual-luciferase assay in *Nicotiana benthamiana* leaves. (**A**) Diagrams of the reporter and effector vectors. (**B**) BrTCP7 activates *BrOPR3* and *BrRCCR* promoters. Data are means ± S.D. of six independent biological replicates. Asterisks indicate significant differences by student's *t*-test (** $p < 0.01$).

Similar to BrTCP7, Arabidopsis class II CINCINNATA (CIN)/TCP TFs, such as TCP4, positively influence JA biosynthesis by directly activating the transcription of a JA biosynthetic gene *LOX2* [32]. Conversely, the class I TCP members TCP9 and TCP20 are negative regulators of leaf senescence via repressing *LOX2* expression [32]. These findings highlight the complexity of TCPs in modulating JA-promoted leaf senescence. Auxin usually promotes leaf senescence and functions downstream of class I and CIN TCPs by regulating the same auxin-responsible TFs, indicating the possible involvement of TCPs in auxin-modulated leaf senescence [39,40]. A class I TCP TF GhTCP19 from gladiolus was reported to positively regulate corm dormancy release by suppressing the transcription of an ABA biosynthetic gene *GhNCED* while activating the transcription of cytokinin biosynthetic gene *GhIPT* and signal transduction gene *GhARR* [29]. As ABA and cytokinins have been implicated to accelerate and delay leaf senescence in Chinese flowering cabbage, respectively [4,37], investigating whether BrTCP7 or other TCP members like GhTCP19 have regulatory roles in ABA-, cytokinins-, and auxin-mediated leaf senescence will be required.

Transcriptional activity of TFs on their targets is finely controlled by many factors, such as homo-/hetero-dimerization, post-transcriptional/translational modification, and protein–protein interactions [41]. For example, miR319a/JAGGED targets class II TCPs and subsequently represses *LOX2* expression, which attenuates JA-induced senescence [26,30]. A study in bananas showed that MaTCP20 is associated with MaTCP5 or MaTCP19 to form complexes, which influence the regulation of cell wall-modifying genes during fruit ripening [42]. An ERF TF BrERF72 was found to be associated with MeJA-promoted leaf senescence in Chinese flowering cabbage by enhancing JA accumulation through upregulating the expression of three JA biosynthetic genes: *BrLOX4*, *BrAOC3*, and *BrOPR3* [5]. Thus, elucidating whether BrTCP7 and BrERF72 can form a complex to affect the expression of JA biosynthetic genes, as well as identifying other interaction proteins and regulatory factors, will provide more detailed information about the BrTCP7-mediated regulation of gene networks operating MeJA-promoted leaf senescence of Chinese flowering cabbage. Targeted transgenic research is required to unravel the biological function of BrTCP7 in regulating MeJA-promoted leaf senescence in Chinese flowering cabbage.

3. Materials and Methods

3.1. Plant Materials, MeJA Treatment, and Growth Conditions

Chinese flowering cabbages (*Brassica rapa* var. parachinensis) grown in a local commercial vegetable plantation near Guangzhou, Southern China, were harvested after 40 days of growth, and were immediately transported to laboratory under low temperature. Uniform size cabbages without appearance defects were selected and randomly divided into two groups for control and MeJA treatment. MeJA (100 µM) and distilled water (control treatment) were foliar-sprayed onto cabbages as described previously [5]. Subsequently, both control and MeJA-treated cabbages were stored at 15 °C in incubators with a 16-h light/8-h dark cycle for 5 days. On days 0, 1, 3, and 5 of storage, the third leaves from the bottom of 10 cabbages were collected for physiological and molecular analysis. All samples were immediately frozen in liquid nitrogen and stored at −80 °C for further assays.

Tobacco (*Nicotiana benthamiana*) plants were planted in a growth chamber set at 22 °C and a 16-h photoperiod. Leaves from four-week-old tobacco plants were used for *Agrobacterium-tumefaciens*-mediated transient expression assays.

3.2. Physiological Measurements of Leaf Senescence

Leaf total chlorophyll content and photochemical efficiency (Fv/Fm) are physiological parameters commonly used as indicators of leaf senescence. Total chlorophyll content was measured by extracting approximately 0.1 g of leaves in 10 mL in 80% acetone in the dark for 24 h and measuring the absorbance of extracts at 663 nm and 645 nm using a spectrophotometer as described earlier [5,6]. *Fv/Fm* was determined non-invasively after leaves were adapted to dark conditions for 30 min. A chlorophyll fluorometer (Imaging-PAM-M series, Heinz Walz GmbH, Effeltrich, Germany) equipped with a charge-coupled device (CCD) camera to capture high-resolution digital images of the emitted fluorescence from the dark-adapted leaves.

3.3. RNA Extraction, Gene Cloning, and Bioinformatic Analysis

A Quick RNA Isolation Kit (Huayueyang, Beijing, China) was used to extract total RNA from cabbage leaves. RNA was then treated with DNase I to remove contaminating DNA and reversely transcribed to cDNA using the reverse transcriptase M-MLV (TaKaRa, Shiga, Japan) using 1 µg RNA template, following the manufacturer's protocol. The full-length BrTCP7 gene was isolated from our transcriptome database. Theoretical isoelectric points (*p*I) and mass values were calculated using the Compute *p*I/Mw tool (http://web.expasy.org/compute_pi/). Sequence alignment and phylogenetic analysis of TCP proteins were conducted with the CLUSTALW program (version 1.83, The Conway Institute of Biomolecular and Biomedical Research, University College Dublin, Dublin, Ireland). A phylogenetic tree was created with the MEGA5.0 program (Institute of Molecular Evolutionary Genetics, The Pennsylvania State University, PA, USA), using the maximum likelihood inference method.

3.4. Gene Expression via qRT-PCR

cDNA from each sample were quantified by quantitative real-time PCR (qRT-PCR) assays. qRT-PCR was performed using the step one plus a CFX96 Touch™ real-time PCR detection system (Bio-Rad, Hercules, CA, USA) and GoTaq qPCR master mix kit (Promega, Madison, WI, USA). The expression levels of target genes were normalized according to the cycle threshold (Ct) value using *BrActin1* [43] as the reference gene.

3.5. Subcellular Localization Assay

The BrTCP7 coding sequence (CDS) fragment without the stop codon was amplified and inserted into the pEAQ-GFP vector [44] to produce the fusion protein BrTCP7-GFP. Then BrTCP7-GFP and the control pEAQ-GFP constructs were transformed to *A. tumefaciens* strain EHA105. Overnight cultures of *Agrobacteria* were collected by centrifugation, resuspended in MES buffer to 0.8–1.0 OD_{600}, incubated at room temperature for 2 h before infiltration. *Agrobacteria* suspension collected in a 1-mL syringe (without the metal needle) was carefully press-infiltrated manually onto healthy leaves of 4-week-old *Nicotiana benthamiana* as described previously [45]. Two days after infiltration, the GFP fluorescent signals in the epidermal cells of leaves were directly observed and images were captured by using a Zeiss fluorescence microscope (Carl Zeiss AG, Oberkochen, Germany).

3.6. Recombinant Protein Induction, Purification, and EMSA Assay

For protein expression and purification, the BrTCP7 coding sequence was recombined into the vector pGEX-4T-1 and transferred to *Escherichia coli* strain Transetta (DE3). When the transformed DE3 cell density reached 0.6 (OD_{600}), protein expression in 1-L cultures maintained at 37 °C was induced by the addition of 0.3 mM isopropyl thio-β-D-galactoside during a 3 h course. Cells were harvested by centrifugation. The cell pellet was resuspended in phosphate-buffered saline and incubated on ice for 30 min. Following sonication, the lysate was cleared by centrifugation. Glutathione-Superflow Resin (Clontech, Mountain View, CA, USA) was used for purification of GST-tagged protein by gravity-flow chromatography according to the manufacturer's protocol, followed by SDS-PAGE and Coomassie Brilliant Blue staining to confirm protein size and purity.

Electrophoretic mobility shift assay (EMSA) was performed as described previously [46]. The synthetic nucleotides (~60 bp) derived from the 5′ UTR of *BrOPR3* and *BrRCCR* oligonucleotides, which contain the consensus binding site (GGNCCCAC) of TCPs, were biotin-labeled at the 5′ end. The purified recombinant BrTCP7 protein (1 μg) was incubated with biotin-labeled probe (2×10^{-6} μmol) in binding buffer for 25 min at 30 °C. Competitions were conducted by adding excess amount (250-fold) of cold probe (unlabeled DNA fragment) or mutant probe. The reaction mixture was electrophoresed on a 6% native polyacrylamide gels, and then transferred onto a positively charged nylon membrane, followed by cross-linking though illumination under an ultraviolet lamp. The signals from the labeled DNA were detected by using the LightShift chemiluminescent EMSA kit (Thermo Scientific, Rockford, IL, USA) in a ChemiDoc™ MP Imaging System (Bio-Rad, Hercules, CA, USA).

3.7. Transient Transcription Dual-Luciferase (Dual-LUC) Assays

To investigate the transcriptional ability of BrTCP7 in vivo, its full length was inserted into pBD [45] to construct pBD-BrTCP7 as an effector. The positive control (pBD-VP16) was constructed by fusing VP16, a herpes simplex virus-encoded transcriptional activator, to pBD. pBD itself was used as a negative control. The GAL4 plasmid with the *firefly luciferase* (*LUC*) gene was used as a reporter [45], and the *renilla luciferase* (*REN*) gene in the same plasmid was used as an internal control.

To determine the activation of *BrOPR3* and *BrRCCR* by BrTCP7, the promoter fragments of *BrOPR3* and *BrRCCR* were amplified and cloned into the transient dual luciferase expression vector pGreenII 0800-LUC [47] as reporter constructs. To generate the 35S::BrTCP7 effector construct, the BrTCP7 coding sequence was amplified by PCR and inserted into pEAQ vector [44]. The empty vector was included as a control.

Transient transcription dual-LUC assays were performed using *Nicotiana benthamiana* plants as described [46]. The reporter and effector constructs mentioned above were co-infiltrated into tobacco leaves. After 2 days of infiltration, the luciferase activity of tobacco leaf extract was quantified by a Luminoskan Ascent Microplate Luminometer (Thermo Fisher Scientific, Rockford, IL, USA), using commercial dual-luciferase reporter assay kit according to the manufacturer's instruction (Promega, Madison, WI, USA). The trans-activation ability of BrTCP7 was indicated by the LUC/REN ratio.

3.8. Statistical Analysis

Data are represented as means ± SD of three or six biological replicates. Statistical differences of two treatments were examined using the Student's *t*-test. Data are considered significant as follows: * $p < 0.05$, ** $p < 0.01$.

3.9. Primers

All primers used in this research are listed in Table S1.

4. Conclusions

In summary, exogenous MeJA treatment promotes leaf senescence of Chinese flowering cabbage. A class I TCP member BrTCP7, which is MeJA-upregulated and acts as a nuclear-localized transcriptional activator, was identified. BrTCP7 targets the promoters of a JA biosynthetic gene *BrOPR3* and a *CCG* gene *BrRCCR*, leading to their transcriptional activation. These findings expand our understanding of TCP TFs' functions and shed light on the transcriptional mechanism operating JA-mediated leaf senescence, and also the molecular mechanism(s) involved in maintaining postharvest quality of an important leafy vegetable, Chinese flowering cabbage.

Supplementary Materials: Supplementary materials can be found at http://www.mdpi.com/1422-0067/20/16/3963/s1.

Author Contributions: J.-w.C. and J.-y.C. conceived and supervised the study; Y.-m.X, X.-m.X., Z.-x.Z., and X.-l.T. conducted the experiments; Y.-m.X, X.-m.X., Z.-x. Z., X.-l.T., Z.-l.L., J.-w.C., X.-g.S., and J.-y.C. analyzed data; Y.-m.X, J.-w.C., and J.-y.C. wrote the manuscript. All authors read and approved the manuscript.

Funding: This study was funded by a grant from the National Natural Science Foundation of China (31671897) and Natural Science Foundation of Guangdong Province, China (2018A030313457).

Acknowledgments: We are grateful to George P. Lomonossoff (Department of Biological Chemistry, John Innes Centre, Norwich Research Park) for providing the pEAQ vectors. We are grateful to Prakash Lakshmanan (Sugarcane Research Institute, Guangxi Academy of Agricultural Sciences) for his critical English language editing of the manuscript.

Conflicts of Interest: The authors declare no conflict of interest.

References

1. Fan, Z.Q.; Tan, X.L.; Shan, W.; Kuang, J.F.; Lu, W.J.; Chen, J.Y. BrWRKY65, a WRKY transcription factor, is involved in regulating three leaf senescence-associated genes in Chinese flowering cabbage. *Int. J. Mol. Sci.* **2017**, *18*, 1228.

2. Fan, Z.Q.; Tan, X.L.; Shan, W.; Kuang, J.F.; Lu, W.J.; Chen, J.Y. Characterization of a transcriptional regulator, BrWRKY6, associated with gibberellin-suppressed leaf senescence of Chinese flowering cabbage. *J. Agric. Food Chem.* **2018**, *66*, 1791–1799. [CrossRef] [PubMed]

3. Ombra, M.N.; Cozzolino, A.; Nazzaro, F.; d'Acierno, A.; Tremonte, P.; Coppola, R.; Fratianni, F. Biochemical and biological characterization of two Brassicaceae after their commercial expiry date. *Food Chem.* **2017**, *218*, 335–340. [CrossRef] [PubMed]

4. Zhang, X.; Zhang, Z.; Li, J.; Wu, L.; Guo, J.; Ouyang, L.; Xia, Y.; Huang, X.; Pang, X. Correlation of leaf senescence and gene expression/activities of chlorophyll degradation enzymes in harvested Chinese flowering cabbage (*Brassica rapa* var. *parachinensis*). *J. Plant Physiol.* **2011**, *168*, 2081–2087. [CrossRef] [PubMed]

5. Tan, X.L.; Fan, Z.Q.; Shan, W.; Yin, X.R.; Kuang, J.F.; Lu, W.J.; Chen, J.Y. Association of BrERF72 with methyl jasmonate-induced leaf senescence of Chinese flowering cabbage through activating JA biosynthesis-related genes. *Hortic. Res.* **2018**, *5*, 22. [CrossRef]

6. Fan, Z.Q.; Tan, X.L.; Chen, J.W.; Liu, Z.L.; Kuang, J.F.; Lu, W.J.; Shan, W.; Chen, J.Y. BrNAC055, a novel transcriptional activator, regulates leaf senescence in Chinese flowering cabbage by modulating reactive oxygen species production and chlorophyll degradation. *J. Agric. Food Chem.* **2018**, *66*, 9399–9408. [CrossRef]

7. Jibran, R.; Hunter, D.; Dijkwel, P. Hormonal regulation of leaf senescence through integration of developmental and stress signals. *Plant Mol. Biol.* **2013**, *82*, 547–561. [CrossRef]

8. Schippers, J.H.; Schmidt, R.; Wagstaff, C.; Jing, H.C. Living to die and dying to live: The survival strategy behind leaf senescence. *Plant Physiol.* **2015**, *169*, 914–930. [CrossRef]

9. Schippers, J.H. Transcriptional networks in leaf senescence. *Curr. Opin. Plant Biol.* **2015**, *27*, 77–83. [CrossRef]

10. Kim, J.; Woo, H.R.; Nam, H.G. Toward systems understanding of leaf senescence: An integrated multi-omics perspective on leaf senescence research. *Mol. Plant* **2016**, *9*, 813–825. [CrossRef]

11. Woo, H.R.; Kim, H.J.; Lim, P.O.; Nam, H.G. Leaf senescence: Systems and dynamics aspects. *Annu. Rev. Plant Biol.* **2019**, *70*, 347–376. [CrossRef]

12. Zhang, H.; Zhou, C. Signal transduction in leaf senescence. *Plant Mol. Biol.* **2013**, *82*, 539–545. [CrossRef]

13. Huang, H.; Liu, H.; Liu, L.; Song, S. Jasmonate action in plant growth and development. *J. Exp. Bot.* **2017**, *68*, 1349–1359. [CrossRef]

14. Hu, Y.; Jiang, Y.; Han, X.; Wang, H.; Pan, J.; Yu, D. Jasmonate regulates leaf senescence and tolerance to cold stress: Crosstalk with other phytohormones. *J. Exp. Bot.* **2017**, *68*, 1361–1369. [CrossRef] [PubMed]

15. He, Y.; Fukushige, H.; Hildebrand, D.F.; Gan, S. Evidence supporting a role of jasmonic acid in Arabidopsis leaf senescence. *Plant Physiol.* **2002**, *128*, 876–884. [CrossRef]

16. Yan, Y.; Christensen, S.; Isakeit, T.; Engelberth, J.; Meeley, R.; Hayward, A.; Emery, R.J.; Kolomiets, M.V. Disruption of OPR7 and OPR8 reveals the versatile functions of jasmonic acid in maize development and defense. *Plant Cell* **2012**, *24*, 1420–1436. [CrossRef] [PubMed]

17. Lee, S.H.; Sakuraba, Y.; Lee, T.; Kim, K.W.; An, G.; Lee, H.Y.; Paek, N.C. Mutation of Oryza sativa CORONATINE INSENSITIVE 1b (OsCOI1b) delays leaf senescence. *J. Integr. Plant Biol.* **2015**, *57*, 562–576. [CrossRef] [PubMed]

18. Zhu, X.; Chen, J.; Xie, Z.; Gao, J.; Ren, G.; Gao, S.; Zhou, X.; Kuai, B. Jasmonic acid promotes degreening via MYC2/3/4- and ANAC019/055/072-mediated regulation of major chlorophyll catabolic genes. *Plant J.* **2015**, *84*, 597–610. [CrossRef] [PubMed]

19. Qi, T.; Wang, J.; Huang, H.; Liu, B.; Gao, H.; Liu, Y.; Song, S.; Xie, D. Regulation of jasmonate-induced leaf senescence by sntagonism between bHLH subgroup IIIe and IIId factors in Arabidopsis. *Plant Cell* **2015**, *27*, 1634–1649. [CrossRef]

20. Jiang, Y.; Liang, G.; Yang, S.; Yu, D. Arabidopsis WRKY57 functions as a node of convergence for jasmonic acid- and auxin-mediated signaling in jasmonic acid-induced leaf senescence. *Plant Cell* **2014**, *26*, 230–245. [CrossRef] [PubMed]

21. Cubas, P.; Lauter, N.; Doebley, J.; Coen, E. The TCP domain: A motif found in proteins regulating plant growth and development. *Plant J. Cell Mol. Biol.* **1999**, *18*, 215–222. [CrossRef]

22. Kosugi, S.; Ohashi, Y. PCF1 and PCF2 specifically bind to cis elements in the rice proliferating cell nuclear antigen gene. *Plant Cell* **1997**, *9*, 1607–1619. [PubMed]

23. Doebley, J.; Stec, A.; Hubbard, L. The evolution of apical dominance in maize. *Nature* **1997**, *386*, 485–488. [CrossRef] [PubMed]

24. Parapunova, V.; Busscher, M.; Busscher-Lange, J.; Lammers, M.; Karlova, R.; Bovy, A.G.; Angenent, G.C.; de Maagd, R.A. Identification, cloning and characterization of the tomato TCP transcription factor family. *BMC Plant Biol.* **2014**, *14*, 157. [CrossRef] [PubMed]

25. Liu, Y.; Guan, X.; Liu, S.; Yang, M.; Ren, J.; Guo, M.; Huang, Z.; Zhang, Y. Genome-wide identification and analysis of TCP transcription factors involved in the formation of leafy head in Chinese cabbage. *Int. J. Mol. Sci.* **2018**, *19*, 847. [CrossRef]

26. Nicolas, M.; Cubas, P. TCP factors: New kids on the signaling block. *Curr. Opin. Plant Biol.* **2016**, *33*, 33–41. [CrossRef]

27. Lucero, L.E.; Uberti-Manassero, N.G.; Arce, A.L.; Colombatti, F.; Alemano, S.G.; Gonzalez, D.H. TCP15 modulates cytokinin and auxin responses during gynoecium development in Arabidopsis. *Plant J.* **2015**, *84*, 267–282. [CrossRef]

28. Mukhopadhyay, P.; Tyagi, A.K. OsTCP19 influences developmental and abiotic stress signaling by modulating ABI4-mediated pathways. *Sci. Rep.* **2015**, *5*, 9998. [CrossRef]

29. Wu, J.; Wu, W.; Liang, J.; Jin, Y.; Gazzarrini, S.; He, J.; Yi, M. GhTCP19 transcription factor regulates corm dormancy release by repressing *GhNCED* expression in gladiolus. *Plant Cell Physiol.* **2019**, *60*, 52–62. [CrossRef]

30. Schommer, C.; Palatnik, J.F.; Aggarwal, P.; Chételat, A.; Cubas, P.; Farmer, E.E.; Nath, U.; Weigel, D. Control of jasmonate biosynthesis and senescence by miR319 targets. *PLoS Biol.* **2008**, *6*, e230. [CrossRef]

31. Breeze, E.; Harrison, E.; McHattie, S.; Hughes, L.; Hickman, R.; Hill, C.; Kiddle, S.; Kim, Y.S.; Penfold, C.A.; Jenkins, D.; et al. High-resolution temporal profiling of transcripts during Arabidopsis leaf senescence reveals a distinct chronology of processes and regulation. *Plant Cell* **2011**, *23*, 873–894. [CrossRef] [PubMed]

32. Danisman, S.; van der Wal, F.; Dhondt, S.; Waites, R.; de Folter, S.; Bimbo, A.; van Dijk, A.; Muino, J.; Cutri, L.; Dornelas, M.; et al. Arabidopsis class I and class II TCP transcription factors regulate jasmonic acid metabolism and leaf development antagonistically. *Plant Physiol.* **2012**, *159*, 1511–1523. [CrossRef] [PubMed]

33. Kim, J.; Kim, J.H.; Lyu, J.I.; Woo, H.R.; Lim, P.O. New insights into the regulation of leaf senescence in Arabidopsis. *J. Exp. Bot.* **2018**, *69*, 787–799. [CrossRef] [PubMed]

34. Zheng, X.; Yang, J.; Lou, T.; Zhang, J.; Yu, W.; Wen, C. Transcriptome profile analysis reveals that CsTCP14 induces susceptibility to foliage diseases in cucumber. *Int. J. Mol. Sci.* **2019**, *20*, 2582. [CrossRef] [PubMed]

35. Martin-Trillo, M.; Cubas, P. TCP genes: A family snapshot ten years later. *Trends Plant Sci.* **2010**, *15*, 31–39. [CrossRef]

36. Du, J.; Hu, S.; Yu, Q.; Wang, C.; Yang, Y.; Sun, H.; Yang, Y.; Sun, X. Genome-wide identification and characterization of BrrTCP transcription factors in *Brassica rapa ssp. rapa*. *Front Plant Sci.* **2017**, *8*, 1588. [CrossRef] [PubMed]

37. Fan, H.M.; Sun, C.H.; Wen, L.Z.; Liu, B.W.; Ren, H.; Sun, X.; Ma, F.F.; Zheng, C.S. CmTCP20 plays a key role in nitrate and auxin signaling-regulated lateral root development in *Chrysanthemum*. *Plant Cell Physiol.* **2019**, *60*, 1581–1594. [CrossRef]

38. Tan, X.L.; Fan, Z.Q.; Kuang, J.F.; Lu, W.J.; Reiter, R.J.; Lakshmanan, P.; Su, X.G.; Zhou, J.; Chen, J.Y.; Shan, W. Melatonin delays leaf senescence of Chinese flowering cabbage by suppressing ABFs-mediated abscisic acid biosynthesis and chlorophyll degradation. *J. Pineal. Res.* **2019**, *67*, e12570. [CrossRef]

39. Uberti-Manassero, N.G.; Lucero, L.E.; Viola, I.L.; Vegetti, A.C.; Gonzalez, D.H. The class I protein AtTCP15 modulates plant development through a pathway that overlaps with the one affected by CIN-like TCP proteins. *J. Exp. Bot.* **2012**, *63*, 809–823. [CrossRef]

40. Koyama, T.; Sato, F.; Ohme-Takagi, M. Roles of miR319 and TCP transcription factors in leaf development. *Plant Physiol.* **2017**, *175*, 874–885. [CrossRef]

41. Shaikhali, J. GIP1 protein is a novel cofactor that regulates DNA-binding affinity of redox-regulated members of bZIP transcription factors involved in the early stages of Arabidopsis development. *Protoplasma* **2015**, *252*, 867–883. [CrossRef] [PubMed]

42. Song, C.B.; Shan, W.; Yang, Y.Y.; Tan, X.L.; Fan, Z.Q.; Chen, J.Y.; Lu, W.J.; Kuang, J.F. Heterodimerization of MaTCP proteins modulates the transcription of MaXTH10/11 genes during banana fruit ripening. *BBA-Gene Regul. Mech.* **2018**, *1861*, 613–622. [CrossRef] [PubMed]

43. Qi, J.N.; Yu, S.C.; Zhang, F.L.; Shen, X.Q.; Zhang, X.Z.; Yu, Y.J.; Zhang, D.S. Reference gene selection for real-time quantitative polymerase chain reaction of mRNA transcript levels in Chinese Cabbage (*Brassica rapa* L. ssp. pekinensis). *Plant Mol. Biol. Rep.* **2010**, *28*, 597–604. [CrossRef]

44. Sainsbury, F.; Thuenemann, E.C.; Lomonossoff, G.P. pEAQ: Versatile expression vectors for easy and quick transient expression of heterologous proteins in plants. *Plant Biotechnol. J.* **2009**, *7*, 682–693. [CrossRef] [PubMed]

45. Wei, W.; Cheng, M.N.; Ba, L.J.; Zeng, R.X.; Luo, D.L.; Qin, Y.H.; Liu, Z.L.; Kuang, J.F.; Lu, W.J.; Chen, J.Y.; et al. Pitaya HpWRKY3 is associated with fruit sugar accumulation by transcriptionally modulating sucrose metabolic genes *HpINV2* and *HpSuSy1*. *Int. J. Mol. Sci.* **2019**, *20*, 1890. [CrossRef]

46. Fan, Z.Q.; Ba, L.J.; Shan, W.; Xiao, Y.Y.; Lu, W.J.; Kuang, J.F.; Chen, J.Y. A banana R2R3-MYB transcription factor MaMYB3 is involved in fruit ripening through modulation of starch degradation by repressing starch degradation-related genes and *MabHLH6*. *Plant J.* **2018**, *96*, 1191–1205. [CrossRef] [PubMed]

47. Hellens, R.P.; Allan, A.C.; Friel, E.N.; Bolitho, K.; Grafton, K.; Templeton, M.D.; Karunairetnam, S.; Gleave, A.P.; Laing, W.A. Transient expression vectors for functional genomics, quantification of promoter activity and RNA silencing in plants. *Plant Methods* **2005**, *1*, 13. [CrossRef]

© 2019 by the authors. Licensee MDPI, Basel, Switzerland. This article is an open access article distributed under the terms and conditions of the Creative Commons Attribution (CC BY) license (http://creativecommons.org/licenses/by/4.0/).

International Journal of
Molecular Sciences

Article

Fertility of Pedicellate Spikelets in Sorghum Is Controlled by a Jasmonic Acid Regulatory Module

Nicholas Gladman [1,2], Yinping Jiao [1,2], Young Koung Lee [2,3], Lifang Zhang [2], Ratan Chopra [1,4],
Michael Regulski [2], Gloria Burow [1], Chad Hayes [1], Shawn A. Christensen [5],
Lavanya Dampanaboina [1], Junping Chen [1], John Burke [1], Doreen Ware [2,6,*] and Zhanguo Xin [1,*]

[1] Plant Stress and Germplasm Development Unit, Cropping Systems Research Laboratory, U.S. Department of
Agriculture-Agricultural Research Service, Lubbock, TX 79415, USA; gladman@cshl.edu (N.G.);
yjiao@cshl.edu (Y.J.); rchopra@umn.edu (R.C.); gloria.burow@ars.usda.gov (G.B.);
chad.hayes@ars.usda.gov (C.H.); lavanya.dampanaboina@ttu.edu (L.D.); junping.chen@ars.usda.gov (J.C.);
john.burke@ars.usda.gov (J.B.)
[2] Cold Spring Harbor Laboratory, Cold Spring Harbor, NY 11724, USA; leeyk@nfri.re.kr (Y.K.L.);
zhangl@cshl.edu (L.Z.); regulski@cshl.edu (M.R.)
[3] Plasma Technology Research Center, National Fusion Research Institute, 37, Dongjangsan-ro, Gunsan-si,
Jeollabuk-do 54004, Korea
[4] Current address: Department of Agronomy and Plant Genetics, University of Minnesota, St. Paul,
MN 55108, USA
[5] Chemistry Research Unit, USDA-ARS, 1700 S.W. 23RD DRIVE, Gainesville, FL 32608, USA;
shawn.christensen@ars.usda.gov
[6] U.S. Department of Agriculture-Agricultural Research Service, NEA Robert W. Holley Center for Agriculture
and Health, Cornell University, Ithaca, NY 14853, USA
[*] Correspondence: ware@cshl.edu or Doreen.Ware@ARS.USDA.GOV (D.W.);
Zhanguo.Xin@ARS.USDA.GOV (Z.X.)

Received: 13 September 2019; Accepted: 2 October 2019; Published: 8 October 2019

Abstract: As in other cereal crops, the panicles of sorghum (*Sorghum bicolor* (L.) Moench) comprise two types of floral spikelets (grass flowers). Only sessile spikelets (SSs) are capable of producing viable grains, whereas pedicellate spikelets (PSs) cease development after initiation and eventually abort. Consequently, grain number per panicle (GNP) is lower than the total number of flowers produced per panicle. The mechanism underlying this differential fertility is not well understood. To investigate this issue, we isolated a series of ethyl methane sulfonate (EMS)-induced *multiseeded* (*msd*) mutants that result in full spikelet fertility, effectively doubling GNP. Previously, we showed that MSD1 is a TCP (Teosinte branched/Cycloidea/PCF) transcription factor that regulates jasmonic acid (JA) biosynthesis, and ultimately floral sex organ development. Here, we show that *MSD2* encodes a lipoxygenase (LOX) that catalyzes the first committed step of JA biosynthesis. Further, we demonstrate that MSD1 binds to the promoters of *MSD2* and other JA pathway genes. Together, these results show that a JA-induced module regulates sorghum panicle development and spikelet fertility. The findings advance our understanding of inflorescence development and could lead to new strategies for increasing GNP and grain yield in sorghum and other cereal crops.

Keywords: transcriptional regulators; plant development; jasmonic acid signaling; gene expression

---

1. Significance

Through a single base pair mutation, grain number can be increased by ~200% in the globally important crop *Sorghum bicolor*. This mutation affects the expression of an enzyme, MSD2, that catalyzes the jasmonic acid pathway in developing floral meristems. The global gene expression profile in this enzymatic mutant is similar to that of a transcription factor mutant, *msd1*, indicating that disturbing any component of this regulatory module disrupts a positive feedback loop that occurs normally due to regular developmental perception of jasmonic acid. Additionally, the MSD1 transcription factor is able to regulate *MSD2* in addition to other jasmonic acid pathway genes, suggesting that it is a primary transcriptional regulator of this hormone signaling pathway in floral meristems.

2. Introduction

Sorghum (*Sorghum bicolor* (L.) Moench) is a crop plant domesticated in northern Africa ~6000 years ago [1,2]. A C_4 grass with robust tolerance to drought, heat, and high-salt conditions, sorghum is the fifth most agriculturally important crop in terms of global dedicated acreage and production quantity. It also serves as a useful model for crop research due to its completely sequenced compact genome (~730 Mb) [3] and similarity to the functional genomics capabilities of maize, sugarcane, and other bioenergy grasses comprising more convoluted genomes.

Increasing grain yield has always been a high priority for breeders. Increasing grains per panicle (GNP) and optimizing panicle architecture represent feasible goals for modern gene editing in sorghum [4,5]. GNP and seed head architecture are related traits, with origins in early stages of inflorescence development [6,7]. Sorghum forms a determinant panicle that manifests at the end of the shoot meristem, with nodes regularly extending throughout from which secondary and tertiary branches emanate [8]. A terminal trio of spikelet florets are attached through a pedicel to these branches: one sessile spikelet (SS) that is fertile and two pedicellate spikelets (PSs) that fail to generate mature pistils and sometimes anthers, which results in an inability of PSs to fertilize. They will ultimately senesce during grain filling instead of becoming viable seed [6]. Below this terminal spikelet group, several pairs of SSs and PSs populate the branches down to the nodes.

Jasmonic acid (JA) is a plant hormone derived from α-linolenic acid and shares structural similarities to animal prostaglandins [9,10]. JA plays roles in organ development, as well as biotic and abiotic response signaling mechanisms [7,11–14], spikelet formation in rice [15], and sex determination in maize [13,16,17]. In sorghum, the TCP family transcription factor MSD1 (*multiseeded 1*) controls PS fertility [7]. MSD1 is expressed in a narrow spatiotemporal window within the developing panicle in wild type (WT), BTx623; its expression is dramatically reduced in ethyl methane sulfonate (EMS)-induced *msd1* mutants. Many genes involved in JA biosynthesis, including 12-oxophytodeinoate reductase 3 (*OPR3*) [18,19], allene oxide synthase (*AOS*) [20], cytochrome P450 [21,22], and lipoxygenase (*LOX*) [23], are also downregulated in *msd1* mutants. MSD1 is thought to activate the programmed cell death pathway through activation of JA biosynthesis, which destines the PS to abortion.

In this study, we characterized another *msd* mutant, *msd2*, from the same publicly available sorghum EMS population in which *msd1* was identified [24]. *MSD2* encodes a 13-lipoxygenase that catalyzes the conversion of free α-linolenic acid (18:3) to 13(S)-hydroperoxylinolenic acid (13-Hpot), the first committed step of the JA biosynthetic pathway [11,25]. As with *msd1*, mutants in *msd2* exhibit complete spikelet fertility for both SSs and PSs, resulting in seed formation from both flower types. Multiple independent alleles were discovered for *msd2*, including nonsense and missense mutations within the LOX functional domain. Mutants in *msd1* and *msd2* exhibit similar regulatory network profiles, including downregulation of JA pathway genes and other expression cascades related to developmental and cellular restructuring. Finally, MSD1 is capable of activating *MSD2* expression, as well as regulating other gene networks in trans, leading to the *multiseeded* phenotype. Taken together, our findings demonstrate that MSD2, along with MSD1, modulates the JA pathway during sorghum sex organ development.

3. Results

3.1. MSD2 Encodes a Lipoxygenase in the Jasmonic Acid Biosynthetic Pathway

Sorghum bicolor (L.) Moench plants manifesting the *multiseeded* phenotype were identified from a collection of EMS-induced single nucleotide polymorphisms (SNP) [24]. *MSD1*, which encodes a TCP (*Teosinte branched 1 (TB1), Cycloidea (Cyc), and Proliferating Cell nuclear antigen binding Factor (PCF)*) [26–28] transcription factor, was the first to be characterized, revealing a role in controlling bioactive JA levels in developing floral meristems [7]. To identify additional causative alleles, we subjected seventeen different *msd* mutants to whole-genome sequencing followed by comparative variant calling analysis. Three of these independent alleles, *msd2-1, -2, and -3*, localized to SORBI_3006G095600 (Sb06g018040) [7], which encodes a class II 13-lipoxygenase that shares >95% amino acid identity with the maize *tasselseed 1 (TS1)* gene [16]. SORBI_3006G095600 is syntenic to *TS1* and is the closest related maize orthologue based on a maximum likelihood phylogenetic analysis (Supplemental Figure S1). The *msd2-1* mutant harbors a nonsense mutation (peptide residue Q402*) and *msd2-2* a missense mutation (peptide residue A423V), respectively, within the lipoxygenase (LOX) domain (Figure 1A). The *msd2-3* allele contains the same mutation as *msd2-1*, but the lines are not siblings, as evidenced by the lack of other shared SNPs.

Lipoxygenases catalyze linolenic acid to hydroperoxyoctadecadienoic acid in the initial committed step of the JA biosynthetic pathway [11]. There are 12 LOX paralogs in *Sorghum bicolor* (Figure 1B), which exhibit different patterns of tissue-specific expression in WT BTx623 plants. *MSD2* is expressed at lower levels overall than other LOX genes [29–32], but is more strongly expressed in developing panicles than its 13-LOX paralogs SORBI_3007g210400 and SORBI_3001G483400 (Figure 1C); only SORBI_3004G078600 is more strongly expressed at particular stages. Thus, like *MSD1*, *MSD2* operates under low levels of expression in specific tissues within developing meristems. EMS-induced SNP mutant lines exist for the other 13-LOX paralogs (Supplemental Table S1), including a nonsense mutation in the more highly expressed SORBI_3004G078600, but no *multiseeded* phenotype has been observed in any of these lines. This suggests that MSD2 is a specific and necessary LOX isoform involved in the JA pathway, which controls floral organ progression.

MSD2 mutants display the same floral phenotype as *msd1*: complete development of anthers and ovaries in both PSs and SSs. Electron micrographs of developing floral spikelets revealed that the developmental pattern of *msd2* is similar to that of WT [7] (Figure 1D), but like *msd1*, the end result is complete floral fertility of all spikelets with near 100% grain filling, increasing the GNP of the mutant (Figure 1E), although the *msd2* seeds from both PS and SS are smaller than those of WT. Dissected images of PSs also show that *msd2* has the same full pistil development phenotype as *msd1* in contrast to WT PSs that lack mature gynoecia (Supplemental Figure S1). The only other consistent agronomic difference between *msd2* and WT plants were a slight increase in initial root growth rate in the mutant (Supplemental Figure S1).

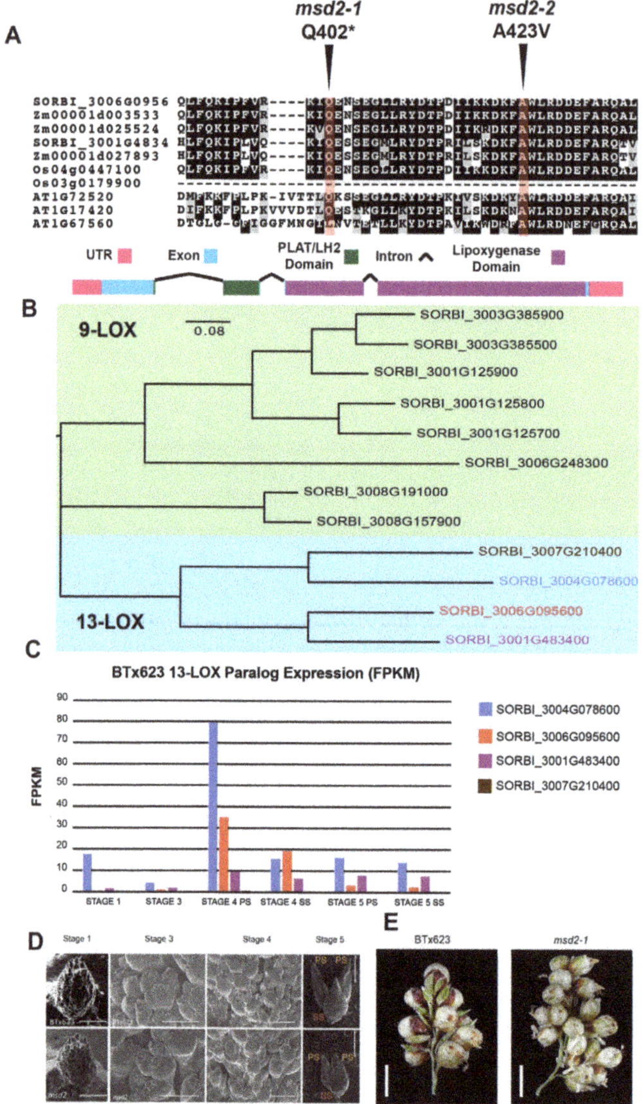

Figure 1. (**A**) Boxshade section of a MUSCLE alignment for MSD2-orthologous sorghum, maize, rice, and Arabdiopsis lipoxygenase (LOX) peptide sequences surrounding the ethyl methane sulfonate (EMS)-induced changes within MSD2. Arrows with red highlights indicate the positions of the *msd2-1* (GLN > premature stop) and *msd2-2* (Ala > Val) mutations. Below the alignment is a diagram of the *MSD2* gene; colored boxes indicate encoded domains of the gene product. The sizes of legend boxes are equivalent to 100 base pairs. (**B**) Phylogenetic tree of sorghum 9- and 13-LOX proteins (MSD2 highlighted in red). (**C**) RNA-seq FPKM expression data of the 13-LOX paralogs across developing panicle tissue stages (colors correspond to panel **B**). (**D**) SEM images of developing inflorescence meristems in WT and *msd2-1*. Scale bars are 1 mm in length for stages 1, 4, and 5, and 500 µm for stage 3. Sessile spikelets (SS) are indicated in red and pedicellate spikelets (PS) in orange. (**E**) A section of a secondary branch of late-dough filling panicles from WT and *msd2-1* lines. White scale bar indicates a length of 0.5 cm.

3.2. The msd2 Phenotype is Rescued by Exogenous Methyl-JA Treatment

To determine whether exogenous application of JA could rescue the *msd2* phenotype, as it does in *msd1* mutants, we pipetted 1 mM methyl-JA (Me-JA) directly down the whorls of young WT and *msd2* mutant plants. This chemical treatment restored PS infertility (Figure 2). Panicle size was reduced in Me-JA treated plants, as was branch length and number. Panicle emergence was also delayed in all genotypes relative to untreated or negative control plants, likely due to other developmental signaling effects and inhibition of cell expansion caused by the introduction of exogenous jasmonates [33,34].

Figure 2. Phenotypic rescue of *msd2* plants with exogenous application of methyl-JA. Wild type (WT) and *msd2* lines were treated every 48 h with either water + 0.05% Tween-20 or 1 mM Me-JA + 0.05% Tween-20.

3.3. MSD2 Regulatory Networks Are Similar to MSD1

Transcriptomic data indicated that many JA biosynthetic pathway genes, including all *LOX* paralogs, were coordinately downregulated in stage 4 SS and PS tissues of developing *msd2* panicles (Figure 3A,B). Within these tissues, the global transcriptomic profile of genes downregulated in *msd2* revealed Gene Ontology (GO) term enrichment for proteins involved in oxylipin biosynthesis, as well as reorganization of cellular structure (Supplemental Figure S2), including members of the glycoside hydrolase, lipid transferase, and cellulose synthase families. Genes upregulated within stage 4 PS and SS tissues of *msd2* were enriched for functions related to system development, an ontological group consisting of transcription factors involved in developmental signaling and progression (Supplemental Figure S3).

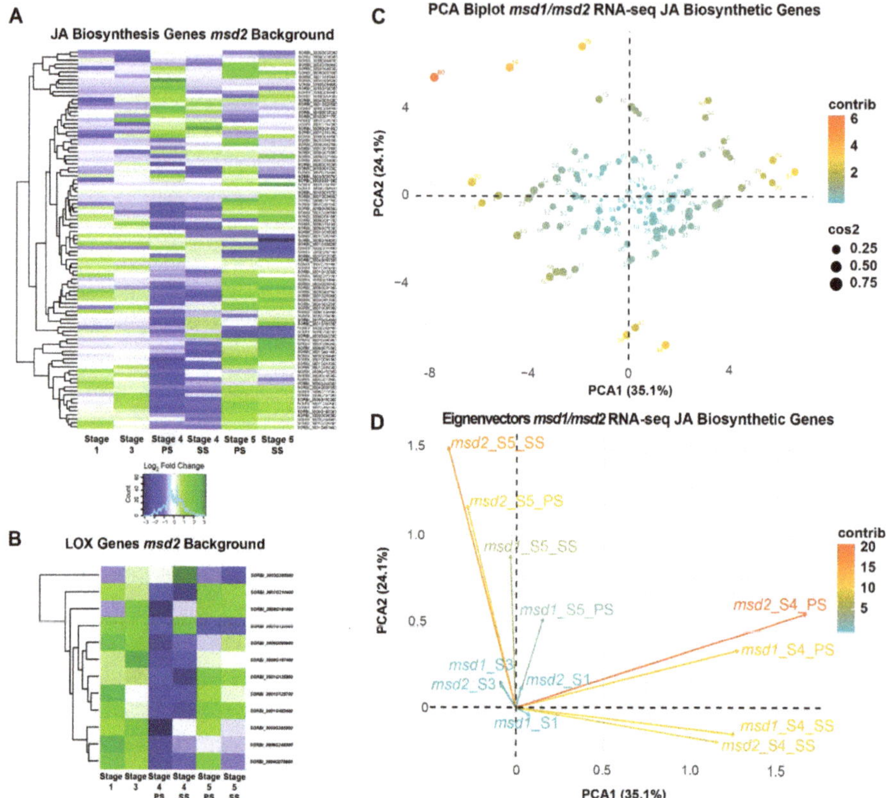

Figure 3. Transcriptomic profile showing the WT: *msd2* log$_2$(fold change) of (**A**) Jasmonic Acid (JA) biosynthetic pathway genes and (**B**) only LOX paralogs (based on homology from *Arabidopsis* and maize orthologs) across various stages of meristem development. (**C**) Principal component analysis (PCA) biplot and (**D**) eigenvectors of meristem stages from *msd1* and *msd2* RNA-seq data for the JA biosynthesis genes. Plot points are colored and sized according to dimensional contribution and quality, respectively. Eigenvectors are colored according to dimensional contribution.

Comparison of *msd1* and *msd2* transcriptomes revealed conserved GO enrichment categories, with little difference in expression of JA biosynthetic and signaling genes between mutants in the TCP transcription factor and lipoxygenase components of the hormone pathway. Principal component analysis (PCA) of JA pathway gene expression in both *msd1* and *msd2* showed that the greatest variance involved particular JA biosynthesis genes, predominantly cytochrome, jasmonate methyl transferases, *OPC-8*, *OPR*, and *LOX* genes (Figure 3C). Early-stage meristems (stage 1 and stage 3) exhibited the least variance between the *msd1* and *msd2* transcriptional profiles, whereas stage 4 and 5 inflorescences and spikelets made the greatest contribution to PCA dimensionality. PCA eigenvectors also indicated that the transcriptional divergence between WT and *multiseeded* plants occurs around stage 4 and continues through maturation in stage 5 (Figure 3D). A set of 149 genes identified by Jiao et al. (2018) as putative regulatory targets of MSD1 was strongly downregulated in *msd2* in either stage 4 or stage 5 tissues (Supplemental Figure S4a,b). Again, dimensional analysis of RNA-seq data revealed little variation between *msd1* and *msd2*, and indicated that stage 4 meristem marks the moment of demarcation between *multiseeded* and WT plants (Supplemental Figure S4d,c).

Motif analysis of JA biosynthetic and signaling genes revealed enrichment for various developmental and environmentally responsive DNA-binding domains, specifically the AP2, BZR (brassinazole-resistant family), bZIP, and WRKY families, as well as TCP proteins (Supplemental Figure S5a). A similar analysis of the 149 candidate MSD1 regulatory targets yielded a significant enrichment for CG-rich motifs strongly recognized by TCP, AP2, MYB, and E2F (specifically Della) transcription factors (Supplemental Figure S5b).

3.4. MSD2 Is Regulated by the TCP Transcription Factor MSD1

To evaluate if MSD1 directly regulates components of the JA biosynthetic pathway, Yeast 1-hybrid (Y1H) analysis was performed to determine whether MSD1 directly regulates *MSD2* in trans. Indeed, MSD1 bound to the sequence upstream of the *MSD2* transcriptional start site (TSS). MSD1 also bound sequence upstream of its own TSS (Figure 4A). These observations are consistent with the idea that MSD1 controls expression of both itself and *MSD2*.

In a less biased investigation of MSD1 regulatory targets, we conducted DNA Affinity Purification sequencing (DAP-seq) [35] analysis using Illumina short-read libraries constructed from developing floral meristem tissues and incubated with bacterially-expressed GST-tagged MSD1 proteins. The full list of 2730 identified peaks with their nearest annotated genes is provided as Supplementary Table S2. Motif analysis of peaks localized near TSSs were only enriched for the canonical TCP DNA binding motif (GTGGGNCC) bound by other plant TCP proteins (Figure 4B) [35–37]. Comparing these peaks with RNA-seq data revealed that 124 of the genes associated with these peaks were at least two-fold downregulated in stage 4 PS or SS tissues of *msd1* mutants, and the size of this candidate gene list increased when we included genes downregulated in any meristem tissue stage. Based on homology to JA pathway genes from other plant species, a subset of these downregulated genes is involved in JA biosynthesis or signaling (Figure 4C), including the *LOX* genes SORBI_3008G157900 and SORBI_3007G210400. Additionally, several genes associated with DAP-seq peaks were among the 149 MSD1 regulatory targets identified by Jiao et al. (2018): they are SORBI_3001G012200 (cytochrome P450), SORBI_3001G202600 (glutamyl-tRNA reductase), SORBI_3002G227700 (lipase), SORBI_3003G061900 (zinc finger), SORBI_3007G004600 (ferredoxin-type iron-sulfur binding domain containing protein), SORBI_3007G035600 (MSP domain containing protein), SORBI_3009G032600 (peroxidase), and SORBI_3009G100500 (WRKY).

The majority of DAP-seq peaks were localized more than 2000 base pairs upstream or downstream of the nearest annotated gene TSS, suggesting that they represent MSD1-targeted enhancer regions. When we applied motif analysis to all 2730 peaks, we identified additional DNA binding sequences. Several of the most significant motifs are recognized by other environmentally responsive and developmental transcription factors, including AP2, WRKY, HOMEOBOX, bZIP, and MYB (Figure 4D). Transcripts downregulated in *msd1* panicles that were also associated with MSD1 DAP-seq peaks included homologs of developmental signaling gene products, such as *Ramosa3* and *Embryonic Flower 1*, as well as several WRKY, AP2/ERF, and ZINC-finger transcription factors.

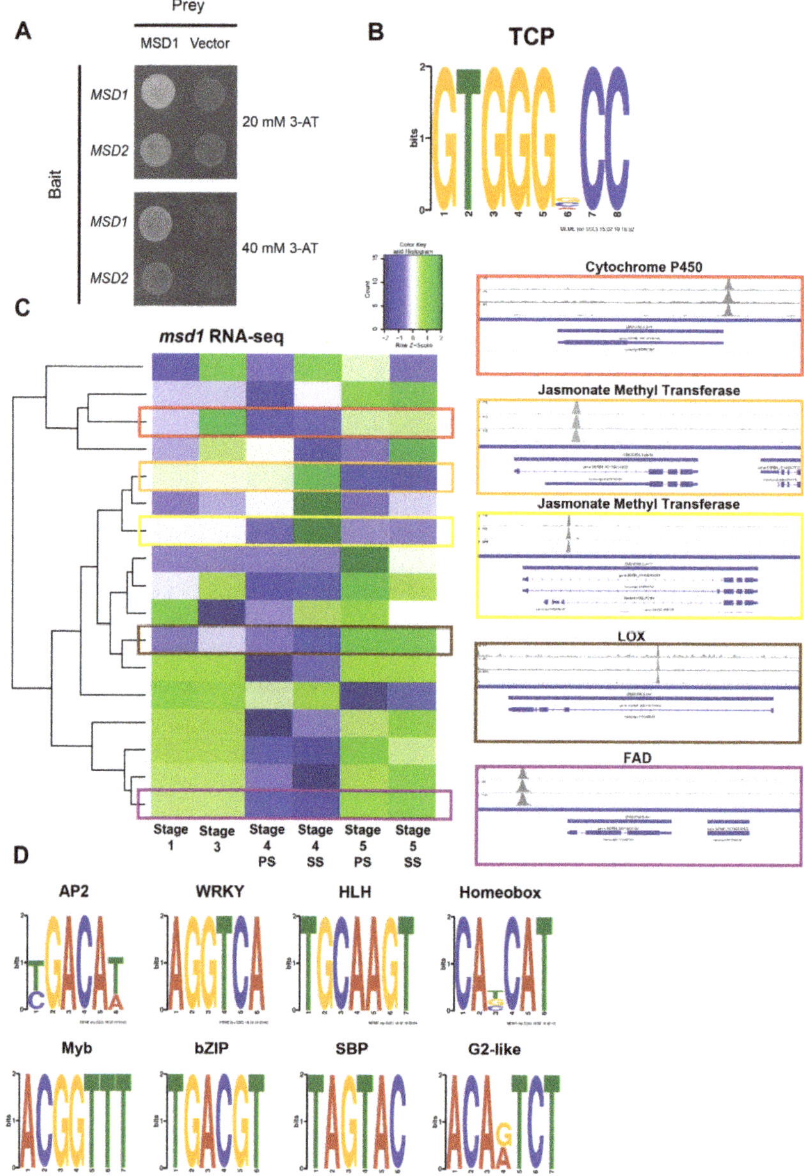

Figure 4. MSD1 as a regulator of developmental signaling genes. (**A**) Yeast 1-Hybrid of MSD1-mediated activation of gene expression by binding to the *MSD1* and *MSD2* upstream promoter regions. (**B**) Teosinte branched/Cycloidea/PCF (TCP) binding motif enriched in the MSD1 DNA Affinity Purification sequencing (DAP-seq) peaks that are localized within 2000 bp of an annotated gene transcriptional start site (TSS). (**C**) RNA-seq data from *msd1* showing downregulation across developing panicles stages in coordination with highlighted DAP-seq peaks localized near the transcriptional start sites (TSSs) of JA pathway genes. (**D**) Enriched DNA-binding motifs from all significant DAP-seq peak sequences, regardless of distance from an annotated gene TSS.

4. Discussion

MSD2 functions as an essential developmental gatekeeper in floral sex organ development; *MSD2*-deficient plants exhibit 100% flower fertility and grain filling, culminating in higher GNP. Mutant *msd2* panicles have similar transcriptomic profiles to mutants in the TCP transcription factor *MSD1*, suggesting that their respective phenotypes are both the result of disrupting an enzymatically-controlled feedback loop. MSD1 has the capacity to bind the promoter of *MSD2*, as well as promoters and more distant genetic elements associated with developmental and JA pathway genes, including those encoding other LOX-domain containing proteins.

JA is integral to environmental responsiveness and developmental progression; regulators of JA, JA biosynthetic genes, and JA signaling proteins influence pest/pathogen sensitivity [38,39], wound response [22,40,41], cell expansion [33], and sex determination and floral organ progression (specifically, anther and pistil development) [7,15–17,22,27,42,43]. Barley, rice, and maize display complex mechanisms of floral development. These modules have been genetically dissected via mutant analyses of homeobox, AP2/ERF, and MADS-box transcription factors as well as JA signaling genes and biosynthetic lipases. Notably, *MSD2* influences pistil progression in sorghum spikelets and the maize ortholog *TS1* controls pistillate determinacy in tassels. Molecular interpretation of these regulatory network ensembles reveals that repression of spikelet fertility in grasses is the norm and is modulated through one or a number of hormonal pathways in a given Poaceae lineage, which include JA and auxin [15,44–46].

Specifically, the role of *MSD2* in regulating floral organ fertility in sorghum is analogous to those of other LOX domain–containing proteins from other plant models [11,16,17,25,47]; multiple paralogs exist and exhibit variable expression through the stages of meristem development, indicating potential redundancy and narrow spatiotemporal expression of key *LOX* genes during development. Despite the higher expression of some LOX paralogs in *msd* meristems, exogenous Me-JA is sufficient to rescue the *multiseeded* phenotype, indicating that *msd2* is specifically responsible for sufficient JA signaling in the meristem cells that differentiate into male and female organs in PS and SS tissues.

Furthermore, the *msd2* RNA-seq data reveals the specific JA module of developmental control within sorghum; the data confirms the previous observation that the lack of a functional lipid enzyme can dismantle a regulatory network, resulting in observable downregulation of other biosynthetic pathway genes [10,48,49]; ablating elements of the JA pathway triggers the disruption of a positive feedback loop that would otherwise progress normally due to regular developmental perception of JA. In the case of *msd2*, this yielded a transcriptomic profile similar to that of the TCP transcription factor mutant *msd1* across developmental time points in immature meristems. Furthermore, protein–protein interaction and GO enrichment analyses of *msd1* and *msd2* gene networks in developing panicles suggest the existence of a distinct molecular avenue that leads to elevated GNP by diverting the expression of cellular restructuring genes and shunting to alternate developmental cascades mediated by other transcription factors and influenced by other hormone pathways.

MSD1 can bind to the *MSD2* promoter and activate gene expression. Consistent with this, expression analysis also revealed reduced levels of *MSD2* transcript in the *msd1* mutants [7]. In addition, DAP-seq analysis showed that MSD1 associates with other JA biosynthetic and signaling genes, including both 9- and 13-LOX paralogs of *MSD2*. Additionally, we identified potential enhancer binding regions for MSD1 that also exhibit enrichment for motifs bound by other developmental and environmental response transcription factors, suggesting that MSD1 participates in a mixed model of enhancer organization throughout the genome [50]. However, further chromatin conservation and architecture analyses will be required to elucidate the complete enhancer profile of this and other TCP proteins. It should be noted that although the sorghum *Ramosa3* ortholog was downregulated, several other trehalose phosphatase genes, in addition to *Clavata3*, were strongly upregulated in the mutants, suggesting that in *msd* mutants there is a diverting of the developmental signaling networks that canonically dictate sex organ determinacy in some plant lineages. The MSD1 DNA-binding data, together with the transcriptomic overlap of *msd1* and *msd2* mutants, provide further support of a model in which JA is responsible for regulating floral sex organ fate in *Sorghum bicolor*, and MSD1 is a major regulator of gene expression in this developmental schema (Supplemental Figure S6).

5. Materials and Methods

5.1. Identification of MSD2 and Phenotypic Evaluation

Seventeen *msd* mutants were isolated from an ethyl methane sulfonate (EMS) population [51] and grown in the field of the USDA-ARS Cropping Systems Research Laboratory at Lubbock, TX (33′39″ N, 101′49″ W). High-quality DNA was extracted [52] from confirmed *msd* lines and submitted for whole-genome sequencing at Beijing Genomic Institute (https://www.bgi.com/us/). Reads were trimmed and aligned to the sorghum reference genome v1.4 with Bowtie2 [53]. SNP calling was carried out on reads with PHRED >20 using SAMtools [54] and BCFtools; read depth was set from 3 to 50. Only homozygous G/C to A/T SNP transitions were filtered through to prediction by the Ensembl variation predictor [55]. Functional annotations of genes along with homology and syntenic analyses were derived from the Gramene database release 39 [56]. Phylogenetic analysis (boxshade and trees) was performed using MUSCLE alignment from MEGA X software. Phenotypic observations including grain number per panicle, root length, and days to emergence were taken from individual plants and seedlings grown in greenhouse or growth chamber conditions (16 h:8 h light:dark photoperiod, 27 °C). Photomicrographs of inflorescence tissues at five stages (from meristem to immature spikelets as described in Jiao et al., 2018) were collected and processed for scanning electron microscopy (SEM).

5.2. Transcriptome Profiling

Sample collection, processing, and transcriptomic profiling was conducted as described in Jiao et al. (2018). Three replicates (ten plants each) at each stage of panicle development were used for tissue collection. The ten plants for each replicate were processed as follows: at stage 1, whole panicles were harvested; at stage 3, differentiated floral organs on the tips of panicles were isolated; and at stages 4 and 5, the SS and PS tissues were isolated. For each replicate, the ten samples for each stage were pooled together. Tissues were immediately frozen in liquid nitrogen and stored at −80 °C prior to RNA extraction.

RNA was extracted using the TRIzol reagent, and then treated with DNase and purified using the RNeasy Mini Clean-up kit (Qiagen). Total RNA quality was examined on 1% agarose gels and RNA Nanochips on an Agilent 2100 Bioanalyzer (Agilent Technologies). Samples with RNA integrity number ≥ 7.0 were used for library preparation. Poly (A)$^+$ selection was applied to RNA via oligo (dT) magnetic beads (Invitrogen 610-02) and eluted in 11 µl of water. RNA-seq library construction was carried out with the ScriptSeq™ v2 kit (Epicentre SSV21124). Final libraries were amplified with 13 PCR cycles. RNA-seq of three biological replicates was executed at the sequencing center of Cold Spring Harbor Laboratory on an Illumina HiSeq2500 instrument.

RNA-seq data from each sample was first aligned to the sorghum version 3.4 reference genome using STAR [57]. Quantification of gene expression levels in each biological replicate was performed using Cufflinks [58]. The correlation coefficient among the three biological replicates for each sample was evaluated by the Pearson test in the R statistical environment. After removal of two low-quality samples, the biological replicates were merged together for differential expression analysis using Cuffdiff [58]. Only genes with at least five reads supported in at least one sample were subjected to differential expression analysis. The cutoff for differential expression was an adjusted FDR of $p < 0.05$. Motif enrichment analysis was performed using the MEME SUITE [59]. GO term analysis was performed with either the agriGO [60] Singular Enrichment Analysis using the hypergeometric statistical test method with significance level set to 0.01, or the GO Enrichment Analysis using PANTHER version 11 with all default presets [61]. All raw FASTQ files have been deposited in the NCBI Sequence Read Archive (see Data Availability statement). Statistical analysis, including PCA biplots (factoextra package), heatmap generation (heatmap2), along with additional figure generation (ggplot2), was performed using RStudio v1.1.463 [62].

5.3. DAP-Seq Analysis

The full length *MSD1* coding sequence (CDS) was cloned into the pDEST15 Gateway vector, and the resultant plasmid was transformed into BL21 competent cells. GST-MSD1 protein was induced by growing cells in Terrific Broth at 28 °C while shaking at 220 rpm; isopropyl-beta-D-thiogalactoside (IPTG) (Goldbio: I2481C25) was added to a final concentration of 0.001 M. GST-MSD1 protein was purified by resuspending cells in 1x PBS + 10 mM phenylmethanesulfonyl fluoride (PMSF), and then sonicating at 4 °C to disrupt cell membranes and plasmid DNA. Soluble cell extracts were added to MagneGST beads (Promega) and incubated and washed as described in Bartlet et al. (2017) [63]. High-purity DNA was isolated from stage 4–5 developing BTx623 meristems and sheared on a Covaris S220 sonicator. Template DNA from three biological replicates was incubated with bead-bound GST-MSD1 protein or GST beads alone (negative control). The MSD1-bound DNA was then washed, eluted, and ligated with Illumina adaptor sequences and quality-controlled using Qubit and Bioanalyzer as described in Bartlet et al. (2017). Sequencing was performed using the mid-output from the Illumina NextSeq platform, multiplexing all six samples, yielding ~16–20 × 10^6 reads per samples. Two separate sequencings runs were performed on experimental samples to increase detection power, with biological replicates undergoing 75-bp and 100-bp paired-end reads. The resultant FASTQ files were aligned and merged as follows: Trimmomatic [64] was used for FASTQ trimming, followed by Bowtie2 [53] alignment and MACS2 [65] peak calling (using the bead-only control for background subtraction), and finally the annotatePeaks program from the Homer [66] package was used to associate peaks with gene models from the version 3.4 *Sorghum bicolor* reference genome files housed by Gramene [67,68]. Sorghum GFF and GTF files were both used for annotatePeaks features functionality; however, the GTF file yielded more total gene-associated peaks than the GFF file. SAMtools was used for various file formatting and manipulation steps, including sorting and merging of the 75-bp and 100-bp paired-end read files. Motif enrichment analysis was performed using the MEME SUITE.

5.4. Phenotypic Rescue of msd2 with Exogenous Methyl-Jasmonate

Phenotypic rescue was performed exactly as described in Jiao et al. (2018). Briefly, BTx623 or *msd2* mutant seeds were germinated and grown at 16-h day cycles at 24 °C in a polyethylene greenhouse in Lubbock, TX. Beginning at leaf stage 7, 1 mL of either 0.05% Tween-20 (polyethylene glycol sorbitan monolaurate) in water (control) or 0.5 mM or 1.0 mM methyl-jasmonate in 0.05% Tween-20 was aspirated directly down the floral whorl. This treatment was repeated every 48 h until the majority of control plants reached the flag leaf stage. At that point, all experimental treatments were halted for that genotype. All plants were allowed to mature to the soft dough stage prior phenotypic rescue evaluation.

6. Data Availability

Sequencing data is available on the National Center for Biotechnology Information Sequence Read Archive (NCBI SRA: https://www.ncbi.nlm.nih.gov/sra). Accession codes for FASTQ files are as follows: DAP-seq, PRJNA550273; RNA-seq, SRP127741 [7] and PRJNA550261. DAP-seq BED files from MACS2 calls are available in Supplementary Data File 1.

Supplementary Materials: Supplementary materials can be found at http://www.mdpi.com/1422-0067/20/19/4951/s1.

Author Contributions: Conceptualization, D.W. and Z.X.; Data curation, N.G. and Y.J.; Formal analysis, N.G., Y.J. and S.A.C.; Funding acquisition, D.W. and Z.X.; Investigation, N.G., Y.K.L., L.Z., R.C., M.R., G.B., C.H., S.A.C., L.D., J.C., J.B. and Z.X.; Methodology, N.G. and Z.X.; Project administration, D.W. and Z.X.; Validation, G.B.; Visualization, N.G.; Writing—original draft, N.G.; Writing—review & editing, N.G., Y.J., Y.K.L., L.Z., R.C., G.B., C.H., S.A.C., L.D., D.W. and Z.X.

Funding: N.G., Y.J., R.C., G.B., J.B., C.H., and Z.X. acknowledge support from the United Sorghum Checkoff program. Z.X. was also partly supported by USDA ARS 3096-21000-019-00-D. Y.J., Y.K.L., N.G., M.R., and D.W. were partly supported by USDA ARS 8062-21000-041-00D. Y.K.L. acknowledges that this work was supported by a grant from the Next-Generation BioGreen 21 Program (Project No. PJ013658032019), Rural Development Administration, Republic of Korea, and R&D Program of 'Plasma Advanced Technology for Agriculture and Food (Plasma Farming)' through the National Fusion Research Institute of Korea (NFRI). NG was supported by an ARS-funded postdoc fellowship. The APC was funded by USDA ARS 8062-21000-041-00D.

Acknowledgments: The authors wish to thank all the support staff, students, and farmers that have made this work possible.

Conflicts of Interest: The authors declare no conflict of interest.

References

1. Dillon:, S.L.; Shapter, F.M.; Henry, R.J.; Cordeiro, G.; Izquierdo, L.; Lee, L.S. Domestication to crop improvement: Genetic resources for *Sorghum* and *Saccharum* (Andropogoneae). *Ann. Bot.* **2007**, *100*, 975–989. [CrossRef] [PubMed]

2. de Wet, J.M.J.; Huckabay, J.P. The origin of *Sorghum bicolor*. II. Distribution and domestication. *Evolution* **1967**, *21*, 787–802. [CrossRef] [PubMed]

3. Paterson, A.H.; Bowers, J.E.; Bruggmann, R.; Dubchak, I.; Grimwood, J.; Gundlach, H.; Haberer, G.; Hellsten, U.; Mitros, T.; Poliakov, A.; et al. The *Sorghum bicolor* genome and the diversification of grasses. *Nature* **2009**, *457*, 551–556. [CrossRef] [PubMed]

4. Jiang, W.; Zhou, H.; Bi, H.; Fromm, M.; Yang, B.; Weeks, D.P. Demonstration of CRISPR/Cas9/sgRNA-mediated targeted gene modification in Arabidopsis, tobacco, sorghum and rice. *Nucleic Acids Res.* **2013**, *41*, e188. [CrossRef] [PubMed]

5. Ding, Y.; Li, H.; Chen, L.L.; Xie, K. Recent advances in genome editing using CRISPR/Cas9. *Front. Plant Sci.* **2016**, *7*, 703. [CrossRef] [PubMed]

6. Brown, P.J.; Klein, P.E.; Bortiri, E.; Acharya, C.B.; Rooney, W.L.; Kresovich, S. Inheritance of inflorescence architecture in sorghum. *Theor. Appl. Genet.* **2006**, *113*, 931–942. [CrossRef]

7. Jiao, Y.; Lee, Y.K.; Gladman, N.; Chopra, R.; Christensen, S.A.; Regulski, M.; Burow, G.; Hayes, C.; Burke, J.; Ware, D.; et al. MSD1 regulates pedicellate spikelet fertility in sorghum through the jasmonic acid pathway. *Nat. Commun.* **2018**, *9*, 822. [CrossRef]

8. Dalberg, J. *Classification and Characterization of Sorghum. Sorghum: Origin, History, Technology, and Production*; John Wiley & Sons, Inc.: New York, NY, USA, 2000; pp. 99–130.

9. Yuan, Z.; Zhang, D. Roles of jasmonate signalling in plant inflorescence and flower development. *Curr. Opin. Plant Biol.* **2015**, *27*, 44–51. [CrossRef]

10. Wasternack, C.; Hause, B. Jasmonates: Biosynthesis, perception, signal transduction and action in plant stress response, growth and development. An update to the 2007 review in annals of botany. *Ann. Bot.* **2013**, *111*, 1021–1058. [CrossRef]

11. Lyons, R.; Manners, J.M.; Kazan, K. Jasmonate biosynthesis and signaling in monocots: A comparative overview. *Plant Cell Rep.* **2013**, *32*, 815–827. [CrossRef]

12. Robert-Seilaniantz, A.; Grant, M.; Jones, J.D. Hormone crosstalk in plant disease and defense: More than just jasmonate-salicylate antagonism. *Annu. Rev. Phytopathol.* **2011**, *49*, 317–343. [CrossRef] [PubMed]

13. Yan, Y.; Christensen, S.; Isakeit, T.; Engelberth, J.; Meeley, R.; Hayward, A.; Emery, R.J.; Kolomiets, M.V. Disruption of OPR7 and OPR8 reveals the versatile functions of jasmonic acid in maize development and defense. *Plant Cell* **2012**, *24*, 1420–1436. [CrossRef] [PubMed]

14. Ye, M.; Luo, S.M.; Xie, J.F.; Li, Y.F.; Xu, T.; Liu, Y.; Song, Y.Y.; Zhu-Salzman, K.; Zeng, R.S. Silencing COI1 in rice increases susceptibility to chewing insects and impairs inducible defense. *PLoS ONE* **2012**, *7*, e36214. [CrossRef] [PubMed]

15. Cai, Q.; Yuan, Z.; Chen, M.; Yin, C.; Luo, Z.; Zhao, X.; Liang, W.; Hu, J.; Zhang, D. Jasmonic acid regulates spikelet development in rice. *Nat. Commun.* **2014**, *5*, 3476. [CrossRef] [PubMed]

16. Acosta, I.F.; Laparra, H.; Romero, S.P.; Schmelz, E.; Hamberg, M.; Mottinger, J.P.; Moreno, M.A.; Dellaporta, S.L. Tasselseed1 is a lipoxygenase affecting jasmonic acid signaling in sex determination of maize. *Science* **2009**, *323*, 262–265. [CrossRef]

17. DeLong, A.; Calderon-Urrea, A.; Dellaporta, S.L. Sex determination gene TASSELSEED2 of maize encodes a short-chain alcohol dehydrogenase required for stage-specific floral organ abortion. *Cell* **1993**, *74*, 757–768. [CrossRef]

18. Schaller, F.; Biesgen, C.; Mussig, C.; Altmann, T.; Weiler, E.W. 12-Oxophytodienoate reductase 3 (OPR3) is the isoenzyme involved in jasmonate biosynthesis. *Planta* **2000**, *210*, 979–984. [CrossRef]

19. Stintzi, A.; Browse, J. The Arabidopsis male-sterile mutant, opr3, lacks the 12-oxophytodienoic acid reductase required for jasmonate synthesis. *Proc. Natl. Acad. Sci. USA* **2000**, *97*, 10625–10630. [CrossRef]

20. Park, J.H.; Halitschke, R.; Kim, H.B.; Baldwin, I.T.; Feldmann, K.A.; Feyereisen, R. A knock-out mutation in allene oxide synthase results in male sterility and defective wound signal transduction in arabidopsis due to a block in jasmonic acid biosynthesis. *Plant J.* **2002**, *31*, 1–12. [CrossRef]

21. Koo, A.J.; Cooke, T.F.; Howe, G.A. Cytochrome P450 CYP94B3 mediates catabolism and inactivation of the plant hormone jasmonoyl-L-isoleucine. *Proc. Natl. Acad. Sci. USA* **2011**, *108*, 9298–9303. [CrossRef]

22. Lunde, C.; Kimberlin, A.; Leiboff, S.; Koo, A.J.; Hake, S. *Tasselseed5* overexpresses a wound-inducible enzyme, ZmCYP94B1, that affects jasmonate catabolism, sex determination, and plant architecture in maize. *Commun. Biol.* **2019**, *2*, 114. [CrossRef] [PubMed]

23. Caldelari, D.; Wang, G.; Farmer, E.E.; Dong, X. Arabidopsis *lox3 lox4* double mutants are male sterile and defective in global proliferative arrest. *Plant Mol. Biol.* **2011**, *75*, 25–33. [CrossRef] [PubMed]

24. Jiao, Y.; Burke, J.; Chopra, R.; Burow, G.; Chen, J.; Wang, B.; Hayes, C.; Emendack, Y.; Ware, D.; Xin, Z. A sorghum mutant resource as an efficient platform for gene discovery in grasses. *Plant Cell* **2016**, *28*, 1551–1562. [CrossRef] [PubMed]

25. Feussner, I.; Wasternack, C. The lipoxygenase pathway. *Annu. Rev. Plant Biol.* **2002**, *53*, 275–297. [CrossRef] [PubMed]

26. Luo, D.; Carpenter, R.; Vincent, C.; Copsey, L.; Coen, E. Origin of floral asymmetry in antirrhinum. *Nature* **1996**, *383*, 794–799. [CrossRef] [PubMed]

27. Doebley, J.; Stec, A.; Hubbard, L. The evolution of apical dominance in maize. *Nature* **1997**, *386*, 485–488. [CrossRef]

28. Kosugi, S.; Ohashi, Y. PCF1 and PCF2 specifically bind to cis elements in the rice proliferating cell nuclear antigen gene. *Plant Cell* **1997**, *9*, 1607–1619.

29. Davidson, R.M.; Gowda, M.; Moghe, G.; Lin, H.; Vaillancourt, B.; Shiu, S.H.; Jiang, N.; Robin Buell, C. Comparative transcriptomics of three poaceae species reveals patterns of gene expression evolution. *Plant J.* **2012**, *71*, 492–502. [CrossRef]

30. Makita, Y.; Shimada, S.; Kawashima, M.; Kondou-Kuriyama, T.; Toyoda, T.; Matsui, M. Morokoshi: Transcriptome database in *Sorghum bicolor*. *Plant Cell Physiol.* **2015**, *56*, e6. [CrossRef]

31. Petryszak, R.; Burdett, T.; Fiorelli, B.; Fonseca, N.A.; Gonzalez-Porta, M.; Hastings, E.; Huber, W.; Jupp, S.; Keays, M.; Kryvych, N.; et al. Expression atlas update—A database of gene and transcript expression from microarray- and sequencing-based functional genomics experiments. *Nucleic Acids Res.* **2014**, *42*, D926–D932. [CrossRef]

32. Olson, A.; Klein, R.R.; Dugas, D.V.; Lu, Z.; Regulski, M.; Klein, P.E.; Ware, D. Expanding and vetting *Sorghum bicolor* gene annotations through transcriptome and methylome sequencing. *Plant Genome* **2014**, *7*. [CrossRef]

33. Huang, H.; Liu, B.; Liu, L.; Song, S. Jasmonate action in plant growth and development. *J. Exp. Bot.* **2017**, *68*, 1349–1359. [CrossRef] [PubMed]

34. Zhang, Y.; Turner, J.G. Wound-induced endogenous jasmonates stunt plant growth by inhibiting mitosis. *PLoS ONE* **2008**, *3*, e3699. [CrossRef] [PubMed]

35. O'Malley, R.C.; Huang, S.C.; Song, L.; Lewsey, M.G.; Bartlett, A.; Nery, J.R.; Galli, M.; Gallavotti, A.; Ecker, J.R. Cistrome and epicistrome features shape the regulatory DNA landscape. *Cell* **2016**, *166*, 1598. [CrossRef] [PubMed]

36. Li, W.; Li, D.D.; Han, L.H.; Tao, M.; Hu, Q.Q.; Wu, W.Y.; Zhang, J.B.; Li, X.B.; Huang, G.Q. Genome-wide identification and characterization of tcp transcription factor genes in upland cotton (*Gossypium hirsutum*). *Sci. Rep.* **2017**, *7*, 10118. [CrossRef] [PubMed]

37. Weirauch, M.T.; Yang, A.; Albu, M.; Cote, A.G.; Montenegro-Montero, A.; Drewe, P.; Najafabadi, H.S.; Lambert, S.A.; Mann, I.; Cook, K.; et al. Determination and inference of eukaryotic transcription factor sequence specificity. *Cell* **2014**, *158*, 1431–1443. [CrossRef] [PubMed]

38. Stitz, M.; Baldwin, I.T.; Gaquerel, E. Diverting the flux of the ja pathway in *Nicotiana attenuata* compromises the plant's defense metabolism and fitness in nature and glasshouse. *PLoS ONE* **2011**, *6*, e25925. [CrossRef]

39. Ahmad, P.; Rasool, S.; Gul, A.; Sheikh, S.A.; Akram, N.A.; Ashraf, M.; Kazi, A.M.; Gucel, S. Jasmonates: Multifunctional roles in stress tolerance. *Front. Plant Sci.* **2016**, *7*, 813. [CrossRef]

40. Li, C.; Schilmiller, A.L.; Liu, G.; Lee, G.I.; Jayanty, S.; Sageman, C.; Vrebalov, J.; Giovannoni, J.J.; Yagi, K.; Kobayashi, Y.; et al. Role of beta-oxidation in jasmonate biosynthesis and systemic wound signaling in tomato. *Plant Cell* **2005**, *17*, 971–986. [CrossRef]

41. Yan, L.; Zhai, Q.; Wei, J.; Li, S.; Wang, B.; Huang, T.; Du, M.; Sun, J.; Kang, L.; Li, C.B.; et al. Role of Tomato Lipoxygenase D in Wound-Induced Jasmonate Biosynthesis and Plant Immunity to Insect Herbivores. *PLoS Genet.* **2013**, *9*, e1003964. [CrossRef]

42. Dobritzsch, S.; Weyhe, M.; Schubert, R.; Dindas, J.; Hause, G.; Kopka, J.; Hause, B. Dissection of jasmonate functions in tomato stamen development by transcriptome and metabolome analyses. *BMC Biol.* **2015**, *13*, 28. [CrossRef] [PubMed]

43. Niwa, T.; Suzuki, T.; Takebayashi, Y.; Ishiguro, R.; Higashiyama, T.; Sakakibara, H.; Ishiguro, S. Jasmonic acid facilitates flower opening and floral organ development through the upregulated expression of SlMYB21 transcription factor in tomato. *Biosci. Biotechnol. Biochem.* **2018**, *82*, 292–303. [CrossRef] [PubMed]

44. Bull, H.; Casao, M.C.; Zwirek, M.; Flavell, A.J.; Thomas, W.T.B.; Guo, W.; Zhang, R.; Rapazote-Flores, P.; Kyriakidis, S.; Russell, J.; et al. Barley SIX-ROWED SPIKE3 encodes a putative jumonji C-type H3K9me2/me3 demethylase that represses lateral spikelet fertility. *Nat. Commun.* **2017**, *8*, 936. [CrossRef] [PubMed]

45. Ishiguro, S.; Kawai-Oda, A.; Ueda, J.; Nishida, I.; Okada, K. The defective in anther dehiscence gene encodes a novel phospholipase A1 catalyzing the initial step of jasmonic acid biosynthesis, which synchronizes pollen maturation, anther dehiscence, and flower opening in Arabidopsis. *Plant Cell* **2001**, *13*, 2191–2209. [CrossRef] [PubMed]

46. Shuai, B.; Reynaga-Pena, C.G.; Springer, P.S. The lateral organ boundaries gene defines a novel, plant-specific gene family. *Plant Physiol.* **2002**, *129*, 747–761. [CrossRef] [PubMed]

47. Yang, X.Y.; Jiang, W.J.; Yu, H.J. The expression profiling of the lipoxygenase (LOX) family genes during fruit development, abiotic stress and hormonal treatments in cucumber (*Cucumis sativus* L.). *Int. J. Mol. Sci.* **2012**, *13*, 2481–2500. [CrossRef] [PubMed]

48. Tiwari, G.J.; Liu, Q.; Shreshtha, P.; Li, Z.; Rahman, S. RNAi-mediated down-regulation of the expression of OsFAD2-1: Effect on lipid accumulation and expression of lipid biosynthetic genes in the rice grain. *BMC Plant Biol.* **2016**, *16*, 189. [CrossRef] [PubMed]

49. De Geyter, N.; Gholami, A.; Goormachtig, S.; Goossens, A. Transcriptional machineries in jasmonate-elicited plant secondary metabolism. *Trends Plant Sci.* **2012**, *17*, 349–359. [CrossRef]

50. Long, H.K.; Prescott, S.L.; Wysocka, J. Ever-changing landscapes: Transcriptional enhancers in development and evolution. *Cell* **2016**, *167*, 1170–1187. [CrossRef]

51. Xin, Z.; Wang, M.L.; Barkley, N.A.; Burow, G.; Franks, C.; Pederson, G.; Burke, J. Applying genotyping (TILLING) and phenotyping analyses to elucidate gene function in a chemically induced sorghum mutant population. *BMC Plant Biol.* **2008**, *8*, 103. [CrossRef] [PubMed]

52. Xin, Z.; Chen, J. A high throughput DNA extraction method with high yield and quality. *Plant Methods* **2012**, *8*, 26. [CrossRef]

53. Langmead, B.; Salzberg, S.L. Fast gapped-read alignment with Bowtie 2. *Nat. Methods* **2012**, *9*, 357–359. [CrossRef] [PubMed]

54. Li, H.; Handsaker, B.; Wysoker, A.; Fennell, T.; Ruan, J.; Homer, N.; Marth, G.; Abecasis, G.; Durbin, R.; Genome Project Data Processing Subgroup. The sequence alignment/map format and SAMtools. *Bioinformatics* **2009**, *25*, 2078–2079. [CrossRef] [PubMed]

55. McLaren, W.; Pritchard, B.; Rios, D.; Chen, Y.; Flicek, P.; Cunningham, F. Deriving the consequences of genomic variants with the Ensembl API and SNP Effect Predictor. *Bioinformatics* **2010**, *26*, 2069–2070. [CrossRef] [PubMed]

56. Monaco, M.K.; Stein, J.; Naithani, S.; Wei, S.; Dharmawardhana, P.; Kumari, S.; Amarasinghe, V.; Youens-Clark, K.; Thomason, J.; Preece, J.; et al. Gramene 2013: Comparative plant genomics resources. *Nucleic Acids Res.* **2014**, *42*, D1193–D1199. [CrossRef] [PubMed]

57. Dobin, A.; Davis, C.A.; Schlesinger, F.; Drenkow, J.; Zaleski, C.; Jha, S.; Batut, P.; Chaisson, M.; Gingeras, T.R. Star: Ultrafast universal RNA-seq aligner. *Bioinformatics* **2013**, *29*, 15–21. [CrossRef] [PubMed]

58. Trapnell, C.; Williams, B.A.; Pertea, G.; Mortazavi, A.; Kwan, G.; van Baren, M.J.; Salzberg, S.L.; Wold, B.J.; Pachter, L. Transcript assembly and quantification by RNA-seq reveals unannotated transcripts and isoform switching during cell differentiation. *Nat. Biotechnol.* **2010**, *28*, 511–515. [CrossRef] [PubMed]

59. Bailey, T.L.; Boden, M.; Buske, F.A.; Frith, M.; Grant, C.E.; Clementi, L.; Ren, J.; Li, W.W.; Noble, W.S. Meme suite: Tools for motif discovery and searching. *Nucleic Acids Res.* **2009**, *37*, W202–W208. [CrossRef] [PubMed]

60. Du, Z.; Zhou, X.; Ling, Y.; Zhang, Z.; Su, Z. AgriGo: A GO analysis toolkit for the agricultural community. *Nucleic Acids Res.* **2010**, *38*, W64–W70. [CrossRef]

61. Mi, H.; Huang, X.; Muruganujan, A.; Tang, H.; Mills, C.; Kang, D.; Thomas, P.D. Panther version 11: Expanded annotation data from gene ontology and reactome pathways, and data analysis tool enhancements. *Nucleic Acids Res.* **2017**, *45*, D183–D189. [CrossRef]

62. RStudio Team. *RStudio: Integrated Development for R*; Rstudio, Inc.: Boston, MA, USA, 2015.

63. Bartlett, A.; O'Malley, R.C.; Huang, S.C.; Galli, M.; Nery, J.R.; Gallavotti, A.; Ecker, J.R. Mapping genome-wide transcription-factor binding sites using DAP-seq. *Nat. Protoc.* **2017**, *12*, 1659–1672. [CrossRef] [PubMed]

64. Bolger, A.M.; Lohse, M.; Usadel, B. Trimmomatic: A flexible trimmer for illumina sequence data. *Bioinformatics* **2014**, *30*, 2114–2120. [CrossRef] [PubMed]

65. Zhang, Y.; Liu, T.; Meyer, C.A.; Eeckhoute, J.; Johnson, D.S.; Bernstein, B.E.; Nusbaum, C.; Myers, R.M.; Brown, M.; Li, W.; et al. Model-based analysis of ChIP-seq (macs). *Genome Biol.* **2008**, *9*, R137. [CrossRef] [PubMed]

66. Heinz, S.; Benner, C.; Spann, N.; Bertolino, E.; Lin, Y.C.; Laslo, P.; Cheng, J.X.; Murre, C.; Singh, H.; Glass, C.K. Simple combinations of lineage-determining transcription factors prime cis-regulatory elements required for macrophage and B cell identities. *Mol. Cell* **2010**, *38*, 576–589. [CrossRef] [PubMed]

67. Tello-Ruiz, M.K.; Naithani, S.; Stein, J.C.; Gupta, P.; Campbell, M.; Olson, A.; Wei, S.; Preece, J.; Geniza, M.J.; Jiao, Y.; et al. Gramene 2018: Unifying comparative genomics and pathway resources for plant research. *Nucleic Acids Res.* **2018**, *46*, D1181–D1189. [CrossRef] [PubMed]

68. McCormick, R.F.; Truong, S.K.; Sreedasyam, A.; Jenkins, J.; Shu, S.; Sims, D.; Kennedy, M.; Amirebrahimi, M.; Weers, B.D.; McKinley, B.; et al. The *Sorghum bicolor* reference genome: Improved assembly, gene annotations, a transcriptome atlas, and signatures of genome organization. *Plant J.* **2018**, *93*, 338–354. [CrossRef] [PubMed]

© 2019 by the authors. Licensee MDPI, Basel, Switzerland. This article is an open access article distributed under the terms and conditions of the Creative Commons Attribution (CC BY) license (http://creativecommons.org/licenses/by/4.0/).

International Journal of
Molecular Sciences

MDPI

Article

Sorghum *MSD3* Encodes an ω-3 Fatty Acid Desaturase that Increases Grain Number by Reducing Jasmonic Acid Levels

Lavanya Dampanaboina [1], Yinping Jiao [1,2], Junping Chen [1], Nicholas Gladman [1,2], Ratan Chopra [1,3], Gloria Burow [1], Chad Hayes [1], Shawn A. Christensen [4], John Burke [1], Doreen Ware [2,5,*] and Zhanguo Xin [1,*]

[1] Plant Stress and Germplasm Development Unit, Cropping Systems Research Laboratory, U.S. Department of Agriculture-Agricultural Research Service, Lubbock, TX 79415, USA; lavanya.dampanaboina@ttu.edu (L.D.); yjiao@cshl.edu (Y.J.); Junping.chen@ars.usda.gov (J.C.); gladman@cshl.edu (N.G.); rchopra@umn.edu (R.C.); Gloria.burow@ars.usda.gov (G.B.); Chad.Hayes@ars.usda.gov (C.H.); John.Burke@ars.usda.gov (J.B.)

[2] Cold Spring Harbor Laboratory, Cold Spring Harbor, New York, NY 11724, USA

[3] Current address: Department of Agronomy and Plant Genetics, University of Minnesota, St. Paul, MN 55108, USA

[4] Chemistry Research Unit, USDA-ARS, 1700 S.W. 23rd Drive, Gainesville, FL 32608, USA; Shawn.Christensen@ars.usda.gov

[5] U.S. Department of Agriculture-Agricultural Research Service, NEA Robert W. Holley Center for Agriculture and Health, Cornell University, Ithaca, New York, NY 14853, USA

* Correspondence: ware@cshl.edu or doreen.ware@ars.usda.gov (D.W.); zhanguo.xin@ars.usda.gov (Z.X.)

Received: 16 September 2019; Accepted: 26 October 2019; Published: 28 October 2019

Abstract: Grain number per panicle is an important component of grain yield in sorghum (*Sorghum bicolor* (L.)) and other cereal crops. Previously, we reported that mutations in multi-seeded 1 (*MSD1*) and *MSD2* genes result in a two-fold increase in grain number per panicle due to the restoration of the fertility of the pedicellate spikelets, which invariably abort in natural sorghum accessions. Here, we report the identification of another gene, *MSD3*, which is also involved in the regulation of grain numbers in sorghum. Four bulked F_2 populations from crosses between BTx623 and each of the independent *msd* mutants p6, p14, p21, and p24 were sequenced to 20× coverage of the whole genome on a HiSeq 2000 system. Bioinformatic analyses of the sequence data showed that one gene, Sorbi_3001G407600, harbored homozygous mutations in all four populations. This gene encodes a plastidial ω-3 fatty acid desaturase that catalyzes the conversion of linoleic acid (18:2) to linolenic acid (18:3), a substrate for jasmonic acid (JA) biosynthesis. The *msd3* mutants had reduced levels of linolenic acid in both leaves and developing panicles that in turn decreased the levels of JA. Furthermore, the *msd3* panicle phenotype was reversed by treatment with methyl-JA (MeJA). Our characterization of *MSD1*, *MSD2*, and now *MSD3* demonstrates that JA-regulated processes are critical to the *msd* phenotype. The identification of the *MSD3* gene reveals a new target that could be manipulated to increase grain number per panicle in sorghum, and potentially other cereal crops, through the genomic editing of *MSD3* functional orthologs.

Keywords: jasmonic acid; fatty acid desaturase; multiseeded; *msd*; grain number; MutMap; sorghum

---

1. Introduction

Grain number per panicle is a major determinant of grain yield in sorghum (*Sorghum bicolor* (L.) Moench) and other cereal crops [1–6]. Increased grain number and grain size, which are directly related to improved grain yield, are common goals during domestication of cereal crops, resulting in selection of genetic stocks with greater grain number and larger seeds [7]. Genetic, physiological, and agronomic studies performed in different environments have revealed that grain number per panicle is strongly correlated with grain yield per hectare [3,4,6,8,9]. Thus, understanding the mechanisms that govern the genetic determination of grain number per panicle and the effect of genetic manipulation of this trait have great potential to increase grain yield.

Currently, however, very little is known about how grain number per panicle is determined. Several features of the inflorescence contribute to the final grain number, including the number and size of the primary and secondary flower branches and fertility of spikelets (grass flowers). In sorghum, the inflorescence or panicle consists of a main rachis on which many primary branches are developed. Secondary branches and tertiary branches are developed from the primary branches [10,11]. The main inflorescence—primary, secondary, and tertiary branches—all end with a terminal triplet of spikelets, consisting of one sessile bisexual spikelet and two lateral pedicellate spikelets [12]. Below the terminal spikelets, one or more spikelet pairs can develop. These adjacent spikelet pairs consist of one sessile and one pedicellate spikelet.

In natural accessions of sorghum, only the sessile spikelets are fertile and capable of producing viable grains. The pedicellate spikelets occasionally develop anthers, but do not develop ovaries and eventually abort. Recently, we isolated and characterized a series of sorghum mutants in which both the sessile and pedicellate spikelets are fertile [11], and in which the number and size of the primary inflorescence branches are increased. These mutants were designated as multiseeded (*msd*) because their panicles are capable of producing more than 200% of the normal grain number per panicle relative to the nonmutated BTx623 [11]. Previously, we reported two *MSD* genes, *MSD1* and *MSD2*, which determine the fertility of the pedicellate spikelets. The *MSD1* gene encodes a TCP (teosinte branched/cycloidea/proliferating cell nuclear antigen)-domain plant-specific transcription factor that increases the expression of enzymes involved in the biosynthesis of jasmonic acid (JA) during panicle development [13]. The *MSD2* encodes a lipoxygenase (LOX) that catalyzes the first committed step of JA biosynthesis [14]. The elevated levels of JA in the wild type panicle may activate programmed cell death, leading to the arrest of the pedicellate spikelets in the wild type BTx623 [13,15]. Because the increase in JA is blocked in the *msd1* and *msd2* mutants, the pedicellate spikelets continue to develop into viable grains.

Here, we describe the identification of the *MSD3* gene, defined by a new *msd* mutant locus. *MSD3* encodes a major plastidial ω-3 fatty acid desaturase that catalyzes the desaturation of linoleic acid (18:2, 18 carbon chain with two double bonds) to linolenic acid (18:3, 18 carbon chain with three double bonds). The *msd3* mutants have dramatically reduced levels 18:3 fatty acid, as well as reduced levels of JA. JAs are lipid-derived cyclopentanone compounds that functionally resemble animal prostaglandins. Similar to the eicosanoid pathway of animals, JA is synthesized from linolenic acid (18:3) and hexadecatrienoic acid (16:3) through a series of steps of cyclization, reduction, and oxidation [16,17]. The reduced levels of 18:3 fatty acid in *msd3* mutants may lead to lower levels of endogenous JA. Indeed, the *msd3* phenotype can be reverted to the wild-type phenotype by application of MeJA. Our results indicate that the *msd* phenotype observed in the *msd3* mutants is caused by the reduction of endogenous levels of JA.

2. Results

2.1. Phenotype of the msd3 Mutants

Here, we characterized a new group of multi-seeded (*msd*) mutants that include p6 (putative mutant #6), p14, p21, and p24. Similar to the *msd1* and *msd2* mutants, the *msd3* mutants also exhibit increases in number and size of the primary inflorescence branches (Figure S1) [11,13,14], and both their sessile and pedicellate spikelets are fertile. These coordinated phenotypic changes led to a ~2-fold increase in grain number per panicle. Complementation tests revealed that these *msd* mutants p6, p14, p21, and p24 represented a new locus that is distinct from the *msd1* and *msd2* loci (Table 1). These mutants were designated as *msd3*. Similar to the *msd1* and *msd2* mutants, the increase in grain number in *msd3* mutants was also associated with reduction in grain size [11,13,14]. However, the *msd3* mutants had larger grains than the *msd1* and *msd2* mutants, (Figure 1), making *msd3* potentially useful trait for improving grain yield and grain appeal.

Table 1. Complementation tests between *msd3*, *msd2*, and *msd1* mutants. The *msd3* mutants were crossed to each other and to *msd1-1* and *msd2-1* mutants by hand emasculation and manual pollination. The panicle phenotypes of the F_1 plants were evaluated during the grain filling stage.

Mutant	*msd1-1 (p12)*	*msd2-1 (p4)*	*msd3-1 (p24)*	*msd3-2 (p14)*	*msd3-3 (p6)*
msd1-1 (p12)					
msd2-1 (p4)	WT				
msd3-1 (p24)	WT	WT			
msd3-2 (p14)	WT	WT	*msd*		
msd3-3 (p6)	WT	WT	*msd*	*msd*	
msd3-4 (p37)	WT	WT	*msd*	*msd*	*msd*

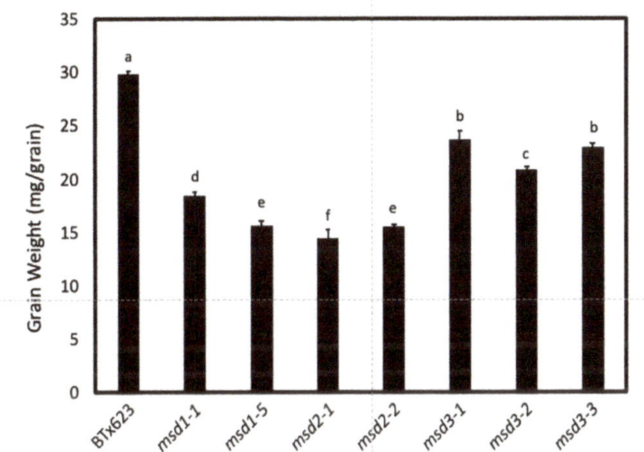

Figure 1. Comparison of grain weight (mg/grain) of *msd3* mutants vs. BTx623 and the *msd1* and *msd2* mutants. Grain weight was determined from plants planted on the Farm of Texas Tech University on Quaker Avenue in 2017. Four samples of 100 grains were weighed from each line. One-way ANOVA analysis revealed that grain weight varied significantly with an F-value of 276 and p-value of 1.8^{-15}. Grain weights labeled with different letters are significantly different.

2.2. Identification of the MSD3 Gene

The *MSD3* gene was identified through bulk segregant analysis of whole-genome sequencing data with an in-house bioinformatics pipeline, as described in rice and sorghum [18,19]. Previously, we crossed p6, p14, p21, and p24 to BTx623 and derived four F_2 populations. After bioinformatics analysis of the four bulked F_2 pools, we identified only one gene (Sorbi_3001G407600) that carried homozygous mutations in all four bulked F_2 pools (Figure 2). The genomic sequence of *MSD3* is 3132 bp in length, with a CDS of 1356 bp that encodes a protein of 451 amino acids. Two putative *msd* mutants, p21 and p24, harbored a mutation that converted the G residue at Chr01_69163608 to A, creating a splice site mutation at the junction of the third exon and the third intron (Figure 3). Both were renamed as *msd3-1*. The mutant p14, renamed *msd3-2*, harbored a G-to-A transition at Chr01_69163762 that created a nonsynonymous mutation, R240W, in the MSD3 protein. Mutant p6, renamed *msd3-3*, harbored a G-to-A transition at Chr01_69164229, resulting in a premature stop codon (W321*). Subsequently, we identified another mutation in the *MSD3* gene from the sequenced mutant library (ASR106, 25M2-1370) at Chr01_69165175, which created a splice site mutation at the junction of the seventh exon and the seventh intron [20]. The homozygous mutant (p37) exhibited the expected *msd* phenotype and was named as *msd3-4*.

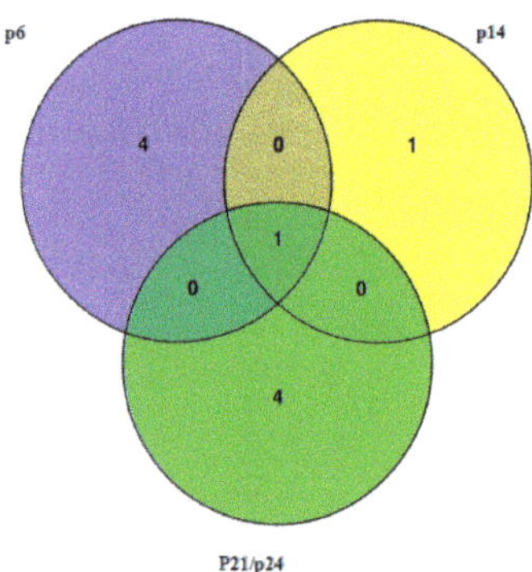

Figure 2. Identification of the *Msd3* gene by MutMap. Four F2 populations, BTx623*p6, BTx623*p14, BTx623*p21, and BTx623*p24, were subjected to sequencing of 20 bulked F2 mutants. The number in each circle represents the number of homozygous deleterious mutations. Only one gene, Sorbi_3001G407600, was commonly mutated in all four populations. In p21 and p24, the *MSD3* gene was mutated at the same position.

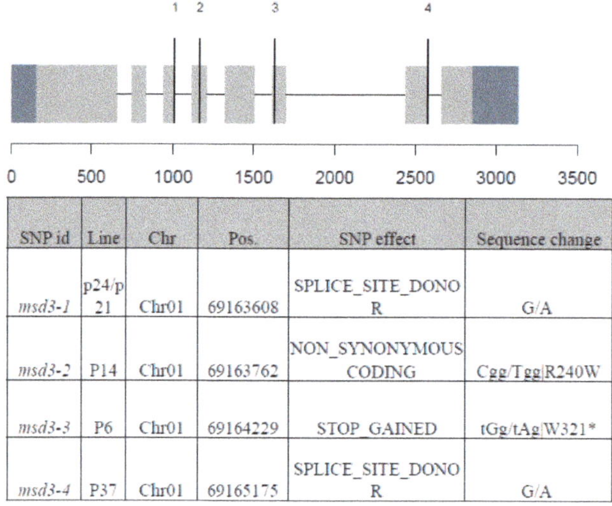

SNP id	Line	Chr	Pos.	SNP effect	Sequence change
msd3-1	p24/p21	Chr01	69163608	SPLICE_SITE_DONOR	G/A
msd3-2	P14	Chr01	69163762	NON_SYNONYMOUS CODING	Cgg/Tgg|R240W
msd3-3	P6	Chr01	69164229	STOP_GAINED	tGg/tAg|W321*
msd3-4	P37	Chr01	69165175	SPLICE_SITE_DONOR	G/A

Figure 3. *MSD3* gene model and mutations. The *MSD3* gene, Sorbi_3001G407600, spanning a genomic sequence of 3132 bp, with a CDS of 1356 bp, encodes a protein of 451 amino acids. The gene has eight exons (grey boxes) and seven introns (black line). The vertical lines indicate mutation sites as described in the table above.

To determine whether these mutations in the *MSD3* gene resulted in the *msd* phenotype, we made pairwise crosses among these four *msd3* mutants; all F_1 plants from these crosses exhibited the *msd* phenotype (Table 1). By contrast, F_1 plants resulting from crosses between *msd3* and *msd1*, *msd3* and *msd2*, or *msd2* and *msd1* exhibited the wild type panicle phenotype. Together, these results indicate that Sorbi_3001G407600 is the *MSD3* gene. Furthermore, we tested the co-segregation of the *msd3-4* mutation by Kompetitive allele-specific PCR (KASP) (Figure 4). Among 63 individual F_2 plants derived from the cross of BTx623 * *msd3-4* (p37), 15 plants were scored as AA at the mutation site, and all 15 lines exhibited the expected *msd3* phenotype. Fifteen plants were scored as GG, and 33 were scored as GA; these two genotypic classes, GG (wild-type, WT) and GA (heterozygote), exhibited that the wild-type panicle structure indicating that the *msd3* mutation is recessive. These results also showed that the causal mutation in the *msd3-4* co-segregated perfectly with the *msd* phenotype.

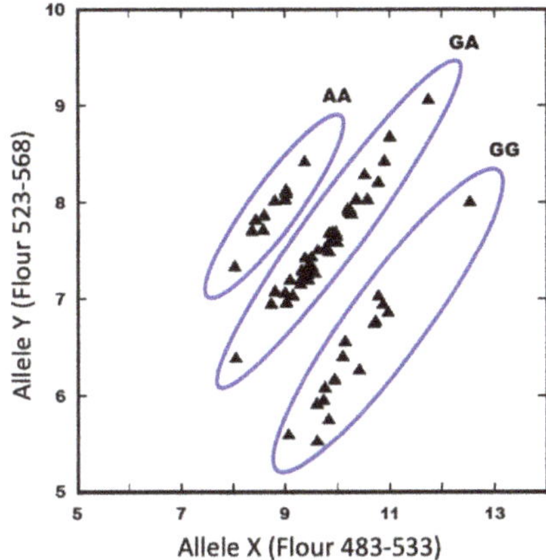

Figure 4. Genotyping of *msd3-4* F₂ backcrossed population by KASP (Kompetitive Allele-Specific PCR) SNP (single nucleotide polymorphism) analysis. The x-axis represents the endpoint fluorescence data at 483–533 nm (allele X or wild type allele) and the y-axis represents the endpoint fluorescence data at 523–568 nm (allele Y or mutant allele). Genotype for individual F₂ plants was determined by cluster analysis of the endpoint KASP assay. AA, homozygous *msd3-4*; GG, homozygous wild-type (WT); and GA, heterozygous WT.

2.3. MSD3 is FAD7, a Major Plastidial ω-3 Fatty Acid Desaturase

Sequencing analysis indicates that the *MSD3* gene (Sorbi_3001G407600) encodes a plastidial fatty acid desaturase that adds a double bond to the ω-3 carbon of linoleic acid (18:2) to convert it to linolenic acid (18:3). The sorghum genome has two plastid-targeted ω-3 fatty acid desaturases, Sorbi_3001G407600 and Sorbi_3002G430100. However, in many plants, *FAD7* is the major plastidial linoleic acid desaturase [21–25]. Based on the panicle phenotype of the *msd3* mutant, we reasoned that *MSD3* could be *FAD7* rather than *FAD8*. To determine whether *MSD3* is *FAD7* or *FAD8*, we constructed a matrix of identity between the two sorghum genes using the annotated *FAD7* and *FAD8* genes from rice and maize (Figure S2 and Table S1). *MSD3* exhibits 84.8% identity with rice *FAD7* (OsFAD7_LOC_Os03g18070.1) and 74.6% identity with rice *FAD8* (OsFAD8_LOC_Os07g49310.1), whereas the other linoleic acid desaturase (Sorbi_3002g430100) shows 76.2% identity with rice *FAD7* but 81.5% with rice *FAD8*. Therefore, we concluded that *MSD3* is *FAD7* not *FAD8*.

FAD7 appears to be the major ω-3 fatty acid desaturase in sorghum. To determine the relative transcript abundance of the two plastidial ω-3 fatty acid desaturases in the developing panicle of the wild type BTx623, we compared the transcript levels of *FAD7* and *FAD8* using qPCR with three biological and three technical replicates. Comparing with the internal gene control EIF4α, the average cycle threshold (Ct) to detect *FAD8* transcript was 24.2 cycles, while the Ct number to detect *FAD7* transcript was 19.8. This difference in Ct translated to a relative abundance of *FAD7* transcript 21-fold over *FAD8* according to the calculation method as described previously [26]. This relative abundance was largely consistent with the trends of online data (Figure S3).

To determine which of two plastidial ω-3 fatty acid desaturases were affected in the *msd3* mutants, we compared the abundance of *FAD7* and *FAD8* transcripts in the *msd3-1* and *msd3-3* mutants relative to the wild type BTx623 (Figure 5). The abundance of *FAD7* was reduced to 10% in *msd3-1* and 30%

in *msd3-3* in comparison to the wild type *FAD7* levels. Because the primers of *FAD7* and *FAD8* were designed based on the 3′-UTR sequences, which lie outside of the mutation sites, the low levels of the *FAD7* and *FAD8* transcripts may be due to truncated RNAs that do not encode active proteins. On the other hand, the transcript levels of *FAD8* were increased by 3.8-fold and 1.5-fold in the *msd3-1* and *msd3-3* mutants relative to the levels of BTx623, respectively, indicating that only the transcript abundance of the *FAD7* gene was significantly reduced in the *msd3* mutants. The slight increase in the abundance of *FAD8* gene suggested that the loss of function mutations in *FAD7* gene may be partially compensated by the elevated expression of *FAD8* (Figure 5). This result suggested that expression of *FAD7*, but not *FAD8*, was greatly reduced in the *msd3* mutants.

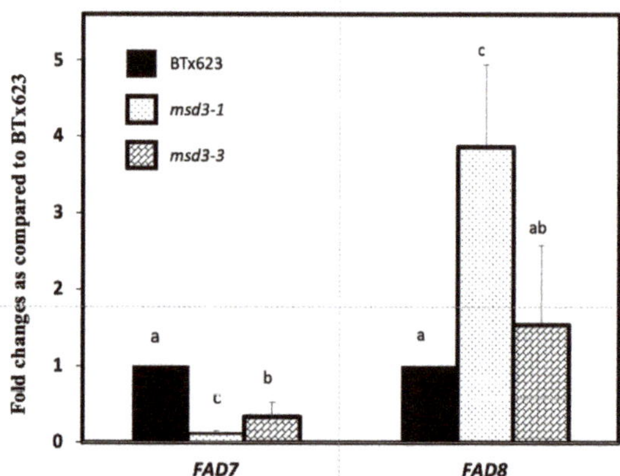

Figure 5. Expression of *FAD7* and *FAD8* genes in developing panicles. Relative change in *FAD7* and *FAD8* mRNA levels in *msd3* mutants in comparison to BTx623 (normalized as 1).

In addition to the two plastidial ω-3 fatty acid desaturases, the sorghum genome encodes two microsomal ω-3 fatty acid desaturases, Sorbi_3005G002800 and Sorbi_3008G003200, homologous to the Arabidopsis *FAD3* gene [25,27]. We searched our mutation database from the 256 sequenced mutant lines [20] and identified nonsynonymous mutations in all three genes (Table S2). However, none of the sequenced lines that harbored a mutation in one of the three ω-3 fatty acid desaturase genes exhibited a phenotype similar to the *msd* mutants.

2.4. Mutation in MSD3 Gene Dramatically Reduced the Levels of Linolenic Acid

To understand how mutations in the *MSD3* (*FAD7*) gene lead to the dramatic change in panicle architecture and restore the fertility of the pedicellate spikelets, we first assessed the effect of the *msd3* mutations on lipid composition of leaves and panicles of the *msd3-1* mutant. In WT BTx623 leaves, 18-carbon fatty acids account for over 92% of total leaf lipids (Figure 6, Table S2). Levels of seven lipid species with molar percentage >1% are plotted in Figure 6. Galactolipids, including monogalactosyl diacylglycerol (MGDG) and digalactosyl diacylglycerol (DGDG), are the major lipids in leaves, accounting for over 90% of the total lipids. The lipid species with 36 carbon and six double bonds (36:6) consist of two linolenic acid molecules, whereas those with four double bonds (36:4) consist of two linoleic acid molecules, and species with five double bonds (36:5) consist of one linoleic acid and one linolenic acid. Sorghum has very low levels of 16-carbon fatty acids, and hexadecatrienoic acid (16:3) was not detectable in either leaves or panicles (Table S3).

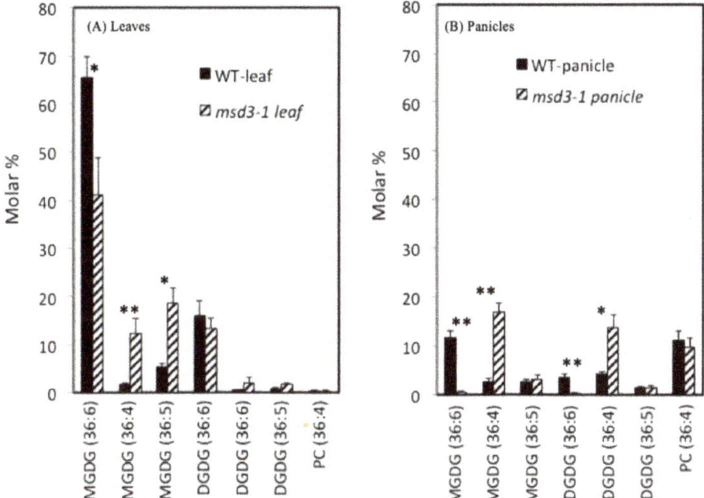

Figure 6. Lipid profiles of leaves and panicles from BTx623 and *msd3-3* (p6) mutants. Lipid profiles of the youngest mature leaves and panicles were carried out by the Lipidomics Center at Kansas State University. Numbers in parenthesis indicate the length and number of double bonds of the fatty acid moiety: (36:6) represents lipid species with two linolenic acids (18:3), (36:4) represents lipid species with two linoleic acids (18:2), and (36:5) represents lipid species with one linolenic acid and one linoleic acid moiety. Minor lipid species (molar percentage <1%) were not plotted, but can be found in Table S3. The difference between WT and *msd3-1* in each lipid species was analyzed by t-test. * $p < 0.05$; ** $p < 0.01$.

As can be seen from Figure 6, mutation in *MSD3* decreased the levels of 36:6 lipid species and concomitantly increased the levels of 36:4 species. The ratio of linolenic acid to linoleic acid was 13.43 in the WT BTx623 leaves, falling to 2.48 in *msd3-3* leaves (Table S3). In general, panicles contained much less linolenic acid than in leaves. In WT BTx623 panicles, the ratio of linolenic acid to linoleic acid was 0.62, falling to 0.08 in the *msd3-3* mutant panicles. Thus, *msd3-3* panicles contained very little linolenic acid (7% vs. 38% in WT panicles). This result confirms that *MSD3* is an ω-3 fatty acid desaturase that catalyzes the desaturation of linoleic acid to linolenic acid in both leaves and panicles. Moreover, the dramatic effect of *msd3* mutations on linoleic acid desaturation was consistent with the idea that *MSD3* is sorghum *FAD7*, as *FAD7* has major effects on linoleic acid desaturation in other plant species [22–24].

2.5. The msd3 Phenotype Was Reverted by Treatment with MeJA

Because linolenic acid (18:3) is a substrate for JA biosynthesis [28], we next asked whether the *msd* phenotype of the *msd3* mutants was due to a deficiency of JA. For this purpose, we then measured the level of JA in developing panicles of *msd3-1* and BTx623. The JA level was reduced from 709 ng/g fresh weight (FW) in the wild type to 409 ng/g FW in *msd3-1* (Figure S4).

To further demonstrate if the decrease in JA led to the *msd* phenotype, we applied MeJA to the whorls of BTx623 and the *msd3-1* mutant every other day, starting from when the plants had eight fully expanded leaves until the plants had 10 fully expanded leaves. Because the pedicellate spikelets never produce viable grains in BTx623 and other natural accessions, the reversion of the pedicellate spikelets from grain-bearing perfect flowers to sterile spikelets is a reliable indicator to demonstrate the effect of JA treatment. As shown in Figure 7, the pedicellate spikelets of the wild type BTx623 were sterile while that of *msd3-3* mutant were fertile and produced viable grain. After treatment with MeJA, the pedicellate spikelets of *msd3-3* mutants became sterile spikelets as in the wild type BTx623.

Furthermore, the grain size of the *msd3* mutant also restored to similar size as the wild type BTx623 (Table S4). This result indicates that the lack of sufficient JA in the *msd3* panicle resulted in full fertility in the pedicellate spikelets.

Figure 7. Reversion of the *msd* panicle phenotype by jasmonic acid (JA). 0.05% Tween-20 in water or 1 mM methyl-JA (MeJA) in 0.05% Tween-20 was directly applied to the central whorl of BTx623 and *msd3-3* (p6) plants every other day, starting from when the plants had eight fully expanded leaves and continuing until they had 10 fully expanded leaves. Shown in the figure were a few spikelets to demonstrate the reversion of the fertile pedicellate spikelets of the *msd3* mutant to sterile spikelets by the addition of MeJA. The effect of JA treatment on an entire primary inflorescence branch was shown in Figure S5.

3. Discussion

In this study, we characterized a new group of *msd* mutants, *msd3*, that exhibits increases in the size and number of the primary branches and fertile pedicellate spikelets, similar to the panicle structure previously reported in the *msd1* and *msd2* mutants [13,14]. Using bulk segregant analysis of four independent F_2 populations, we identified that all causal mutations are located within the *FAD7* (Sorbi_3001G407600) gene. Because *FAD7* is a major plastidial fatty acid desaturase, loss-of-function mutations in the *FAD7* gene resulted in a dramatic reduction in the content of linolenic acid, a substrate of JA, in both leaves and panicles. The reduction in JA levels in *msd3* panicles and the reversion of the *msd3* phenotype to the wild type by application of MeJA indicate that the *msd* phenotype in *msd3* mutants is probably due to insufficient levels of JA to activate the programmed cell death pathway leading to the arrest of the pedicellate spikelets. This conclusion is consistent with the previously reported mechanisms in *msd1* and *msd2* mutants [13,14].

MSD3 (*FAD7*) appears to the predominant form of ω-3 fatty acid desaturase in sorghum. Most plants, including sorghum, have multiple forms of microsomal and plastidial ω-3 fatty acid desaturases that catalyze the formation of linolenic acids from linoleic acid. Arabidopsis has one microsomal fatty acid desaturase, originally known as *FAD3*, and two closely related plastidial ω-3 desaturases, *FAD7* and *FAD8* (plastidial) [21–24,29]. Mutations in all three genes, *FAD3*, *FAD7*, and *FAD8*, are required to reduce the linolenic acid levels sufficiently to affect JA-regulated processes such as male sterility and the defense response to insect attack [29–31]. However, loss-of-function mutation in *FAD7* alone blocks the JA-induced responses in tomato [22]. The sorghum genome encodes two microsomal ω-3 fatty acid desaturases, Sorbi_3005G002800 and Sorbi_3008G003200, which are homologous to Arabidopsis *FAD3* [25,27], and two plastidial fatty acid ω-3 desaturases, Sorbi_3001G407600 and

Sorbi_3002g430100, which are homologous to Arabidopsis *FAD7* and *FAD8* [27,32]. Loss-of-function mutations in *MSD3* (*FAD7*) alone decreased the ratio of linolenic acid to linoleic acid from 13.43 to 2.48 in leaves, and from 0.62 to 0.08 in panicles, resulting in dramatic changes in panicle architecture, and restoration of the fertility of the pedicellate spikelets. As noted above, *MSD3* encodes one of the four ω-3 fatty acid desaturases, but the roles of the other three enzymes in fatty acid desaturation remains unclear. Based on their relative expression levels (Figure S3), these genes may not contribute much to the desaturation of linoleic acid except at specific developmental stages. Thus, *MSD3* (*FAD7*) plays a major role in the desaturation of linoleic acid to linolenic acid in sorghum.

Linolenic acid (18:3) and hexadecatrienoic acid (16:3) are the substrates for the lipid-derived JAs [16,17]. Because sorghum has non-detectable levels of hexadecatrienoic acid, linolenic acid serves as the main substrate for JA biosynthesis. We propose that the dramatic decrease in linolenic acid in the *msd3* mutants may result in deficiency of JA, which failed to arrest the development of the pedicellate spikelets in sorghum. Thus, we determined the JA levels in the developing panicles of BTx623 and *msd3-1* mutant. Although JA level in the *msd3-1* mutant panicle was significantly reduced to 409 ng/g FW from 709 ng/g fresh weight (FW) of the wild type, the *msd3* panicles still contained substantial levels of JA. Next, we tested if the *msd3* phenotype, such as fertile pedicellate spikelets, can be reversed by JA treatment. As shown in Figure 7 and Figure S5, the fertile pedicellate spikelets on the *msd3* panicle was reverted to sterile spikelets as in the BTx623. Thus, it is more likely that the dramatic reduction in linolenic acid in *msd3* mutants may lead to lower levels of JA, which failed to trigger the programmed cell death pathway and allowed the pedicellate spikelets to develop viable grains. This conclusion is also supported by our previous reports on *msd1* and *msd2* mutants [13,14].

The role of JA in the mediation of the *msd* phenotype in sorghum may be closely related to its role in maize tassel development by functioning at the level of sexual determination. Maize is monoecious with separate staminate (male) and pistillate (female) inflorescences called the tassel and the ear, respectively. During early floral development, the florets in both the tassel and the ear spikelets are bisexual. The monoecious nature is conferred by the selective abortion of pistillate organs in tassel florets [15]. A large number of sex-determination mutants (tasselseed) have been identified in maize [33]. The *TS1* encodes a lipoxygenase that catalyze the conversion of free linolenic acid to 13(S)-hydroperoxylinolenic acid, the first committed step in JA biosynthesis. The sorghum *MSD2* gene is an ortholog of the maize *TS2* gene [14]. The maize *TS2* gene encodes a short-chain alcohol dehydrogenase, probably involved in JA biosynthesis [34]. Another tasselseed mutant was created by knocking out the duplicated orthologs of OPR3, a major OPR (12-oxo-phytodienoic acid reductase) gene in Arabidopsis that acts in JA biosynthesis [35]. The resulting maize *opr7opr8* double mutant is phenotypically similar to *ts1* and *ts2* mutants and can be reverted to normal tassel phenotype by treatment with MeJA [36]. The maize *TS5* gene encodes a wound-inducible gene, CY94B, which serves as an enzyme that inactivates the biologically active JA-isoleucine [33]. Although the causal mutation in the *TS5* gene has not been identified, the mutation of *ts5* is dominant and leads to the overexpression of CY94B, decreased JA-isoleucine, and a concomitant increase in the deactivated 12OH-JA-isoleucine. As with the maize *ts1* and *ts2* mutants, the tasselseed phenotype was reverted by JA treatment. It is well-established that JA plays a critical role in the maintenance of the staminate state of the tassel by suppressing the development of female flower organs through JA-induced programmed cell death [15].

Sorghum is hermaphroditic and produces perfect flowers with both male and female floral organs. Like many other grasses, sorghum produces two types of spikelets depending on the mode of attachment to the inflorescence axis. The sessile spikelets, directly attached to the inflorescence axis, contain perfect flowers and produce viable grains. The pedicellate spikelets, attached to the inflorescence axis through a short petiole called pedicel, occasionally produce anthers but never produce mature ovaries, and thus, become sterile spikelets in mature panicles. In sorghum, *msd* mutants produce both sessile and pedicellate spikelets that contain perfect flowers and can develop viable grains. Sorghum may employ similar mechanisms to control the fertility of the pedicellate spikelets as reported in maize in controlling the ovary development in tassels. In the wild type BTx623, both sessile and pedicellate spikelets have both male and female floral organs initially but the floral organs, especially the ovary, in the pedicellate spikelets become arrested in a late development stage [13]. The abortion of the ovary has been shown to be regulated by JA-induced programmed cell death as in maize tassels [13,15]. As previously reported, the *MSD1* gene encodes a TCP-transcription factor that activates many genes involved in JA biosynthesis and *MSD2* encodes a lipoxygenase that catalyzes the first committed step of JA biosynthesis from linolenic acid [13,14]. The discovery of the *MSD3* gene as a predominant form of ω-3 fatty acid desaturase responsible for the conversion of linoleic acid to linolenic acid highlights the importance of the JA regulatory module(s) in the control of the fertility of the pedicellate spikelets in sorghum.

The mechanisms determining the staminate state of tassels in maize and that suppressing the development of the pedicellate spikelets in sorghum may also differ. All maize tasselseed mutants are male sterile due to the ectopic growth of the silks. However, the sorghum *msd* mutants develop both anthers and ovaries. Male sterility was never observed in the sorghum *msd* mutants. In addition, JA may also contribute to panicle development in sorghum, because all *msd* mutants exhibit increases in the number and size of the primary inflorescence branches [11,13,14].

4. Materials and Methods

4.1. Plant Materials

Sorghum (*Sorghum bicolor* (L.) Moench) *msd* mutants were identified from a pedigreed sorghum mutant library that was developed by mutagenizing the sorghum inbred line BTx623 seeds with the chemical mutagen ethyl methane sulfonate (EMS) [37]. The wild type BTx623, the *msd* mutants, and their backcrossed F$_2$ populations were planted in the Agricultural Experiment Station at Lubbock, TX (33′39″ N, 101′49″ W) of the Agricultural Research Service of the United States Department of Agriculture (USDA-ARS) in May 2017. During the late grain-filling stage, when the *msd* phenotype could be easily observed, leaf samples were collected from each of the confirmed homozygous *msd* mutants to prepare genomic DNA as described previously [38].

The *msd* mutants isolated from the library harbored a high density of EMS-induced mutations [20]. To decrease the number of background mutations, these mutants were backcrossed (BC) to WT BTx623 for three generations before use in this study. Moreover, the new *msd* mutants were also crossed to the known *msd1* and *msd2* mutants, as well as to each other, to determine whether they represented alleles of known *msd* loci or new locus.

4.2. Gene Identification by Next-Generation Sequencing

Homozygous *msd* mutants were identified from the backcrossed F$_2$ populations of p6, p14, p21, and p24 during the grain-filling stage, during which the *msd* phenotype could be easily scored. Genomic DNA pooled from the homozygous *msd* mutants from each F$_2$ population was sequenced to >20× coverage on an Illumina HiSeq 2000 through a service from BGI (https://www.bgi.com/us/). Causal mutations in these populations were identified by an inhouse bioinformatic pipeline described previously [19]. Briefly, low-quality reads, adapter sequences, and contamination were excluded from the raw reads, and then the clean reads were aligned to the sorghum reference genome v3.0 with Bowtie2 [39]. SNP (single nucleotide polymorphism) calling was performed by Samtools and Bcftools using only reads with mapping and sequencing quality greater than Q20 [40]. The read depth for true SNPs was set from 3 to 50. Because EMS induces only G/C-to-A/T transition mutations [41], only homozygous G/C to A/T SNPs were used for prediction of effects on gene function by the Ensembl Variant Effect Predictor [42]. Homology analysis and functional annotation of candidate genes were obtained from Gramene database release 39 [43] (http://www.gramene.org).

4.3. Confirmation of the Causal Mutation with KASP

The BC_3F_2 population consisting of 63 F_2 offspring was utilized for this study. Based on putative SNPs discovered through the sequencing analysis described above, three types of SNP–KASP primers were designed [44,45] and synthesized by Integrated DNA Technologies (Coralville, IA, USA). Genotyping analyses were performed using extracted genomic DNA from leaf tissues of sorghum plants collected from the field. SNP genotype in each offspring was determined by Kompetitive Allele Specific PCR (KASP) chemistry (www.lgcgenomics.com) with some modifications [44]. Analysis of genotypes was conducted at the Plant Stress and Germplasm Development Unit at Lubbock, TX, USA. Phenotype data for the population were gathered at the full maturity stage of the panicle. Co-segregation analyses were conducted based on the correspondence of genotype to phenotype and chi-square test.

4.4. Lipidomic Assay

Samples were obtained from ~5 cm^2 from the middle lamina of the first mature leaf (the youngest leaf with a visible leaf collar) and from developing panicles (approximately 3 cm long). The samples were quickly immersed in liquid nitrogen and stored at −80 °C until use. Lipids were extracted and analyzed following the procedure established at the Lipidomic Center of Kansas State University (https://www.k-state.edu/lipid/). The original protocol was modified as follows: instead of cutting leaves into small pieces, the leaf tissue was completely ground into a fine powder with liquid nitrogen chilled mortars and pestles. The plant material was then treated with hot isopropanol and 0.01% butylated hydroxytoluene (preheated to 75 °C). The resultant lipid extract was dissolved in chloroform and stored at −80 °C. Lipid extracts were dried with a stream of nitrogen gas, packed in dry ice, and shipped to the Lipidomic Center of Kansas State University for lipid analysis. The tissue used for extraction was dried in an oven at 65 °C for 1 week to estimate the total dry weight of tissue used for extraction of each lipid sample.

4.5. Jasmonic Acid Determination

Panicles at stage 4, when the fate of the pedicellate spikelets is determined [13,14], were collected and immediately frozen in liquid nitrogen. Five biological replicates were solvent-extracted, methylated, collected on a polymeric adsorbent using vapor-phase extraction, and analyzed using GC/isobutene chemical ion mass spectrometry (GC/CI-MS) [46]. For metabolite quantification, d_5JA (Sigma–Aldrich, St. Louis, MO, USA) was used as an internal standard. The JA level in each sample was normalized to the mass of the panicle, and expressed as ng/g fresh weight (FW).

4.6. Quantitative Gene Expression Analysis of FAD7 and FAD8

Sorghum *msd3* and BTx623 plants were grown in a greenhouse with temperatures maintained at 28 °C (12 h day)/25 °C (12 h night) cycle. Panicle samples at stage 4, about 3 cm long, were collected, flash frozen in liquid nitrogen, and stored at −80 °C until use. Total RNA was extracted from a single panicle with Trizol reagent (Thermo–Fisher Scientific, Waltham, MA, USA), followed by column purification using the RNA Extraction kit from Sigma–Aldrich (Sigma–Aldrich, St. Louis, MO, USA). After purification, genomic DNA contaminants were removed by On-Column DNase I Digestion supplied in the Sigma kit (Sigma, St Louis, MO, USA). RNA concentration was measured using a BioSpectrophotometer® Kinetic (Eppendorf, Hamburg, Germany). RNA quality was checked by electrophoresis on a 1% agarose gel. First-strand cDNA was synthesized from 1 µg RNA per sample using the Superscript 2 Reverse Transcriptase kit (Invitrogen, Carlsbad, CA, USA). The reaction mixture was incubated at 42 °C for 50 min followed by heat inactivation at 70 °C for 15 min. The resultant cDNA is used as a template for quantitative analysis of *FAD7* and *FAD8* genes with control eukaryotic translation initiation factor.

Real-time polymerase chain reaction (PCR) analysis was used to determine which plastidial fatty acid desaturase was affected in the *msd3* mutants. The Primer3 software was used to design primers for the *FAD7* and *FAD8* genes from sorghum. Combinations of forward and reverse primers with Tm above 60 °C were selected for real-time PCR analysis, and the annealing temperature was set to 53 °C. Because of the high sequence similarity among fatty acid desaturase genes, primers were designed within the 3′-UTR region to prevent non-specific binding to genes in the same family. 3′-UTR regions unique to *FAD7* (*FAD7* forward primer: 5′ TCC CTC AAA TCC CAC ATT 3′, *FAD7* reverse primer: 5′ GAA GAG CAC CCG ACT TCT TT 3′) and *FAD8* (*FAD8* forward primer: 5′ TGC ATG GAG GTT CAT ATA CTG C 3′, *FAD8* reverse primer: 5′ AAT TCT GTT CTG TTT GGT TGG TG 3′) were used to design primer pairs that amplified a product of ~100 base pairs. Internal control primers were designed against the gene encoding eukaryotic translation initiation factor 4 α (*EIF4α*) (SB_EIF4α Forward: 5′ CAA CTT TGT CAC CCG CGA TGA 3′ SB_EIF4αReverse: 5′ TCC AGA AAC CTT AGC AGC CCA 3′).

The cDNA was diluted five times with nuclease-free water, mixed with Roche Fast Start SYBR Green® (Roche, San Francisco, CA, USA) along with specific primers and run on a Roche LightCycler 96. Three biological replicates (control and mutant samples) were used for gene expression analysis. The melt curve from the Light Cycler was analyzed by exporting the ΔCt and $\Delta\Delta$Ct values to an Excel file. *EIF4α* was used as an internal control for analyzing the relative expression level differences of *FAD7* and *FAD8* genes.

4.7. Reversion of the msd3 Panicle Structure by MeJA

Treatment with MeJA was performed as described [13,14]. Briefly, 1 mL of 1 mM MeJA in 0.05% Tween-20 in water was applied to the central whorl every other day, starting when the plants had eight fully expanded leaves until the plants had 10 fully expanded leaves. Controls were treated with 1 mL of 0.05% Tween-20.

4.8. Statistical Analyses

Statistical analyses, such as one-way ANOVA and t-test, were run in Excel with the XRealStats add-in (http://www.real-statistics.com). Option 'Tukey's HSD' was used for multiple comparisons.

5. Conclusions

We identified *MSD3* as *FAD7* (Sorbi_3001G407600), a major plastid-targeted ω-3 fatty acid desaturase. Loss-of-function mutations in *MSD3* (as in *msd3-1*, *msd3-3*, and *msd3-4*) result in low levels of linolenic acid, the substrate for JA biosynthesis. The *msd* panicle architecture was restored to the wild-type panicle phenotype by treatment with MeJA. The discovery that *MSD3* is *FAD7* provides further support for the idea that JA-regulated modules are responsible for suppressing the development of pedicellate spikelets in sorghum, as described previously in our characterization of the *msd1* and *msd2* mutants. Compared with *msd1* and *msd2* mutants, *msd3* mutants have higher grain weight indicating that *msd3* mutants may be better adapted to increasing sink capacity, and therefore may have better potential to increase grain yield. These findings of this study provide a new approach for genetic manipulation of grain numbers per panicle with the goal of increasing grain yield.

Supplementary Materials: Supplementary materials can be found at http://www.mdpi.com/1422-0067/20/21/5359/s1.

Author Contributions: Z.X. and D.W. conceived the study. L.D. conducted the lipidomics study and characterization of the *MSD3* expression. Y.J. analyzed genomic data. J.C. isolated the *msd3* mutants. R.C. and G.B. conducted genetic and co-segregation analyses. S.A.C. performed J.A. experiments. L.D., I.C., N.G., G.B., R.C., C.H., S.C., J.C. and J.B. participated in the characterization of the *msd3* mutants. L.D. and Z.X. wrote the manuscript. All authors participated in the revision and agreed with the final manuscript.

Funding: This research was funded in part by the United Sorghum Checkoff. ZX was also partly supported by USDA–ARS 3096-21000-020-00-D, and DW was partly supported by USDA–ARS 8062-21000-041-00D.

Acknowledgments: The authors are grateful to Lan Liu-Gitz, Halee Hughes, and Jing Wang from the Plant Stress and Germplasm Development Research Unit, USDA–ARS for their technical support. We thank the two peer reviewers for their valuable input, which helped us to improve the manuscript.

Conflicts of Interest: The authors declare no conflicts of interest pertaining to this work.

Disclaimer: Mention of trade names or commercial products in this article is solely for the purpose of providing specific information and does not imply recommendation or endorsement by the U.S. Department of Agriculture. USDA is an equal opportunity provider and employer.

References

1. Saeed, M.; Francis, C.A.; Clegg, M.D. Yield component analysis in grain sorghum. *Crop Sci.* **1986**, *26*, 346–351. [CrossRef]

2. Ashikari, M.; Sakakibara, H.; Lin, S.; Yamamoto, T.; Takashi, T.; Nishimura, A.; Angeles, E.R.; Qian, Q.; Kitano, H.; Matsuoka, M. Cytokinin oxidase regulates rice grain production. *Science* **2005**, *309*, 741–745. [CrossRef] [PubMed]

3. Duggan, B.L.; Domitruk, D.R.; Fowler, D.B. Yield component variation in winter wheat grown under drought stress. *Can. J. Plant Sci.* **2000**, *80*, 739–745. [CrossRef]

4. Reynolds, M.; Foulkes, M.J.; Slafer, G.A.; Berry, P.; Parry, M.A.J.; Snape, J.W.; Angus, W.J. Raising yield potential in wheat. *J. Exp. Bot.* **2009**, *60*, 1899–1918. [CrossRef] [PubMed]

5. Richards, R.A. Selectable traits to increase crop photosynthesis and yield of grain crops. *J. Exp. Bot.* **2000**, *51*, 447–458. [CrossRef] [PubMed]

6. Boyles, R.E.; Cooper, E.A.; Myers, M.T.; Brenton, Z.; Rauh, B.L.; Morris, G.P.; Kresovich, S. Genome-Wide Association Studies of Grain Yield Components in Diverse Sorghum Germplasm. *Plant Genome* **2016**, *9*. [CrossRef] [PubMed]

7. Zohary, D.; Hopf, M.; Weiss, E. *Domestication of Plants in the Old World: The Origin and Spread of Cultivated Plants in West Asia, Europe, and the Mediterranean Basin*, 4th ed.; Oxford University Press: Oxford, UK, 2012; p. 280.

8. Dolferus, R.; Ji, X.; Richards, R. Abiotic stress and control of grain number in cereals. *Plant Sci. Int. J. Exp. Plant Biol.* **2011**, *181*, 331–341. [CrossRef]

9. Sreenivasulu, N.; Schnurbusch, T. A genetic playground for enhancing grain number in cereals. *Trends Plant Sci.* **2012**, *17*, 91–101. [CrossRef]

10. Brown, P.J.; Klein, P.E.; Bortiri, E.; Acharya, C.B.; Rooney, W.L.; Kresovich, S. Inheritance of inflorescence architecture in sorghum. *Theor. Appl. Genet.* **2006**, *113*, 931–942. [CrossRef]

11. Burow, G.; Xin, Z.; Hayes, C.; Burke, J. Characterization of a multiseeded mutant of sorghum for increasing grain yield. *Crop Sci.* **2014**, *54*, 2030–2037. [CrossRef]

12. Walters, D.R.; Keil, D.J. *Vascular Plant Taxonomy*, 4th ed.; Kendall/Hunt Pub. Co.: Dubuque, IA, USA, 1988; p. 607.

13. Jiao, Y.; Lee, Y.K.; Gladman, N.; Chopra, R.; Christensen, S.A.; Regulski, M.; Burow, G.; Hayes, C.; Burke, J.; Ware, D.; et al. MSD1 regulates pedicellate spikelet fertility in sorghum through the jasmonic acid pathway. *Nat. Commun.* **2018**, *9*, 822. [CrossRef] [PubMed]

14. Gladman, N.; Jiao, Y.; Lee, Y.K.; Zhang, L.; Chopra, R.; Regulski, M.; Burow, G.; Hayes, C.; Christensen, S.A.; Dampanaboina, L.; et al. Fertility of Pedicellate Spikelets in Sorghum Is Controlled by a Jasmonic Acid Regulatory Module. *Int. J. Mol. Sci.* **2019**, *20*, 4951. [CrossRef] [PubMed]

15. Acosta, I.F.; Laparra, H.; Romero, S.P.; Schmelz, E.; Hamberg, M.; Mottinger, J.P.; Moreno, M.A.; Dellaporta, S.L. tasselseed1 is a lipoxygenase affecting jasmonic acid signaling in sex determination of maize. *Science* **2009**, *323*, 262–265. [CrossRef] [PubMed]

16. Ghasemi Pirbalouti, A.; Sajjadi, S.E.; Parang, K. A Review (Research and Patents) on Jasmonic Acid and Its Derivatives. *Arch. Der Pharm.* **2014**, *347*, 229–239. [CrossRef]

17. Wasternack, C.; Hause, B. Jasmonates: Biosynthesis, perception, signal transduction and action in plant stress response, growth and development. An update to the 2007 review in Annals of Botany. *Ann. Bot.* **2013**, *111*, 1021–1058. [CrossRef]

18. Abe, A.; Kosugi, S.; Yoshida, K.; Natsume, S.; Takagi, H.; Kanzaki, H.; Matsumura, H.; Mitsuoka, C.; Tamiru, M.; Innan, H.; et al. Genome sequencing reveals agronomically important loci in rice using MutMap. *Nat. Biotechnol.* **2012**, *30*, 174–178. [CrossRef]

19. Jiao, Y.; Burow, G.; Gladman, N.; Acosta-Martinez, V.; Chen, J.; Burke, J.; Ware, D.; Xin, Z. Efficient identification of causal mutations through sequencing of bulked f2 from two allelic bloomless mutants. *Front. Plant Sci.* **2018**, *8*, 2267. [CrossRef]

20. Jiao, Y.; Burke, J.J.; Chopra, R.; Burow, G.; Chen, J.; Wang, B.; Hayes, C.; Emendack, Y.; Ware, D.; Xin, Z. A sorghum mutant resource as an efficient platform for gene discovery in grasses. *Plant Cell* **2016**, *28*, 1551–1562. [CrossRef]

21. Berberich, T.; Harada, M.; Sugawara, K.; Kodama, H.; Iba, K.; Kusano, T. Two maize genes encoding ω-3 fatty acid desaturase and their differential expression to temperature. *Plant Mol. Biol.* **1998**, *36*, 297–306. [CrossRef]

22. Li, C.; Liu, G.; Xu, C.; Lee, G.I.; Bauer, P.; Ling, H.-Q.; Ganal, M.W.; Howe, G.A. The Tomato Suppressor of prosystemin-mediated responses2 Gene Encodes a Fatty Acid Desaturase Required for the Biosynthesis of Jasmonic Acid and the Production of a Systemic Wound Signal for Defense Gene Expression. *Plant Cell* **2003**, *15*, 1646–1661. [CrossRef]

23. Roman, A.; Hernandez, M.L.; Soria-Garcia, A.; Lopez-Gomollon, S.; Lagunas, B.; Picorel, R.; Martinez-Rivas, J.M.; Alfonso, M. Non-redundant Contribution of the Plastidial FAD8 omega-3 Desaturase to Glycerolipid Unsaturation at Different Temperatures in Arabidopsis. *Mol. Plant* **2015**, *8*, 1599–1611. [CrossRef] [PubMed]

24. Tovuu, A.; Zulfugarov, I.S.; Wu, G.; Kang, I.S.; Kim, C.; Moon, B.Y.; An, G.; Lee, C.H. Rice mutants deficient in omega-3 fatty acid desaturase (FAD8) fail to acclimate to cold temperatures. *Plant Physiol. Biochem.* **2016**, *109*, 525–535. [CrossRef] [PubMed]

25. Arondel, V.; Lemieux, B.; Hwang, I.; Gibson, S.; Goodman, H.; Somerville, C. Map-based cloning of a gene controlling omega-3 fatty acid desaturation in Arabidopsis. *Science* **1992**, *258*, 1353–1355. [CrossRef] [PubMed]

26. Rao, X.; Huang, X.; Zhou, Z.; Lin, X. An improvement of the 2^(−delta delta CT) method for quantitative real-time polymerase chain reaction data analysis. *Biostat. Bioinforma. Biomath.* **2003**, *3*, 71–85.

27. Iba, K.; Gibson, S.; Nishiuchi, T.; Fuse, T.; Nishimura, M.; Arondel, V.; Hugly, S.; Somerville, C. A gene encoding a chloroplast omega-3 fatty acid desaturase complements alterations in fatty acid desaturation and chloroplast copy number of the fad7 mutant of Arabidopsis thaliana. *J. Biol. Chem.* **1993**, *268*, 24099–24105.

28. Lyons, R.; Manners, J.M.; Kazan, K. Jasmonate biosynthesis and signaling in monocots: A comparative overview. *Plant Cell Rep.* **2013**, *32*, 815–827. [CrossRef]

29. McConn, M.; Browse, J. The Critical Requirement for Linolenic Acid Is Pollen Development, Not Photosynthesis, in an Arabidopsis Mutant. *Plant Cell* **1996**, *8*, 403–416. [CrossRef]

30. McConn, M.; Creelman, R.A.; Bell, E.; Mullet, J.E.; Browse, J. Jasmonate is essential for insect defense in Arabidopsis. *Proc. Natl. Acad. Sci. USA* **1997**, *94*, 5473–5477. [CrossRef]

31. Vijayan, P.; Shockey, J.; Levesque, C.A.; Cook, R.J.; Browse, J. A role for jasmonate in pathogen defense of arabidopsis. *Proc. Natl. Acad. Sci. USA* **1998**, *95*, 7209–7214. [CrossRef]

32. McConn, M.; Hugly, S.; Browse, J.; Somerville, C. A Mutation at the fad8 Locus of Arabidopsis Identifies a Second Chloroplast [omega]-3 Desaturase. *Plant Physiol.* **1994**, *106*, 1609–1614. [CrossRef]

33. Lunde, C.; Kimberlin, A.; Leiboff, S.; Koo, A.J.; Hake, S. Tasselseed5 overexpresses a wound-inducible enzyme, ZmCYP94B1, that affects jasmonate catabolism, sex determination, and plant architecture in maize. *Commun. Biol.* **2019**, *2*, 114. [CrossRef] [PubMed]

34. DeLong, A.; Calderon-Urrea, A.; Dellaporta, S.L. Sex determination gene TASSELSEED2 of maize encodes a short-chain alcohol dehydrogenase required for stage-specific floral organ abortion. *Cell* **1993**, *74*, 757–768. [CrossRef]

35. Stintzi, A.; Browse, J. The Arabidopsis male-sterile mutant, opr3, lacks the 12-oxophytodienoic acid reductase required for jasmonate synthesis. *Proc. Natl. Acad. Sci. USA* **2000**, *97*, 10625–10630. [CrossRef] [PubMed]

36. Yan, Y.; Christensen, S.; Isakeit, T.; Engelberth, J.; Meeley, R.; Hayward, A.; Emery, R.J.; Kolomiets, M.V. Disruption of OPR7 and OPR8 reveals the versatile functions of jasmonic acid in maize development and defense. *Plant Cell* **2012**, *24*, 1420–1436. [CrossRef]

37. Xin, Z.; Wang, M.L.; Barkley, N.A.; Burow, G.; Franks, C.; Pederson, G.; Burke, J. Applying genotyping (TILLING) and phenotyping analyses to elucidate gene function in a chemically induced sorghum mutant population. *BMC Plant Biol.* **2008**, *8*, 103. [CrossRef]

38. Xin, Z.; Chen, J. A high throughput DNA extraction method with high yield and quality. *Plant Methods* **2012**, *8*, 26. [CrossRef]

39. Langmead, B.; Salzberg, S.L. Fast gapped-read alignment with Bowtie 2. *Nat. Methods* **2012**, *9*, 357–359. [CrossRef]

40. Li, H.; Handsaker, B.; Wysoker, A.; Fennell, T.; Ruan, J.; Homer, N.; Marth, G.; Abecasis, G.; Durbin, R.; Genome Project Data Processing, S. The Sequence Alignment/Map format and SAMtools. *Bioinformatics* **2009**, *25*, 2078–2079. [CrossRef]

41. Greene, E.A.; Codomo, C.A.; Taylor, N.E.; Henikoff, J.G.; Till, B.J.; Reynolds, S.H.; Enns, L.C.; Burtner, C.; Johnson, J.E.; Odden, A.R.; et al. Spectrum of chemically induced mutations from a large-scale reverse-genetic screen in Arabidopsis. *Genetics* **2003**, *164*, 731–740.

42. McLaren, W.; Pritchard, B.; Rios, D.; Chen, Y.; Flicek, P.; Cunningham, F. Deriving the consequences of genomic variants with the Ensembl API and SNP Effect Predictor. *Bioinformatics* **2010**, *26*, 2069–2070. [CrossRef]

43. Monaco, M.K.; Stein, J.; Naithani, S.; Wei, S.; Dharmawardhana, P.; Kumari, S.; Amarasinghe, V.; Youens-Clark, K.; Thomason, J.; Preece, J.; et al. Gramene 2013: Comparative plant genomics resources. *Nucleic Acids Res.* **2014**, *42*, D1193–D1199. [CrossRef] [PubMed]

44. Burow, G.; Chopra, R.; Sattler, S.; Burke, J.; Acosta-Martinez, V.; Xin, Z. Deployment of SNP (CAPS and KASP) markers for allelic discrimination and easy access to functional variants for brown midrib genes bmr6 and bmr12 in Sorghum bicolor. *Mol. Breed.* **2019**, *39*, 115. [CrossRef]

45. Rosas, J.E.; Bonnecarrère, V.; Pérez de Vida, F. One-step, codominant detection of imidazolinone resistance mutations in weedy rice (Oryza sativa L.). *Electron. J. Biotechnol.* **2014**, *17*, 95–101. [CrossRef]

46. Schmelz, E.A.; Engelberth, J.; Tumlinson, J.H.; Block, A.; Alborn, H.T. The use of vapor phase extraction in metabolic profiling of phytohormones and other metabolites. *Plant J.* **2004**, *39*, 790–808. [CrossRef] [PubMed]

© 2019 by the authors. Licensee MDPI, Basel, Switzerland. This article is an open access article distributed under the terms and conditions of the Creative Commons Attribution (CC BY) license (http://creativecommons.org/licenses/by/4.0/).

International Journal of
Molecular Sciences

MDPI

Article

Comprehensive Identification of PTI Suppressors in Type III Effector Repertoire Reveals that *Ralstonia solanacearum* Activates Jasmonate Signaling at Two Different Steps

Masahito Nakano [1,2,*] and Takafumi Mukaihara [1,*]

[1]   Research Institute for Biological Sciences, Okayama (RIBS), 7549-1 Yoshikawa, Kibichuo-cho, Okayama 716-1241, Japan
[2]   Graduate School of Environmental and Life Science, Okayama University, 1-1-1 Tsushima-naka, Kita-ku, Okayama 700-8530, Japan
*   Correspondence: nakanom@okayama-u.ac.jp (M.N.); mukaihara@bio-ribs.com (T.M.); Tel.: +81-866-56-9452 (T.M.)

Received: 15 November 2019; Accepted: 26 November 2019; Published: 28 November 2019

Abstract: *Ralstonia solanacearum* is the causative agent of bacterial wilt in many plants. To identify *R. solanacearum* effectors that suppress pattern-triggered immunity (PTI) in plants, we transiently expressed *R. solanacearum* RS1000 effectors in *Nicotiana benthamiana* leaves and evaluated their ability to suppress the production of reactive oxygen species (ROS) triggered by flg22. Out of the 61 effectors tested, 11 strongly and five moderately suppressed the flg22-triggered ROS burst. Among them, RipE1 shared homology with the *Pseudomonas syringae* cysteine protease effector HopX1. By yeast two-hybrid screening, we identified jasmonate-ZIM-domain (JAZ) proteins, which are transcriptional repressors of the jasmonic acid (JA) signaling pathway in plants, as RipE1 interactors. RipE1 promoted the degradation of JAZ repressors and induced the expressions of JA-responsive genes in a cysteine–protease-activity-dependent manner. Simultaneously, RipE1, similarly to the previously identified JA-producing effector RipAL, decreased the expression level of the salicylic acid synthesis gene that is required for the defense responses against *R. solanacearum*. The undecuple mutant that lacks 11 effectors with a strong PTI suppression activity showed reduced growth of *R. solanacearum* in *Nicotiana* plants. These results indicate that *R. solanacearum* subverts plant PTI responses using multiple effectors and manipulates JA signaling at two different steps to promote infection.

Keywords: *Ralstonia solanacearum*; type III effector; jasmonic acid; salicylic acid; *Nicotiana* plants

---

1. Introduction

Plants are exposed to various abiotic and biotic stresses during their life cycle. To combat pathogens, plants have developed a specialized surveillance system, the so-called pattern-triggered immunity (PTI), to reject or attenuate infection by potential pathogens [1]. In PTI, plants sense evolutionarily conserved molecules from diverse pathogens, namely, pathogen/microbe-associated molecular patterns (PAMPs), such as flagellin, cold shock protein, and chitin, through pattern-recognition receptors (PRRs) on the plasma membrane [2]. The recognition of PAMPs by PRRs activates a large set of physiological responses including ion-flux changes, generation of reactive oxygen species (ROS), phosphorylation of mitogen-activated protein kinases, deposition of callose, production of phytohormones, and transcriptional reprogramming of defense-related genes, conferring disease resistance to a wide variety of pathogens.

Phytohormones act as signaling molecules that are required for immune responses against attacks from pathogens. Salicylic acid (SA) mediates defense responses against biotrophic and hemibiotrophic pathogens, whereas jasmonic acid (JA) controls defense responses against necrotrophic pathogens [3,4]. In many cases, their signaling network shows an antagonistic relationship with each other to induce appropriate immune responses against various pathogens with different infection strategies. During the coevolutionary arms race between pathogens and their host plants, pathogens acquired various virulence strategies to manipulate host hormonal signaling networks to accelerate successful infection [5]. One well-known example is the polyketide toxin coronatine (COR) produced by the hemibiotrophic bacterial pathogen *Pseudomonas syringae* pv. *tomato* (Pto) DC3000 [6]. COR is composed of two moieties, coronafacic acid and coronamic acid, and functions as a structural mimic of an active isoleucine conjugate of JA (JA-Ile). In the presence of COR, the F-box protein coronatie-insensitive1 (COI1) can promote the degradation of jasmonate-ZIM-domain (JAZ) proteins that repress the JA signaling pathway, resulting in the activation of JA signaling [7,8]. Upon Pto infection, the activation of JA signaling by COR antagonistically suppresses the SA-mediated signaling pathway, leading to the inhibition of stomatal closure and callose deposition to promote bacterial infection [9–11].

Many plant pathogenic bacteria have evolved a series of secretary proteins called effector proteins and inject them into plant cells via the Hrp type III secretion system to subvert plant immune responses [12]. Pathogen effectors often localize to specific organelles and exert their virulence functions in the early stage of infection. For example, AvrPtoB from Pto DC3000 degrades *Arabidopsis* PRR FLS2 through the E3 ubiquitin ligase activity to suppress PTI responses [13]. HopM1 localizes to endosomes and induces the proteasomal degradation of its target protein, AtMIN7, which is involved in PTI responses [14].

Ralstonia solanacearum is a Gram-negative phytopathogenic bacterium that causes bacterial wilt disease in more than 200 plant species, such as tomato, potato, banana, and eggplant [15]. The pathogen injects approximately 70 type III effectors into plant cells through the Hrp type III secretion system [16,17]. To date, several studies have clarified the biochemical functions of *R. solanacearum* effectors in PTI suppression. RipP2 suppresses the expressions of defense-related genes by acetylating WRKY transcription factors [18]. RipAY suppresses PTI by degrading glutathione in plant cells [19,20]. RipAR and RipAW suppress PTI responses through their E3 ubiquitin ligase activity [21]. RipAK inhibits the activity of host catalases and suppresses a hypersensitive response [22]. RipAL suppresses the SA signaling pathway by activating JA production in plant cells [23]. RipN suppresses PTI and alters the NADH/NAD$^+$ ratio in plant cells through its ADP-ribose/NADH pyrophosphorylase activity [24]. However, the functions of other effectors are as yet largely unknown.

To expand our knowledge of *R. solanacearum* effectors in PTI suppression, in this study, we comprehensively screened for *R. solanacearum* RS1000 effectors with the ability to suppress flg22-triggered ROS burst in *N. benthamiana*. We identified 16 effectors that show PTI suppression activity. The detailed functional analysis of one of the effectors, RipE1, revealed that *R. solanacearum* manipulates the plant JA signaling pathway at two different steps to suppress SA-mediated defense responses. We also show that these PTI suppressors collectively contribute to bacterial virulence in *Nicotiana* plants.

2. Results

2.1. Identification of R. solanacearum Effectors that Suppress Flg22-Triggered ROS Burst in N. benthamiana

We previously identified a type III effector repertoire of *R. solanacearum* strain RS1000 [16] and constructed the binary vectors expressing effector genes under the control of a constitutive promoter [25]. To identify RS1000 effectors that affect plant PTI responses, we transiently expressed each effector protein in *N. benthamiana* leaves by agroinfiltration and evaluated its ability to suppress ROS burst triggered by flg22 treatment. Among the effector repertoire of RS1000, three effectors, namely, RipB, RipP1, and RipAA, were excluded from the screening because these effectors act as avirulence determinants and induce rapid effector-triggered immunity (ETI) responses in *N. benthamiana* [26]. In this screening, two type III effectors, AvrPtoB and HopM1, from Pto DC3000 with the ability to suppress PTI [13,14] were used as a positive control and an empty vector (EV) as a negative control. Out of the 61 Rip effectors tested, 11 (RipA5, RipE1, RipI, RipQ, RipAC, RipAL, RipAP, RipAR, RipAU, RipAW, and RipAY) strongly (≤50%) suppressed flg22-triggered ROS burst compared with the EV control (Figure 1). Among them, four effectors (RipAL, RipAR, RipAW, and RipAY) were previously shown to suppress plant PTI responses [19,21,23], indicating that our screening worked effectively. We also identified five effectors that weakly (51–70%) suppress ROS burst in *N. benthamiana*. By the screening, we identified a total of 16 effectors that affect flg22-triggered ROS burst.

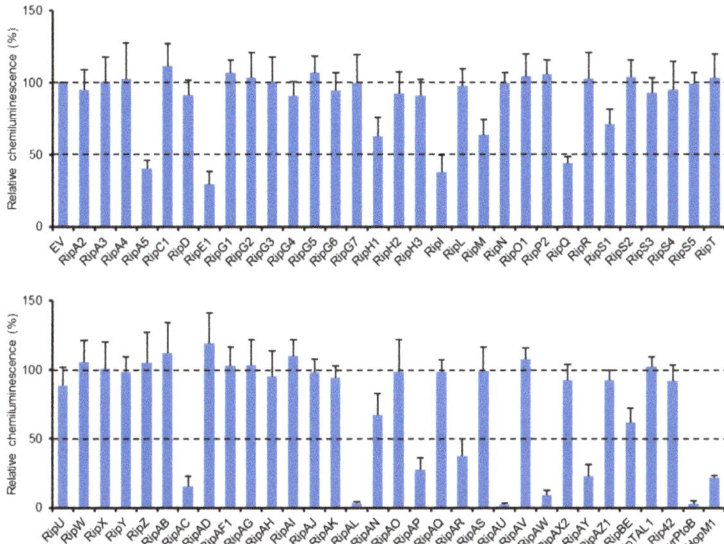

Figure 1. Identification of *R. solanacearum* effectors that suppress flg22-triggered ROS burst. Leaves of *N. benthamiana* were infiltrated with *A. tumefaciens* harboring the binary vector expressing the effector or empty vector (EV). Leaf disks were treated with the flg22 elicitor 2 days after agroinfiltration, and ROS production was monitored as photon counts for 120 min. Total photon counts are shown. Values are means ±SD of four replicates.

2.2. RipE1 is a Member of the HopX Family and Suppresses PTI through its Cysteine Protease Activity

Among the seven newly identified effectors that strongly suppress flg22-triggered ROS burst, we focused on RipE1 because it showed sequence similarity to *P. syringae* effectors belonging to the HopX family, such as HopX1 from *P. syringae* pv. *tabaci* 11528 (HopX1$_{Pta}$) (21% identity and 69% similarity; Figure S1). HopX1$_{Pta}$ has been shown to activate plant JA signaling by directly degrading JAZ repressors with its cysteine protease activity [27]. Although the entire sequence homology between RipE1 and HopX1$_{Pta}$ is low, the cysteine, histidine, and aspartic acid residues corresponding to the catalytic triad of HopX1$_{Pta}$ were conserved in RipE1 as C172, H203, and D222, respectively (Figure S1). We previously showed that the *R. solanacearum* effector RipAL activates JA signaling by inducing JA production to suppress SA-mediated defense responses in plant cells [23]. It has been proposed that RipAL targets chloroplast lipids and releases JA precursors to induce JA production because RipAL contains a putative lipase domain similar to that of the *Arabidopsis* DAD1 lipase that is involved in JA biosynthesis through the production of JA precursors [28]. To elucidate whether RipE1 manipulates and activates JA signaling by a mechanism different from that used by RipAL, we tested the cysteine protease activity of RipE1. We transiently expressed hemagglutinin (HA)-tagged RipE1 in *N. benthamiana* leaves by agroinfiltration (Figure S2A) and purified the recombinant RipE1 proteins using an anti-HA affinity resin. A protease activity assay using the fluorescein-labeled casein substrate revealed that RipE1 showed a clear protease activity in vitro compared with the buffer control (Figure S3). We also constructed two RipE1 mutants, in which the cysteine at position 172 was changed to alanine (RipE1^{C172A}), and the histidine at position 203 was also changed to alanine (RipE1^{H203A}). The recombinant RipE1^{C172A} and RipE1^{H203A} proteins lost their protease activity in vitro. This finding indicates that RipE1 has cysteine protease activity and that the two putative catalytic residues are essential for its protease activity.

To clarify the role of the protease activity of RipE1 in PTI suppression, we transiently expressed HA-tagged RipE1, RipE1^{C172A}, and RipE1^{H203A} in *N. benthamiana* leaves by agroinfiltration. The expression of RipE1, but not those of RipE1^{C172A} and RipE1^{H203A}, suppressed the flg22-triggered ROS burst (Figure 2A,B) and the expressions of PTI marker genes (Figure 2C) compared with the EV control. We confirmed that the expression of RipE1 induced no visible changes in colors and ion leakage levels in the infiltrated leaves at least 3 days after agroinfiltration (Figure S2B,C). These findings show that RipE1 suppresses PTI through its cysteine protease activity.

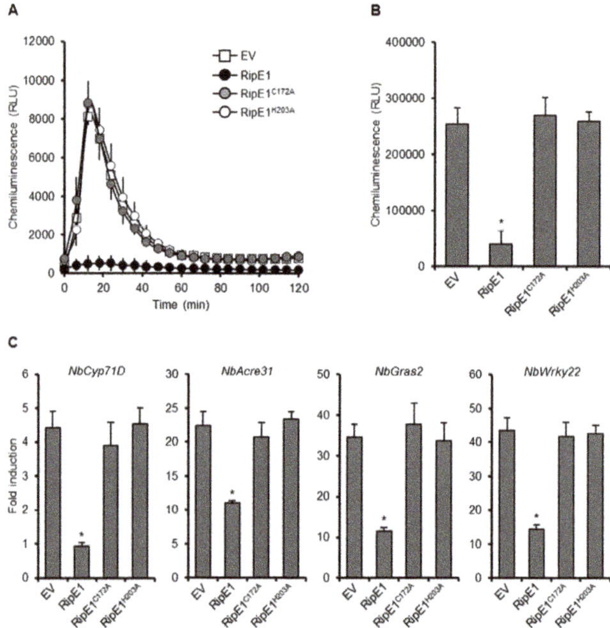

Figure 2. Effect of RipE1 expression on PTI responses in *N. benthamiana*. Leaves were infiltrated with *A. tumefaciens* harboring the binary vector expressing N-terminal HA-tagged RipE1, RipE1^{C172A}, or RipE1^{H203A}, or EV. (**A**) Flg22-triggered reactive oxygen species (ROS) burst in leaves expressing RipE1 and its catalytic site mutants. Leaf disks were treated with the flg22 elicitor 2 days after agroinfiltration, and ROS production was monitored as photon counts for 120 min. Values are means ±SD of six replicates. (**B**) Total photon counts in (**A**). The asterisk (*) denotes a statistically significant difference compared with the EV control ($p < 0.01$, Student's t-test). (**C**) Expression levels of PTI marker genes in leaves expressing RipE1 and its mutants. Leaves were treated with the flg22 elicitor 2 days after agroinfiltration, and total RNA was isolated from leaves 60 min after the treatment. Expression levels were determined by qRT-PCR analysis and normalized to that in the water treatment. Values are means ±SD of three replicates. Asterisks (*) denote statistically significant differences compared with the EV control ($p < 0.01$, Student's t-test). All experiments were repeated three times with similar results, and representative results are shown.

2.3. RipE1 Localizes to Nucleocytoplasm in Plant Cells

To examine the subcellular localization of RipE1 in plant cells, we constructed binary vectors expressing green fluorescent protein (GFP)-tagged RipE1, RipE1^{C172A}, and RipE1^{H203A} under the control of the β-estradiol-inducible promoter. When the GFP-RipE1 fusions were transiently coexpressed with the nucleocytoplasm marker mCherry in *N. benthamiana* leaves by agroinfiltration (Figure 3A), the fluorescence signals of GFP-RipE1, GFP-RipE1^{C172A}, and GFP-RipE1^{H203A} were observed in the cytoplasm of mesophyll cells as well as that of mCherry (Figure 3B). On the other hand, line-scanning analysis revealed that the fluorescence intensity of GFP-RipE1 or GFP-RipE1^{C172A}, but not GFP-RipE1^{H203A}, completely overlapped with that of mCherry in the nucleus. To verify this finding, we isolated total and nuclear protein fractions from *N. benthamiana* leaves expressing HA-tagged RipE1, RipE1^{C172A}, and RipE1^{H203A} and analyzed the accumulation levels of RipE1 and its mutants by immunoblotting. RipE1^{H203A} was detected in the total fraction, but not in the nuclear fraction, whereas RipE1 and RipE1^{C172A} were detected in both fractions (Figure 3C). These observations indicate that RipE1 localizes to the nucleocytoplasm of *N. benthamiana* cells and that the H203A mutation affects not only the protease activity but also the nuclear localization of RipE1.

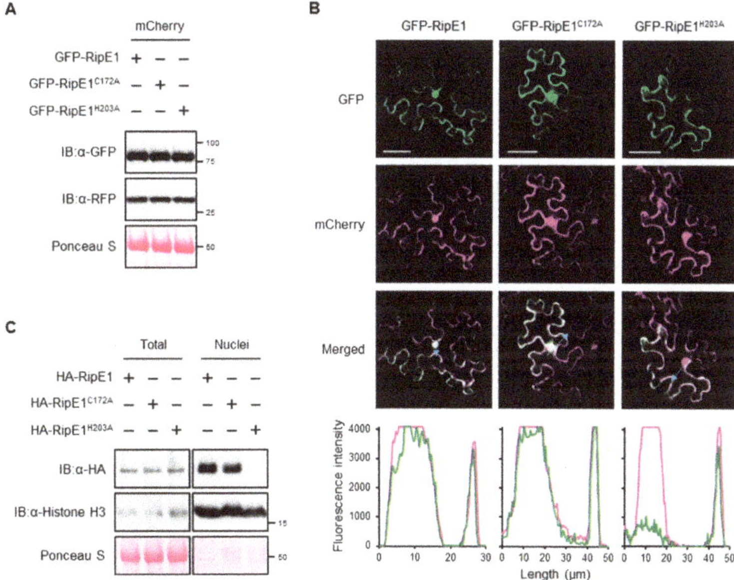

Figure 3. Subcellular localization of RipE1 in *N. benthamiana* cells. (**A**) Immunoblot analysis of green fluorescent protein (GFP)-RipE1 fusions. Leaves were coinfiltrated with *A. tumefaciens* harboring the binary vector expressing N-terminal GFP-tagged RipE1, RipE1^{C172A}, or RipE1^{H203A} and the binary vector expressing mCherry under the control of the β-estradiol-inducible promoter. Leaves were treated with β-estradiol 1 day after agroinfiltration. Total protein was extracted from the leaves 1 day after the treatment and subjected to immunoblot analysis using an anti-GFP or anti-RFP antibody. The membrane was stained with Ponceau S as the loading control. (**B**) Subcellular localization of GFP-RipE1 and its mutants. Fluorescence was observed 1 day after the treatment by confocal microscopy. The overlay of fluorescence was monitored by scanning fluorescence intensities in the regions indicated by blue arrows. Bars, 50 μm. (**C**) Immunoblot analysis of RipE1 in the nuclear fraction. Leaves were infiltrated with *A. tumefaciens* harboring the binary vector expressing N-terminal HA-tagged RipE1, RipE1^{C172A}, or RipE1^{H203A}. Total and nuclear fractions were extracted from leaves 2 days after agroinfiltration and subjected to immunoblot analysis using an anti-HA or anti-histone H3 antibody. The membrane was stained with Ponceau S as the loading control.

2.4. RipE1 Interacts with JAZ Proteins in Yeast and Plant Cells

To identify plant proteins that interact with RipE1, we performed yeast two-hybrid screening using RipE1 as the bait. Yeast cells expressing *ripE1* showed no significant difference in their growth (Figure 4A), indicating that RipE1 is not toxic to yeast cells. Upon screening approximately 2 × 10^6 transformants with an *A. thaliana* cDNA-derived prey library, we identified JAZ4 as a candidate plant target of RipE1. The JAZ family consists of 12 members in *A. thaliana* [4,29]. We examined the interaction of RipE1 with the JAZ proteins and found that RipE1 interacts with JAZ4, JAZ9, and JAZ10 in yeast cells (Figure 4B), particularly strongly with JAZ4 and JAZ9 (Figure 4C). The interaction with JAZ proteins was not affected by the C172A and H203A mutations of RipE1 (Figure 4B,C).

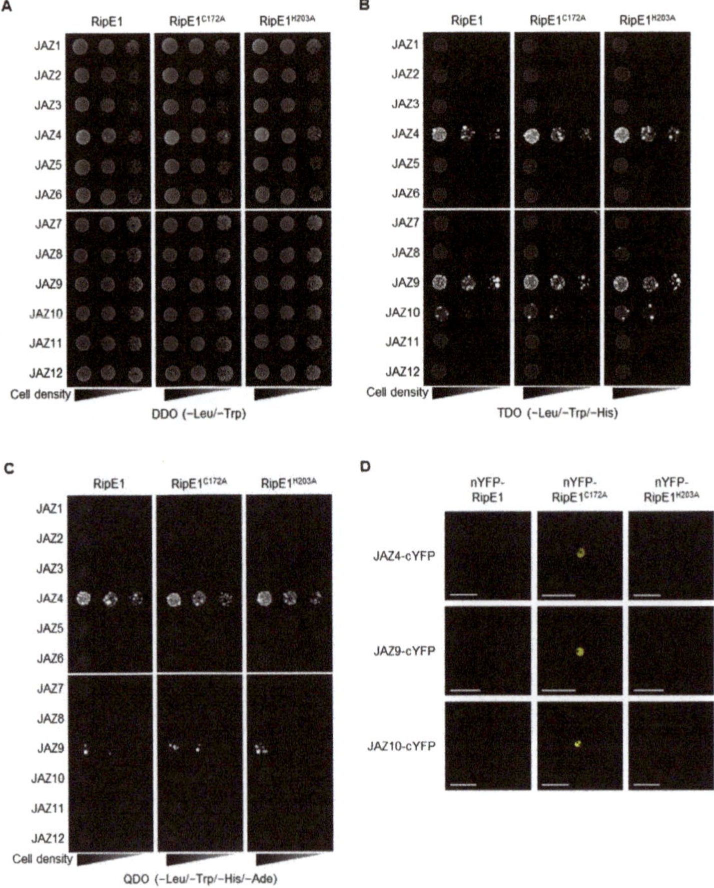

Figure 4. Interaction of RipE1 and jasmonate-ZIM-domain (JAZ) family. (**A–C**) Interaction of RipE1 and JAZ family in yeast. Serial dilutions of diploid cells harboring the bait vector expressing RipE1 and the prey vector expressing JAZs were spotted on DDO (**A**), TDO moderately selective (**B**), and QDO stringently selective (**C**) plates. Photographs were taken after 5 days of incubation. (**D**) Interaction of RipE1 and JAZ proteins in planta. Leaves of *N. benthamiana* were coinfiltrated with *A. tumefaciens* harboring the binary vector expressing N-terminal nYFP-tagged RipE1, RipE1^{C172A}, or RipE1^{H203A} and the binary vector expressing C-terminal cYFP-tagged JAZ4, JAZ9, or JAZ10. Fluorescence was observed 2 days after agroinfiltration by confocal microscopy. Bars, 50 μm. All experiments were repeated three times with similar results, and representative results are shown.

JAZ proteins are mainly localized in the plant nucleus and act as repressors in the JA signaling pathway [4]. Indeed, fluorescence signals of GFP-tagged JAZ4, JAZ9, and JAZ10 were observed in the nucleus of *N. benthamiana* cells (Figure S4). To elucidate whether RipE1 interacts with JAZs in the plant nucleus, we performed a bimolecular fluorescence complementation (BiFC) assay using *N. benthamiana* leaves by agroinfiltration. Notably, when the catalytically inactive nYFP-RipE1^{C172A}, but not the catalytically active nYFP-RipE1, was coexpressed with JAZ4-cYFP, JAZ9-cYFP, or JAZ10-cYFP, fluorescence signals were observed in the nucleus of *N. benthamiana* cells (Figure 4D). No fluorescence signal was observed when the catalytically inactive but non-nucleus-localized nYFP-RipE1^{H203A} was used in the assay. Collectively, these observations indicate that RipE1 interacts with JAZ4, JAZ9, and JAZ10, mainly in plant nuclei.

2.5. RipE1 Degrades JAZ Repressors to Activate JA Signaling and Simultaneously Suppresses SA Signaling

In the aforementioned BiFC experiments, no fluorescence signal was detected in the leaves coexpressing nYFP-RipE1 and JAZs-cYFP fusions (Figure 4D). To test whether RipE1 degrades JAZ proteins, we transiently coexpressed HA-tagged RipE1 and GFP-tagged JAZs in *N. benthamiana* leaves by agroinfiltration. The expression of RipE1 markedly decreased the accumulation of JAZ4-GFP, JAZ9-GFP, and JAZ10-GFP in the infiltrated leaves (Figure 5A). On the other hand, the expression of RipE1^{C172A} and RipE1^{H203A} did not affect the accumulation of JAZs-GFP fusions. These results show that RipE1 degrades at least three JAZ repressors through its cysteine protease activity in plant cells.

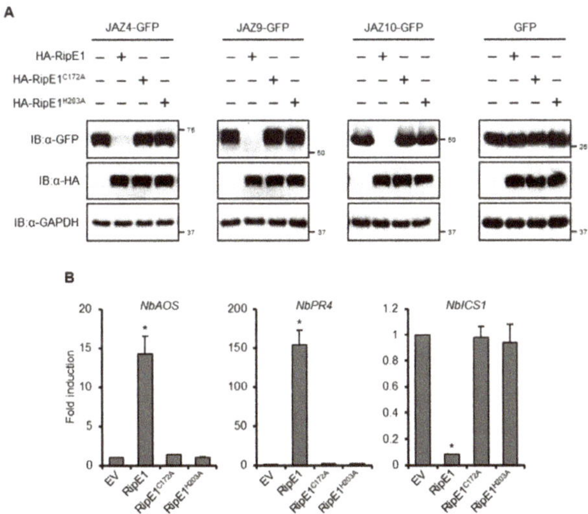

Figure 5. Effect of RipE1 expression on the accumulation of JAZs and defense-related gene expression in *N. benthamiana*. (**A**) Leaves were coinfiltrated with *A. tumefaciens* harboring the binary vector expressing N-terminal HA-tagged RipE1, RipE1^{C172A}, or RipE1^{H203A} and the binary vector expressing C-terminal GFP-tagged JAZ4, JAZ9, JAZ10, or GFP alone. Total protein was extracted from the leaves 2 days after agroinfiltration and subjected to immunoblot analysis using an anti-HA or anti-GFP antibody. The membrane was subjected to immunoblot analysis using an anti-GAPDH antibody as the endogenous and loading control. (**B**) Expression levels of jasmonic acid (JA) and salicylic acid (SA) marker genes in leaves expressing RipE1 and its mutants. Leaves were infiltrated with *A. tumefaciens* harboring the binary vector expressing N-terminal HA-tagged RipE1, RipE1^{C172A}, or RipE1^{H203A}, or EV. Total RNA was isolated from leaves 2 days after agroinfiltration. Expression levels were determined by qRT-PCR analysis and normalized to that of EV. Values are means ±SD of three replicates. Asterisks (*) denote statistically significant differences compared with the EV control ($p < 0.01$, Student's *t*-test). All experiments were repeated three times with similar results, and representative results are shown.

We observed that the prolonged expression of HA-tagged RipE1, but not RipE1^{C172A} or RipE1^{H203A}, induced leaf chlorosis, and a reduction in the chlorophyll content of the leaves (Figure S2D,E), which are hallmark events accompanied by the activation of JA signaling [30,31]. Notably, the transient expression of RipE1, but not RipE1^{C172A} or RipE1^{H203A}, induced the expressions of JA signaling marker genes, such as *NbAOS* and *NbPR4*, in *N. benthamiana* (Figure 5B). In contrast to JA signaling, the transient expression of RipE1 greatly decreased the expression level of the SA signaling marker gene *NbICS1* encoding the SA-producing enzyme. This finding indicates that RipE1 activates JA signaling by degrading JAZ repressors and simultaneously suppresses the antagonistic SA signaling pathway in *N. benthamiana*.

2.6. RipE1 can Complement the Impaired Virulence Phenotype of the COR-Deficient Mutant of Pto in Arabidopsis Plants

Pto DC3000, the causative agent of bacterial speck disease in tomato, could also infect *A. thaliana*. The *A. thaliana*–Pto interaction has been used as a model pathosystem for determining the contribution of effector proteins from other pathogens to virulence [21,24,32]. Therefore, we generated transgenic *A. thaliana* plants expressing GFP-RipE1, GFP-RipE1^{C172A}, and GFP-RipE1^{H203A} under the control of the β-estradiol-inducible promoter (Figure 6A). The growth of Pto in the transgenic plant leaves expressing GFP-RipE1 showed no significant difference compared with that in the parental Col-0 leaves (Figure 6B). Pto produces the phytotoxin COR, which is a functional mimic of JA-Ile that activates JA signaling to suppress SA-mediated defense responses in plants [6,11]. The growth of the COR-deficient mutant of Pto (Pto *cor*$^-$) decreased 100-fold compared with that of the wild-type strain in the Col-0 leaves (Figure 6B). Notably, the growth of the Pto *cor*$^-$ mutant increased 100-fold and reached the wild-type level in the transgenic plants expressing GFP-RipE1, but not GFP-RipE1^{C172A} and GFP-RipE1^{H203A} compared with the parental Col-0 plants 2 days after inoculation. Moreover, the development of disease symptoms caused by Pto *cor*$^-$ was accelerated in the transgenic plant leaves expressing GFP-RipE1, but not GFP-RipE1^{C172A} and GFP-RipE1^{H203A} (Figure 6C). These observations indicate that RipE1 can complement the impaired virulence phenotype of Pto *cor*$^-$ through the activation of JA signaling.

2.7. Multiple Deletions of Effector Genes that Show a Strong PTI Suppression Activity Affect the Growth of R. solanacearum in Nicotiana Plants

To evaluate the contribution of RipE1 to bacterial virulence, we generated a Δ*ripE1* mutant of *R. solanacearum* strain RS1002, a nalidixic acid-resistant derivative of RS1000, and inoculated the Δ*ripE1* mutant into solanaceous host plants. However, the Δ*ripE1* mutant showed no significant difference in symptom development in the inoculated plants compared with the wild-type strain (Figure S5), probably owing to functional redundancy among the effector repertoire. We next generated a Δ*ripA5* Δ*ripE1* Δ*ripI* Δ*ripQ* Δ*ripAC* Δ*ripAL* Δ*ripAP* Δ*ripAR* Δ*ripAU* Δ*ripAW* Δ*ripAY* undecuple mutant, in which all of the 11 effector genes that strongly suppressed flg22-triggered ROS burst in *N. benthamiana* were deleted. We monitored the growth of the wild-type strain and the undecuple mutant in vitro and *in planta*. Although the growth rates of the two strains showed no significant difference in a rich medium (Figure 7A), the undecuple mutant showed reduced bacterial growth in susceptible *N. sylvestris* leaves compared with the wild-type strain (Figure 7B). Furthermore, the undecuple mutant showed a significant growth defect in resistant *N. benthamiana* leaves. These findings indicate that the 11 effectors collectively contribute to the growth of *R. solanacearum* in *Nicotiana* plants.

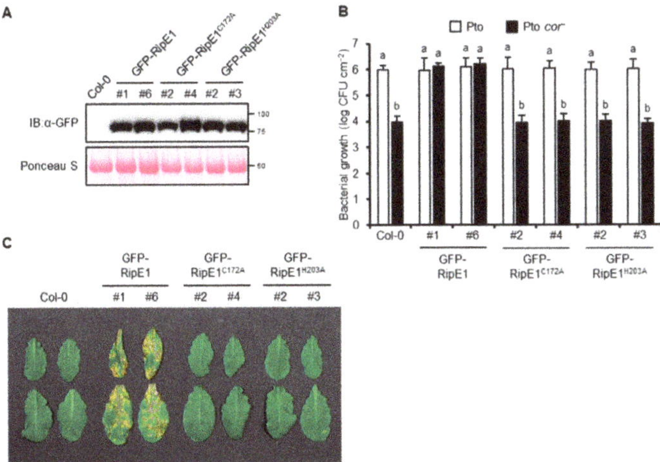

Figure 6. The phenotype of transgenic *A. thaliana* expressing RipE1. *A. thaliana* Col-0 and transgenic plants expressing N-terminal GFP-tagged RipE1, RipE1^{C172A}, and RipE1^{H203A} under the control of the β-estradiol-inducible promoter were treated with β-estradiol. (**A**) Immunoblot analysis of GFP-RipE1 fusions. Total protein was extracted from the leaves 3 days after the treatment and subjected to immunoblot analysis using an anti-GFP antibody. The membrane was stained with Ponceau S as the loading control. (**B**) Growth of Pto and Pto *cor⁻* in *A. thaliana* Col-0 and transgenic plants expressing GFP-RipE1 and its mutants. Leaves treated with β-estradiol for 1 day were sprayed with the bacterial suspension, and the bacterial population was determined 2 days after inoculation. Values are means ±SD of three replicates. Different letters denote statistically significant differences ($p < 0.05$, one-way ANOVA with Tukey–Kramer HSD test). (**C**) Disease symptoms caused by Pto *cor⁻* in *A. thaliana* Col-0 and transgenic plants expressing GFP-RipE1 and its mutants. One day after treatment with β-estradiol, the leaves were sprayed with the bacterial suspension. Photographs were taken 1 week after inoculation. All experiments were repeated three times with similar results, and representative results are shown.

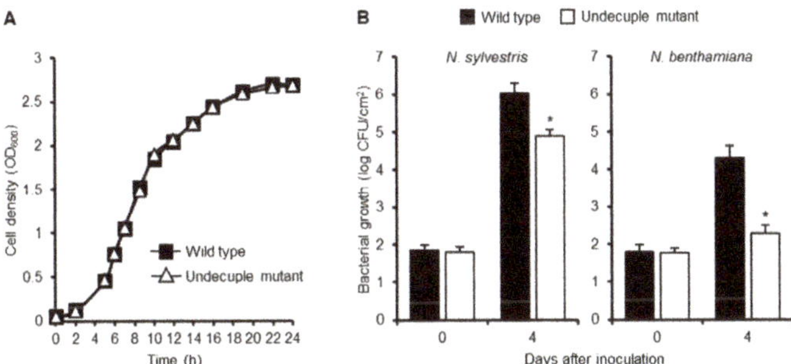

Figure 7. Contribution of PTI suppressors to virulence of *R. solanacearum* in *Nicotiana* plants. (**A**) Growth of *R. solanacearum* mutants in vitro. *R. solanacearum* wild-type strain and the Δ*ripA5* Δ*ripE1* Δ*ripI* Δ*ripQ* Δ*ripAC* Δ*ripAL* Δ*ripAP* Δ*ripAR* Δ*ripAU* Δ*ripAW* Δ*ripAY* undecuple mutant were diluted in BG medium to OD$_{600}$ = 0.05. The cultures were incubated at 28 °C with shaking, and cell density was spectrometrically measured at the indicated time points. Values are means ±SD of three replicates. (**B**) Growth of *R. solanacearum* mutants *in planta*. Leaves of *N. sylvestris* and *N. benthamiana* were inoculated with the suspension of the wild type or undecuple mutant of *R. solanacearum*. The bacterial population was determined on the indicated days after inoculation. Values are means ±SD of four replicates. Asterisks (*) denote statistically significant differences compared with the wild type ($p < 0.01$, Student's *t*-test). All experiments were repeated three times with similar results, and representative results are shown.

3. Discussion

In this study, we screened *R. solanacearum* RS1000 effectors for their ability to suppress plant PTI responses using an *Agrobacterium*-mediated transient expression system. Out of the 61 effectors tested, 16 (26%) were found to suppress flg22-triggered ROS burst in *N. benthamiana* (Figure 1). Similar comprehensive screenings for bacterial effectors that suppress PTI have been performed in Pto and *Xanthomonas euvesicatoria*. In Pto, seven out of the 22 effectors tested (32%) can suppress flg22-triggered ROS burst and expression of PTI marker genes in *N. benthamiana* [33]. In *X. euvesicatoria*, 17 out of the 33 effectors tested (52%) show the ability to suppress flg22-triggered PTI signaling in *N. benthamiana* [34]. Our finding corresponds well to the previous studies revealing that many effectors in the effector repertoire of a plant pathogenic bacterium can suppress plant PTI responses.

Out of the 16 effectors identified by our screening, 11 strongly suppressed flg22-triggered ROS burst in *N. benthamiana* (Figure 1). Among them, RipE1 was found to share low homology with the cysteine protease effector HopX1$_{Pta}$ from *P. syringae* pv. *tabaci* 11528 (Figure S1). We showed that RipE1 suppresses flg22-triggered ROS burst and reduces the expression levels of PTI marker genes in a protease-activity-dependent manner (Figure S3 and Figure 2). However, the functions of HopX family effectors vary among the members. For example, HopX1 from Pto DC3000 (HopX1$_{Pto}$) does not show the protease activity in *N. benthamiana* [27]. For PTI suppression, HopX1$_{Pta}$ and HopX1$_{Pto}$ do not suppress flg22-induced ROS burst and expressions of PTI marker genes in *N. benthamiana* [33]. *X. euvesicatoria* possesses two HopX1 family effectors, XopE1 and XopE2 [35]. XopE2, but not XopE1, has been shown to suppress flg22-triggered PTI in *N. benthamiana*, although their enzymatic activity remains unclear [34]. In this study, we clearly demonstrated that cysteine protease activity is indispensable for the PTI suppression activity of RipE1.

For the function of RipE1 in plant cells, we obtained the following results. (i) RipE1 localized to the nucleocytoplasm of plant cells (Figure 3). (ii) RipE1 interacted with JAZ4, JAZ9, and JAZ10 in yeast and plant cells (Figure 4) and degraded the JAZ repressors in the nucleus when coexpressed in plant cells (Figures 4D and 5A). (iii) RipE1 activated JA signaling and simultaneously suppressed SA signaling (Figure 5B). (iv) The expression of RipE1 complemented the reduced growth phenotype of the Pto *cor⁻* mutant in *Arabidopsis* plants (Figure 6). It has been shown that HopX1$_{Pta}$ degrades JAZ repressors through its cysteine protease activity and activates JA signaling to suppress SA signaling [27]. Except for the ability to suppress plant PTI responses, RipE1 functions similarly to HopX1$_{Pta}$.

It is widely known that the JA and SA signaling pathways antagonize each other to fine-tune proper defense responses to combat against pathogens with different infection strategies. Upon activation of JA signaling, the MYC2 transcription factor released by the degradation of JAZ repressors induces the expression of NAC transcription factors that repress the expression of the SA synthesis enzyme ICS1, leading to the downregulation of SA signaling [11]. We showed that RipE1 activates JA signaling and simultaneously suppresses the expression of *NbICS1* in *N. benthamiana* (Figure 5). It has been shown that the virulence of *R. solanacearum* is enhanced in *NbICS1*-silenced plants [23]. Our work provides new evidence that *R. solanacearum* exploits antagonistic interactions between the SA and JA signaling pathways to suppress SA signaling in plants and promote successful infection. We previously showed that RipAL targets chloroplasts and induces JA production to activate JA signaling and simultaneously suppress SA-mediated defense responses in plants [23]. Our findings indicate that *R. solanacearum* uses at least two effectors that target different organelles and activate host JA signaling in two different steps, JA production and JAZ degradation, probably giving a synergistic effect (Figure S6).

We observed that RipE1 preferentially interacts with JAZ4, JAZ9, and JAZ10 among the JAZ members in the yeast two-hybrid system (Figure 4) and proteolytically degrades them in plant cells (Figure 5A). It has been reported that the loss-of-function mutant *jaz4-1* or *jaz10-1* in *A. thaliana* makes the plant hypersusceptible to Pto DC3000 and enhances pathogen growth in infected leaves compared with the wild-type plants [36,37]. No other single-gene mutations of 12 JAZ members affect the *in planta* growth of Pto DC3000, indicating that JAZ4 and JAZ10 play important roles in the defense response to the bacterial pathogen. It is noteworthy that RipE1 interacts most strongly with JAZ4 (Figure 4C) since the *jaz4-1* mutation showed the most striking effect in terms of the enhanced growth of the bacterial pathogen in mutant plants [37]. JAZ4 and JAZ9 belong to the same subgroup of JAZ members based on their amino acid sequences [29]; therefore, it is reasonable to consider that RipE1 interacts with JAZ9 (Figure 4). The roles of JAZ4, JAZ9, and JAZ10 in defense responses against *R. solanacearum* should be clarified in future work.

Generally, a single mutation in effector genes does not affect the virulence of plant pathogenic bacteria, probably owing to its functional redundancy among the effector repertoire. Therefore, we generated the undecuple mutant that lacks 11 effectors with a strong PTI suppression activity. Notably, the growth of the undecuple mutant was reduced in susceptible *N. sylvestris* leaves and extremely decreased in resistant *N. benthamiana* leaves (Figure 7). This finding clearly shows that Rip effectors with PTI suppression activity collectively promote the growth of *R. solanacearum* in plants. The greater inhibition of bacterial growth in resistant plants might be explained by the more rapid and stronger induction of defense responses in resistant plants. From another viewpoint, it has recently been suggested that one effector targets multiple plant factors to subvert immune responses. For example, the acetyltransferase effector HopZ1a from *P. syringae* degrades not only JAZ proteins but also GmHID1, which is involved in isoflavone biosynthesis [38,39]. Therefore, RipE1 or another effector(s) in the 11 effectors deleted in the undecuple mutant might suppress defense responses other than PTI, for example, ETI, in resistant plants. It would be interesting to determine whether any of the 11 effectors show the ability to affect ETI in *N. benthamiana*.

4. Materials and Methods

4.1. Plant, Yeast, and Bacterial Growth Conditions

N. benthamiana, *N. sylvestris*, and *A. thaliana* were grown in a controlled environment room, as described previously [21]. The bacterial strains used in this study are listed in Table S1. The growth conditions, media, and antibiotics used for the *Escherichia coli*, *Saccharomyces cerevisiae*, *R. solanacearum*, Pto, and *A. tumefaciens* strains were described previously [19,21].

4.2. Agrobacterium-Mediated Transient Expression (Agroinfiltration)

ripE1 and *JAZs* were cloned into the *NcoI*-*Eco*RV sites of the pENTER4 vector (Invitrogen, Carlsbad, CA, USA) using an In-Fusion HD cloning kit. *ripE1^{C172A}* and *ripE1^{H203A}* were produced by PCR-based site-directed mutagenesis. The resultant entry clones were subcloned into pGWB5 and pGWB15 vectors [40] using a Gateway LR Clonase II enzyme mix (Invitrogen). *A. tumefaciens* cells harboring the resultant plasmids were suspended in infiltration buffer [10 mM $MgCl_2$, 10 mM MES (pH 5.6)] supplemented with 150 μM acetosyringone. The inoculum preparations were spectrometrically adjusted to OD_{600} = 1.0 and incubated at 30 °C for 3 h with shaking before infiltration. For coexpression assays, *A. tumefaciens* cells harboring pDGBα2-35S:P19 [41], pGWB5 derivatives, and pGWB15 derivatives were mixed at a ratio of 1:3:3 and infiltrated into the leaves of *N. benthamiana*. The primer sets used for plasmid construction are listed in Table S2.

4.3. Measurement of ROS

ROS were measured using a chemiluminescence probe L-012 (Wako, Osaka, Japan) as described previously [21]. Leaf disks were floated on water overnight. Then, the water was replaced with 0.5 mM L-012 solution (10 mM MOPS-KOH, pH 7.4) containing 100 nM flg22 (Funakoshi, Tokyo, Japan). Chemiluminescence was continuously monitored using a microplate reader (SH-8000Lab, Corona Electric, Ibaraki, Japan).

4.4. Protein Extraction and Immunoblotting

Leaf disks (60 mg) were collected, frozen in liquid nitrogen, and ground to a fine powder. Proteins were extracted in 100 μL of extraction buffer (0.35 M Tris-HCl (pH 6.8), 30% glycerol, 10% SDS, 0.6 M

DTT, 0.012% bromophenol blue). The fraction of nuclear proteins was extracted from the leaves using a CelLytic PN Isolation/Extraction kit (Sigma–Aldrich, St. Louis, MO, USA) in accordance with the manufacturer's instructions. Total protein (5 μL) was separated by 10% SDS-PAGE. Separated proteins were transferred onto a membrane and incubated with an HRP-conjugated anti-GFP (1:5000; Miltenyi Biotec, Bergisch Gladbach, Germany), anti-HA (1:2000; MBL, Aichi, Japan, #561-7), anti-RFP (1:2000; MBL, #PM005-7), anti-GAPDH (1:5000; Proteintech, Rosemont, IL, USA), or anti-histone H3 (1:2000; MBL) antibody. Immunodetection was performed using an ECL Prime Western blotting detection reagent (GE Healthcare, Marlborough, MA, USA) or Clarity Max Western ECL Substrate (Bio-rad, Hercules, CA, USA).

4.5. Protease Activity Assays

Protease activity assays were performed as described previously [27] with slight modifications. Briefly, leaves of *N. benthamiana* plants transiently expressing HA-tagged RipE1, RipE1^{C172A}, and RipE1^{H203A} were homogenized in an extraction buffer containing 100 mM Tris-HCl (pH 7.5), 150 mM NaCl, 5 mM EDTA, 5% glycerol, 10 mM DTT, 1 mM PMSF, 0.5% Triton X-100, and a protease inhibitor cocktail (Roche, Mannheim, Germany). HA-tagged proteins were purified from the homogenate using an anti-HA affinity resin (Pierce, Rockford, IL, USA). Protease activity was determined using a Protease Fluorescent Detection kit (Sigma–Aldrich) in accordance with the manufacturer's instructions. Fluorescence intensity was measured at 485 nm excitation and 535 nm emission using a microplate reader.

4.6. Measurement of Ion Leakage and Chlorophyll Content

The severity of cell death was quantified by the degree of electrolyte leakage from leaves. Leaf disks (8 mM in diameter) were immersed in 1 mL of water for 2 h with gentle shaking. The ion conductivity of water was measured using a conductivity meter (LAQUAtwin, Horiba, Kyoto, Japan).

The chlorophyll content of the leaves was spectrophotometrically measured using a chlorophyll meter (SPAD-502, Konica Minolta, Tokyo, Japan) in accordance with the manufacturer's instructions.

4.7. Real-Time PCR Analysis

Real-time PCR was performed as described previously [42]. Briefly, total RNA was extracted from leaves using an RNeasy Plant Mini Kit (Qiagen, Hilden, Germany), and cDNA was synthesized using a High Capacity cDNA reverse transcription kit (Applied Biosystems, Foster, CA, USA). Quantitative PCR was performed using a Power SYBR Green PCR master mix (Applied Biosystems). Expression levels of target genes were normalized to those of multiple endogenous control genes, such as *NbEF1α*, *NbNQO*, and *NbF-box*. Gene-specific primer sets used for the real-time PCR analysis are listed in Table S3.

4.8. Microscopic Analyses

ripE1 and its derivative genes were cloned into the *Bam*HI site of the SUPERR:sXVE:GFP$_N$:Hyg vector [43] using an In-Fusion HD cloning kit. The resultant plasmids were used for *Agrobacterium*-mediated transient expression in *N. benthamiana* leaves. *A. tumefaciens* cells harboring SUPERR:sXVE:mCherry$_N$:Bar vector were coinfiltrated and used as a nucleocytoplasm marker. Leaves were treated with 50 μM β-estradiol (Wako) 1 d after the agroinfiltration. The fluorescence of GFP and RFP was observed 1 day after the treatment with β-estradiol under a laser scanning microscope (FV1200, Olympus, Tokyo, Japan).

For BiFC analysis, the entry clones of *ripE1* and its derivatives and *JAZs* were, respectively, subcloned into the pB5nYGW and pB5GWcY vectors [44] using a Gateway LR Clonase II enzyme mix. *A. tumefaciens* cells harboring pDGBα2-35S:P19, pB5nYGW derivatives, and pB5GWcY derivatives were mixed at a ratio of 1:3:3 and infiltrated into the leaves of *N. benthamiana*.

4.9. Yeast Two-Hybrid Analysis

Yeast two-hybrid analysis was performed using the Matchmaker Gold yeast two-hybrid system (Clontech, Palo Alto, CA, USA) in accordance with the manufacturer's instructions. *ripE1* and its derivatives were cloned into the *Eco*RI-*Bam*HI sites of the pGBKT7 vector and transformed into *S. cerevisiae* Y2HGold as the bait. For the identification of RipE1 targets, yeast two-hybrid screening was performed using Make Your Own Mate & Plate Library System (Clontech) in accordance with the manufacturer's instructions. Briefly, total RNA was extracted from *A. thaliana* leaves infiltrated with *R. solanacearum* RS1002 (1×10^8 CFU mL^{-1}) and 1 μM flg22. A mixture of total RNA was used to synthesize the cDNA library and ligated into the pGADT7-Rec vector. The resultant plasmids were transformed into *S. cerevisiae* Y187 as prey. Positive clones were selected on QDO (SD/−Ade/−His/−Leu/−Trp) stringent selective plates by mating the Y2HGold and Y187 strains.

For the validation of RipE1 and JAZs interaction, the entry clones for *JAZs* were obtained from ABRC and RIKEN BRC through the National Bio-Resource Project of MEXT/AMED, Japan. The entry clones were subcloned into the pGADT7-GW vector [45] using a Gateway LR Clonase II enzyme mix. The resultant plasmids were transformed into Y187. Diploid cells harboring pGBKT7 and pGADT7 derivatives were selected on DDO (SD/−Leu/−Trp) plates. Positive interactions of RipE1s and JAZs were tested on TDO (SD/−His/−Leu/−Trp) moderate selective plates and QDO plates.

4.10. Generation of R. solanacearum Mutants

Each fragment (0.6 kb) upstream and downstream of the effector coding region was tandemly inserted into the plasmid pK18*mobsacB* [46] using an In-Fusion HD cloning kit (Takara, Shiga, Japan) The resultant plasmids were used to generate *R. solanacearum* mutants by the marker-exchange method using *E. coli* S17-1 [46].

4.11. Bacterial Virulence Assay

The virulence of *R. solanacearum* and Pto in plants was assayed, as described previously [21]. For measuring the growth of *R. solanacearum* strains, the inoculum (1×10^4 CFU mL^{-1}) was infiltrated into the leaves of *Nicotiana* plants using a needleless syringe. For measuring the growth of Pto, *A. thaliana* was treated with 100 μM β-estradiol supplemented with 0.01% Silwet L-77 for 1 day and then inoculated by spraying the inoculum (1×10^8 CFU mL^{-1}) containing 10 mM MgCl$_2$ and 0.01% Silwet L-77. Leaf disks were taken from the inoculated leaves and homogenized in water. Serial dilutions of the homogenate were spread on BG plates containing nalidixic acid for *R. solanacearum* and King's B plates containing rifampicin for Pto.

Supplementary Materials: Supplementary materials can be found at http://www.mdpi.com/1422-0067/20/23/5992/s1.

Author Contributions: Conceptualization, M.N. and T.M.; Formal analysis, M.N.; Investigation, M.N.; Writing—original draft, M.N. and T.M.; Writing—review and editing, M.N. and T.M.

Funding: This research was funded by JSPS KAKENHI, grant number JP18K05666 (Takafumi Mukaihara) and JP18J02213 and JP19K15847 (Masahito Nakano). Masahito Nakano was funded by the Ichimura Foundation for New Technology, the Yamazaki Spice Promotion Foundation, the Wesco Scientific Promotion Foundation, and the Institute for Fermentation, Osaka.

Acknowledgments: We would like to thank Tsuyoshi Nakagawa and Shoji Mano for providing the Gateway binary vectors and BiFC vectors, respectively.

Conflicts of Interest: The authors declare no conflict of interest.

Abbreviations

PTI	pattern-triggered immunity
PAMPs	pathogen/microbe-associated molecular patterns
PRRs	pattern-recognition receptors
ROS	reactive oxygen species

SA	salicylic acid
JA	jasmonic acid
COR	coronatine
Pto	*Pseudomonas syringae* pv. *tomato*
COI1	Coronatine-insensitive1
JAZ	Jasmonate-ZIM-domain
ETI	effector-triggered immunity
HA	hemagglutinin
GFP	green fluorescent protein
BiFC	bimolecular fluorescence complementation

References

1. Jones, J.D.; Dangl, J.L. The plant immune system. *Nature* **2006**, *444*, 323–329. [CrossRef] [PubMed]
2. Couto, D.; Zipfel, C. Regulation of pattern recognition receptor signalling in plants. *Nat. Rev. Immunol.* **2016**, *16*, 537–552. [CrossRef] [PubMed]
3. Pieterse, C.M.; Van der Does, D.; Zamioudis, C.; Leon-Reyes, A.; Van Wees, S.C. Hormonal modulation of plant immunity. *Annu. Rev. Cell Dev. Biol.* **2012**, *28*, 489–521. [CrossRef] [PubMed]
4. Wasternack, C.; Hause, B. Jasmonates: Biosynthesis, perception, signal transduction and action in plant stress response, growth and development. An update to the 2007 review in Annals of Botany. *Ann. Bot.* **2013**, *111*, 1021–1058. [CrossRef] [PubMed]
5. Zhang, L.; Zhang, F.; Melotto, M.; Yao, J.; He, S.Y. Jasmonate signaling and manipulation by pathogens and insects. *J. Exp. Bot.* **2017**, *68*, 1371–1385. [CrossRef]
6. Bender, C.L.; Alarcón-Chaidez, F.; Gross, D.C. *Pseudomonas syringae* phytotoxins: Mode of action, regulation, and biosynthesis by peptide and polyketide synthetases. *Microbiol. Mol. Biol. Rev.* **1999**, *63*, 266–292.
7. Katsir, L.; Schilmiller, A.L.; Staswick, P.E.; He, S.Y.; Howe, G.A. COI1 is a critical component of a receptor for jasmonate and the bacterial virulence factor coronatine. *Proc. Natl. Acad. Sci. USA* **2008**, *105*, 7100–7105. [CrossRef]
8. Melotto, M.; Mecey, C.; Niu, Y.; Chung, H.S.; Katsir, L.; Yao, J.; Zeng, W.; Thines, B.; Staswick, P.; Browse, J.; et al. A critical role of two positively charged amino acids in the Jas motif of Arabidopsis JAZ proteins in mediating coronatine- and jasmonoyl isoleucine-dependent interactions with the COI1 F.-box protein. *Plant J.* **2008**, *55*, 979–988. [CrossRef]
9. Brooks, D.M.; Bender, C.L.; Kunkel, B.N. The *Pseudomonas syringae* phytotoxin coronatine promotes virulence by overcoming salicylic acid-dependent defences in *Arabidopsis thaliana*. *Mol. Plant Pathol.* **2005**, *6*, 629–639. [CrossRef]
10. Melotto, M.; Underwood, W.; Koczan, J.; Nomura, K.; He, S.Y. Plant stomata function in innate immunity against bacterial invasion. *Cell* **2006**, *126*, 969–980. [CrossRef]
11. Zheng, X.Y.; Spivey, N.W.; Zeng, W.; Liu, P.P.; Fu, Z.Q.; Klessig, D.F.; He, S.Y.; Dong, X. Coronatine promotes *Pseudomonas syringae* virulence in plants by activating a signaling cascade that inhibits salicylic acid accumulation. *Cell Host Microbe* **2012**, *11*, 587–596. [CrossRef] [PubMed]
12. Büttner, D. Behind the lines-actions of bacterial type III effector proteins in plant cells. *FEMS Microbiol. Rev.* **2016**, *40*, 894–937. [CrossRef] [PubMed]
13. Göhre, V.; Spallek, T.; Häweker, H.; Mersmann, S.; Mentzel, T.; Boller, T.; de Torres, M.; Mansfield, J.W.; Robatzek, S. Plant pattern-recognition receptor FLS2 is directed for degradation by the bacterial ubiquitin ligase AvrPtoB. *Curr. Biol.* **2008**, *18*, 1824–1832. [CrossRef]
14. Nomura, K.; Debroy, S.; Lee, Y.H.; Pumplin, N.; Jones, J.; He, S.Y. A bacterial virulence protein suppresses host innate immunity to cause plant disease. *Science* **2006**, *313*, 220–223. [CrossRef] [PubMed]
15. Hayward, A.C. Biology and epidemiology of bacterial wilt caused by *Pseudomonas solanacearum*. *Annu. Rev. Phytopathol.* **1991**, *29*, 65–87. [CrossRef]
16. Mukaihara, T.; Tamura, N.; Iwabuchi, M. Genome-wide identification of a large repertoire of *Ralstonia solanacearum* type III effector proteins by a new functional screen. *Mol. Plant Microbe Interact* **2010**, *23*, 251–262. [CrossRef]

17. Peeters, N.; Carrère, S.; Anisimova, M.; Plener, L.; Cazalé, A.C.; Genin, S. Repertoire, unified nomenclature and evolution of the Type III effector gene set in the *Ralstonia solanacearum* species complex. *BMC Genomics* **2013**, *14*, 859. [CrossRef]

18. Le Roux, C.; Huet, G.; Jauneau, A.; Camborde, L.; Tremousaygue, D.; Kraut, A.; Zhou, B.; Levaillant, M.; Adachi, H.; Yoshioka, H.; et al. A receptor pair with an integrated decoy converts pathogen disabling of transcription factors to immunity. *Cell* **2015**, *161*, 1074–1088. [CrossRef]

19. Mukaihara, T.; Hatanaka, T.; Nakano, M.; Oda, K. *Ralstonia solanacearum* type III effector RipAY is a glutathione-degrading enzyme that is activated by plant cytosolic thioredoxins and suppresses plant immunity. *MBio* **2016**, *7*, e00359-16. [CrossRef]

20. Sang, Y.; Wang, Y.; Ni, H.; Cazalé, A.C.; She, Y.M.; Peeters, N.; Macho, A.P. The *Ralstonia solanacearum* type III effector RipAY targets plant redox regulators to suppress immune responses. *Mol. Plant Pathol.* **2018**, *19*, 129–142. [CrossRef]

21. Nakano, M.; Oda, K.; Mukaihara, T. *Ralstonia solanacearum* novel E3 ubiquitin ligase (NEL) effectors RipAW and RipAR suppress pattern-triggered immunity in plants. *Microbiology* **2017**, *163*, 992–1002. [CrossRef] [PubMed]

22. Sun, Y.; Li, P.; Deng, M.; Shen, D.; Dai, G.; Yao, N.; Lu, Y. The *Ralstonia solanacearum* effector RipAK suppresses plant hypersensitive response by inhibiting the activity of host catalases. *Cell Microbiol.* **2017**, *19*, e12736. [CrossRef] [PubMed]

23. Nakano, M.; Mukaihara, T. *Ralstonia solanacearum* type III effector RipAL targets chloroplasts and induces jasmonic acid production to suppress salicylic acid-mediated defense responses in plants. *Plant Cell Physiol.* **2018**, *59*, 2576–2589. [CrossRef] [PubMed]

24. Sun, Y.; Li, P.; Shen, D.; Wei, Q.; He, J.; Lu, Y. The *Ralstonia solanacearum* effector RipN suppresses plant PAMP-triggered immunity, localizes to the endoplasmic reticulum and nucleus, and alters the NADH/NAD. *Mol. Plant Pathol.* **2019**, *20*, 533–546. [CrossRef] [PubMed]

25. Nahar, K.; Matsumoto, I.; Taguchi, F.; Inagaki, Y.; Yamamoto, M.; Toyoda, K.; Shiraishi, T.; Ichinose, Y.; Mukaihara, T. *Ralstonia solanacearum* type III secretion system effector Rip36 induces a hypersensitive response in the nonhost wild eggplant *Solanum torvum*. *Mol. Plant Pathol.* **2014**, *15*, 297–303. [CrossRef]

26. Nakano, M.; Mukaihara, T. The type III effector RipB from *Ralstonia solanacearum* RS1000 acts as a major avirulence factor in *Nicotiana benthamiana* and other *Nicotiana* species. *Mol. Plant Pathol.* **2019**, *20*, 1237–1251. [CrossRef] [PubMed]

27. Gimenez-Ibanez, S.; Boter, M.; Fernandez-Barbero, G.; Chini, A.; Rathjen, J.P.; Solano, R. The bacterial effector HopX1 targets JAZ transcriptional repressors to activate jasmonate signaling and promote infection in *Arabidopsis*. *PLoS Biol.* **2014**, *12*, e1001792. [CrossRef]

28. Ishiguro, S.; Kawai-Oda, A.; Ueda, J.; Nishida, I.; Okada, K. The *DEFECTIVE IN ANTHER DEHISCIENCE1* gene encodes a novel phospholipase A1 catalyzing the initial step of jasmonic acid biosynthesis, which synchronizes pollen maturation, anther dehiscence, and flower opening in Arabidopsis. *Plant Cell.* **2001**, *13*, 2191–2209. [CrossRef]

29. Chung, H.S.; Cooke, T.F.; Depew, C.L.; Patel, L.C.; Ogawa, N.; Kobayashi, Y.; Howe, G.A. Alternative splicing expands the repertoire of dominant JAZ repressors of jasmonate signaling. *Plant J.* **2010**, *63*, 613–622. [CrossRef]

30. Creelman, R.A.; Mullet, J.E. Jasmonic acid distribution and action in plants: Regulation during development and response to biotic and abiotic stress. *Proc. Natl. Acad. Sci. USA* **1995**, *92*, 4114–4119. [CrossRef]

31. He, Y.; Fukushige, H.; Hildebrand, D.F.; Gan, S. Evidence supporting a role of jasmonic acid in Arabidopsis leaf senescence. *Plant Physiol.* **2002**, *128*, 876–884. [CrossRef] [PubMed]

32. Fabro, G.; Steinbrenner, J.; Coates, M.; Ishaque, N.; Baxter, L.; Studholme, D.J.; Körner, E.; Allen, R.L.; Piquerez, S.J.; Rougon-Cardoso, A.; et al. Multiple candidate effectors from the oomycete pathogen *Hyaloperonospora arabidopsidis* suppress host plant immunity. *PLoS Pathog.* **2011**, *7*, e1002348. [CrossRef] [PubMed]

33. Gimenez-Ibanez, S.; Hann, D.R.; Chang, J.H.; Segonzac, C.; Boller, T.; Rathjen, J.P. Differential suppression of *Nicotiana benthamiana* innate immune responses by transiently expressed *Pseudomonas syringae* type III effectors. *Front. Plant Sci.* **2018**, *9*, 688. [CrossRef] [PubMed]

34. Popov, G.; Fraiture, M.; Brunner, F.; Sessa, G. Multiple *Xanthomonas euvesicatoria* type III effectors inhibit flg22-triggered immunity. *Mol. Plant Microbe Interact* **2016**, *29*, 651–660. [CrossRef] [PubMed]

35. Thieme, F.; Szczesny, R.; Urban, A.; Kirchner, O.; Hause, G.; Bonas, U. New type III effectors from *Xanthomonas campestris* pv. *vesicatoria* trigger plant reactions dependent on a conserved N-myristoylation motif. *Mol. Plant Microbe Interact.* **2007**, *20*, 1250–1261. [PubMed]

36. De Torres Zabala, M.; Zhai, B.; Jayaraman, S.; Eleftheriadou, G.; Winsbury, R.; Yang, R.; Truman, W.; Tang, S.; Smirnoff, N.; Grant, M. Novel JAZ co-operativity and unexpected JA dynamics underpin *Arabidopsis* defence responses to *Pseudomonas syringae* infection. *New Phytol.* **2016**, *209*, 1120–1134. [CrossRef]

37. Oblessuc, P.R.; Obulareddy, N.; DeMott, L.; Matiolli, C.C.; Thompson, B.K.; Melotto, M. JAZ4 is involved in plant defense, growth, and development in Arabidopsis. *Plant J.* **2019**. Epub ahead of print. [CrossRef]

38. Zhou, H.; Lin, J.; Johnson, A.; Morgan, R.L.; Zhong, W.; Ma, W. *Pseudomonas syringae* type III effector HopZ1 targets a host enzyme to suppress isoflavone biosynthesis and promote infection in soybean. *Cell Host Microbe* **2011**, *9*, 177–186. [CrossRef]

39. Jiang, S.; Yao, J.; Ma, K.W.; Zhou, H.; Song, J.; He, S.Y.; Ma, W. Bacterial effector activates jasmonate signaling by directly targeting JAZ transcriptional repressors. *PLoS Pathog.* **2013**, *9*, e1003715. [CrossRef]

40. Nakagawa, T.; Kurose, T.; Hino, T.; Tanaka, K.; Kawamukai, M.; Niwa, Y.; Toyooka, K.; Matsuoka, K.; Jinbo, T.; Kimura, T. Development of series of gateway binary vectors, pGWBs, for realizing efficient construction of fusion genes for plant transformation. *J. Biosci. Bioeng.* **2007**, *104*, 34–41. [CrossRef]

41. Sarrion-Perdigones, A.; Vazquez-Vilar, M.; Palací, J.; Castelijns, B.; Forment, J.; Ziarsolo, P.; Blanca, J.; Granell, A.; Orzaez, D. GoldenBraid 2.0: A comprehensive DNA assembly framework for plant synthetic biology. *Plant Physiol.* **2013**, *162*, 1618–1631. [CrossRef] [PubMed]

42. Nakano, M.; Nishihara, M.; Yoshioka, H.; Takahashi, H.; Sawasaki, T.; Ohnishi, K.; Hikichi, Y.; Kiba, A. Suppression of DS1 phosphatidic acid phosphatase confirms resistance to *Ralstonia solanacearum* in *Nicotiana benthamiana*. *PLoS ONE* **2013**, *8*, e75124. [CrossRef] [PubMed]

43. Schlücking, K.; Edel, K.H.; Köster, P.; Drerup, M.M.; Eckert, C.; Steinhorst, L.; Waadt, R.; Batistic, O.; Kudla, J. A new β-estradiol-inducible vector set that facilitates easy construction and efficient expression of transgenes reveals CBL3-dependent cytoplasm to tonoplast translocation of CIPK5. *Mol. Plant* **2013**, *6*, 1814–1829. [CrossRef]

44. Kamigaki, A.; Nito, K.; Hikino, K.; Goto-Yamada, S.; Nishimura, M.; Nakagawa, T.; Mano, S. Gateway vectors for simultaneous detection of multiple protein-protein interactions in plant cells using bimolecular fluorescence complementation. *PLoS ONE* **2016**, *11*, e0160717. [CrossRef] [PubMed]

45. Lu, Q.; Tang, X.; Tian, G.; Wang, F.; Liu, K.; Nguyen, V.; Kohalmi, S.E.; Keller, W.A.; Tsang, E.W.; Harada, J.J.; et al. Arabidopsis homolog of the yeast TREX-2 mRNA export complex: Components and anchoring nucleoporin. *Plant J.* **2010**, *61*, 259–270. [CrossRef]

46. Schäfer, A.; Tauch, A.; Jäger, W.; Kalinowski, J.; Thierbach, G.; Pühler, A. Small mobilizable multi-purpose cloning vectors derived from the *Escherichia coli* plasmids pK18 and pK19: Selection of defined deletions in the chromosome of *Corynebacterium glutamicum*. *Gene* **1994**, *145*, 69–73. [CrossRef]

© 2019 by the authors. Licensee MDPI, Basel, Switzerland. This article is an open access article distributed under the terms and conditions of the Creative Commons Attribution (CC BY) license (http://creativecommons.org/licenses/by/4.0/).

International Journal of
Molecular Sciences

Article

PatJAZ6 Acts as a Repressor Regulating JA-Induced Biosynthesis of Patchouli Alcohol in *Pogostemon Cablin*

Xiaobing Wang [1], Xiuzhen Chen [1], Liting Zhong [1], Xuanxuan Zhou [1], Yun Tang [1], Yanting Liu [1], Junren Li [1], Hai Zheng [2], Ruoting Zhan [1] and Likai Chen [1,*]

[1] Joint Laboratory of National Engineering Research Center for the Pharmaceutics of Traditional Chinese Medicines, Key Laboratory of Chinese Medicinal Resource from Lingnan, Ministry of Education, Research Center of Chinese Herbal Resource Science and Engineering, Guangzhou University of Chinese Medicine, Guangzhou University of Chinese Medicine, Guangzhou 510006, China; 13570332896@163.com (X.W.); xiuzhenchan@163.com (X.C.); 13726857667@163.com (L.Z.); 15521166528@163.com (X.Z.); TY1216569833@163.com (Y.T.); sonicflyt@163.com (Y.L.); leejunren@163.com (J.L.); zhanrt@gzucm.edu.cn (R.Z.)

[2] School of Pharmaceutical Sciences, Guangdong Food and Drug Vocational College, Guangzhou 510520, China; zhenghai2007@163.com

* Correspondence: chenlk@gzucm.edu.cn; Tel.: +020-3935-8066

Received: 12 October 2019; Accepted: 27 November 2019; Published: 30 November 2019

Abstract: The JASMONATE ZIM DOMAIN (JAZ) proteins act as negative regulators in the jasmonic acid (JA) signaling pathways of plants, and these proteins have been reported to play key roles in plant secondary metabolism mediated by JA. In this study, we firstly isolated one JAZ from *P. cablin*, PatJAZ6, which was characterized and revealed based on multiple alignments and a phylogenic tree analysis. The result of subcellular localization indicated that the PatJAZ6 protein was located in the nucleus of plant protoplasts. The expression level of *PatJAZ6* was significantly induced by the methyl jasmonate (MeJA). Furthermore, by means of yeast two-hybrid screening, we identified two transcription factors that interact with the PatJAZ6, the PatMYC2b1 and PatMYC2b2. Virus-induced gene silencing (VIGS) of *PatJAZ6* caused a decrease in expression abundance, resulting in a significant increase in the accumulation of patchouli alcohol. Moreover, we overexpressed *PatJAZ6* in *P. cablin*, which down-regulated the patchoulol synthase expression, and then suppressed the biosynthesis of patchouli alcohol. The results demonstrate that PatJAZ6 probably acts as a repressor in the regulation of patchouli alcohol biosynthesis, contributed to a model proposed for the potential JA signaling pathway in *P. cablin*.

Keywords: PatJAZ6; jasmonic acid (JA) signaling pathway; *Pogostemon cablin*; patchouli alcohol; biosynthesis

1. Introduction

Pogostemon cablin (*P. cablin*), the medicinal part of which is dry whole grass, is a kind of Labiatae plant that has long been considered an important Chinese herbal medicine in Lingnan [1]. The main medicine produced from *P. cablin* is patchouli alcohol, which has been reported to have antibacterial [2], anti-inflammatory [3] and vasodilatory [4] properties, among others. On the one hand, patchouli alcohol may be mainly used in the perfume and cosmetics industry, and on the other hand, it may be frequently used for medical treatment. Due to the large market demand [5], an increasing number of scholars have begun to study *P. cablin*, especially the molecular synthesis mechanism of patchouli alcohol. Many investigations have focused on pharmacological effects in *P. cablin*; however, current knowledge on patchouli alcohol biosynthetic pathways is limited. In cultivation and production, the quality of patchouli is often related to planting light conditions, ambient temperature and different abiotic and biotic stresses. Previous studies in our research group found that these factors affected the expression of the patchouli alcohol synthase gene and caused significant differences in patchouli alcohol accumulation [6]. Subsequently, various exogenous hormone treatment experiments showed that patchouli alcohol synthesis was specifically induced by Methyl jasmonate (MeJA). A large number of studies have shown that different environmental signals stimulate plant synthesis of jasmonic acid [7], which affects the synthesis and accumulation of important secondary metabolites and their chemical reactions through JA signaling. Therefore, we predict that the biosynthesis process of patchouli alcohol may be highly dependent on the JA signaling. The pattern of JA signaling regulating the synthesis of secondary metabolites in medicinal plants has progressed in *Catharanthus roseus*, the main component of which is vinblastine [8]. In addition, this signaling also exists in tobacco (nicotine) [9].

JA is one of the most important signaling molecules in plants, governing responses to abiotic and biotic stresses, as well as in secondary metabolite biosynthesis, signal transduction, stress response [10], growth and development [11]. The JASMONATE ZIM DOMAIN (JAZ) proteins are not only an important component of the JA signaling pathway but also a node that links different signaling pathways in plants [12]. The N-terminus of the JAZ contains a weakly conserved N-terminal (NT) domain. The ZIM domain contains a conserved TIF[F/Y]XG (TIFY) motif, and the Jas domain at the C-terminus is highly conserved and can interact with many proteins [13], such as various transcription factors (TFs). In addition, there are nuclear localization signals in the Jas domain, which causes JAZ proteins to have nuclear localization properties [14]. Thirteen JAZ proteins have been found in *Arabidopsis thaliana*, playing roles in growth, defense, and reproductive output [15], these proteins are regarded as repressors in the JA signaling pathway, and they interact with the MYC2 transcription factor and repress its function. MYC2 is the initial transcription factor of the JA response gene [16], but the JA induction of plants is controlled by CORONATINE INSENSITIVE 1 (COI1). JAZ family proteins have been identified as COI1 targets and repressors of MYC2 [17]. To date, JAZ proteins have been proven to repress the activity of TFs, and function as repressors in the JA pathway to mediate many developmental processes, including secondary metabolite synthesis.

As a pivotal regulator in the plant JA signaling pathway, JAZ proteins should play important regulatory roles in the biosynthesis of patchouli alcohol in *P. cablin*. However, little is known about the unambiguous roles of JAZ proteins in patchouli alcohol biosynthesis in *P. cablin*. In our previous work, we obtained 82,335 raw data of transcriptome using next-generation sequencing (NGS) technology from leaves of *P. cablin* treated with MeJA [18], 12 unigenes were recognized and identified as JAZ family genes, based on the expression levels of 12 genes under the induction of 300 μM MeJA and our previous screening experiment, we ultimately chose *PatJAZ6* (Unigene48011) as the research object, which may participate in the regulation of biosynthesis of patchouli alcohol.

In this study, the *PatJAZ6* gene from *P. cablin* was cloned and characterized, and the relative expression pattern analysis was performed. Then, the subcellular localization study of PatJAZ6 was performed. To further research on the interaction between PatJAZ6 and PatMYC2b1/PatMYC2b2, which was cloned and identified by our laboratory [18], we performed a Yeast two hybrid (Y2H) assay and Firefly Luciferase Complementation Imaging Assay (LCI) verification. Furthermore, to illuminate the roles of PatJAZ6 in patchouli alcohol biosynthesis, gene silencing induced by viruses and overexpression of *PatJAZ6* in plants was examined and analyzed. We ultimately elucidated a molecular conduction model between PatJAZ6 and PatMYC2b1/PatMYC2b2 in patchouli alcohol biosynthesis. The present study is the first to analyze the function of the *PatJAZ6* gene in *P. cablin*, which may be instructive for the study of JA signaling molecular mechanisms and secondary metabolites in *P. cablin*.

2. Results

2.1. Bioinformatics Analysis of PatJAZ6 from P. cablin

To determine the basic bioinformatics of *PatJAZ6*, several software and websites were used in this work. Analysis of the sequence revealed that *PatJAZ6* consists of 1375 bp with a 972-bp open reading frame (ORF). Amino acid alignment confirmed that PatJAZ6 is a member of the *JAZ* family. JAZ proteins contain three conserved domains, NT, ZIM, and Jas (Figure 1A).Multiple alignments of PatJAZ6 with 12 *JAZ* proteins from *Arabidopsis thaliana* showed that PatJAZ6 contained two conserved domains: the ZIM domain (TIFY motif), which is located near the *N*-terminus, and the Jas domain, which is near the *C*-terminal (Figure 1B). To further identify the characteristics of PatJAZ6, a phylogenetic tree was constructed using MEGA 7 software. The result illustrated that PatJAZ6 was highly homologous to AtJAZ9 (Figure 1C).

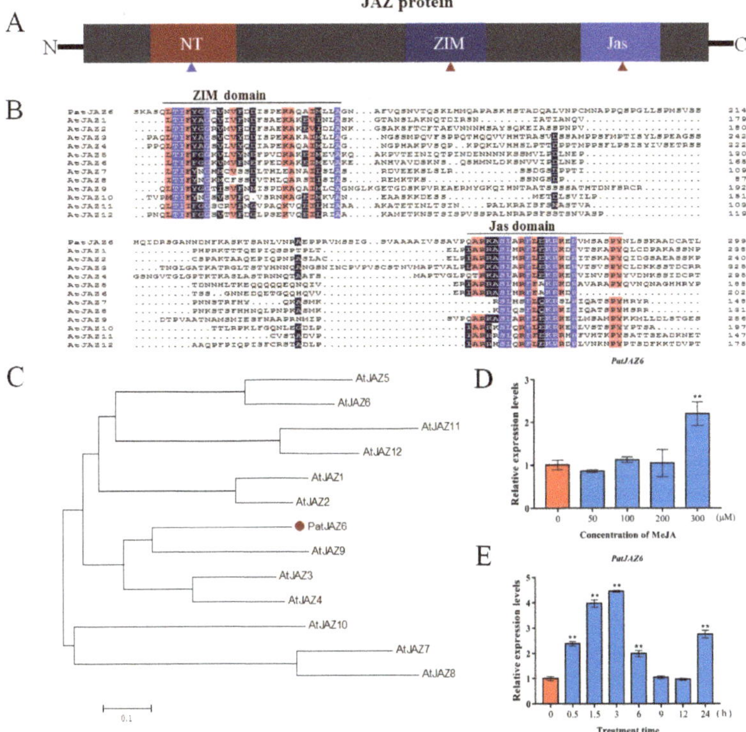

Figure 1. Bioinformatics analysis and expression profiles of *PatJAZ6*. (**A**) A schematic diagram of JAZ (Jasmonate ZIM domain, JAZ) protein. Red triangles represent conserved domains, blue triangle represents a weakly conserved NT domain. (**B**) Multiple alignments of PatJAZ6 with 12 JAZ proteins from *Arabidopsis thaliana*. AtJAZ1 (NP_564075.1), AtJAZ2 (NP_565096.1), AtJAZ3 (NP_974330.1), AtJAZ4 (NP_175283.2), AtJAZ5 (NP_001320905.1), AtJAZ6 (NP_001321693.1), AtJAZ7(NP_181007.1), AtJAZ8(NP_564349.1), AtJAZ9(AAL32593.1), AtJAZ10(NP_568287.1), AtJAZ11(NP_001190007.1), AtJAZ12 (NP_197590.1). (**C**) Phylogenic tree of PatJAZ6 with 12 JAZ proteins from *A. thaliana* was built using MEGA 7. The red solid dot represents PatJAZ6. (**D**) The relative expression of *PatJAZ6* was calculated after treatment with different concentrations of MeJA for 8 h, and the results were calculated according to the expression of *PatJAZ6* at 0 μM MeJA. (**E**) The relative expression of *PatJAZ6* was analyzed at different time points under 300 μM MeJA treatment. The results were calculated based on the expression of *PatJAZ6* at 0 h. (One-way ANOVA test; ** $p < 0.01$).

2.2. Expression Profiles of PatJAZ6 under MeJA Treatments

MeJA plays an important role in regulating secondary metabolite synthesis in a variety of plants. To detect whether *PatJAZ6* responds to MeJA, the relative expression of *PatJAZ6* was analyzed by qRT-PCR after treatment with different concentrations of exogenous MeJA for 8 h (Figure 1D). This analysis showed that 300 μM MeJA effectively induced the expression of *PatJAZ6* in *P. cablin*; therefore, 300 μM MeJA was selected as the lowest effective concentration to detect the expression of *PatJAZ6*. Different time points (0, 0.5, 1.5, 3, 6, 9, 12, and 24 h) were set after MeJA treatment to detect *PatJAZ6* expression. The results showed that MeJA effectively induced the expression of *PatJAZ6* in *P. cablin* (Figure 1E). Within 3 h of MeJA treatment, the expression levels of *PatJAZ6* increased rapidly and reached a maximum at 3 h, subsequently decreased at 6 h, and returned to the initial levels at 9 h and 12 h. After 24 h, the expression of *PatJAZ6* began to increase again, which is possible that expression of *PatJAZ6* is driven by circadian rhythms.

2.3. Subcellular Localization of PatJAZ6

To determine the subcellular localization of PatJAZ6, the ORF of *PatJAZ6* without a termination codon was inserted into the *N*-terminus of the Green Fluorescent Protein (GFP) tag in vector PAN580. The recombinant plasmid was transformed into *A. thaliana* protoplasts by the polyethylene glycol (PEG-mediated method [19]. Subcellular localization results showed that PatJAZ6 was localized in the nucleus (Figure 2). We can infer that PatJAZ6 is highly likely to play functional roles in the nucleus, such as regulating transcription factors in the JA signaling pathway.

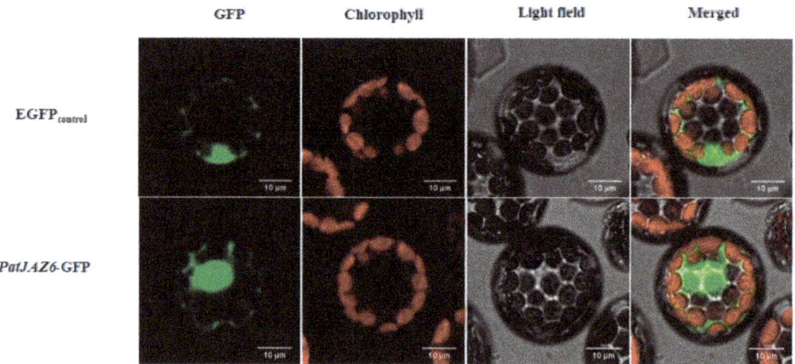

Figure 2. Subcellular localization of PatJAZ6 in *Arabidopsis* protoplasts. The open reading frame (ORF) without a termination codon was inserted into the vector named PAN580, Enhanced Green Fluorescent Protein (EGFP) was used as a control.

2.4. PatJAZ6 Protein Interacts with PatMYC2b1 and PatMYC2b2

Based on the reported research that JAZ proteins interact with some TFs [20], such as MYC2, which plays a regulated role in the JA signaling pathway, the ORFs of *PatJAZ6* and TFs (*PatMYC2b1/PatMYC2b2*) used in this study were cloned by the gene primers (Table S1) and fused to digested vectors PGBKT7 and pGADT7, respectively, in our research group. To explore whether PatJAZ6 can interact with PatMYC2b1 and PatMYC2b2, a Y2H screen was chosen to confirm the possible interaction relationship between them. Our screen results showed that pGBKT7-*PatJAZ6* + pGADT7-*PatMYC2b1* and pGBKT7-*PatJAZ6*+pGADT7-*PatMYC2b2* could grow normally on three types of screening plates and turned blue on SD/-Trp/-Leu/-His/-Ade/ plates containing X-α-Gal, which is consistent with the positive control, indicating that there is an interaction between PatJAZ6 and PatMYC2b1 /PatMYC2b2 in yeast systems, which implied that a relationship between PatJAZ6 and PatMYC2b1 /PatMYC2b2 may have existed in plants to regulate the biosynthesis of secondary metabolites (Figure 3A).

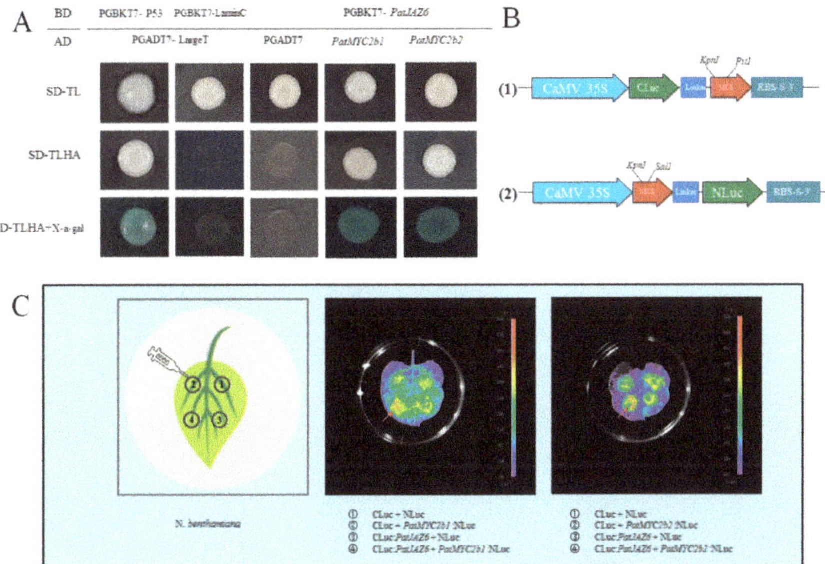

Figure 3. Protein interaction verification of PatJAZ6. (**A**) Yeast two-hybrid assay. Plasmids PGADT7-LargeT and pGBKT7-P53 were cotransformed into AH109 as the positive control. Plasmids PGADT7-LargeT and PGBKT7-LaminC were cotransformed into AH109 as the negative control. PGBKT7-*PatJAZ6* and pGADT7-TFs were cotransformed into *Saccharomyces cerevisiae* AH109 competent cells. The blue colonies represent the positive results. (**B**) (1) The complete ORF of *PatJAZ6* was inserted into the vector named PCAMBIA1300CLuc with the restriction enzyme sites *KpnI* and *PstI*. (2) The complete ORFs of *PatMYC2b1* and *PatMYC2b2* were inserted into the vector named PCAMBIA1300NLuc with the restriction enzyme sites *KpnI* and *SalI*, respectively. (**C**) Firefly Luciferase Complementation Imaging Assay. LCI images of *N. benthamiana* leaves coinfiltrated with the *Agrobacterial* GV3101-Psoup-p19 strains containing *PatMYC2b1/PatMYC2b2*:NLuc and CLuc:*PatJAZ6*. Arrow positions indicate where the signal is strongest.

To verify whether this interaction exists in plants, further confirming the reliability of the Y2H results, a LCI was performed in *Nicotiana benthamiana* leaves by injecting *A. tumefaciens* GV3101-Psoup-p19 cultures containing recombinant constructs (Figure 3C). Injection positions ①, ② and ③, which represent different plasmid combinations, were set as negative controls. Injection positions ④ in LCI images showed large red areas, whereas ①, ② and ③ had almost no red areas, indicating that the signal at position ④ was significantly stronger than the control, revealing that PatJAZ6 and PatMYC2b1 /PatMYC2b2 have a strong interaction in *N. benthamiana*, implying a relationship between them most likely existed in *P. cablin*.

2.5. *Effect on Patchouli Alcohol Biosynthesis by the Virus Induced PatJAZ6 Silencing*

To investigate the roles of the PatJAZ6 protein in the JA signaling pathway affecting the synthesis of patchouli alcohol, Virus-induced gene silencing (VIGS) was selected to silence *PatJAZ6* to explore the effects of *PatJAZ6* on related genes and patchouli alcohol. The buffer containing a 1:1 ratio of PTRV1 and PTRV2 was set as a control. Leaf tissues were collected after *PatJAZ6* was silenced 14 days, which was used for qRT-PCR and Gas Chromatography-Mass Spectrometer (GC-MS) analysis. In the virus-induced *PatJAZ6* silencing, the relative expression level of *PatJAZ6* was clearly downregulated in comparison to the control (Figure 4B), while the expression of patchoulol synthase (*PTS*), which is the key enzyme for patchouli synthesis, was upregulated by approximately 80%. Moreover, the relative expression of *PatMYC2b1* and *PatMYC2b2* interacting with *PatJAZ6* were all increased, especially *PatMYC2b2*. The results of GC-MS showed that the content of patchouli alcohol in the VIGS-JAZ6 group (3.67 mg/g Fresh weight (FW)) was significantly higher than that in CK (2.5 mg/g FW), exhibiting an increase of 32% (Figure 4C). GC-MS chromatograms of samples from the standard of patchouli alcohol (top panel), CK (middle panel) and VIGS-JAZ6 (bottom panel) leaves showing abundance of patchouli alcohol (Figure 4D). The expression tendency of *PatJAZ6* was contrary to *PatMYC2b1* and *PatMYC2b2*, and based on the results of Y2H and LCI, we can speculate that PatJAZ6 plays a role as a transcriptional repressor in *P. cablin*. In addition, when *PatJAZ6* was silenced, patchouli alcohol synthesis was increased, which may be the result of a common increase in TFs and *PTS* gene expression.

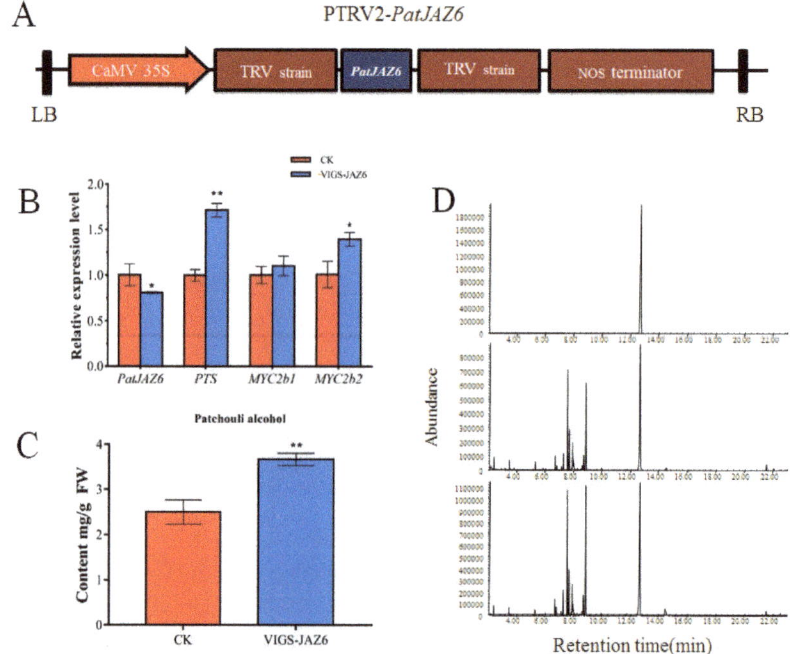

Figure 4. Analysis of virus-induced *PatJAZ6* silencing. (**A**) The *PatJAZ6* gene fragment (less than 500 bp) was cloned into the PTRV2 vector to form the PTRV2-*PatJAZ6*. (**B**) The corresponding mRNA expression level of VIGS-JAZ6 analyzed by real-time q-PCR. (**C**) The content of patchouli alcohol detected in control check (CK) and VIGS-JAZ6 leaves. (**D**) Gas Chromatography-Mass Spectrometer (GC-MS) chromatograms of samples from the standard of patchouli alcohol (top panel), CK (middle panel) and VIGS-JAZ6 (bottom panel) leaves showing abundance of patchouli alcohol. Asterisks indicate a significant difference from the control. (Student's *t*-test; ** $p < 0.01$, * $p < 0.05$).

2.6. Effect on Patchouli Alcohol Accumulation by the Overexpression of PatJAZ6

To confirm that PatJAZ6 plays a role as a transcriptional repressor in *P. cablin*, based on the VIGS-PatJAZ6 experiment, we hypothesized that overexpression of *PatJAZ6* may show opposite results to gene silencing of *PatJAZ6*. The empty PJLTRBO vector was transformed into GV3101-Psoup-p19 as CK. In the *PatJAZ6*-overexpressing *P. cablin* leaves, *PatJAZ6* expression was upregulated in comparison to the CK, while the transcripts of *PTS*, *PatMYC2b1* and *PatMYC2b2* were all reduced at different levels. Among these genes, *PatMYC2b2* showed the most significant decrease, exhibiting a nearly 70% reduction (Figure 5B). The content of patchouli alcohol in PJLTRBO-PatJAZ6 was determined by GC-MS, consistent with the gene expression profile of *PTS*, *PatMYC2b1* and *PatMYC2b2*. The accumulation of patchouli alcohol in *PatJAZ6* overexpressing appeared to decrease. Overexpression of *PatJAZ6* produced lower levels of patchouli alcohol (5.04 mg/g FW) compared with the control (6.76 mg/g FW) (Figure 5C). The results described above confirmed our conjecture, indicating that PatJAZ6 may repress the biosynthesis of patchouli alcohol.

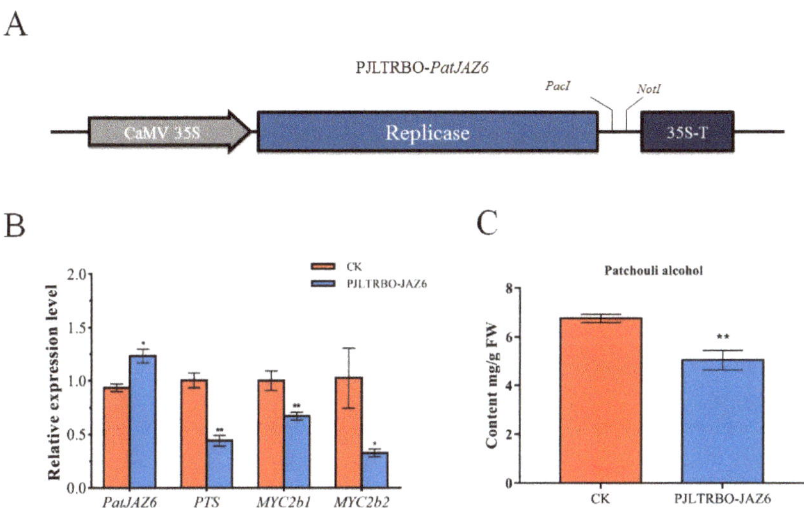

Figure 5. Overexpression analysis of *PatJAZ6*. (**A**) The *PatJAZ6* gene fragment was cloned into the PJLTRBO vector to form the PJLTRBO-*PatJAZ6* construct with the restriction enzyme sites *PacI* and *NotI*. (**B**) The corresponding mRNA expression level of PJLTRBO-*PatJAZ6* analyzed by real-time q-PCR. (**C**) The content of patchouli alcohol detected in CK and PJLTRBO-*PatJAZ6* leaves. Asterisks indicate a significant difference from the control (Student's *t*-test; ** $p < 0.01$, * $p < 0.05$).

3. Discussion

Patchouli alcohol, a natural tricyclic sesquiterpene compound that is a bioactive ingredient in *P. cablin* [21], is used worldwide in decorative cosmetics, toilet soaps, perfume industries [22] and medical treatments [23]. With the increasing market demand in the world, a metabolic engineering approach has been considered an effective approach to increase useful metabolites in medical plants, and several strategies have been reported to promote the industrialization process of patchouli alcohol production using *Saccharomyces cerevisiae* [24], and gene isolation and cloning in MVA and MEP pathways participating in patchouli alcohol biosynthesis have been reported [25]. In addition, full-length transcriptome data reported provide a valuable genetic resource in *P. cablin* [18]. Despite in-depth research on genes related to patchouli alcohol synthesis, little is known concerning the regulation of patchouli alcohol biosynthesis.

MeJA is an important plant endogenous hormone that is widely present in plants and regulates metabolic and developmental processes in plants [26]. Exogenous application of MeJA can stimulate the expression of defense genes and induce chemical defense in plants, including stimulating the synthesis of a series of secondary metabolites [27]. A growing number of reports indicate that the synthesis of many secondary metabolites in medicinal plants is increased under MeJA induction [28]. Our previous experiments showed that MeJA treatment on *P. cablin* leaves can significantly increase the accumulation of patchouli alcohol (Figure S1), but the specific molecular mechanisms involved have not been elucidated. We hypothesize that this effect may be related to JA signaling in plants and that the key factors of JA signaling are JAZ proteins. In our current research, the *PatJAZ6* gene was cloned from *P. cablin* and functionally identified as a repressor involved in patchouli alcohol biosynthesis. Bioinformatics analysis revealed that the *PatJAZ6* gene showed high homology with 12 JAZs from *A.*

thaliana Arabidopsis and contained highly conserved ZIM and Jas domains, indicating that PatJAZ6 may have similar effects to previously reported JAZ proteins. The expression profiles under MeJA revealed that 300 μM MeJA was the lowest effective concentration to detect the expression of *PatJAZ6*, this concentration is higher in comparison with other JAZs, such as *NtJAZ* in tobacco [29] and *SmJAZ* in *Salvia miltiorrhiza* [30], that response to 100 μM MeJA. Subcellular localization results showed that PatJAZ6 was localized in the nucleus (Figure 2). This result is consistent with the characteristics of the Jas motif with nuclear localization. Previous studies on the subcellular localization of other JAZs also support this result [31].

JAZ proteins belong to ZIM-domain proteins, are located near the *C*-terminus, and have a highly conserved Jas motif of 26 amino acids. Studies have now determined that Jas motifs are involved in protein-protein interactions with MYC and COI1 [32]. In our present study, it was found that the PatJAZ6 protein can interact with PatMYC2b1 and PatMYC2b2 through a yeast two-hybrid (Y2H) screening method, which was consistent with previous reports that JAZ proteins interact with the MYC2 transcription factor [33]. Furthermore, an LCI assay was performed in *N. benthamiana* leaves, which further confirmed the interaction between PatJAZ6 and PatMYC2b1/PatMYC2b2 separately. The above data suggest that PatMYC2b1 and PatMYC2b2 transcription factors in *P. cablin* may be targets of the PatJAZ6 protein in *P. cablin* and play key regulatory roles in the accumulation of patchouli alcohol. Of course, there are other transcription factors involved in this synthesis process, which warrants further investigation.

VIGS is widely used to downregulate target genes in a majority of plants [34]. Tobacco rattle virus (TRV) is one of the most widely used viruses in VIGS technology [35]. The VIGS system constructed by this virus is rapidly applied to the model plant *N. benthamiana* [36] and such crops as pepper [37], cotton [38], and tomato [39]. For the silencing effect, some studies have shown that the silencing effect of the target gene fragment between 300 and 500 bp is the best [40]. Not long ago, the efficient VIGS system in *P. cablin* was established by our own laboratory (Figure S2A). According to other research reports, gene silencing in wild tobacco revealed that *NaJAZi* functions as a flower-specific jasmonate repressor that regulates JAs, TPIs, (E)-α-bergamotene and a defensin. Flowers silenced in *NaJAZi* are more resistant to tobacco budworm attack [41]. In addition, there are reports that knockdown of *AsJAZ1* expression through RNA interference led to decreased number of nodules, abnormal development of bacteroids, accumulation of poly-x-hydroxybutyrate (PHB) and loss of nitrogenase activity in legumes–rhizobia symbiosis [42]. However, in our experiments, gene silencing of *PatJAZ6* in *P. cablin* leaves did not exhibit a distinct phenotype, except for slight curling of the leaves, but an increase in the expression of *PTS*, *PatMYC2b1* and *PatMYC2b2* was observed, resulting in a significant increase in the accumulation of patchouli alcohol in the VIGS-JAZ6 group (3.67 mg/g FW) compared with the control in CK (2.5 mg/g FW).

Research on JAZs in cash crops, including *Oryza sativa* [43], *Glycine soja* [44] and *Gossypium hirsutum* [45], has progressed rapidly in the past several years. Overexpression of JAZs in these crops produces different phenotypes. For example, overexpression of *GsJAZ2* in soybean can increase the sensitivity of plants to salinity; overexpression of *GhJAZ2* in cotton impairs the sensitivity to JA, decreases the expression level of JA-response genes (*GhPDF1.2* and *GhVSP*) and enhances the susceptibility to *V. dahliae* and insect herbivory. However, we did not observe a significant phenotype in the *P. cablin* plants that overexpressed *PatJAZ6*, but we observed a decrease in the expression of *PTS*, *PatMYC2b1* and *PatMYC2b2*, resulting in a lower level of patchouli alcohol (5.04 mg/g FW) in PJLTRBO-JAZ6 compared with the control (6.76 mg/g FW). JAZ proteins have different functions, which may be due to the different roles of transcription factors interacting with JAZ proteins. Since it has been reported that JAZ protein may be involved in the development of glandular trichomes which, in turn, affects the synthesis of secondary metabolites in *Artemisia annua* [46]; therefore, whether the gene silencing or overexpression of *PatJAZ6* also affects the development of glandular trichomes in *P. cablin* requires further experimental confirmation.

In *S. miltiorrhiza* hairy roots, SmJAZ8 acts as a core repressor regulating JA-induced biosynthesis of salvianolic acids and tanshinones [30]. According to the results of our study, it is reasonable to speculate that PatJAZ6 may act as a repressor in patchouli alcohol biosynthesis. Based on the above experimental results, we first present a model for the JA signaling pathway in *P. cablin* (Figure 6).

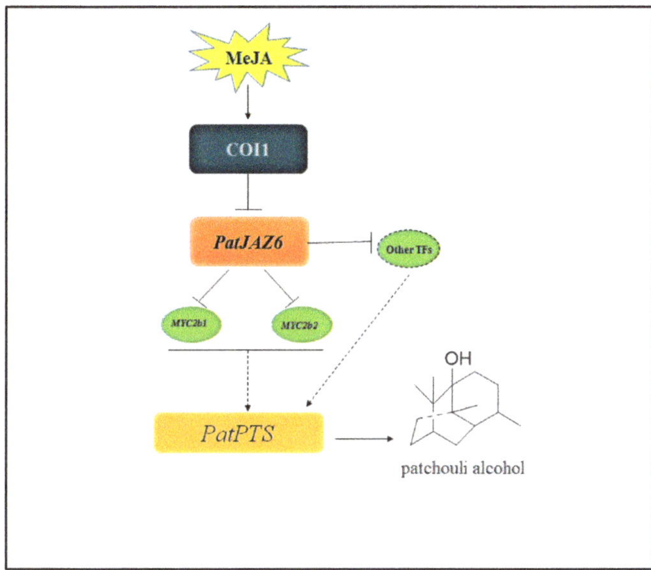

Figure 6. Model describing the function of *PatJAZ6* in JA-induced biosynthesis of patchouli alcohol. *PatJAZ6* acts as a repressor regulating JA-induced biosynthesis of patchouli alcohol in *Pogostemon cablin*.

The model clearly illustrates the connection of how *PatJAZ6* acts as a transcription factor suppressor regulating JA-induced biosynthesis of patchouli alcohol in *P. cablin*. However, further research is needed to determine whether PatJAZ6 interacts with other transcription factors and whether other JAZ proteins are involved in this signaling pathway. We also need to further explore the COI-JAZ-TFs model and how these proteins interact to regulate the synthesis of patchouli alcohol in *P. cablin*. Taken together, the results of this study help to elucidate the molecular regulation of JA signal-induced patchouli alcohol biosynthesis. Our work indicated that PatJAZ6 acts as a repressor in the regulation of patchouli alcohol biosynthesis. The discovery of the PatJAZ6 function points out a direction for the JA signaling pathway molecular mechanism and patchouli alcohol production in *P. cablin*.

4. Materials and Methods

4.1. Experimental Materials and Total RNA Extraction

The *P. cablin* plants were gathered from Yangjiang city, Guangdong Province, China. The cutting propagation method was used to obtain more seedlings that were used for the analysis of the expression patterns of *PatJAZ6* and content of patchouli alcohol in leaves. The seeds of *N. benthamiana* were kept in our laboratory and grown in flower pots in a growth chamber. Well-growing plant materials, which were cultured under a constant environment at 25 °C with a 16/8 h photoperiod treatment, were selected for our experiments. The vector plasmids, *E. coli* competent cells DH5α and *A. tumefaciens* competent cells used in this study were all kept in our own laboratory. Total RNA was extracted from *P. cablin* leaves with the GeneMark Plant Total RNA Purification Kit (GeneMarkBio, Taichung, Taiwan), and then a spectrophotometer (IMPLEN GNBH, Germany) was used to determine the RNA concentration and purity. cDNA synthesis was performed via oligo dT and stored at −20 °C for subsequent experiments.

4.2. MeJA Treatments

MeJA was purchased from Sigma-Aldrich, St. Louis, Missouri, The United States of America, USA, dissolved in ethanol, formulated into 50 mM mother liquor for later use. To screen for the best response concentration, *P. cablin* plants were sprayed with MeJA solution at 0, 50, 100, 200 and 300 μM concentrations containing 0.1% Tween-80, and leaf samples were collected at 8 h after MeJA treatments.

To investigate the effect of MeJA on *PatJAZ6* expression, *P. cablin* plants were sprayed in the morning with MeJA solution at a 300 μM concentration. Leaf samples were collected at time intervals of 0, 0.5, 1.5, 3, 6, 9, 12, and 24 h after MeJA treatment. All leaf samples were frozen with liquid nitrogen and stored in a −80 °C refrigerator for subsequent RNA extraction.

4.3. Bioinformatics Analysis of PatJAZ6

Bioinformatics analysis of *PatJAZ6* was performed using several bioinformatics software and websites. The ORF of *PatJAZ6* was determined using ORF finder (http://www.bioinformatics.org/sms2/orf_find.html), and 12 JAZ proteins from *A. thaliana* were searched from NCBI (https://www.ncbi.nlm.nih.gov/). MEGA v.7 software was used to construct the phylogenetic tree, and DNAMAN software was used to perform multiple sequence alignment.

4.4. Expression Patterns of PatJAZ6 by qRT-PCR

The expression patterns of *PatJAZ6* and related genes under different treatments were quantified by qRT-PCR. Plant tissues were gathered after various treatments, and total RNA was isolated. cDNA synthesis was performed using HiScript II QRT SuperMix for qPCR (Vazyme R222-01, Nanjing, China), and the CFX96TM Real-Time System was selected to carry out qRT-PCR analysis with ChamQ Universal SYBR qPCR Master Mix (Vazyme, Q711-02/03). qRT-PCR conditions were as follows: 95 °C for 3 min for one cycle, followed by 40 cycles of 95 °C for 10 s and 60 °C for 30 s. The relative expression levels of *PatJAZ6* and related genes were calculated based on the $2^{-\Delta\Delta Ct}$ method.

4.5. Subcellular Localization of PatJAZ6

The ORF of *PatJAZ6* without a termination codon was fused to the *N*-terminus of the vector pAN580-GFP tag with the restriction enzyme sites *SpeI* and *BamHI*. The recombinant plasmids PAN580-*PatJAZ6* were transformed into *Arabidopsis* protoplasts. The empty vector pAN580 was used as a negative control. The result of subcellular localization of PatJAZ6 was observed with ZEISS LSM 800 with Airyscan (ZEISS, Jena city, Germany).

4.6. Yeast Two-Hybrid Assays

The yeast two-hybrid (Y2H) screen was chosen to confirm possible TFs interacting with PatJAZ6. The ORF of *PatJAZ6* was cloned into the pGBKT7 vector to generate pGBKT7-*PatJAZ6*. Both the ORFs of *PatMYC2b1* and *PatMYC2b2* were inserted into the PGADT7 vector to form pGADT7-*PatMYC2b1* and pGADT7-*PatMYC2b2*, respectively. Plasmids PGADT7-LargeT and pGBKT7-P53 were cotransformed into *S. cerevisiae* AH109 competent cells as the positive control, while PGADT7-LargeT and PGBKT7-LaminC were cotransformed into AH109 as the negative control. Recombinant plasmids (pGBKT7-*PatJAZ6*+pGADT7-*PatMYC2b1*, pGBKT7-*PatJAZ6*+ pGADT7-*PatMYC2b2* were cotransformed separately into AH109 cells and cultured on SD/-Trp/-Leu medium. Then, transformants were vaccinated on SD/-Trp/-Leu/-His/-Ade and SD/-Trp/-Leu/-His/-Ade/X-α-Gal to observe the interaction situation of *PatJAZ6* with *PatMYC2b1* and *PatMYC2b2*. All medium plates were incubated in an incubator at 29 °C for 3 days.

4.7. Firefly Luciferase Complementation Imaging Assay

To further understand the interaction between PatJAZ6 and PatMYC2b1/PatMYC2b2 in living plants, Firefly Luciferase Complementation Imaging Assay (LCI) was performed in *N. benthamiana* leaves. The complete ORF of *PatJAZ6* was inserted into the vector PCAMBIA1300CLuc with the restriction enzyme sites *KpnI* and *PstI*, while ORFs of *PatMYC2b1* and *PatMYC2b2* were inserted into the vector PCAMBIA1300NLuc with the restriction enzyme sites *KpnI* and *SalI*, respectively (Figure 3B). The *A. tumefaciens* strain GV3101-Psoup-p19 was transformed with recombinant constructs by the freeze–thaw method. The cultured bacteria solution was mixed at a ratio of 1:1, centrifuged and resuspended in buffer, adjusted optical density (OD) to 0.8–1.0, then placed at room temperature for 2–4 h, injected the back of *N. benthamiana* leaves with a needle-free syringe (Figure 3C), incubated for 12 h at 23 °C in the dark, and moisturized for 2–4 days at room temperature. The leaves were placed in MS solid medium with the leaves facing up, sprayed with 100 mM D-luciferin potassium salt, and kept in the dark for 6 min. A Berthold Technologies (LB983 NC100, Germany) was used to capture the images with an exposure time of 2 min.

4.8. Virus-Induced PatJAZ6 Silencing

The pTRV1 and pTRV2 vectors were kept in our laboratory and used in this study. A 418-bp fragment (Figure S2B) from the *PatJAZ6* ORF was cloned into the *EcoRI* and *BamHI* sites of the pTRV2 vector (Figure 4A). The resulting pTRV2-*PatJAZ6* constructs, PTRV1 and PTRV2, were transformed into *A. tumefaciens* GV3101. The mixture of *A. tumefaciens* cultures containing a 1:1 ratio of PTRV1 and PTRV2 or pTRV2-*PatJAZ6* was harvested by centrifugation and resuspended in infiltration buffer to obtain an OD of 1.0 and then incubated at room temperature for 2–4 h. Six-leaf-staged *P. cablin* plants were selected to infect with 1 mL needleless syringe on the abaxial side of leaves, and two to three leaves per plant needed infiltration. When *PatJAZ6* was silenced for 14 days, leaf tissues were collected and frozen for later use.

4.9. Overexpression Analysis

The complete ORF fragment of *PatJAZ6* was cloned into the PJLTRBO vector to form the PJLTRBO-*PatJAZ6* construct with the restriction enzyme sites *PacI* and *NotI* (Figure 5A). The recombinant plasmids PJLTRBO-*PatJAZ6* were transformed into GV3101-Psoup-p19, and the empty PJLTRBO vector was transformed into the same strain as the control. *A. tumefaciens* cultures were harvested and resuspended in infiltration buffer. For *P. cablin* leaf infiltration, the same injection method described in 4.8 was adopted. Samples were collected 4 days after plants were injected and frozen for later use.

4.10. Patchouli Alcohol Extraction and GC-MS Analysis

200 mg leaf tissues were ground frizzed in liquid nitrogen, 1.5 mL hexane was added into centrifuge tubes, ultrasonic for 30 min with 60 Hz and then heated under a 56 °C water bath for 1 h. After centrifugation, the supernatant was taken and passed through a 0.22-μm organic membrane as the test solution for GC-MS analysis using Agilent 7890B Gas Chromatograph with 5977A inert Mass Selective Detector (Agilent, California, USA). The gas chromatograph was equipped with an HP-5MS capillary column (30 m × 250 mm × 0.25 mm). The instrument was set to an initial temperature of 50 °C and maintained for 0 min. Then, the oven temperature was increased to 130 °C at a rate of 20 °C/min and then increased to 150 °C at a rate of 2 °C/min. The temperature was maintained at 150 °C for 5 min and later increased to 230 °C at a rate of 20 °C/min. The injection volume was 1 μL, and the injection port temperature was 230 °C. In addition, patchouli alcohol standards were purchased from NanTong FeiYu, China. All reagents used were analytical grade.

4.11. Agrobacterium Culture and Buffer Formulation

A. tumefaciens cultures were grown in the shaker with shaking (200 rpm/min) at 28 °C for 20–24 h. The infiltration buffer formulation as follows: 1 M 2-(4-morpholino)-ethane sulfonic acid; 1 M $MgCl_2$ and 200 mM acetosyringone, dissolved in dimethyl sulfoxide.

4.12. Statistical Analysis

Statistical significance was determined by student's *t*-test and one-way ANOVA.

Supplementary Materials: Supplementary materials can be found at http://www.mdpi.com/1422-0067/20/23/6038/s1. Figure S1. MeJA treatment on *P. cablin* leaves can significantly increase the accumulation of patchouli alcohol. Figure S2. (A) Virus induced *PatPDS* silencing in *P. cablin*. (B) Electropherogram of cloning a 418 bp fragment of *PatJAZ6* into pTRV2 vector. Table S1: List of primers used in this study.

Author Contributions: L.C. and X.W. designed the research. Y.T. and X.C. performed the experiments and collected the data. X.Z., L.Z., Y.L., X.W. and J.L. analyzed the data and wrote the manuscript. R.Z., H.Z. and L.C. edited the manuscript and provided guidance during experimentation.

Funding: This research was supported by grants from the National Natural Science Foundation of China (81803657).

Acknowledgments: We appreciate Rui He (Guangzhou University of Chinese Medicine) for helping improve our manuscript writing.

Conflicts of Interest: The authors declare no conflict of interest.

References

1. Chen, M.; Zhang, J.; Lai, Y.; Wang, S.; Li, P.; Xiao, J.; Fu, C.; Hu, H.; Wang, Y. Analysis of Pogostemon cablin from pharmaceutical research to market performances. *Expert Opin. Investig. Drugs* **2013**, *22*, 245–257. [CrossRef] [PubMed]

2. Xu, Y.F.; Lian, D.W.; Chen, Y.Q.; Cai, Y.F.; Zheng, Y.F.; Fan, P.L.; Ren, W.K.; Fu, L.J.; Li, Y.C.; Xie, J.H.; et al. In Vitro and In Vivo Antibacterial Activities of Patchouli Alcohol, a Naturally Occurring Tricyclic Sesquiterpene, against Helicobacter pylori Infection. *Antimicrob. Agents Chemother.* **2017**, *61*, e00122-17. [CrossRef] [PubMed]

3. Lian, D.W.; Xu, Y.F.; Ren, W.K.; Fu, L.J.; Chen, F.J.; Tang, L.Y.; Zhuang, H.L.; Cao, H.Y.; Huang, P. Unraveling the Novel Protective Effect of Patchouli Alcohol Against Helicobacter pylori-Induced Gastritis: Insights Into the Molecular Mechanism in vitro and in vivo. *Front. Pharm.* **2018**, *9*, 1347. [CrossRef] [PubMed]

4. Hu, G.Y.; Peng, C.; Xie, X.F.; Xiong, L.; Zhang, S.Y.; Cao, X.Y. Patchouli alcohol isolated from Pogostemon cablin mediates endothelium-independent vasorelaxation by blockade of Ca(2+) channels in rat isolated thoracic aorta. *J. Ethnopharmacol.* **2018**, *220*, 188–196. [CrossRef] [PubMed]

5. Swamy, M.K.; Sinniah, U.R. Patchouli (Pogostemon cablin Benth.): Botany, agrotechnology and biotechnological aspects. *Ind. Crop. Prod.* **2016**, *87*, 161–176. [CrossRef]

6. Li, J.; Chen, X.; Zhong, L.; Wang, X.; Zhou, X.; Tang, Y.; Liu, Y.; Zheng, H.; Zhan, R.; Chen, L. Comparative iTRAQ-based proteomic analysis provides insight into a complex regulatory network of Pogostemon cablin in response to exogenous MeJA and Ethrel. *Ind. Crop. Prod.* **2019**, *140*, 111661. [CrossRef]

7. Rahnamaie-Tajadod, R.; Goh, H.H.; Mohd Noor, N. Methyl jasmonate-induced compositional changes of volatile organic compounds in Polygonum minus leaves. *J. Plant Physiol.* **2019**, *240*, 152994. [CrossRef]

8. Van Moerkercke, A.; Steensma, P.; Gariboldi, I.; Espoz, J.; Purnama, P.C.; Schweizer, F.; Miettinen, K.; Vanden Bossche, R.; De Clercq, R.; Memelink, J.; et al. The basic helix-loop-helix transcription factor BIS2 is essential for monoterpenoid indole alkaloid production in the medicinal plant Catharanthus roseus. *Plant J.* **2016**, *88*, 3–12. [CrossRef]

9. Shoji, T.; Hashimoto, T. Tobacco MYC2 regulates jasmonate-inducible nicotine biosynthesis genes directly and by way of the NIC2-locus ERF genes. *Plant Cell Physiol.* **2011**, *52*, 1117–1130. [CrossRef]

10. Reinbothe, C.; Springer, A.; Samol, I.; Reinbothe, S. Plant oxylipins: Role of jasmonic acid during programmed cell death, defence and leaf senescence. *FEBS J.* **2009**, *276*, 4666–4681. [CrossRef]

11. Wasternack, C.; Hause, B. Jasmonates: Biosynthesis, perception, signal transduction and action in plant stress response, growth and development. An update to the 2007 review in Annals of Botany. *Ann. Bot.* **2013**, *111*, 1021–1058. [CrossRef] [PubMed]

12. Song, S.; Qi, T.; Huang, H.; Ren, Q.; Wu, D.; Chang, C.; Peng, W.; Liu, Y.; Peng, J.; Xie, D. The Jasmonate-ZIM domain proteins interact with the R2R3-MYB transcription factors MYB21 and MYB24 to affect Jasmonate-regulated stamen development in Arabidopsis. *Plant Cell* **2011**, *23*, 1000–1013. [CrossRef] [PubMed]

13. Pauwels, L.; Goossens, A. The JAZ Proteins: A Crucial Interface in the Jasmonate Signaling Cascade. *Plant Cell* **2011**, *23*, 3089–3100. [CrossRef] [PubMed]

14. Grunewald, W.; Vanholme, B.; Pauwels, L.; Plovie, E.; Inzé, D.; Gheysen, G.; Goossens, A. Expression of the Arabidopsis jasmonate signalling repressor JAZ1/TIFY10A is stimulated by auxin. *EMBO Rep.* **2009**, *10*, 923–928. [CrossRef] [PubMed]

15. Guo, Q.; Yoshida, Y.; Major, I.T.; Wang, K.; Sugimoto, K.; Kapali, G.; Havko, N.E.; Benning, C.; Howe, G.A. JAZ repressors of metabolic defense promote growth and reproductive fitness inArabidopsis. *Proc. Natl. Acad. Sci. USA* **2018**, *115*, E10768–E10777. [CrossRef]

16. Lorenzo, O.; Chico, J.M.; Sanchez-Serrano, J.J.; Solano, R. JASMONATE-INSENSITIVE1 encodes a MYC transcription factor essential to discriminate between different jasmonate-regulated defense responses in Arabidopsis. *Plant Cell* **2004**, *16*, 1938–1950. [CrossRef]

17. Chini, A.; Boter, M.; Solano, R. Plant oxylipins: COI1/JAZs/MYC2 as the core jasmonic acid-signalling module. *FEBS J.* **2009**, *276*, 4682–4692. [CrossRef]

18. Chen, X.; Li, J.; Wang, X.; Zhong, L.; Tang, Y.; Zhou, X.; Liu, Y.; Zhan, R.; Zheng, H.; Chen, W.; et al. Full-length transcriptome sequencing and methyl jasmonate-induced expression profile analysis of genes related to patchoulol biosynthesis and regulation in Pogostemon cablin. *BMC Plant Biol.* **2019**, *19*, 266. [CrossRef]

19. Cao, Y.; Li, H.; Pham, A.Q.; Stacey, G. An Improved Transient Expression System Using Arabidopsis Protoplasts. *Curr. Protoc. Plant Biol.* **2016**, *1*, 285–291. [CrossRef]

20. Fonseca, S.; Fernandez-Calvo, P.; Fernandez, G.M.; Diez-Diaz, M.; Gimenez-Ibanez, S.; Lopez-Vidriero, I.; Godoy, M.; Fernandez-Barbero, G.; Van Leene, J.; De Jaeger, G.; et al. bHLH003, bHLH013 and bHLH017 are new targets of JAZ repressors negatively regulating JA responses. *PLoS ONE* **2014**, *9*, e86182. [CrossRef]

21. Hu, G.; Peng, C.; Xie, X.; Zhang, S.; Cao, X. Availability, Pharmaceutics, Security, Pharmacokinetics, and Pharmacological Activities of Patchouli Alcohol. *Evid. Based Complement. Altern. Med.* **2017**, *2017*, 4850612. [CrossRef] [PubMed]

22. Bhatia, S.P.; Letizia, C.S.; Api, A.M. Fragrance material review on patchouli alcohol. *Food Chem. Toxicol.* **2008**, *46*, S255–S256. [CrossRef] [PubMed]

23. Swamy, M.K.; Sinniah, U.R. A Comprehensive Review on the Phytochemical Constituents and Pharmacological Activities of Pogostemon cablin Benth: An Aromatic Medicinal Plant of Industrial Importance. *Molecules* **2015**, *20*, 8521–8547. [CrossRef] [PubMed]

24. Ma, B.; Liu, M.; Li, Z.H.; Tao, X.; Wei, D.Z.; Wang, F.Q. Significantly Enhanced Production of Patchoulol in Metabolically Engineered Saccharomyces cerevisiae. *J. Agric. Food Chem.* **2019**, *67*, 8590–8598. [CrossRef]

25. Tang, Y.; Zhong, L.; Wang, X.; Zheng, H.; Chen, L. Molecular identification and expression of sesquiterpene pathway genes responsible for patchoulol biosynthesis and regulation in *Pogostemon cablin*. *Bot. Stud.* **2019**, *60*, 11. [CrossRef]

26. Cao, J.; Li, M.; Chen, J.; Liu, P.; Li, Z. Effects of MeJA on Arabidopsis metabolome under endogenous JA deficiency. *Sci. Rep.* **2016**, *6*, 37674. [CrossRef]

27. Gundlach, H.; Müller, M.J.; Kutchan, T.M.; Zenk, M.H. Jasmonic acid is a signal transducer in elicitor-induced plant cell cultures. *Proc. Natl. Acad. Sci. USA* **1992**, *89*, 2389–2393. [CrossRef]

28. Yu, H.; Guo, W.; Yang, D.; Hou, Z.; Liang, Z. Transcriptional Profiles of SmWRKY Family Genes and Their Putative Roles in the Biosynthesis of Tanshinone and Phenolic Acids in *Salvia miltiorrhiza*. *Int. J. Mol. Sci.* **2018**, *19*, 1593. [CrossRef]

29. Zhang, H.; Li, W.; Niu, D.; Wang, Z.; Yan, X.; Yang, X.; Yang, Y.; Cui, H. Tobacco transcription repressors NtJAZ: Potential involvement in abiotic stress response and glandular trichome induction. *Plant Physiol. Biochem.* **2019**, *141*, 388–397. [CrossRef]

30. Pei, T.; Ma, P.; Ding, K.; Liu, S.; Jia, Y.; Ru, M.; Dong, J.; Liang, Z. SmJAZ8 acts as a core repressor regulating JA-induced biosynthesis of salvianolic acids and tanshinones in Salvia miltiorrhiza hairy roots. *J. Exp. Bot.* **2018**, *69*, 1663–1678. [CrossRef]

31. Li, W.; Xia, X.C.; Han, L.H.; Ni, P.; Yan, J.Q.; Tao, M.; Huang, G.Q.; Li, X.B. Genome-wide identification and characterization of JAZ gene family in upland cotton (*Gossypium hirsutum*). *Sci. Rep.* **2017**, *7*, 2788. [CrossRef] [PubMed]

32. Zhang, F.; Yao, J.; Ke, J.; Zhang, L.; Lam, V.Q.; Xin, X.F.; Zhou, X.E.; Chen, J.; Brunzelle, J.; Griffin, P.R.; et al. Structural basis of JAZ repression of MYC transcription factors in jasmonate signalling. *Nature* **2015**, *525*, 269–273. [CrossRef] [PubMed]

33. Zhou, Y.; Sun, W.; Chen, J.; Tan, H.; Xiao, Y.; Li, Q.; Ji, Q.; Gao, S.; Chen, L.; Chen, S.; et al. SmMYC2a and SmMYC2b played similar but irreplaceable roles in regulating the biosynthesis of tanshinones and phenolic acids in Salvia miltiorrhiza. *Sci. Rep.* **2016**, *6*, 22852. [CrossRef] [PubMed]

34. Lange, M.; Yellina, A.L.; Orashakova, S.; Becker, A. Virus-induced gene silencing (VIGS) in plants: An overview of target species and the virus-derived vector systems. *Methods Mol. Biol.* **2013**, *975*, 1–14.

35. Ratcliff, F.; Martin-Hernandez, A.M.; Baulcombe, D.C. Tobacco rattle virus as a vector for analysis of gene function by silencing. *Plant J.* **2001**, *25*, 237–245. [CrossRef]

36. Senthil-Kumar, M.; Mysore, K.S. Tobacco rattle virus-based virus-induced gene silencing in Nicotiana benthamiana. *Nat. Protoc.* **2014**, *9*, 1549–1562. [CrossRef]

37. Li, C.; Hirano, H.; Kasajima, I.; Yamagishi, N.; Yoshikawa, N. Virus-induced gene silencing in chili pepper by apple latent spherical virus vector. *J. Virol. Methods* **2019**, *273*, 113711. [CrossRef]

38. Li, X.; Liu, N.; Sun, Y.; Wang, P.; Ge, X.; Pei, Y.; Liu, D.; Ma, X.; Li, F.; Hou, Y. The cotton GhWIN2 gene activates the cuticle biosynthesis pathway and influences the salicylic and jasmonic acid biosynthesis pathways. *BMC Plant Biol.* **2019**, *19*, 379. [CrossRef]

39. Liu, Y.L.; Schiff, M.; Dinesh-Kumar, S.P. Virus-induced gene silencing in tomato. *Plant J.* **2002**, *31*, 777–786. [CrossRef]

40. Burch-Smith, T.M.; Anderson, J.C.; Martin, G.B.; Dinesh-Kumar, S.P. Applications and advantages of virus-induced gene silencing for gene function studies in plants. *Plant J.* **2004**, *39*, 734–746. [CrossRef]

41. Li, R.; Wang, M.; Wang, Y.; Schuman, M.C.; Weinhold, A.; Schafer, M.; Jimenez-Aleman, G.H.; Barthel, A.; Baldwin, I.T. Flower-specific jasmonate signaling regulates constitutive floral defenses in wild tobacco. *Proc. Natl. Acad. Sci. USA* **2017**, *114*, E7205–E7214. [CrossRef] [PubMed]

42. Li, Y.; Xu, M.; Wang, N.; Li, Y. A JAZ Protein in Astragalus sinicus Interacts with a Leghemoglobin through the TIFY Domain and Is Involved in Nodule Development and Nitrogen Fixation. *PLoS ONE* **2015**, *10*, e0139964. [CrossRef] [PubMed]

43. Hori, Y.; Kurotani, K.; Toda, Y.; Hattori, T.; Takeda, S. Overexpression of the JAZ factors with mutated jas domains causes pleiotropic defects in rice spikelet development. *Plant Signal. Behav.* **2014**, *9*, e970414. [CrossRef] [PubMed]

44. Zhu, D.; Cai, H.; Luo, X.; Bai, X.; Deyholos, M.K.; Chen, Q.; Chen, C.; Ji, W.; Zhu, Y. Over-expression of a novel JAZ family gene from Glycine soja, increases salt and alkali stress tolerance. *Biochem. Biophys. Res. Commun.* **2012**, *426*, 273–279. [CrossRef] [PubMed]

45. He, X.; Zhu, L.; Wassan, G.M.; Wang, Y.; Miao, Y.; Shaban, M.; Hu, H.; Sun, H.; Zhang, X. GhJAZ2 attenuates cotton resistance to biotic stresses via the inhibition of the transcriptional activity of GhbHLH171. *Mol. Plant Pathol.* **2018**, *19*, 896–908. [CrossRef] [PubMed]

46. Yan, T.; Chen, M.; Shen, Q.; Li, L.; Fu, X.; Pan, Q.; Tang, Y.; Shi, P.; Lv, Z.; Jiang, W.; et al. HOMEODOMAIN PROTEIN 1 is required for jasmonate-mediated glandular trichome initiation in Artemisia annua. *New Phytol.* **2017**, *213*, 1145–1155. [CrossRef] [PubMed]

© 2019 by the authors. Licensee MDPI, Basel, Switzerland. This article is an open access article distributed under the terms and conditions of the Creative Commons Attribution (CC BY) license (http://creativecommons.org/licenses/by/4.0/).

International Journal of
Molecular Sciences

Article

Interruption of Jasmonic Acid Biosynthesis Causes Differential Responses in the Roots and Shoots of Maize Seedlings against Salt Stress

Ramala Masood Ahmad [1,2], Cheng Cheng [1,2], Jia Sheng [1,2], Wei Wang [3], Hong Ren [3], Muhammad Aslam [4] and Yuanxin Yan [1,2,*]

[1] State Key Laboratory of Crop Genetics and Germplasm Enhancement, Nanjing Agricultural University, Nanjing 210095, China; 2016201089@njau.edu.cn (R.M.A.); 2016201062@njau.edu.cn (C.C.); 2017101141@njau.edu.cn (J.S.)

[2] Jiangsu Collaborative Innovation Center for Modern Crop Production, Nanjing 210095, China

[3] Guizhou Institute of Upland Food Crops, Guizhou Academy of Agricultural Sciences, Guiyang 550006, China; ww1980666@126.com (W.W.); rhong666@163.com (H.R.)

[4] Department of Plant Breeding and Genetics, University of Agriculture Faisalabad, Faisalabad 38000, Pakistan; aslampbg@gmail.com

* Correspondence: yuanxin.yan@njau.edu.cn

Received: 6 November 2019; Accepted: 6 December 2019; Published: 9 December 2019

Abstract: Jasmonates (JAs) together with jasmonic acid and its offshoots are lipid-derived endogenous hormones that play key roles in both developmental processes and different defense responses in plants. JAs have been studied intensively in the past decades for their substantial roles in plant defense comebacks against diverse environmental stresses among model plants. However, the role of this phytohormone has been poorly investigated in the monocotyledonous species against abiotic stresses. In this study, a JA biosynthesis mutant *opr7opr8* was used for the investigation of JA roles in the salt stress responses of maize seedlings, whose roots were exposed to 0 to 300 mM NaCl. Foliar stomatal observation showed that *opr7opr8* had a larger stomatal aperture than wild type (WT) (B73) under salinity stress, indicating that JA positively regulates guard cell movement under salt stress. The results regarding chlorophyll content and leaf senescence showed that *opr7opr8* exhibited delayed leaf senescence under salt stress as compared to WT, indicating that JA plays a role in salt-inducing cell death and subsequent leaf senescence. Moreover, the morphological parameters, including the length of the shoots and roots, and the fresh and dry weights of the shoots and roots, showed that after 7 days of salt treatment, *opr7opr8* had heavier and longer shoots than WT but slighter and shorter roots than WT. In addition, ion analysis showed that *opr7opr8* accumulated less sodium but more potassium in the leaves than WT but more sodium and less potassium in the roots than WT, suggesting that JA deficiency causes higher salt stress to the roots but less stress to the leaves of the seedlings. Reactive oxygen species (ROS) analysis showed that *opr7opr8* produced less H_2O_2 than WT in the leaves but more H_2O_2 in the roots under salt treatment, and correspondingly, ROS-scavenging enzymes superoxide dismutase (SOD), catalase (CAT), and ascorbate peroxidase (APX) showed a similar variation, i.e., *opr7opr8* has lower enzymatic activities in the shoots but higher activities in the roots than WT under salt treatment. For osmotic adjustment, *opr7opr8* produced less proline in the shoots at 100 and 300 mM NaCl treatments but more in the roots than the WT roots under all salt treatments. In addition, the gene expression for abscisic acid (ABA) biosynthesis under salt stress was investigated. Results showed that the expression levels of four key enzymes of ABA biosynthesis, *ZEP1*, *NCED5*, *AO1*, and *VP10*, were significantly downregulated in the shoots as compared to WT under salt treatment. Putting all the data together, we concluded that JA-deficiency in maize seedlings reduced the salt-stress responses in the shoots but exaggerated the responses in the roots. In addition, endogenous JA acted as a positive regulator for the transportation of sodium ions from the roots to the shoots because the mutant *opr7opr8* had a higher level of sodium in the roots but a significantly lower

Int. J. Mol. Sci. **2019**, *20*, 6202

level in the shoots than WT. Furthermore, JA may act as a positive regulator for ABA biosynthesis in the leaves under salt stress.

Keywords: jasmonate; salt response; *Zea mays*; ROS; proline; ABA biosynthesis

---

1. Introduction

Salt stress is one of the most serious abiotic stresses restraining the production of agricultural crops worldwide, specifically in arid and semi-arid areas. Grounded on the FAO/UNESCO report, 397 million hectares (approximately 3.1% of the world's total land area) is affected by salt stress [1]. In addition, land degradation due to soil salinization has become a major global issue for maintainable agriculture in arid and semi-arid regions. Salt stress affects almost all aspects of plant growth and development including seed germination and the vegetative and reproductive growth development of plants [2]. A high salinity causes ionic toxicity, macro and micro nutrient (Na, K, P, Ca, Fe, Zn, etc.) deficiencies, and limits water uptake from the soil, thus reducing photosynthesis and metabolic processes under oxidative stress [2]. To deal with the biotic and abiotic stresses, a number of plants have developed multifaceted mechanisms to survive in adverse conditions including the saline soils. Salt tolerance either by salt elimination or accumulation within the cells is an economic trait for crops that helps them to produce a relatively high yield under saline soil conditions.

Phytohormones like abscisic acid (ABA), gibberellins (GA), ethylene (ET), salicylic acid (SA), jasmonates (JA), auxins (IAA), cytokinins (CK), brassinosteroids (BR), and strigolactones (SL) play positive roles in improving the tolerance of crops against abiotic stresses [3]. Some of them, such as abscisic acid, have been identified as stress hormones. Classically, abscisic acid is a hormone responsible against abiotic stresses such as drought, salt, cold, heat, and high-temperature stresses [4]. ABA upregulates the turgor pressure in cells, synthesizes osmoprotectants, and regulates the activity of antioxidants conferring dehydration tolerance. ABA activates the expression of a number of responsive genes including the genes of late embryogenesis abundant (LEA) proteins, dehydrins (DHNs), and other defensive proteins that play a fundamental protective role for membranes, organelles, and metabolic processes during water limitation [5]. Moreover, ABA closely interacts with several other stress-response hormones including SA, ET, and JA during the protective response to abiotic stresses [6]. Interestingly, evidence is increasing that growth-promoting hormones including IAA, GA, and CK play an integral part in plant responses to heat, salt, cold, and other stresses [7–9]. In general, ABA is regarded as the universal stress hormone. However, the hormonal crosstalk of ABA with other hormones is crucial to fine-tune plant defense responses against abiotic stresses or combinations of abiotic with biotic stresses.

Jasmonic acid (JA) and its offshoots, such as methyl jasmonate (MeJA), jasmonoyl-L-isoleucine (JA-Ile), and jasmonoyl-L-Tryptophan (JA-Trp), are collectively stated as jasmonates (JAs). These are fatty-acid-derived cyclopentanone compounds that occur ubiquitously and entirely in the plant kingdom [10,11] and serve as natural growth regulators in plant species [12]. These compounds play crucial roles in many plant biological processes, i.e., seed maturing, reproductive development, leaf senescence, root development, trichome and tendril formation, and the biosynthesis of many secondary metabolites in response to environmental stresses [13,14]. In model plants tomato and Arabidopsis, jasmonic acid has been deeply studied for their defensive role against insect and pest attacks. JA mutants, such as *fad3/7/8* [15], *opr3/dde1* [16,17], *aos/dde2* [18], and *coronatine insensitive1* (*coi1*) [19], are all male sterile, suggesting that JA is required for anther/pollen development in plants. All the above mutants except *opr3* have been shown to be susceptible to necrotrophic pathogens and insect pests, indicating that JA is an important element for plant responses to biotic stresses [20]. Interestingly, other JA signaling mutants, such as *jar1* [21] and *jin1/myc2* [22], are fertile but still susceptible to pathogens [22]. Similar results for defense responses have been obtained in tomato. In tomato, the systemin perception mutant *spr1* [23], JA biosynthesis mutant *spr2* [24], and JA perception mutant *spr6/jai1* [25] are impaired in the expression of wound-induced proteinase inhibitors (PIs) and are susceptible to insects and pests [24]. In addition to biotic stress, a number of studies have shown that JAs have taken significant protective responses against abiotic stresses including heavy metals [26], salt [27–29], drought [30], heat [31], and cold stress [32]. Salinity is indisputably a foremost abiotic stress factor that limits crop production by initiating ionic and osmotic stresses [33]. For salinity, JAs have been intensively studied as the positive regulators of salt tolerance [29,34]. For example, the foliar spray of MeJA can effectually lessen salt toxicity symptoms in soybean seedlings [35]. In grapevine, the foliar application of jasmonic acid can save plant growth in the salt-sensitive cell lines [34]. The exogenous application of JAs under saline stress improved the performance of safflower by a collective increase in chlorophyll a, b, photosystem II (Fv/Fm), leaf area index (LAI) [36]. In addition, the foliar application of JA to the seedlings of strawberries regulated enzymatic and non-enzymatic antioxidant activities, reduced lipid peroxidation, and increased the potassium content under salt stress [37]. The foliar spray of JA to the soybean seedlings enhanced the soluble protein content, antioxidant enzyme activity, and membrane stability index of the leaves [38]. In common wheat (*Triticum aestivum*), the *TaAOC1* gene encodes an allene oxide cyclase (AOC) enzyme of the JA biosynthesis pathway, and the over-expression of *TaAOC1* in Arabidopsis elevates jasmonic acid levels and promotes saline tolerance, suggesting that jasmonic acid positively regulates the salt tolerance in wheat [35]. However, there are also several reports available that suggest a negative role of JAs for the salt tolerance of plants. For example, rice mutants *cpm2* and *hebiba* are impaired in the function of allene oxide cyclase (AOC) of their JA biosynthesis pathway. These mutants were resistant to salt and drought stress. Interestingly, both mutants showed better scavenging of reactive oxygen species (ROS) under stress conditions [39]. In wild soybean (*Glycine soja*), the expression level of the GsJAZ2 gene was induced by varied abiotic stresses, and over-expression of GsJAZ2 in Arabidopsis enhanced its tolerance to saline stress [40]. Maize is an important crop for global food security and its production is restricted by environmental stresses especially by drought and soil salinity. Information about maize plant tolerance to salt stress is very limited so far. In this study, the molecular bases of the jasmonate-regulating salt tolerance of maize plants were investigated using the JA-deficient mutant *opr7opr8*. We found that JA-deficiency in maize seedlings reduced the salt-stress responses in the shoots but exaggerated the responses in the roots because JA takes an essential role in Na$^+$ transportation from the root to shoots and JA positively regulates ABA biosynthesis in the leaves under salt stress.

2. Results

2.1. Jasmonate Is a Required Signal for Stomata Closure under Salt Stress

Stomata closure is the early response of plants to water stresses. Stomata closure largely reduces water loss by transpiration during the water stress period. In this study, we noted that JA-deficient mutant *opr7opr8* [41] has delayed "wilting," a water loss symptom of plants under drought and salt-stresses, in comparison to wild type (WT), indicating that the stomata response to water stress of *opr7opr8* could be different from that of WT. We counted the stomatal density in the leaves of *opr7opr8* and WT, and the results showed that *opr7opr8* has a lower stomatal density in their leaves than WT after salt stress (Figure 1d), indicating that endogenous JA positively regulates stomata formation during leaf development. Stomatal visualization under salt stress at 0, 100, 200, and 300 mM salt stress showed that *opr7opr8* has fewer stomata in their leaves than wild type (WT), implying that *opr7opr8* may lose water less during water stress than WT. Stomatal closure during salt stress was studied. The results showed that after 24 h of salt treatment, the pore aperture index (PAI) and stomatal aperture index (SAI) of *opr7opr8* were higher than those of WT under salt stress treatments (100, 200, and 300 mM) (Figure 1a,b), indicating that the JA deficiency in *opr7opr8* slows down the stomatal closure during salt stress in comparison with WT. Figure 1c shows that the stomatal opening in *opr7opr8* is wider than in B73 after 24 h of 200 mM NaCl application. The stomatal visualization of leaves under 0, 100, and

300 mM NaCl treatments is shown in Figure S1. At later hours of salt treatment, *opr7opr8* and WT keep their stomata close, and there is water loss through the epidermal cells in both genotypes.

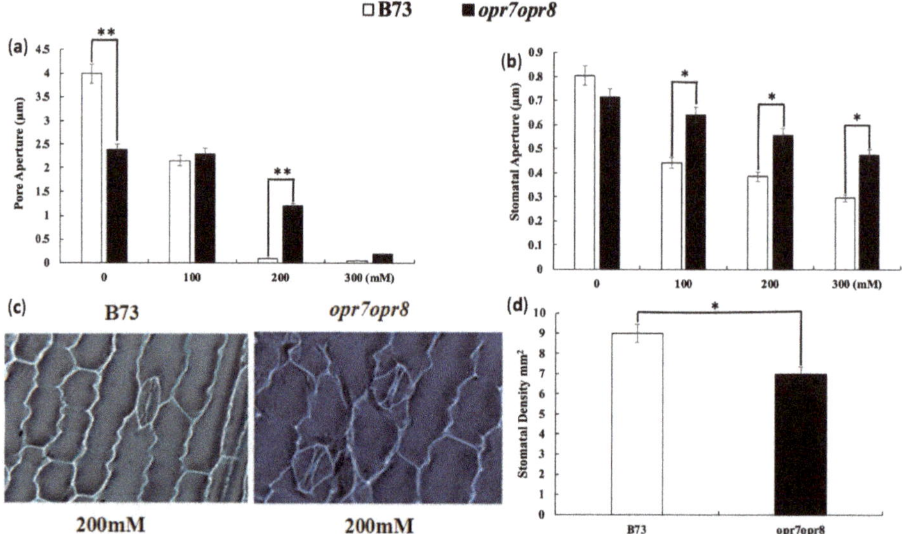

Figure 1. *opr7opr8* displays a higher stomatal aperture than wild type (B73) under salt (NaCl) treatment. (a) Pore aperture index of stomata. (b) Stomatal aperture index (SAI) of B73 and *opr7opr8* after 24 h salt treatment with 0 to 300 mM NaCl. (c) Stomata image of B73 and *opr7opr8* after 24 h under 200 mM NaCl treatment at 40x magnification. (d) Stomatal density on the third leaf of B73 and *opr7opr8*. The asterisks denote significant differences between B73 and the *opr7opr8* mutant at $p < 0.05$ (*) or $p < 0.01$ (**) by analysis of variance.

2.2. opr7opr8 Mutant Showed Delayed Leaf Senescence under Salt Stress

To dissect JA roles in leaf senescence under salt stress, the seedlings of *opr7opr8* and WT (at V3-stage) were applied with different concentrations of NaCl (0, 100, 200, and 300 mM) in a hydroponic system. Data were collected for different physiological parameters, i.e., fresh root length (FRL), fresh shoot length (FSL), fresh root weight (FRW), fresh shoot weight (FSW), dry root weight (DRW), and dry shoot weight (DSW), in these salt treatment experiments. We observed that the leaves of *opr7opr8* were less wilted and necrotic as compared to B73 under salt stress. At treatment of 100 mM sodium chloride, the tips of the leaves of *opr7opr8* and B73 were becoming yellowish after 7 days of salt treatment (Figure 2a), but no substantial differences in leaf senescence were experienced among the two genotypes under 100 mM sodium chloride treatment. The treatment of 200 mM sodium chloride strongly activated leaf senescence of B73 and *opr7opr8*, but the symptoms of leaf rolling and necrosis in *opr7opr8* developed slower than those in B73. At 7 days of treatment, all the leaves of B73 were dried and yellow, but the leaves of *opr7opr8* were rolled and green (Figure 2a). At treatment of 300 mM sodium chloride, the symptoms of leaf rolling, yellowing, and drying were similar to the treatment of 200 mM sodium chloride, but the symptoms under 300 mM sodium chloride developed 2 days earlier than those under 200 mM sodium chloride. At 5 days of 300 mM sodium chloride treatment, all seedlings of B73 were dried, but the new leaves (third and fourth leaf) of *opr7opr8* seedlings remained green (Figure 2a). We measured the chlorophyll contents of the shoots of B73 and *opr7opr8* at 2 days of salt treatment (Figure 2b,c). The result showed that *opr7opr8* had a significantly higher chlorophyll A content than B73 at 200 and 300 mM sodium chloride treatments and significantly higher chlorophyll B at all the salt concentrations including 0 mM. All the results indicated that JA-deficient mutant *opr7opr8* underwent delayed leaf senescence upon salt stress as compared to WT, suggesting that endogenous JA acts as a negative regulator for salt stress response in maize.

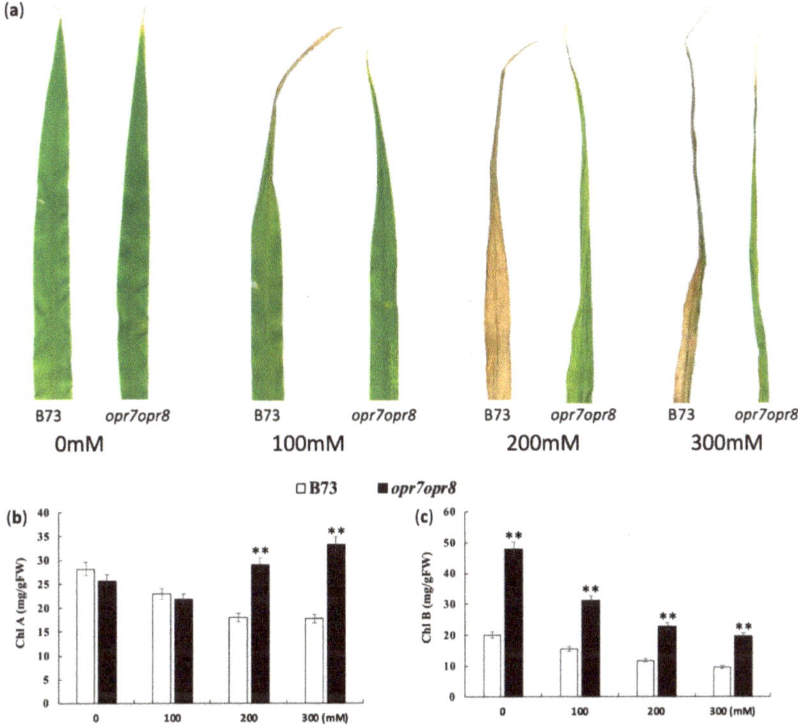

Figure 2. *opr7opr8* undergoes delayed leaf senescence upon salt (NaCl) stress compared to B73. (**a**) Chlorosis symptom of leaves of B73 and *opr7opr8* seedlings whose roots were treated with 0–300 mM NaCl in the hydroponic system. The third leaves of the two genotypes were used to take the pictures at 7 days of salt treatment. (**b**) Measurement of chlorophyll A content in the leaves of B73 and *opr7opr8* seedlings at 2 days of salt stress. (**c**) Measurement of chlorophyll B content in the leaves of B73 and *opr7opr8* seedlings at 2 days of salt stress. The asterisks denote significant differences between wild type (WT) and *opr7opr8* at $p < 0.01$ (**) by analysis of variance.

2.3. opr7op8 Displayed Better Growth in the Shoots but Worse Growth in the Roots Than B73 under Salt Stress

In this study, we investigated the growth inhibition difference under salt stress between *opr7opr8* and WT. The shoot and root length of *opr7opr8* and WT showed growth inhibition under salt treatments (Figure 3a,b). As the salt concentration increased, the effects of growth inhibition were exaggerated. However, we observed that the shoot length of *opr7opr8* was longer than that of WT after 7 days of 100 and 200 mM salt treatments, but the root length of *opr7opr8* was significantly shorter than that of WT. With the concentration of NaCl increasing, the root length of both genotypes decreased, but this decrease was sharper in *opr7opr8* than in B73. Our result indicates that the growth inhibition to *opr7opr8* shoots was slighter than that of B73, but to the roots, it was stronger than that of B73 (Figure 3a,b), suggesting that to JA-deficient mutant *opr7opr8*, salt treatments cause stronger damage to the roots but slighter damage to the leaves as compared to B73. The *opr7opr8* plants showed less shoot fresh weight and shoot dry weight under salt stress as compared to B73 (Figure 3c,e). However, for the roots, *opr7opr8* showed a higher shoot fresh weight and shoot dry weight in comparison with WT

(Figure 3d,f). For WT plants, as the salt concentration increased, the shoot fresh and dry weights and root fresh and dry weights increased (Figure 3c–f).

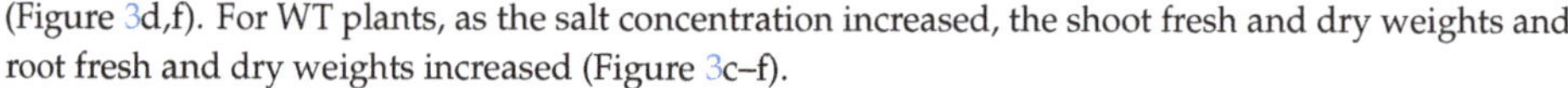
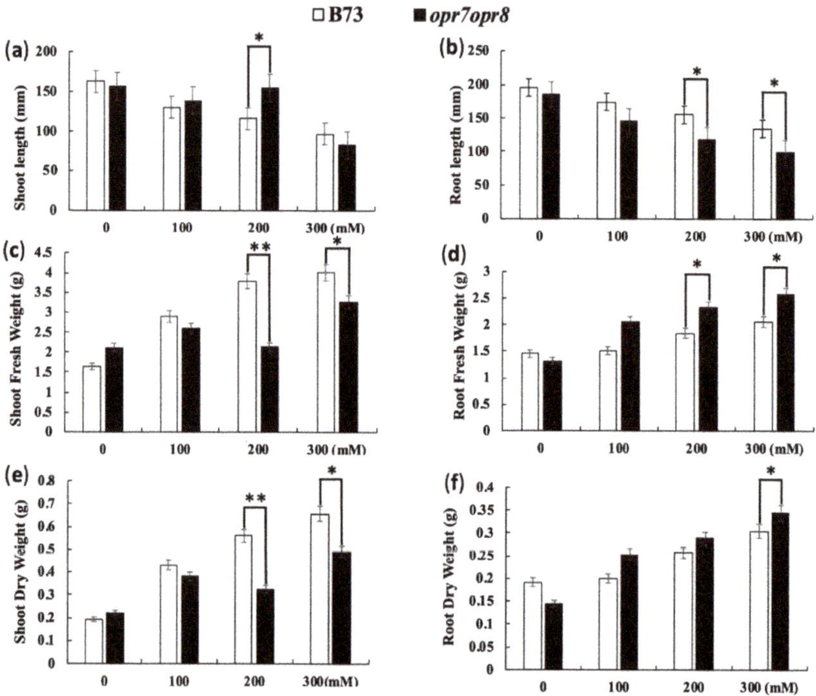

Figure 3. Salt treatments inhibit the growth of shoots and roots of B73 and *opr7opr8* seedlings. The V3-stage plants of B73 and *opr7opr8* were treated with 0–300 mM NaCl in the hydroponic system, and the (**a**) shoot length, (**b**) root length, (**c**) shoot fresh weight, (**d**) root fresh weight, (**e**) shoot dry weight, and (**f**) root dry weight were measured after 7 days of salt treatments. The shoot length is the distance from the first node (the coleoptiles node) to the tip of the leaves. The root length was measured from the first node to the far-end of the root system. The asterisks show significant differences for WT and the mutant at $p \leq 0.05$ (*) or $p \leq 0.01$ (**) by analysis of variance.

2.4. opr7opr8 Accumulated Less Sodium in the Leaves but More Sodium in the Roots Than WT under Salt Stress

In this study, we analyzed sodium and potassium accumulation in the roots and leaves of B73 and *opr7opr8* plants treated at 0, 100, 200, and 300 mM NaCl. Our results showed that the higher the concentration of NaCl applied, the higher the content of Na^+ in the leaves and roots of B73 and *opr7opr8* plants detected (Figure 4a,c). However, B73 and *opr7opr8* had different levels of salt content in the leaves and roots. In the leaves, *opr7opr8* accumulated significantly fewer Na^+ than WT (Figure 4a), indicating that in *opr7opr8*, the Na^+ transportation from the roots to the leaves was reduced during the salt stress. In the roots, *opr7opr8* had a higher Na^+ content than WT (Figure 4c), indicating that the mutant was compromised to exclude Na^+ out of the roots or to transport Na^+ from the roots to the shoot. Homeostasis of potassium ions and sodium ions plays a crucial role in plant development under salt stress conditions. A reduced K^+/Na^+ ratio is the most commonly observed physiological feature of plants challenged by salt stress [42]. In this study, we observed that the K^+ content declined in the roots and leaves of both genotypes with the increase in salt solution concentration (Figure 4b,d). Both genotypes showed different K^+ contents in the shoots and roots under salt stress. At the level of 100 mM NaCl, K^+ contents in the shoots and roots in WT were higher than those in the mutant.

At the level of 200 and 300 mM NaCl, *opr7opr8* retained higher K$^+$ contents in the shoot than WT, but in the roots, *opr7opr8* accumulated lower contents than WT. Putting the sodium and potassium results together, we concluded that the *opr7opr8* mutant retained more sodium in the roots but less sodium in the leaves under salt stress, and that for potassium was the opposite to sodium, suggesting that JA plays an important role in maize plants regarding Na$^+$ ion uptake and transportation under the salinity condition. Salinity may cause other mineral nutrient deficiencies or imbalances due to the accumulation of Na$^+$ ions in the roots and shoots.

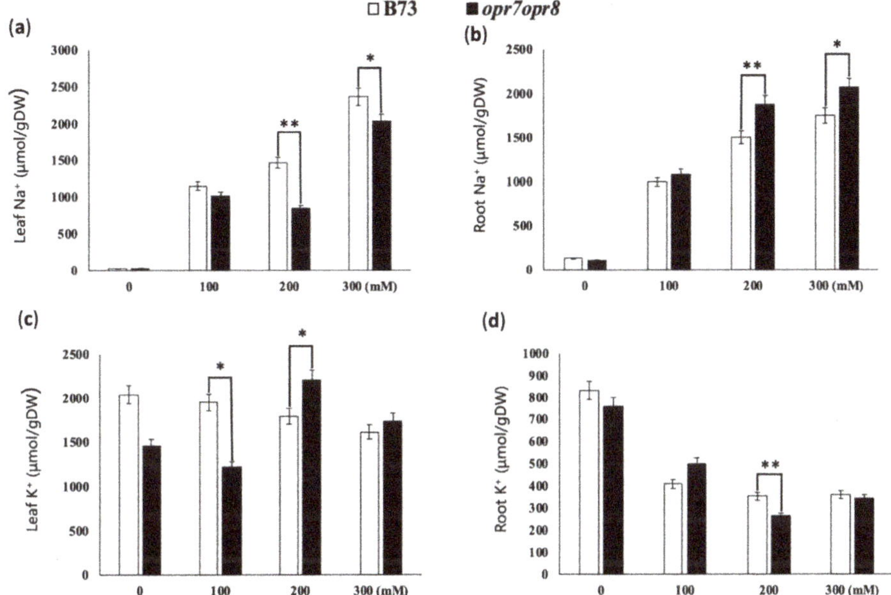

Figure 4. The leaves and roots of B73 and *opr7opr8* seedlings accumulate sodium and potassium under NaCl treatments. The samples were taken at 7 days after salt treatment. (**a**) Na$^+$ ion content in the leaf of B73 and *opr7opr8* plants. (**b**) Na$^+$ ion content in the roots. (**c**) K$^+$ ion content in the leaves. (**d**) K$^+$ ion content in the roots. The asterisks denote significant differences between WT and the mutant at $p \leq 0.05$ (*) or $p \leq 0.01$ (**) by analysis of variance.

2.5. opr7opr8 and WT Accumulated a Different Level of ROS under Salt Stress

To investigate the JA roles involved in the ROS production under salt stress, a ROS visualization experiment was performed in plant roots by using 2′,7′-dichlorofluorescin diacetate (H2DCFDA). Root samples were collected 4 h after salt stress and immediately underwent ROS detection treatment. The meristematic zone of the root tips of nine individual plants for each treatment was used as the samples. ROS production in the meristematic zone of the roots was detected under a confocal microscope for 0, 100, 200, and 300 mM NaCl treatments (Figure 5a). The relative fluorescence quantification of ROS production in the root meristematic zone is shown in Figure 5b. Our results show that the relative fluorescence value of ROS in *opr7opr8* roots was significantly higher than in WT (Figure 5a,b) at 200 and 300 mM salt treatments, suggesting that the roots of *opr7opr8* are highly sensitive to salt damage under salt stress.

In this study, the H$_2$O$_2$ level was also determined in the roots and leaves of both genotypes under salt treatments. H$_2$O$_2$ causes plant cell death as a result of environmental stresses, so its regulation is crucial in the growth and developmental events [43]. In the leaves and roots of B73 and *opr7opr8*, the production of H$_2$O$_2$ increased with the increase in salt concentration (Figure 6c,d). H$_2$O$_2$ production was significantly higher in the leaves of B73 at 200 mM NaCl treatment, whereas there was no significant increase in H_2O_2 production among the leaves of *opr7opr8*. Our study showed that *opr7opr8* produced significantly more H_2O_2 in its roots as compared to its leaves. Salt treatments increased H_2O_2 levels in the roots and leaves of both genotypes (Figure 6c,d), and the genotypes had significantly different H_2O_2 levels. In the roots, *opr7opr8* showed a higher H_2O_2 level than WT, and in the leaves, *opr7opr8* produced a lower H_2O_2 level than WT.

Oxygen-free radicals cause lipid peroxidation in an organism. An upsurge in free radicals under stress conditions causes overproduction of malondialdehyde (MDA). The MDA level is known as a marker of oxidation stress and the antioxidant activity in plants under stress conditions. In this study, MDA contents were detected in both genotypes under the salt treatments. The salt treatments resulted in MDA accumulation in B73 and *opr7opr8*. The MDA levels increased in the leaves and roots of both genotypes with the increase in salt concentration (Figure 6a,b). In the roots, the MDA accumulation in *opr7opr8* was significantly higher than that in WT (Figure 6b), whereas in the leaves, the MDA level in *opr7opr8* was significantly lower than that in WT, indicating that *opr7opr8* suffered from a stronger lipid peroxidation in the roots than WT but slighter lipid peroxidation in the leaves than WT under salt stress.

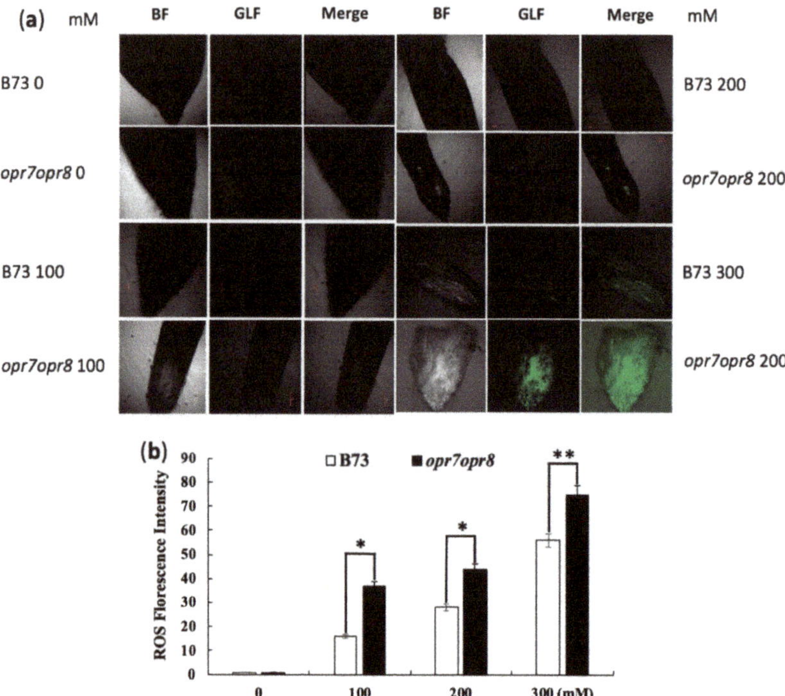

Figure 5. Reactive oxygen species (ROS) production in the meristematic zone of the roots of B73 and *opr7opr8* under 0–300 mM NaCl treatments. (**a**) Visualization detection of ROS production under bright field (BF) and green light (480–550 nm) fluorescence (GLF) with a confocal microscope using 2′,7′-dichlorofluorescin diacetate (H2DCFDA) at 4 h after salt treatments. (**b**) Relative fluorescence quantification of ROS production in root meristematic zones of B73 and *opr7opr8* under salt treatments by software ImageJ scan. The values of B73 and *opr7opr8* were given 1 at 0 mM for relative ROS quantification. The asterisks denote significant differences between WT and the mutant at $p \leq 0.05$ (*) or $p \leq 0.01$ (**) by analysis of variance.

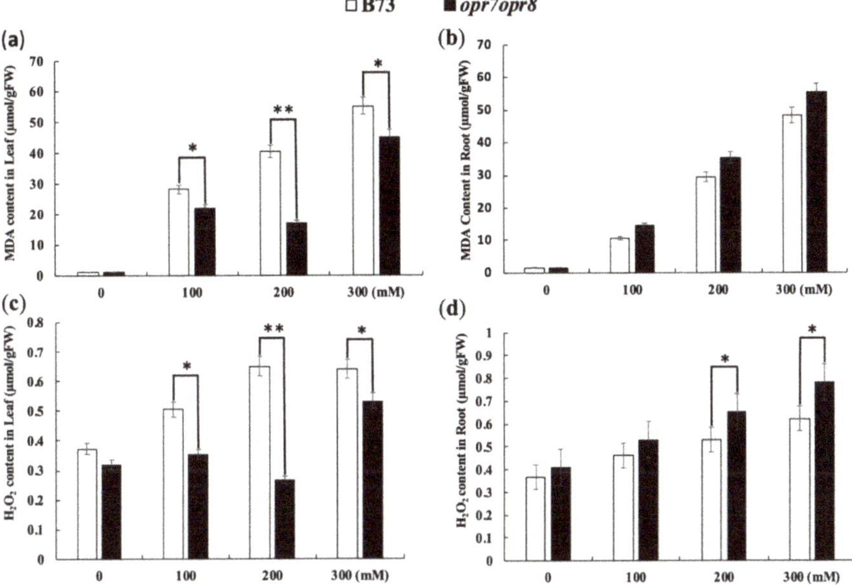

Figure 6. Malondialdehyde (MDA) and H_2O_2 accumulation in *opr7opr8* and B73 under salt stress. The MDA levels in the (**a**) leaves and (**b**) roots were detected after 48 h of 0–300 mM NaCl treatments. The H_2O_2 content in the (**c**) leaves and (**d**) roots was measured after 48 h of salt treatments. The asterisks denote significant differences between WT and the mutant at $p \leq 0.05$ (*) or $p \leq 0.01$ (**) by analysis of variance.

2.6. opr7opr8 Displayed Different Antioxidant Enzyme Activities from WT under Salt Treatment

The enzymatic antioxidant defense system including superoxide dismutase (SOD), peroxidase (POD), catalase (CAT), ascorbate peroxidase (APX), and glutathione peroxidase (GPX) is important for the plant to cope with ROS bursts during salt stress. In this study, the enzymes such as SOD, POD, CAT, and APX were analyzed in the roots and leaves of *opr7opr8* and B73 plants under salt treatments. The activity of SOD increased with the increase in salt concentration in *opr7opr8* and B73 plants. The SOD activity in the leaves of *opr7opr8* was lower than that of WT under 200 and 300 mM NaCl treatments (Figure 7a), but in the roots, the SOD activity of *opr7opr8* was higher than that of WT under 100, 200, and 300 mM NaCl treatments (Figure 7a). For peroxidase (POX) (total class III peroxidases), salt treatments activated POX activity in the roots of B73 and *opr7opr8*. In the roots of B73, the POX activity increased with the increase in salt concentration (Figure 7b). In the roots of *opr7opr8*, the POX activity at 100 mM NaCl treatment was higher than 200 and 300 mM NaCl. *opr7opr8* roots showed a higher POX activity than WT at 200 and 300 mM NaCl treatment. In the leaves, the POX activity was downregulated in WT under salt treatments, but for *opr7opr8* salt, the treatments did not affect the POX activity in the leaves (Figure 7b). For CAT, the fluctuation in enzyme activity under salt treatment was similar to SOD. The salt treatments increased the CAT activity in the roots of B73 and *opr7opr8*, and the mutant had a higher activity than WT (Figure 7c). In the leaves, salt treatments increased the CAT activity in B73 but decreased in *opr7opr8* (Figure 7c). The mutant exhibited a significantly lower level of CAT activity in the leaves than in WT. Ascorbate peroxidase (APX) is a major ROS-scavenging enzyme controlling intracellular ROS levels in varied stresses. The variation in APX activity under salt treatments was similar to the POX activity. In the roots, the APX activity in B73 and *opr7opr8* increased with the increased salt concentration (Figure 7d). At 100 and 200 mM NaCl treatments, in the roots of *opr7opr8*, the APX activity was lower than that of WT. In the leaves, the

APX activity in B73 was elevated when the salt concentration increased, but, in the leaves of *opr7opr8*, the APX activity had no significant change under salt stress (Figure 7d). In the leaves, *opr7opr8* was significantly lower than that of WT for APX activity under salt stress. Putting the four enzyme results together, we saw that *opr7opr8* had a significantly different level of ROS-scavenging enzyme activity in the roots and shoots under the salt treatments as compared to WT, indicating that the JA deficiency in *opr7opr8* plants caused ROS-scavenging enzyme genes to be differentially expressed under salt stress.

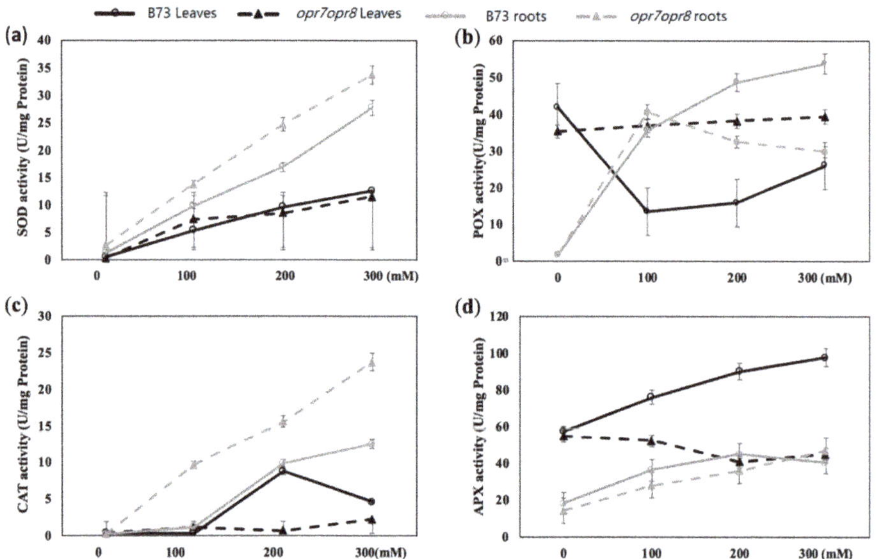

Figure 7. ROS-scavenging enzyme activities in the shoots and roots of *opr7opr8* and B73 seedlings under 0–300 mM NaCl treatments. The samples were taken 2 days after the salt was applied to the hydroponic solution. (**a**) SOD activity in leaves and roots, (**b**) POX activity in leaves and roots, (**c**) CAT activity in leaves and roots, and (**d**) APX activity in leaves and roots.

2.7. opr7opr8 Exhibited Different Levels of Glutathione Reductase (GR) and Glutathione-S-Transferase (GST) Activities from WT under Salt Treatment

Except for SOD, POD, CAT, and APX, glutathione reductase (GR) and glutathione-S-transferase (GST) can also act as the enzymatic antioxidants in plants under abiotic stresses. In this study, the GR and GST activities were analyzed in *opr7op8* and B73 plants under salt treatments. The GR activity increased with the increase in salt stress in the leaves and roots of *opr7op8* and B73 plants (Figure 8c,d). In the roots, the GR activity in *opr7opr8* was higher than that in B73, but in the leaves, *opr7opr8* was lower than WT at salt treatments of 200 and 300 mM NaCl (Figure 8c). For GST activity, in the leaves, salt treatments reduced the GST activity in B73 but activated GST activity in *opr7opr8* (Figure 8a). At 200 and 300 mM NaCl, the GST activity in the leaves of *opr7opr8* was higher than that in WT. In the roots, salt treatments slightly induced the GST activity in WT but slightly inhibited in *opr7opr8* (Figure 8b). At 200 and 300 mM NaCl, WT was significantly higher than *opr7opr8* for GR activity in the roots (Figure 8d).

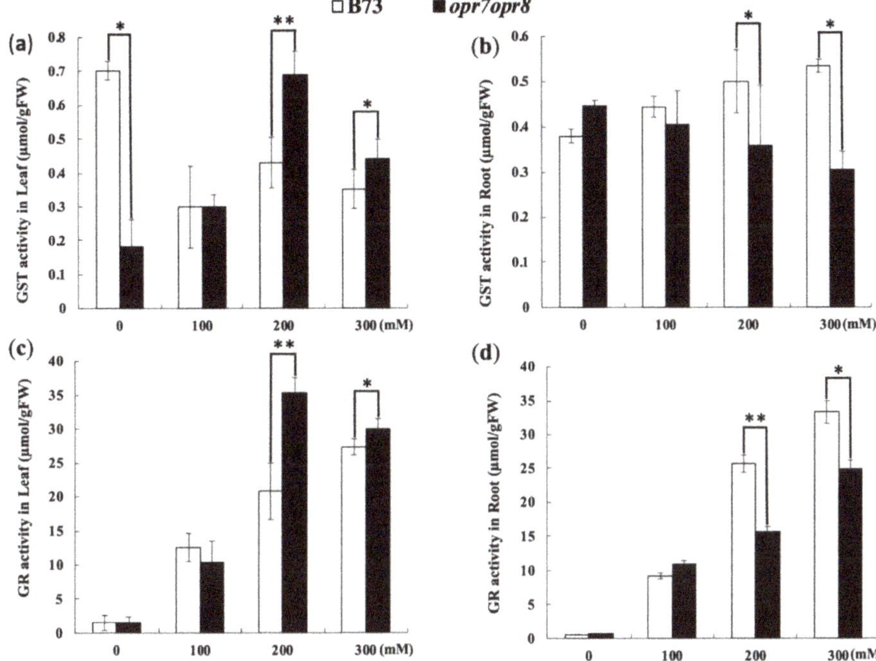

Figure 8. Enzymatic activities of glutathione reductase (GR) and glutathione-S-transferase (GST) in B73 and *opr7opr8* seedlings under 0–300 mM NaCl treatments. The samples were taken 2 days after the salt treatments. GST activity in (**a**) the leaves and (**b**) roots and GR activity in the (**c**) leaves and (**d**) roots. The asterisks denote significant differences between WT and the mutant at $p \leq 0.05$ (*) or $p \leq 0.01$ (**) by analysis of variance.

2.8. opr7opr8 Accumulated More Proline in the Roots but Less in the Leaves Than WT under Salt Stress

Under abiotic stresses, plants tend to accumulate soluble osmotic adjustment substances such as proline to protect the cellular structure and enzyme activity against osmotic and ionic stresses. In this study, the proline content accumulation was tested in the leaves and roots of *opr7opr8* and B73 plants treated with 0, 100, 200, and 300 mM NaCl. The results showed that proline production was highly induced by salt treatments in the leaves and roots in both genotypes (Figure 9a,b). In the leaves, the proline accumulation level in *opr7opr8* was significantly lower than that in WT, except under 200 mM NaCl (Figure 9a). In the roots, *opr7opr8* accumulated a higher proline level than WT (Figure 9b). Our results suggest that JA is involved in osmotic regulation under salt stress.

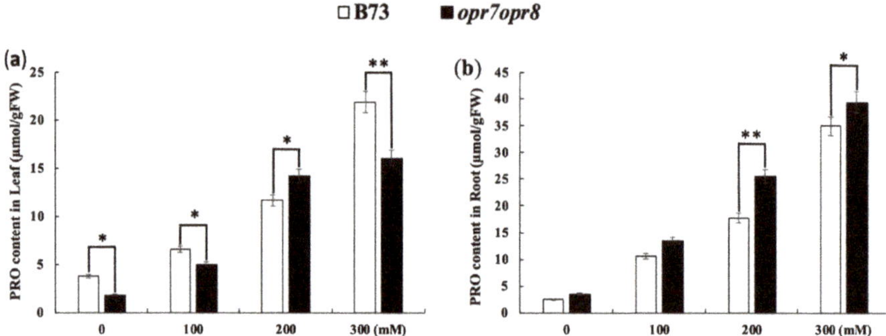

Figure 9. Proline contents were detected in (**a**) the leaves and (**b**) roots at two days after 0–300 mM NaCl was applied to the roots. The asterisks denote significant differences between WT and the mutant at $p \leq 0.05$ (*) or $p \leq 0.01$ (**) by analysis of variance.

2.9. Endogenous JA Production Is Required for Transcriptional Activation of ABA Biosynthesis Genes under Salt Stress

Abscisic acid (ABA) plays an important role for plants to deal with abiotic stresses, including drought and soil salinity. In this study, we tested the expression levels of four key genes, *ZEP1*, *NCED5*, *VP10*, and *AO1*, of the ABA biosynthesis pathway under salt treatment in the leaves and the roots of *opr7opr8* and WT. ZEP1 (zeaxanthin epoxidase1) is the initial enzyme of the ABA biosynthesis pathway in maize. NCED (9-cisepoxycarotenoid dioxygenase) catalyzes the oxidative cleavage of epoxy-carotenoid 9-cis-neoxanthin, the first step of abscisic-acid biosynthesis from carotenoids [44]. *Vp10* (*viviparous10*) encodes the ortholog of Cnx1, which catalyzes the final common step of molybdenum cofactor (MoCo) synthesis. The sulfur-containing form of MoCo, MoCo-S, is a required cofactor of AO1 activity [45]. AO1 (aldehyde oxidase 1), a molybdenum-containing oxidoreductase, catalyzes the final step of ABA biosynthesis, the conversion of abscisic aldehyde to ABA. In our experiments, the expression of all the four genes showed a similar induction pattern, that is, in the leaves of B73, the four genes were highly induced by salt treatment, but not induced or slightly induced in the *opr7opr8* mutant (Figure 10a,c–e). Our results showed that 200 mM NaCl strongly induced the *ZEP1* expression in WT, and the maximum induction was more than 30 times at 6 h of treatment in comparison with that at 0 h (Figure 10d). However, in the mutant *opr7opr8*, the *ZEP1* gene was just slightly induced at the early time points (2 to 12 h of the treatment) and the maximum induction was about 5 times at 2 h of treatment (Figure 10d). Overall, the *ZEP1* gene expression in *opr7opr8* was significantly lower than that of WT at 4 to 72 h of salt treatment (Figure 10d). At 200 mM salt stress, the *NCED5* gene was highly upregulated in the leaves of B73, as compared to mutant *opr7opr8* (Figure 10a). At 24 h of 200 mM salt treatment, the expression level of the *NCED5* gene in B73 was increased to 130 times the expression level at 0 h. Then, it decreased at longer time points. While in the mutant, it upregulated gradually from 2 to 72 h salt stress, the expression level was highest at 72 h (Figure 10a). The *AO1* gene was also highly induced by salt treatment in WT but not in the mutant *opr7opr8* (Figure 10c). The expression level of the *AO1* gene was increased to 50 times at 6 h of salt treatment compared to 0 h of treatment (Figure 10c). In *opr7opr8*, the maximum induction was less than 5 times, which happened at 2 h of salt treatment (Figure 10b). At the timepoints of 2 to 24 h of the treatment, the *AO1* expression level in *opr7opr8* was significantly lower than that in WT (Figure 10c). Salt treatment induced the expression of the *VP10* gene in WT and the mutant (Figure 10e). For WT, the induction peak appeared at 6 h of salt treatment, and for the mutant, it appeared at 24 h (Figure 10e), indicating that *VP10* induction by salt treatment was faster in WT than in the mutant. For all time points, the induction level of *VP10* expression in WT was significantly higher than in the mutant. Putting the data of the four genes together, we saw that four key genes of the ABA biosynthesis pathway, *NCED5*, *ZEP1*, *AO1*, and *VP10*, were strongly induced in the WT plants by salt stress but were significantly less expressed in the mutant with no or a slight induction by salt treatment, indicating that endogenous JA is a positive factor for ABA accumulation under salt stress in maize.

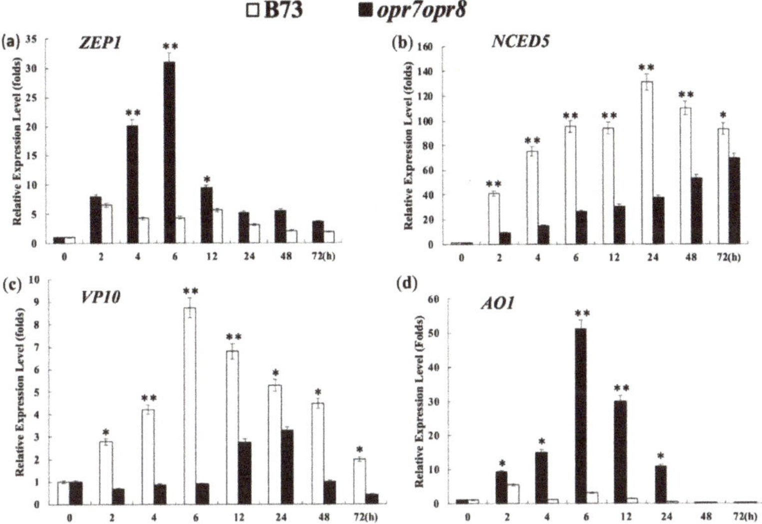

Figure 10. Quantitative real-time polymerase chain reaction (qRT-PCR) expression level analysis of (a) *NCED5*, (b) *AO1*, (c) *VP10*, and (d) *ZEP1* genes in the leaves of the B73 and *opr7opr8* plants under 200 mM NaCl treatment. The relative expression level was calculated according to the expression of the gene at 0 h of treatment. The asterisks denote significant differences between WT and the mutant *opr7opr8* at $p \leq 0.05$ (*) or $p \leq 0.01$ (**) by analysis of variance.

3. Discussion

Salinity is one of the most ubiquitous environmental stresses limiting the yield of agricultural crops with adverse effects on the vegetative and reproductive growth of plants [46]. Growing on saline-alkali soil, the plant has to tolerate the root-absorbed excessive sodium ions that have a damaging effect on biochemical reactions and causes ionic, osmotic, and oxidative stresses to plant cells [46,47]. Phytohormones have long been considered essential endogenous molecules regulating plant development and tolerance to diverse environmental stresses including salinity stress [48]. ABA is well known as the endogenous signal molecules enabling plants to survive the severe adverse environmental conditions such as salt and drought stresses [49]. Increasing evidence supports the idea that jasmonic acid can play relevant functions in the abiotic stress response [50,51]. Up until now, three major lines of evidence have been reported for JA contribution in the adaptive response to salt stress. **(1)** Salt stress-activated JA biosynthesis in plants. Elevated JA levels were detected in a number of plant species such as Arabidopsis [52], tomato [53,54], rice [55,56], maize [57], and *Brassica rapa* [58] when challenged with salt stress, indicating that high levels of JAs accumulated in salt-challenged plants after salt treatment may function as a protection signal in plants against salinity stress. **(2)** The exogenous application of JA or MeJA to leaves or roots enhances the salt tolerance of plants. Here are several examples: The exogenous application of 30 μm JA applied 24 h after salt stress effectively reduced the sodium ion uptake in rice seedlings, especially in the salt-sensitive cultivars rather than the salt-tolerant cultivars [56]. A ten-micrometer solution of MeJA sprayed on rice varieties could effectively alleviate the symptoms of rice varieties to salinity stress [59]. An exogenous spray of 2 mM JA can enhance the tolerance of wheat seedlings to salt stress [28]. The foliar application of 20 to 30 μM MeJA can effectively lessen salinity stress symptoms and change endogenous ABA and GA_4 levels in soybeans [37]. Foliar sprays of 1 mM SA and 0.5 mM JA stimulate the H^+-ATPase activity of tonoplast and salt tolerance of soybean seedlings [60]. In grapevine, 10 to 50 µM JA treatments can save growth in the salt-sensitive cell lines, and the salt stress response of the lines is modulated by JA-signaling components such as JAZ proteins [34]. Exogenous sprays of 100 µM MeJA to the seedlings of *Brassica napus* mitigate the inhibitory effect of all salt treatments [61]. In maize, 10 µM JA applied to the seed before germination alleviates alkaline (Na_2CO_3) stress by improving the ascorbate glutathione cycle and glyoxalase system in the seedlings [62]. (**3**) Alterations of JA biosynthesis or interference of JA signaling affects the salinity tolerance of plants. A couple of JA biosynthesis or signaling mutants have been tested for their tolerance or susceptibility against abiotic stresses so far. The mutants of the JA biosynthesis enzyme AOC, *hebiba*, and *cpm2* in rice showed an increased salt tolerance [51]. Transgenic rice with overexpressed gene *CYP94*, encoding an inactivating JA-Ile catabolic enzyme, displayed enhanced salt tolerance [63]. The suppression of OsJAZ9, a repressor of JA signaling, produced a higher sensitivity to JA and increased sensitivity to salt [64]. The overexpression of OsJAZ8, a JA signaling suppressor, improved the salt tolerance of transgenic rice seedlings [65]. The results of the above four published works indicate that a JA-deficiency or JA signaling depression in transgenic rice causes improved salt tolerance. Putting all the three evidence lines together, we saw that the roles of exogenous JA applications are quite consistent from the different studies: The exogenous application of MeJA or JA can significantly enhance the tolerance of plants to salt stress. However, the roles of endogenous JA signals or JA signaling for salt tolerance can be varied, whose conclusion depends on the plant species. Obviously, more research works using JA biosynthesis or signaling mutants from different plant species are needed to clarify the endogenous JA roles involved in the adaptive response against abiotic stresses in plants.

In this study, *opr7opr8*, a maize JA biosynthesis mutant [41], was applied to identify the roles of endogenous JA in maize plants challenged by salt stress. Our results have shown that the shoots of *opr7opr8* display a number of symptoms weaker than those of WT under salt treatments, including delayed leaf senescence, less sodium accumulation, less ROS production, and less antioxidant enzyme activities in the leaves, among others. From these results, we can conclude that *opr7opr8* shoots are less sensitive to salt stress. Rice JA biosynthesis mutants *hebiba* and *cpm2* showed an increased salt tolerance [51]. Our research results indicated that JA biosynthesis mutant *opr7opr8* [43] resembles the rice JA biosynthesis mutants *hebiba* and *cpm2* [53] in the adaptive response against salt stress. Interestingly, in this study, we found that the roots of *opr7opr8* showed stronger symptoms such as more sodium accumulation and more ROS and antioxidant enzyme activities than WT under salt treatments, indicating that the roots of *opr7opr8* are more susceptible than WT. Comparing the data of shoots and roots, we suggested that endogenous JA differentially regulates salt responses in the shoots and roots in maize seedlings for acclimation to salinity in the soil.

Stomata are the vital organ of plants to control water loss under abiotic stresses. Stomatal closure can sharply reduce transpirational water loss under drought and salt stresses. It has been known that the exogenous application of methyl jasmonate or jasmonic acid elicits stomatal closure in a large number of plant species [66] including Arabidopsis [67], *Olea europaea* [68], and barley [69]. It has also been known that JA negatively regulates stomatal formation in Arabidopsis cotyledons [70]. However, the question of how endogenous JA is involved in stomatal closure under abiotic stresses remains unanswered. In this study, we observed that *opr7op8* had a larger stomatal aperture than WT (B73) under salinity stress (Figure 1), indicating that the JA-deficient mutant *opr7op8* is insensitive to stomatal closure under salt stress. This result implies that endogenous JA positively regulates guard cell movement for stomatal closure under water stress conditions.

A common symptom of damage by salinity stress is the growth inhibition, and leaf senescence will appear afterward during prolonged exposure to salt stress. In this study, we showed that after 7 days of the salt treatments, *opr7opr8* showed late leaf senescence as compared to WT (Figure 2), suggesting that endogenous JA in maize acts as a negative regulator for plant growth and leaf senescence under salt stress. We tested the fresh weight and dry weight and found that *opr7opr8* had a higher fresh/dry weight of shoots than WT, but a smaller fresh/dry weight of roots than WT. These results indicated that JA-deficiency in maize plants causes different stress strengths to the roots and shoots under salt treatments. In addition, we noted that both genotypes have a higher fresh/dry weight under higher NaCl concentrations (Figure 3c–f), suggesting that sodium and potassium accumulation in the leaves and roots could be the major reason of the biomass weight increase under salt stress. We tested the sodium and potassium contents in the shoots and roots, and the results showed that both genotypes have increasing sodium contents in the shoots and roots as the treatment salt concentration increases (Figure 4a,b). For example, under 300 mM NaCl at 7 days, WT leaves contained Na^+ 3000 µmol/gDW (Figure 4a) and K^+ 2000 µmol/gDW (Figure 4c), which is NaCl 0.176 g/gDW and KCl 0.151 g/gDW, indicating that the salts (NaCl and KCl) occupied 32.7% of the dry mass of the WT leaves. Similarly, *opr7opr8* leaves contained Na^+ 2000 µmol/gDW (Figure 4a) and K^+ 1500 µmol/gDW (Figure 4c), that is, salts (NaCl and KCl) occupied 23.0% of the dry mass. This calculation suggests that sodium and potassium accumulation in the leaves and roots promoted the increase in biomass weight under salt treatments.

Salinity induces the formation of reactive oxygen species (ROS) within plant cells, which is a well-known cause of damage to all components of the cell, including proteins, lipids, carbohydrates, and DNA [71]. To scavenge excessive levels of ROS, an effective system composed of non-enzymatic and enzymatic antioxidants is evolved in plants [72]. Non-enzymatic antioxidants include phenolic compounds, flavonoids, alkaloids, tocopherol, carotenoids, ascorbate (ASC), and glutathione (GSH) [73]. Enzymatic antioxidants include superoxide dismutase (SOD), peroxidase (POX), catalase (CAT), ASC peroxidase (APX), guaiacol peroxidase (GPX), glutathione reductase (GR), monodehydroascorbate reductase (MDHAR), and dehydroascorbate reductase (DHAR) [73,74]. A number of previous studies have shown that jasmonate mediates ROS production and antioxidant enzymatic activities under abiotic stresses [75]. An exogenous spray of 1 mM MeJA induced ROS accumulation and activated the activities of CAT, GPX, and APX in *Ricinus communis* leaves [76]. The exogenous application of 50 µM MeJA to sunflower (*Helianthus annuus* L.) elicited a fast increase in ROS content, followed by a marked increase in the activity of H_2O_2-scavenging enzymes such as GPX, APX, and CAT [77]. In grape, salt-tolerant cultivars have higher antioxidant enzyme activities [78]. In this study, we showed that JA-deficient mutant *opr7opr8* had a higher H_2O_2 level in the roots than WT under the salt treatments (Figure 6), indicating that endogenous JA may mediate ROS production or ROS-scavenging under abiotic stress. Meanwhile, we saw that *opr7opr8* had a lower H_2O_2 level in the shoots than WT. This result must indicate that the mutant *opr7opr8* suffered from milder salt stress than WT in the leaves because *opr7opr8* leaves have a significantly lower sodium content than WT (Figure 4). As for ROS-scavengers, *opr7opr8* had lower levels of SOD, CAT, and APX activities in the shoots than WT but higher activities in the roots than WT. This result indicates that JA-deficiency in maize seedlings causes strong ROS production and enzymatic activities of ROS-scavenging in the roots, but an opposite phenomenon in the shoots, suggesting that a differential mechanism of JA is involved in the salt response in the shoots and roots of the maize plant.

In many studies, ABA has been regarded as the most important phytohormone that confers abiotic stress tolerance in plants [79]. However, ABA crosstalk with other hormones is crucial to fine-tune plant responses to varied stresses. It is identified that the JA signaling pathway interacts with the ABA pathway via transcription factors such as MYC2, ABI5, and WRKY57 [80]. In our previous study, we reported that ABA production was dramatically reduced in the senescing leaves of *opr7opr8* compared to the wild type, indicating a significant role for JA in the regulation of ABA biosynthesis during leaf senescence in maize [41]. In this study, we quantified the transcriptional levels of four key genes (*NCED5*, *ZEP1*, *AO1*, and *VP10*) of the ABA biosynthesis pathway by quantitative polymerase chain reaction (PCR) in the mutant and WT. Our results exposed that the expression of these four genes was strongly activated by salt treatment (200 mM) in the leaves of WT but was just slightly induced or non-inducible in the mutant, indicating that *opr7opr8* was insensitive to salt stress for ABA biosynthesis activation.

In this study, we noted that the *opr7opr8* mutant under salt stress showed milder growth inhibition, a lower production of H_2O_2, lower level of ROS-scavenging enzymatic activities of SOD, CAT, and APX, lower production of MDA and proline, and lower expression of ABA biosynthesis genes in the leaves than WT. In the roots, *opr7opr8* under salt stress showed stronger growth inhibition and a much higher responsibility to salt. Our results indicated that endogenous JA in the maize plant might differentially regulate the adaptive response to salt stress in roots and shoots, suggesting that different JA-relevant mechanisms in roots and shoots might work in maize seedlings. Meanwhile, we observed that the *opr7opr8* mutant accumulated less sodium and more potassium in the shoots than WT but more sodium and less potassium in the roots under the salt treatments than WT, indicating that endogenous JA played a role in Na^+ and K^+ ion transportation from the roots to shoots, which might be the primary cause of the differential responsibility of roots and shoots of the mutant to salt treatments.

4. Materials and Methods

4.1. Experimental Material, Planting, and Salt Treatments

JA biosynthesis mutant *opr7opr8*, which carried *opr7-5* and *opr8-2* alleles in *ZmOPR7* and *ZmOPR8*, respectively [41], was crossed with B73 to BC_5-stage, and the double mutant *opr7-5opr7-5/opr8-2opr8-2* (homozygous for both genes) was selected in the self-segregation population for this study. B73 was used as the WT plant. The original mutant *opr7opr8* was provided by Dr. Michael V. Kolomiets (Texas A&M University, USA).

The experiment was conducted in greenhouses in Nanjing Agriculture University, Nanjing, China. Good-quality seeds of two genotypes were surface-sterilized by 20% bleach solution containing sodium hypochlorite ~5% for 10 min followed by three times of washing with sterilized double-distilled water. The seeds were sown in the sand and grew for 10–12 days in a growth room at 28/25 °C with 16/8 h of day/night cycles and ~180 μmol m^{-2} s^{-1} of illumination.

The seedlings of both genotypes at the V3-stage were transferred to an aerated hydroponic system with full nutrients of modified Hoagland solution, which contained 945 mg/L $Ca(NO_3)_24H_2O$, 506 mg/L KNO_3, 80 mg/L NH_4NO_3, 136 mg/L KH_2PO_4, 493 mg/L $MgSO_4$, and 2.5 mL $FeSO_4$. $FeSO_4$ was prepared by $FeSO4.7H2O$ + EDTA-Na. The micro-nutrient solution was 0.83 mg/L KI, 6.2 mg/L H_3BO_3, 22.3 mg/L $MnSO_4$, 8.6 mg/L $ZnSO_4$, 0.25 mg/L $NaMoO_4$, 0.025 mg/L $CuSO_4$, and 0.025 mg/L $CaCL_2$.

The hydroponic solution was replaced every 3rd day to ensure nutrient enrichment and the pH value of the solution. For each 7 L-hydroponic box, nine seedlings of B73 and nine seedlings of *opr7opr8* were planted. The salt treatments were applied to the plant roots 2–3 days after transfer to the hydroponics boxes by replacing the hydroponic solution. The treatment solutions contained the full nutrients of Hoagland and NaCl. The concentrations of NaCl treatments were 0, 100, 200, and 300 mM. The experiment had three repeats for every concentration of salt treatment.

4.2. Analysis of Root and Shoot Elongation

Seven days after salt treatment, the plant's root and shoot were collected and measured instantly after harvesting. Morphological parameters like primary root elongation, shoot length, fresh shoot weight, fresh root weight, dry root-shoot weight, and plant water content were measured by a ruler and an analytical balance. For dry weight (DW) and water content, the samples were dried in an oven at −80 °C for 48 h and measured by an analytical balance.

4.3. Ion Content Profiling

Dry shoot and root tissues of each biological replicate were transferred into digestion tubes (Gerhardt, Brackely, UK), supplemented with 5 mL of concentrated nitric acid (HNO_3), and then vortexed for 6 h. After cooling, the final volume of each sample was adjusted to 10 mL with distilled water and vortexed. Contents of different ions were measured by an inductively coupled plasma optical emission spectrometer (ICP-OES, Perkin Elmer Optima 2100DV) (College of Life Sciences,

Nanjing Agriculture University). Blank samples were prepared by adding 5 mL of concentrated nitric acid to an empty digestion vessel and processed.

4.4. Determination of Enzymatic Antioxidants

Fresh leaves and roots (0.5 g) of plants were collected 48 h after salt stress and stored at −80 °C for the determination of various antioxidant enzymes. Tissues were ground in a tissue homogenizer. CAT, POX, APX, SOD, H_2O_2, and MDA activities were determined by a chemical assay kit (Nanjing Jiancheng Bioengineering Institute, China). One unit of enzyme activity was defined as an absorbance change of 0.01 units per minute, and each enzyme's activity was expressed as a unit per milligram of protein. Activities of proline, GR (glutathione reductase), and GST (glutathione S-transferase) were measured according to the kit protocol of the manufacturer. All chemicals were bought from Nanjing Jiancheng Bioengineering Institute. The soluble proline content was calculated in micromoles of proline per gram of fresh weight according to a standard curve.

4.5. Photosynthetic Pigments

The photosynthetic pigments leaf chlorophyll a (Chl a) and chlorophyll b (Chl b) were measured. Leaf samples were collected 48 h after salt stress. About 30 mg of leaf segment was incubated with 10 mL of acetone (80%) and kept in the dark for 24 h. The absorbance was measured by spectrophotometry at 645 and 663 nm. Chl a and Chl b contents were calculated using MacKinney equations [81].

4.6. Stomatal Imaging and Quantification

Stomatal densities in the maize leaf epidermis were observed after 24 h of salt stress under an optical microscope, Olympus BX53. Stomata on the epidermis layer were imprinted using nail varnish painted on fully expanded leaves at V3-stage plants. The stomatal aperture was calculated using ImageJ software (University of Wisconsin-Madison, Madison, WI, USA). The size was measured in ImageJ from a total of 35 stomata from each genotype of each salt treatment concentration, taken from nine biological replicates. Pore aperture and stomata area were measured from imaged biological replicates. The pore area was calculated from the major axis of the measured aperture length, and the minor axis of the measured aperture width at the center of the pore. The stomatal area was calculated from the axes of the measured guard cell length and the doubled guard cell width at the center of the stoma. The stomata aperture was calculated by dividing the stomata width with the stomata length.

4.7. Reactive Oxygen Species Visualization

The reactive oxygen species (ROS) in the root tips were detected using a TCS-SP2 confocal laser scanning microscope (LSCM 780, Zeiss, Leica Lasertechnik GmbH, Heidelberg, Germany). The excitation was set at 488 nm and the emission was at 500–530 nm. The root tissues were collected four hours after the salt treatments and loaded with 20 μM 2′,7′-dichlorofluorescin diacetate (H2DCFDA, Sigma) in a 20 mM HEPES/NaOH buffer (pH 7.5) for 20 min. Samples were then washed with distilled water three times for each 15 min and detected immediately by the confocal microscope. Nine samples were selected per treatment and measured. The experiment was performed at 25 °C. The Root tip sections are imaged under bright field (BF) and green light (480–550 nm) fluorescence (GLF) mode with a confocal microscope. The relative fluorescence production of reactive oxygen species in the root tips was quantified based on 25 overlapping confocal scopes by using ImageJ software.

4.8. Gene Expression Analysis

Plant leaves were used for gene expression analysis. The total RNA was isolated with the Trizol method (Sigma-Aldrich, Oakville, Ontario, Canada) under RNase-free conditions. The total RNA was isolated from the leaves of control and salt-stressed plants (200 mM NaCl at 0, 2, 4, 6, 12 24, 48, and 72 h). The integrity of isolated RNA samples was examined spectrophotometrically and by gel electrophoresis. RNA samples were quantified using a NanoDrop 2000C spectrophotometer (Nanodrop Technologies, Wilmington, DE, USA).

To eliminate genomic DNA in the total RNA extracted, total RNA samples were treated by DNase I (DNaseI, Invitrogen, Carlsbad, CA, USA) according to the manufacturer instruction. After DNaseI treatment, total RNA samples were tested for genomic DNA (gDNA) residue by the polymerase chain reaction (PCR) using the primer pair for the maize *actin1* gene. No band of PCR amplification of maize *actin1* indicated that the total RNA sample was free of gDNA residue. First-strand cDNA was synthesized from 1 µg of total RNA following the manufacturer's instructions. Quantitative PCR (qPCR) was done with a QuantiTect SYBR Green PCR Kit (QIAGEN China Co., Ltd., Shanghai, China) using an Opticon 2 system (Biorad, CFX96 USA). Specific primers for the amplification of target cDNAs were designated by primer-Blast of NCBI (available online: https://www.ncbi.nlm.nih.gov/tools/primer-blast/index.cgi, accessed on 5 December 2019) based on the target gene sequence. The primers are listed in Table S1. The two-way analysis of variance was employed followed by Duncan's multiple range test to determine the significance of the differences of target gene expression levels among treatments at the level of $p \leq 0.05$ or $p \leq 0.01$.

4.9. Statistical Analysis

All the treatments were arranged in a completely randomized design. Morphological, physiological, and biochemical data were presented as mean ± SD (standard error). The data were analyzed using a statistical package, Statistic 8.1 (Analytical Software, Tallahassee, FL, USA). The data were subjected to the two-way analysis of variance (ANOVA). Significant differences at levels of significance (* $p \leq 0.05$; ** $p \leq 0.01$) are represented by asterisks.

Supplementary Materials: Supplementary materials can be found at http://www.mdpi.com/1422-0067/20/24/6202/s1. Figure S1, Visualization of stomata in B73 and opr7opr8 mutant. The leaves of B73 and opr7opr8 under control, 100 mM, and 300 mM salt stress were used to image stomata 24 h after application of NaCl under a microscope at 40× magnification; Table S1, Primers used for quantitative real time PCR (qRT-PCR) in the study.

Author Contributions: The work presented here was a collaborative study of all the authors. R.M.A. prepared and executed the experiments, and wrote and kept working on the revised version of the manuscript. C.C. and J.S. helped with the experimental performance and data analysis. W.W., H.R., and M.A. helped with seeds and manuscript modification, Y.Y. designed the project, analyzed the data, revised the manuscript and obtained the funds to support this project.

Funding: This work was supported by the National Natural Science Foundation of China (31571580), the Fundamental Research Funds for the Central Universities (KYTZ201402 and KYRC201404), and the Outstanding Scientific Innovation Team Program for Jiangsu Universities (2015).

Acknowledgments: We are grateful to Dr. Michael V. Kolomiets (Texas A&M University, College Station, Texas, USA) who provided the JA-deficient mutant *opr7opr8* for the experiments of this study.

Conflicts of Interest: The authors declare no conflict of interest.

Abbreviations

FRW	Fresh root weight
FSW	Fresh shoot weight
DRW	Dry root weight
DSW	Dry shoot weight
FSL	Fresh shoot length
FRL	Fresh shoot length
Pro	Proline
MDA	Malondialdehyde
Chl	Chlorophyll
CAT	Catalase
SOD	Superoxide dismutase
APX	Ascorbate peroxidase

References

1. Setia, R.; Gottschalk, P.; Smith, P.; Marschner, P.; Baldock, J.; Setia, D.; Smith, J. Soil salinity decreases global soil organic carbon stocks. *Sci. Total Environ.* **2013**, *465*, 267–272. [CrossRef] [PubMed]
2. Shrivastava, P.; Kumar, R. Soil salinity: A serious environmental issue and plant growth promoting bacteria as one of the tools for its alleviation. *Saudi J. Biol. Sci.* **2015**, *22*, 123–131. [CrossRef] [PubMed]
3. Flowers, T.J. Improving crop salt tolerance. *J. Exp. Bot.* **2004**, *55*, 307–319. [CrossRef] [PubMed]
4. Verma, V.; Ravindran, P.; Kumar, P.P. Plant hormone-mediated regulation of stress responses. *BMC Plant. Biol.* **2016**, *16*, 86. [CrossRef]
5. Wani, S.H.; Kumar, V.; Shriram, V.; Sah, S.K. Phytohormones and their metabolic engineering for abiotic stress tolerance in crop plants. *Crop. J.* **2016**, *4*, 162–176. [CrossRef]
6. Ku, Y.S.; Sintaha, M.; Cheung, M.Y.; Lam, H.M. Plant Hormone Signaling Crosstalks between Biotic and Abiotic Stress Responses. *Int. J. Mol. Sci.* **2018**, *19*, 3206. [CrossRef]
7. Popko, J.; Hansch, R.; Mendel, R.R.; Polle, A.; Teichmann, T. The role of abscisic acid and auxin in the response of poplar to abiotic stress. *Plant. Biol.* **2010**, *12*, 242–258. [CrossRef]
8. Shu, K.; Zhou, W.; Chen, F.; Luo, X.; Yang, W. Abscisic Acid and Gibberellins Antagonistically Mediate Plant Development and Abiotic Stress Responses. *Front. Plant. Sci.* **2018**, *9*, 416. [CrossRef]
9. Zwack, P.J.; Rashotte, A.M. Interactions between cytokinin signalling and abiotic stress responses. *J. Exp. Bot.* **2015**, *66*, 4863–4871. [CrossRef]
10. Tohidi, B.; Rahimmalek, M.; Trindade, H. Review on essential oil, extracts composition, molecular and phytochemical properties of Thymus species in Iran. *Ind. Crop. Prod.* **2019**, *134*, 89–99. [CrossRef]
11. Ghasemi Pirbalouti, A.; Sajjadi, S.E.; Parang, K. ChemInform Abstract: A Review (Research and Patents) on Jasmonic Acid and Its Derivatives. *Arch. Der Pharm.* **2014**, *347*. [CrossRef]
12. Farmer, E.E.; Almeras, E.; Krishnamurthy, V. Jasmonates and related oxylipins in plant responses to pathogenesis and herbivory. *Curr. Opin. Plant. Biol.* **2003**, *6*, 372–378. [CrossRef]
13. Browse, J. Jasmonate passes muster: A receptor and targets for the defense hormone. *Annu. Rev. Plant. Biol.* **2009**, *60*, 183–205. [CrossRef]
14. Wasternack, C.; Song, S. Jasmonates: Biosynthesis, metabolism, and signaling by proteins activating and repressing transcription. *J. Exp. Bot.* **2016**, *68*, 1303–1321. [CrossRef]
15. McConn, M.; Browse, J. The Critical Requirement for Linolenic Acid Is Pollen Development, Not Photosynthesis, in an Arabidopsis Mutant. *Plant. Cell* **1996**, *8*, 403–416. [CrossRef]
16. Sanders, P.M.; Lee, P.Y.; Biesgen, C.; Boone, J.D.; Beals, T.P.; Weiler, E.W.; Goldberg, R.B. The Arabidopsis DELAYED DEHISCENCE1 Gene Encodes an Enzyme in the Jasmonic Acid Synthesis Pathway. *Plant. Cell* **2000**, *12*, 1041–1061. [CrossRef]
17. Stintzi, A.; Browse, J. The Arabidopsis male-sterile mutant, opr3, lacks the 12-oxophytodienoic acid reductase required for jasmonate synthesis. *Proc. Natl Acad Sci. USA* **2000**, *97*, 10625–10630. [CrossRef]
18. Park, J.H.; Halitschke, R.; Kim, H.B.; Baldwin, I.T.; Feldmann, K.A.; Feyereisen, R. A knock-out mutation in allene oxide synthase results in male sterility and defective wound signal transduction in Arabidopsis due to a block in jasmonic acid biosynthesis. *Plant. J. Cell Mol. Biol.* **2002**, *31*, 1–12. [CrossRef]
19. Feys, B.; Benedetti, C.E.; Penfold, C.N.; Turner, J.G. Arabidopsis Mutants Selected for Resistance to the Phytotoxin Coronatine Are Male Sterile, Insensitive to Methyl Jasmonate, and Resistant to a Bacterial Pathogen. *Plant. Cell* **1994**, *6*, 751–759. [CrossRef]
20. Browse, J. The power of mutants for investigating jasmonate biosynthesis and signaling. *Phytochemistry* **2009**, *70*, 1539–1546. [CrossRef]
21. Staswick, P.E.; Su, W.; Howell, S.H. Methyl jasmonate inhibition of root growth and induction of a leaf protein are decreased in an Arabidopsis thaliana mutant. *Proc. Natl. Acad Sci. USA* **1992**, *89*, 6837–6840. [CrossRef] [PubMed]
22. Lorenzo, O.; Chico, J.M.; Sanchez-Serrano, J.J.; Solano, R. JASMONATE-INSENSITIVE1 encodes a MYC transcription factor essential to discriminate between different jasmonate-regulated defense responses in Arabidopsis. *Plant. Cell* **2004**, *16*, 1938–1950. [CrossRef] [PubMed]
23. Lee, G.I.; Howe, G.A. The tomato mutant spr1 is defective in systemin perception and the production of a systemic wound signal for defense gene expression. *Plant. J.: Cell Mol. Biol.* **2003**, *33*, 567–576. [CrossRef] [PubMed]

24. Li, L.; Li, C.; Lee, G.I.; Howe, G.A. Distinct roles for jasmonate synthesis and action in the systemic wound response of tomato. *Proc. Natl. Acad Sci. USA* **2002**, *99*, 6416–6421. [CrossRef]

25. Li, L.; Zhao, Y.; McCaig, B.C.; Wingerd, B.A.; Wang, J.; Whalon, M.E.; Pichersky, E.; Howe, G.A. The tomato homolog of CORONATINE-INSENSITIVE1 is required for the maternal control of seed maturation, jasmonate-signaled defense responses, and glandular trichome development. *Plant. Cell* **2004**, *16*, 126–143. [CrossRef]

26. Maksymiec, W.; Wianowska, D.; Dawidowicz, A.L.; Radkiewicz, S.; Mardarowicz, M.; Krupa, Z. The level of jasmonic acid in Arabidopsis thaliana and Phaseolus coccineus plants under heavy metal stress. *J. Plant. Physiol.* **2005**, *162*, 1338–1346. [CrossRef]

27. Dong, W.; Wang, M.; Xu, F.; Quan, T.; Peng, K.; Xiao, L.; Xia, G. Wheat oxophytodienoate reductase gene TaOPR1 confers salinity tolerance via enhancement of abscisic acid signaling and reactive oxygen species scavenging. *Plant. Physiol.* **2013**, *161*, 1217–1228. [CrossRef]

28. Qiu, Z.; Guo, J.; Zhu, A.; Zhang, L.; Zhang, M. Exogenous jasmonic acid can enhance tolerance of wheat seedlings to salt stress. *Ecotoxicol. Environ. Saf.* **2014**, *104*, 202–208. [CrossRef]

29. Zhao, Y.; Dong, W.; Zhang, N.; Ai, X.; Wang, M.; Huang, Z.; Xiao, L.; Xia, G. A wheat allene oxide cyclase gene enhances salinity tolerance via jasmonate signaling. *Plant. Physiol.* **2014**, *164*, 1068–1076. [CrossRef]

30. Brossa, R.; López-Carbonell, M.; Jubany-Marí, T.; Alegre, L. Interplay Between Abscisic Acid and Jasmonic Acid and its Role in Water-oxidative Stress in Wild-type, ABA-deficient, JA-deficient, and Ascorbate-deficient Arabidopsis Plants. *J. Plant. Growth Regul.* **2011**, *30*, 322–333. [CrossRef]

31. Clarke, S.M.; Cristescu, S.M.; Miersch, O.; Harren, F.J.; Wasternack, C.; Mur, L.A. Jasmonates act with salicylic acid to confer basal thermotolerance in Arabidopsis thaliana. *New Phytol.* **2009**, *182*, 175–187. [CrossRef] [PubMed]

32. Sharma, M.; Laxmi, A. Jasmonates: Emerging Players in Controlling Temperature Stress Tolerance. *Front. Plant. Sci.* **2016**, 6. [CrossRef] [PubMed]

33. Golldack, D.; Li, C.; Mohan, H.; Probst, N. Tolerance to drought and salt stress in plants: Unraveling the signaling networks. *Front. Plant. Sci.* **2014**, *5*, 151. [CrossRef] [PubMed]

34. Ismail, A.; Riemann, M.; Nick, P. The jasmonate pathway mediates salt tolerance in grapevines. *J. Exp. Bot.* **2012**, *63*, 2127–2139. [CrossRef] [PubMed]

35. Yoon, J.Y.; Hamayun, M.; Lee, S.-K.; Lee, I.-J. Methyl jasmonate alleviated salinity stress in soybean. *J. Crop. Sci. Biotechnol.* **2009**, *12*, 63–68. [CrossRef]

36. Ghassemi-Golezani, K.; Hosseinzadeh-Mahootchi, A. Improving physiological performance of safflower under salt stress by application of salicylic acid and jasmonic acid. *WALIA J.* **2015**, *31*, 104–109.

37. Faghih, S.; Ghobadi, C.; Zarei, A. Response of Strawberry Plant cv. 'Camarosa' to Salicylic Acid and Methyl Jasmonate Application Under Salt Stress Condition. *J. Plant. Growth Regul.* **2017**, *36*. [CrossRef]

38. Farhangi-Abriz, S.; Ghassemi-Golezani, K. How can salicylic acid and jasmonic acid mitigate salt toxicity in soybean plants? *Ecotoxicol. Environ. Saf.* **2018**, *147*, 1010–1016. [CrossRef]

39. Hazman, M.; Hause, B.; Eiche, E.; Nick, P.; Riemann, M. Increased tolerance to salt stress in OPDA-deficient rice ALLENE OXIDE CYCLASE mutants is linked to an increased ROS-scavenging activity. *J. Exp. Bot.* **2015**, *66*, 3339–3352. [CrossRef]

40. Zhu, D.; Cai, H.; Luo, X.; Bai, X.; Deyholos, M.K.; Chen, Q.; Chen, C.; Ji, W.; Zhu, Y. Over-expression of a novel JAZ family gene from Glycine soja, increases salt and alkali stress tolerance. *Biochem. Biophys. Res. Commun.* **2012**, *426*, 273–279. [CrossRef]

41. Yan, Y.; Christensen, S.; Isakeit, T.; Engelberth, J.; Meeley, R.; Hayward, A.; Emery, R.J.; Kolomiets, M.V. Disruption of OPR7 and OPR8 reveals the versatile functions of jasmonic acid in maize development and defense. *Plant. Cell* **2012**, *24*, 1420–1436. [CrossRef] [PubMed]

42. Wakeel, A. Potassium–sodium interactions in soil and plant under saline-sodic conditions. *J. Plant. Nutr. Soil Sci.* **2013**, *176*, 344–354. [CrossRef]

43. Saxena, I.; Srikanth, S.; Chen, Z. Cross Talk between H2O2 and Interacting Signal Molecules under Plant Stress Response. *Front. Plant. Sci.* **2016**, *7*, 570. [CrossRef] [PubMed]

44. Xiong, L.; Zhu, J.K. Regulation of abscisic acid biosynthesis. *Plant. Physiol.* **2003**, *133*, 29–36. [CrossRef]

45. Porch, T.G.; Tseung, C.W.; Schmelz, E.A.; Settles, A.M. The maize Viviparous10/Viviparous13 locus encodes the Cnx1 gene required for molybdenum cofactor biosynthesis. *Plant. J. Cell Mol. Biol.* **2006**, *45*, 250–263. [CrossRef]

46. Munns, R.; Tester, M. Mechanisms of Salinity Tolerance. *Annu. Rev. Plant. Biol.* **2008**, *59*, 651–681. [CrossRef]
47. Zhu, J.K. Salt and drought stress signal transduction in plants. *Annu. Rev. Plant. Biol.* **2002**, *53*, 247–273. [CrossRef]
48. Ryu, H.; Cho, Y.-G. Plant hormones in salt stress tolerance. *J. Plant. Biol.* **2015**, *58*, 147–155. [CrossRef]
49. Raghavendra, A.S.; Gonugunta, V.K.; Christmann, A.; Grill, E. ABA perception and signalling. *Trends Plant. Sci.* **2010**, *15*, 395–401. [CrossRef]
50. Kazan, K. Diverse roles of jasmonates and ethylene in abiotic stress tolerance. *Trends Plant. Sci.* **2015**, *20*, 219–229. [CrossRef]
51. Riemann, M.; Dhakarey, R.; Hazman, M.; Miro, B.; Kohli, A.; Nick, P. Exploring Jasmonates in the Hormonal Network of Drought and Salinity Responses. *Front. Plant. Sci.* **2015**, *6*, 1077. [CrossRef] [PubMed]
52. Valenzuela, C.E.; Acevedo-Acevedo, O.; Miranda, G.S.; Vergara-Barros, P.; Holuigue, L.; Figueroa, C.R.; Figueroa, P.M. Salt stress response triggers activation of the jasmonate signaling pathway leading to inhibition of cell elongation in Arabidopsis primary root. *J. Exp. Bot.* **2016**, *67*, 4209–4220. [CrossRef] [PubMed]
53. Abdala, G. Jasmonate and octadecanoid occurrence in tomato hairy roots. Endogenous level changes in response to NaCl. *Plant. Growth Regul.* **2003**, *40*, 21–27. [CrossRef]
54. Pedranzani, H.; Racagni, G.; Alemano, S.; Miersch, O.; Ramírez, I.; Pena-Cortes, H.; Taleisnik, E.; Machado, E.; Abdala, G. Salt tolerant tomato plants show increased levels of jasmonic acid. *Plant. Growth Regul.* **2003**, *41*, 149–158. [CrossRef]
55. Moons, A.; Prinsen, E.; Bauw, G.; Van Montagu, M. Antagonistic Effects of Abscisic Acid and Jasmonates on Salt Stress-Inducible Transcripts in Rice Roots. *Plant. Cell* **1997**, *9*, 2243–2259. [CrossRef]
56. Kang, D.-J.; Seo, Y.J.; Lee, J.D.; Ishii, R.; Kim, K.U.; Shin, D.H.; Park, S.K.; Jang, S.W.; Lee, I.J. Jasmonic Acid Differentially Affects Growth, Ion Uptake and Abscisic Acid Concentration in Salt-tolerant and Salt-sensitive Rice Cultivars. *J. Agron. Crop. Sci.* **2005**, *191*, 273–282. [CrossRef]
57. Shahzad, A.N.; Pitann, B.; Ali, H.; Qayyum, M.F.; Fatima, A.; Bakhat, H.F. Maize Genotypes Differing in Salt Resistance Vary in Jasmonic Acid Accumulation During the First Phase of Salt Stress. *J. Agron. Crop. Sci.* **2015**, *201*, 443–451. [CrossRef]
58. Pavlović, I.; Pencík, A.; Novak, O.; Vujčić, V.; Radić, S.; Lepeduš, H.; Strnad, M.; Salopek-Sondi, B. Short-term salt stress in Brassica rapa seedlings causes alterations in auxin metabolism. *Plant. Physiol. Biochem.* **2018**, *125*. [CrossRef]
59. Mahmud, S.; Sharmin, S.; Chowdhury, B.; Hossain, M. Effect of Salinity and Alleviating Role of Methyl Jasmonate in Some Rice Varieties. *Asian J. Plant. Sci.* **2017**, *16*, 87–93. [CrossRef]
60. Ghassemi-Golezani, K.; Farhangi-Abriz, S. Foliar sprays of salicylic acid and jasmonic acid stimulate H(+)-ATPase activity of tonoplast, nutrient uptake and salt tolerance of soybean. *Ecotoxicol. Environ. Saf.* **2018**, *166*, 18–25. [CrossRef]
61. Ahmadi, F.I.; Karimi, K.; Struik, P.C. Effect of exogenous application of methyl jasmonate on physiological and biochemical characteristics of Brassica napus L. cv. Talaye under salinity stress. *South. Afr. J. Bot.* **2018**, *115*, 5–11. [CrossRef]
62. Mir, M.A.; John, R.; Alyemeni, M.N.; Alam, P.; Ahmad, P. Jasmonic acid ameliorates alkaline stress by improving growth performance, ascorbate glutathione cycle and glyoxylase system in maize seedlings. *Sci. Rep.* **2018**, *8*, 2381. [CrossRef]
63. Kurotani, K.; Hayashi, K.; Hatanaka, S.; Toda, Y.; Ogawa, D.; Ichikawa, H.; Ishimaru, Y.; Tashita, R.; Suzuki, T.; Ueda, M.; et al. Elevated levels of CYP94 family gene expression alleviate the jasmonate response and enhance salt tolerance in rice. *Plant. Cell Physiol.* **2015**, *56*, 779–789. [CrossRef]
64. Wu, H.; Ye, H.; Yao, R.; Zhang, T.; Xiong, L. OsJAZ9 acts as a transcriptional regulator in jasmonate signaling and modulates salt stress tolerance in rice. *Plant. Sci. Int. J. Exp. Plant. Biol.* **2015**, *232*, 1–12. [CrossRef]
65. Peethambaran, P.K.; Glenz, R.; Honinger, S.; Shahinul Islam, S.M.; Hummel, S.; Harter, K.; Kolukisaoglu, U.; Meynard, D.; Guiderdoni, E.; Nick, P.; et al. Salt-inducible expression of OsJAZ8 improves resilience against salt-stress. *Bmc Plant. Biol.* **2018**, *18*, 311. [CrossRef]
66. Yastreb, T.; Kolupaev, Y.; Kokorev, A.; Horielova, E.; Dmitriev, A. Methyl Jasmonate and Nitric Oxide in Regulation of the Stomatal Apparatus of Arabidopsis thaliana. *Cytol. Genet.* **2018**, *52*, 400–405. [CrossRef]
67. Hossain, M.A.; Munemasa, S.; Uraji, M.; Nakamura, Y.; Mori, I.C.; Murata, Y. Involvement of endogenous abscisic acid in methyl jasmonate-induced stomatal closure in Arabidopsis. *Plant. Physiol.* **2011**, *156*, 430–438. [CrossRef]

68. Sanz, L.C.; Fernández-Maculet, J.C.; Gómez, E.; Vioque, B.; Olías, J.M. Effect of methyl jasmonate on ethylene biosynthesis and stomatal closure in olive leaves. *Phytochemistry* **1993**, *33*, 285–289. [CrossRef]

69. Metodiev, M.; Tsonev, T.; Popova, L. Effect of jasmonic acid on the stomatal and nonstomatal limitation of leaf photosynthesis in barley leaves. *J. Plant. Growth Regul.* **1996**, *15*, 75–80. [CrossRef]

70. Han, X.; Hu, Y.; Zhang, G.; Jiang, Y.; Chen, X. Jasmonate Negatively Regulates Stomatal Development in Arabidopsis Cotyledons. *Plant Physiol.* **2018**, *176*, 2871–2885. [CrossRef]

71. AbdElgawad, H.; Zinta, G.; Hegab, M.M.; Pandey, R.; Asard, H.; Abuelsoud, W. High Salinity Induces Different Oxidative Stress and Antioxidant Responses in Maize Seedlings Organs. *Front. Plant. Sci.* **2016**, *7*, 276. [CrossRef] [PubMed]

72. Gill, S.S.; Tuteja, N. Reactive oxygen species and antioxidant machinery in abiotic stress tolerance in crop plants. *Plant. Physiol. Biochem. Ppb* **2010**, *48*, 909–930. [CrossRef] [PubMed]

73. Ahmad, P.; Jaleel, C.A.; Salem, M.A.; Nabi, G.; Sharma, S. Roles of enzymatic and nonenzymatic antioxidants in plants during abiotic stress. *Crit. Rev. Biotechnol.* **2010**, *30*, 161–175. [CrossRef] [PubMed]

74. Foyer, C.H.; Noctor, G. Ascorbate and glutathione: The heart of the redox hub. *Plant. Physiol.* **2011**, *155*, 2–18. [CrossRef] [PubMed]

75. Farhangi-Abriz, S.; Ghassemi-Golezani, K. Jasmonates: Mechanisms and functions in abiotic stress tolerance of plants. *Biocatal. Agric. Biotechnol.* **2019**, *20*, 101210. [CrossRef]

76. Soares, A.M.d.S.; Souza, T.F.d.; Jacinto, T.; Machado, O.L.T. Effect of Methyl Jasmonate on antioxidative enzyme activities and on the contents of ROS and H2O2 in Ricinus communis leaves. *Braz. J. Plant. Physiol.* **2010**, *22*, 151–158. [CrossRef]

77. Parra-Lobato, M.; Fernández-García, N.; Olmos, E.; Alvarez-Tinaut, M.; Gomez-Jimenez, M. Methyl jasmonate-induced antioxidant defence in root apoplast from sunflower seedlings. *Environ. Exp. Bot.* **2009**, *66*, 9–17. [CrossRef]

78. Regni, L.; Del Pino, A.M.; Mousavi, S.; Palmerini, C.A.; Baldoni, L.; Mariotti, R.; Mairech, H.; Gardi, T.; D'Amato, R.; Proietti, P. Behavior of Four Olive Cultivars During Salt Stress. *Front. Plant. Sci.* **2019**, *10*. [CrossRef]

79. Sah, S.K.; Reddy, K.R.; Li, J. Abscisic Acid and Abiotic Stress Tolerance in Crop Plants. *Front. Plant. Sci.* **2016**, *7*. [CrossRef]

80. Kundu, S.; Gantait, S. Abscisic acid signal crosstalk during abiotic stress response. *Plant. Gene.* **2017**, *11*, 61–69. [CrossRef]

81. SestÁK, Z.; CatskÝ, J.; Jarvis, P.G. *Plant Photosynthetic Production. Manual of Methods*; Dr. W. Junk NV: The Hague, The Netherlands, 1971; p. 818.

© 2019 by the authors. Licensee MDPI, Basel, Switzerland. This article is an open access article distributed under the terms and conditions of the Creative Commons Attribution (CC BY) license (http://creativecommons.org/licenses/by/4.0/).

International Journal of
Molecular Sciences

Article

JA-Induced Endocytosis of AtRGS1 Is Involved in G-Protein Mediated JA Responses

Li Li [1,†], Bodan Su [1,†], Xueying Qi [1], Xi Zhang [1], Susheng Song [2] and Xiaoyi Shan [1,*

[1]   College of Biological Sciences and Biotechnology, Beijing Forestry University, Beijing 10083, China
[2]   Beijing Key Laboratory of Plant Gene Resources and Biotechnology for Carbon Reduction and Environmental Improvement, College of Life Sciences, Capital Normal University, Beijing 100048, China
*   Correspondence: shanxy@bjfu.edu.cn
†   These authors contributed equally to this article.

Received: 28 June 2019; Accepted: 29 July 2019; Published: 2 August 2019

Abstract: *Arabidopsis* heterotrimeric G proteins regulate diverse plant growth and defense processes by coupling to 7TM AtRGS1 proteins. Although G protein mutants display alterations in response to multiple plant hormones, the underlying mechanism by which G proteins participate in the regulation of hormone responses remains elusive. Here, we show that genetic disruption of Gα and Gβ subunits results in reduced sensitivity to JA treatment. Furthermore, using confocal microscopy, VA-TIRFM, and FRET-FLIM, we provide evidence that stimulation by JA induces phosphorylation- and C-terminus-dependent endocytosis of AtRGS1, which then promotes dissociation of AtRGS1 from AtGPA1. In addition, SPT analysis reveals that JA treatment affects the diffusion dynamics of AtRGS1 and AtRGS1-ΔCt. Taken together, these findings suggest that the JA signal activates heterotrimeric G proteins through the endocytosis of AtRGS1 and dissociation of AtRGS1 from AtGPA1, thus providing valuable insight into the mechanisms of how the G protein system perceives and transduces phytohormone signals.

Keywords: heterotrimeric G proteins; AtRGS1; jasmonates; endocytosis; diffusion dynamics

1. Introduction

Jasmonates (JAs), which include jasmonic acid and its oxylipin derivatives, are synthesized from the octadecanoid/hexadecanoid pathways and are widely distributed throughout the plant kingdom [1]. As one of the important plant hormones, JAs function as growth regulators and defense signals that control various plant developmental processes, such as root growth, anthocyanin accumulation, male fertility, and leaf senescence, as well as mediate plant responses to abiotic and biotic stresses, including insect attack, UV damage, pathogen infections, and wounding [2]. The jasmonate ZIM-domain proteins (JAZs) that consist of 12 members act as repressors to negatively regulate diverse JA responses by directly binding to downstream transcriptional factors (TFs) or interactions with the co-repressor TOPLESS (TPL) via the novel interactor of JAZ (NINJA) [3]. Upon perception of a JA signal, the F-box protein coronatine insensitive 1 (COI1) forms an SCF^{COI1} complex with *Arabidopsis* SKP1 homologue 1 (ASK1)/ASK2, AtCullin1, and *Arabidopsis* Ring-box 1 (AtRbx1) to recruit JAZs for ubiquitination and degradation through the 26S proteasome, which subsequently activates downstream TFs [3].

Signal transduction through a heterotrimeric G protein complex, classically consisting of Gα, Gβ, and Gγ subunits, is an essential plasma membrane (PM) signaling pathway in most eukaryotes [4,5]. In animal cells, the heterotrimeric G proteins are directly regulated by seven transmembrane (7TM) G protein-coupled receptors (GPCRs) that could catalyze the nucleotide exchange from guanosine triphosphate (GTP) to guanosine diphosphate (GDP) on the Gα subunit upon stimulation [6]. Thus, GTP-bound Gα and Gβγ dimers are released to activate downstream effectors, thereby relaying diverse intracellular signals. In *Arabidopsis*, this complex comprises one canonical Gα subunit (AtGPA1), one Gβ subunit (AGB1), and one of three Gβ subunits (AGG1, 2, and 3) [7–10]. In contrast to animal alpha subunits, AtGPA1 spontaneously undergoes the GDP- and GTP-exchange cycle. AtRGS1, a GPCR-like N-terminal 7TM fused to a regulator of G-protein signaling (RGS) protein, keeps the complex in the inactive state by promoting GTP hydrolysis of AtGPA1 [11,12]. D-glucose or other stimuli can induce the dissociation of AtGPA1 from AtRGS1 to maintain its self-active state and trigger effector activities [13]. Genetic analysis of loss-of-function G protein mutants in *Arabidopsis* and rice support the fundamental roles of heterotrimeric G proteins in various biological processes, including morphological development from seed germination to silique development, response to glucose, light stimuli and hormones, stomatal movements, and ion channel regulation, as well as innate immunity [13].

Endocytosis of the 7TM receptors in mammals and plants is considered to be a vital step in the regulation of G protein signaling. In animals, GPCRs are phosphorylated and then endocytosed through a clathrin-dependent pathway to desensitize ligand stimulation [6]. However, phosphorylated AtRGS1 protein is internalized to release its inhibition upon AtGPA1 self-activation, permitting sustained activation of G protein signaling. In the presence of glucose, WITH NO LYSINE kinases (WNKs) can phosphorylate AtRGS1 at the C-terminal domain for endocytosis, which is critical to the activation of G protein-mediated sugar signaling and cell proliferation [14,15]. After induction of a conserved 22-amino acid domain in the N-terminus of flagellin (flg22), AtRGS1 is phosphorylated by BRI1-associated kinase 1 (BAK1) at its C-terminal tail and then internalized [16,17]. Thus, the G protein complex is physically uncoupled from its repressor, and then interacts with its effectors to regulate reactive oxygen species (ROS) production and calcium release [17–19]. In addition, sodium-induced activation of G signaling via AtRGS1 endocytosis plays a key role in plant responses to salt stress [20].

In the present study, we found that the loss of Gα and Gβ subunits conferred hyposensitivity to JA signaling. Furthermore, we demonstrated that JA induced the endocytosis of AtRGS1 in a phosphorylation- and C-terminus-dependent manner, which in turn led to the dissociation of AtRGS1 from AtGPA1 for activation of downstream signaling. In addition, the diffusion dynamics of AtRGS1 and C-terminus-truncated AtRGS1 proteins changed upon JA treatment. Our analyses propose a possible role for JA-induced endocytosis of AtRGS1 in G-protein-mediated JA responses and support the hypothesisthat plant hormone signaling might be relayed, in part, by the AtRGS1-G protein pathway.

2. Results and Discussion

2.1. The Heterotrimeric G Protein Complex Is Involved in JA Signaling

Considering that G protein mutants exhibit various phenotypes during growth and in defense responses, it is possible that the heterotrimeric G protein complex might be involved with JA signaling, which regulates plant development and immunity. To gain insights into the relationship between G proteins and JA signaling, we subjected wild-type (WT, Col-0), Gα-deficient mutant (*gpa1-4*), Gβ-deficient mutant (*agb1-2*), and double mutant *gpa1-4/agb1-2* to a series of JA response assays.

Compared to seedlings without methyl jasmonate (MeJA) treatment, each genotype exhibited JA-induced inhibition of primary root growth when grown on MS medium with different concentrations of MeJA (Figure 1A,B). However, the stunted growth of *gpa1-4*, *agb1-2*, or *gpa1-4/agb1-2* mutants was significantly less than that of WT seedlings (73%, 50%, and 56% for *gpa1-4*, *agb1-2*, and *gpa1-4/agb1-2*, respectively, 44% for WT upon 10 μm MeJA; 58%, 44%, and 47% for *gpa1-4*, *agb1-2*, and *gpa1-4/agb1-2*, respectively, 38% for WT upon 25 μm MeJA) (Figure 1A,B). Consistent with the root length phenotype, the G protein mutants were also hyposensitive to JA-induced anthocyanin accumulation. The anthocyanin content of WT plants was 1.7-, 1.3-, and 1.2-fold greater than that of *gpa1-4*, *agb1-2*, and *gpa1-4/agb1-2*, respectively, in response to JA induction (Figure 1C,D). Moreover, JA treatment could upregulate the expression of *VSP1* and *LOX2* in each genotype, whereas the expression level in *gpa1-4* and *gpa1-4/agb1-2* mutants was significantly less than the WT seedlings (Figure 1E,F). For example, *VSP1* expression was reduced by 36% and 14% in the *gpa1-4* and *gpa1-4/agb1-2* mutant in comparison to that in the WT seedling, respectively (Figure 1E). The JA-induced expression level of *LOX2* was also noticeably decreased in the *agb1-2* plants, which was about half of that in the WT seedlings (Figure 1F). However, the *VSP1* induction pattern was almost identical in WT and *agb1-2* mutants (Figure 1E).

Collectively, knockout mutants of *AtGPA1* and *AGB1* display and share reduced sensitivity to JA in root growth, anthocyanin accumulation, and *LOX2* gene expression (Figure 1). These findings suggest that both Gα and Gβ subunits likely function in some, but not all, JA responses as positive regulators.

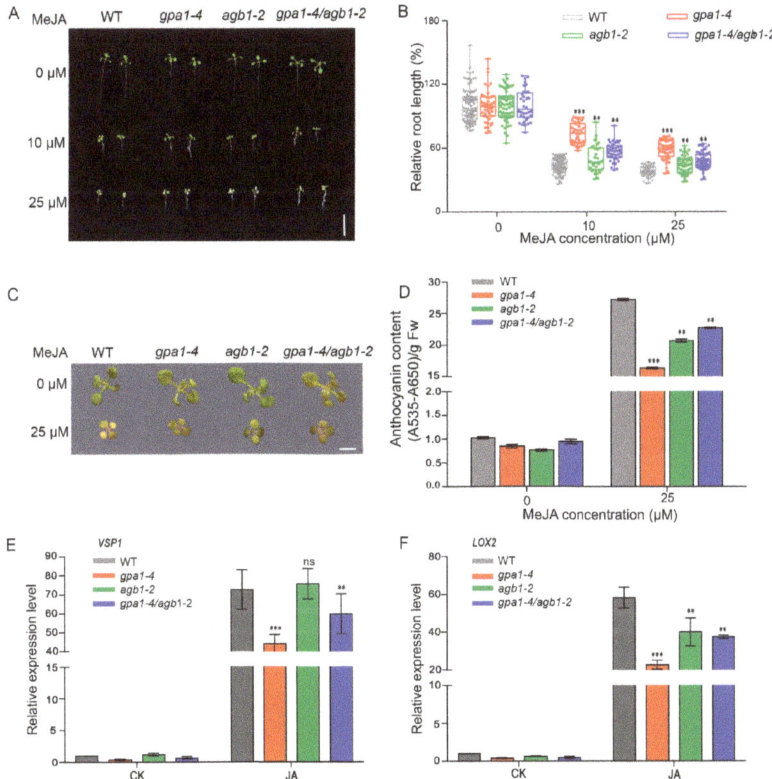

Figure 1. The heterotrimeric G protein complex is involved in JA signaling. (**A**) Root phenotypes of 9-day-old seedlings of WT, *gpa1-4*, *agb1-2*, and *gpa1-4/agb1-2* grown on MS medium containing indicated concentrations of MeJA. Bar = 1 cm. (**B**) Relative root length of 9-day-old seedlings of WT, *gpa1-4*, *agb1-2*, and *gpa1-4/agb1-2* grown on MS medium containing indicated concentrations of MeJA. Data shown are from 34 to 104 plants. Significant differences are denoted by asterisks (** $p < 0.01$, *** $p < 0.001$, Student's *t* test). (**C**) Phenotype of 11-day-old seedlings of WT, *gpa1-4*, *agb1-2*, and *gpa1-4/agb1-2* grown on MS medium containing indicated concentrations of MeJA. Bar = 1 cm. (**D**) Anthocyanin contents of 11-day-old seedlings of WT, *gpa1-4*, *agb1-2*, and *gpa1-4/agb1-2* grown on MS medium containing indicated concentrations of MeJA. Error bars represent SD ($n = 3$). Significant differences are denoted by asterisks (** $p < 0.01$, *** $p < 0.001$, Student's *t* test). (**E,F**), Relative expression level of *VSP1* (**E**) and *LOX2* (**F**) in WT, *gpa1-4*, *agb1-2*, and *gpa1-4/agb1-2* seedlings treated without (CK) or with 100 μm MeJA for 8 h. Error bars represent SD ($n = 3$). Significant differences are denoted by asterisks (** $p < 0.01$, *** $p < 0.001$, ns, no significant difference, Student's *t* test).

2.2. JA Induces AtRGS1 Endocytosis and Dissociation from AtGPA1

One possible explanation for the involvement of the G-protein complex in JA signaling is that JA itself indirectly activates G proteins and their downstream effectors to regulate JA responses. Given that the endocytosis of AtRGS1 is a well-known reporter for G protein activation, plants expressing AtRGS1-YFP (Figure S1) were treated with 100 μm MeJA plus the protein synthesis inhibitor, cycloheximide (CHX), for different times, and AtRGS1-YFP internalization was quantitated. Under steady-state (CK) or with CHX treatment only, AtRGS1-YFP proteins were mainly localized to the PM (Figure 2A). When treated with MeJA for 2 and 4 h, no detectable change in the subcellular location of AtRGS1-YFP was observed (Figure 2A). Notably, distinct YFP-positive vesicles appeared in the cytoplasm upon MeJA treatment for 6 h or more (Figure 2A). By analyzing the percentage of AtRGS1 endocytosis, 2 and 4 h MeJA treatment (12% for both 2 and 4 h) showed no quantitative effect compared to the control conditions (10% for both CK and CHX only) (Figure 2B). A significant increase in internalization (15.8% of AtRGS1) was found at 6 h after MeJA treatment, and by 8 h, more extensive endocytosis was observed (23% of AtRGS1) (Figure 2B), suggesting a time-dependent AtRGS1 internalization upon JA treatment.

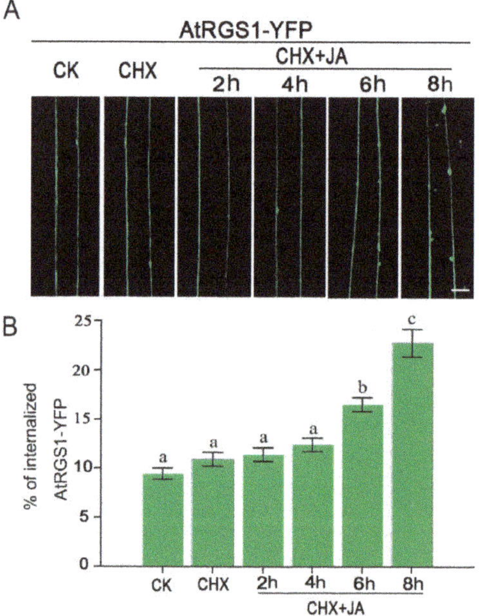

Figure 2. JA induces the endocytosis of AtRGS1. (**A**) Confocal images of AtRGS1-YFP localization without treatment (CK), with CHX treatment and with MeJA plus CHX pretreatment for indicated times in an *Arabidopsis* hypocotyl epidermal cell. Bar = 5 μm. (**B**) Quantification of the internalized AtRGS1-YFP without treatment (CK), with CHX treatment, and with MeJA plus CHX pretreatment for indicated times. Error bars represent SD (n = 5–8). Significant differences are denoted by letters (Duncan multiple-comparisons test; values within the column followed by the same letter show no significant difference ($p > 0.05$), whereas those with the different letter indicate significant difference ($p < 0.05$)).

Using variable-angle total internal reflection fluorescence microscopy (VA-TIRFM), we found that AtRGS1-YFP spots localized to the PM and formed dispersed punctate structures with increased fluorescence intensities (Figure 3B). Sequential images with a 15 s recording in 0.15 s intervals showed that the individual particles remained on the PM with lateral and temporal dynamics (Figure 3C,D).

To obtain more data on JA-induced endocytosis of AtRGS1, we implemented kymograph analysis to investigate the PM residence time of AtRGS1 particles in response to JA treatment. There was no significant difference between seedlings without treatment (CK, t = 2.06 s) and with CHX only (t = 2.02 s), whereas JA treatment led to significantly shorter dwell times (t = 1.56 s) compared to the control seedlings (CK and CHX only) (Figure 6A,B,E). These results provide evidence that JA facilitates rapid internalization of AtRGS1.

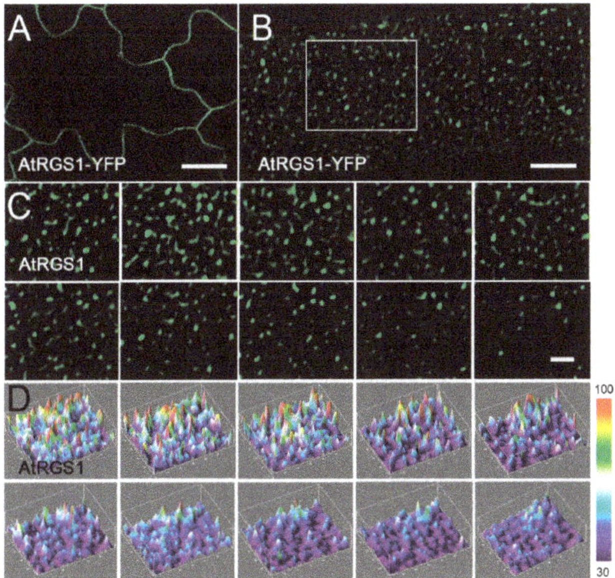

Figure 3. The distribution of AtRGS1 on the PM. (**A**) Confocal images of AtRGS1-YFP in *Arabidopsis* leaf epidermal cells. Bar = 5 µm. (**B**) A typical single-particle image for AtRGS1-YFP at the PM. (**C**) Sequential images of the boxed area in (**B**) with a 15 s recording in 1.5 s intervals by VA-TIRFM. Bar = 2 µm. (**D**) Three-dimensional luminance plots in (**C**) showing varied fluorescence intensity among different spots. Bar = 2 µm.

Endocytosis has been shown to cause physical separation of the Gα subunit from AtRGS1 for sustained G protein-dependent signaling. To test whether JA-induced endocytosis of AtRGS1 also results in its dissociation from AtGPA1, we performed forster resonance energy transfer-fluorescence lifetime imaging microscopy (FRET-FLIM) assay to analyze their interaction by determining the donor fluorescence lifetime (τ) and the FRET frequency with a higher spatial temporal accuracy in living cells. When expressed alone in *Nicotiana benthamiana* leaves, the donor AtRGS1-GFP signal exhibited a generally long fluorescence lifetime (τ = 2.53 ± 0.001 ns) as depicted by the reddish pseudo-color on the heat map, whereas a 'blue-shift' was observed in the AtRGS1-GFP/AtGPA1-mCherry heterologous co-expression system, indicating a strong reduction in GFP lifetime (τ = 2.36 ± 0.005 ns) (Figure 4 and Figure S2). These results confirm that AtRGS1 directly interacts with AtGPA1 under steady-state conditions. In contrast, treatment of MeJA resulted in a marked increase in GFP lifetime in the AtRGS1-GFP/AtGPA1-mCherry co-expressed leaves (τ = 2.50 ± 0.006 ns) (Figure 4). Meanwhile, FRET efficiency significantly decreased from 6.8% to 1.4% (Figure 4B). These observations indicate that JA triggers the dissociation of AtRGS1 and G protein pre-formed complex.

Taken together, our findings reveal that JA can stimulate the endocytosis of AtRGS1 protein and subsequently enable AtRGS1 to move away from AtRGS1. It is reasonable to propose that other potential ligands, such as plant hormones, in addition to D-glucose and flg22, could also activate G signaling as a consequence of AtRGS1 endocytosis. Therefore, the endocytosis of AtRGS1 could be the crux of signal modulation of heterotrimeric G proteins in response to diverse stimuli.

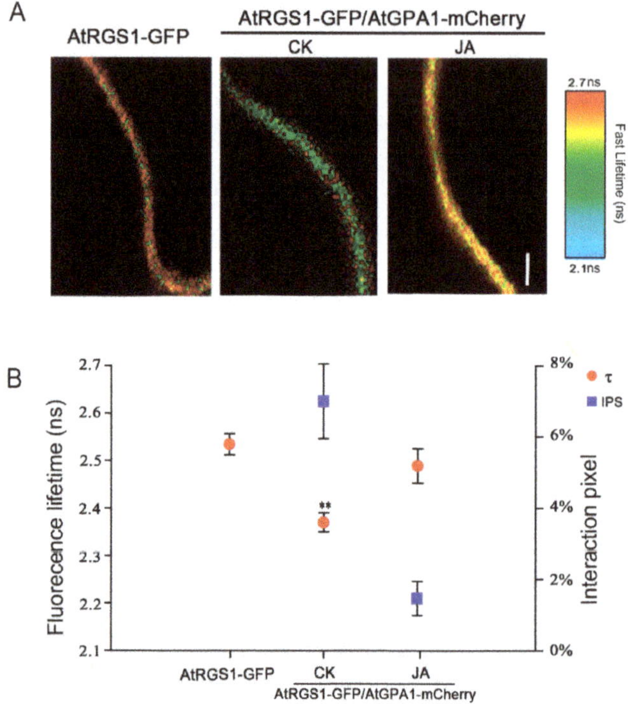

Figure 4. JA induces the dissociation of AtRGS1 from AtGPA1. (**A**) Fluorescence lifetime heat maps of AtRGS1-GFP alone or AtRGS1-GFP/AtGPA1-mCherry in *N. benthamiana* leaves prior and after MeJA treatment for 8 h. Bar = 10 µm. The scale varies from lowest lifetime of 2.1 s to the highest lifetime of 2.7 s. (**B**) The average fluorescence lifetime (τ) and FRET efficiency (IPS) of AtRGS1-GFP alone or AtRGS1-GFP/AtGPA1-mCherry in *N. benthamiana* leaves prior and after MeJA treatment for 8 h. Error bars represent SD (*n* = 9–12). Significant differences are denoted by letters (** *p* < 0.01, Student's *t* test).

2.3. The Phosphorylation and C-Terminal Domain Are Required for JA-Induced AtRGS1 Endocytosis

The C-terminal phosphorylation of AtRGS1 has been shown to play a critical role in its endocytosis. Glucose-induced endocytosis of AtRGS1 is initiated by its transphosphorylation by three WNKs at the C-terminal domain [15]. Stimulation by flg22 induces the phosphorylation of AtRGS1 by the receptor-like kinase BAK1 at the C-terminal domain, leading to its endocytosis [16,17]. Therefore, we were curious as to whether the phosphorylation and C-terminal domain of AtRGS1 are necessary for JA-induced endocytosis.

To determine whether the phosphorylation state of AtRGS1 is required for its endocytosis upon JA treatment, we employed the serine/threonine protein kinases inhibitor, K252a. Similar to our previous observation, AtRGS1-YFP displayed an obvious intracellular accumulation in response to MeJA and CHX co-treatment (Figure 5A). In contrast, a few intracellular puncta were observed in AtRGS1-YFP seedlings after incubation of MeJA with pretreatment of CHX plus K252a (Figure 5A). Statistical analysis of the internalized AtRGS1-YFP fluorescence percentage also showed almost 50% inhibition of AtRGS1-YFP by co-treatment with MeJA, CHX, and K252a compared to that without K252a addition (Figure 5B). We next tested the effect of K252a on the dwell time of AtRGS1 in response to JA treatment by kymograph analysis. With reduced endocytosis, K252a application led to a significantly longer

PM lifetime for AtRGS1-YFP upon JA treatment than that without the specific inhibitor ($t = 1.93$ s vs. $t = 1.56$ s) (Figure 6A,B,E). Thus, phosphorylation is an essential step in JA-induced AtRGS1 endocytosis.

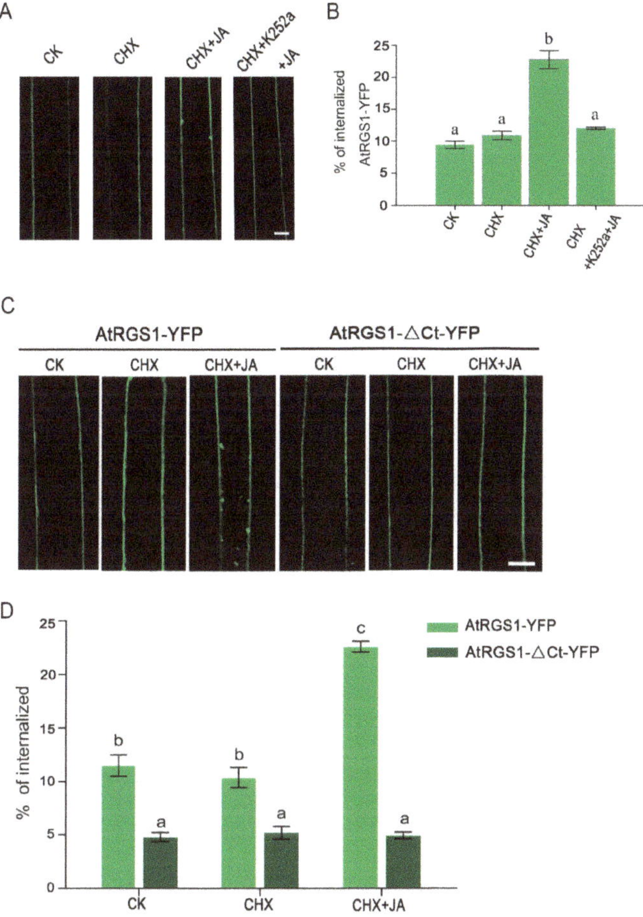

Figure 5. Phosphorylation and the C-terminal domain are required for AtRGS1 endocytosis. (**A**) Confocal images of AtRGS1-YFP localization without treatment (CK), with CHX treatment, with MeJA treatment for 8 h plus CHX pretreatment, and with MeJA treatment for 8 h plus CHX and K252a pretreatment in an *Arabidopsis* hypocotyl epidermal cell. Bar = 10 μm. (**B**) Quantification of the internalized AtRGS1-YFP without treatment (CK), with CHX treatment, with MeJA treatment for 8 h plus CHX pretreatment, and with MeJA treatment for 8 h plus CHX and K252a pretreatment. Error bars represent SD ($n = 5$–8). Significant differences are denoted by letters (Duncan multiple-comparisons test; values within the column followed by the same letter show no significant difference ($p > 0.05$), whereas those with different letter indicate significant difference ($p < 0.05$)). (**C**) Confocal images of AtRGS1-YFP and AtRGS1-ΔCt-YFP without treatment (CK), with CHX treatment, and with MeJA treatment for 8 h plus CHX pretreatment in an *Arabidopsis* hypocotyl epidermal cell. Bar = 10 μm. (**D**) Quantification of the internalized AtRGS1-YFP or AtRGS1-ΔCt-YFP without treatment (CK), with CHX treatment, and with MeJA treatment for 8 h plus CHX pretreatment. Error bars represent SD ($n = 5$–8). Significant differences are denoted by letters (Duncan multiple-comparisons test; values within the column followed by the same letter show no significant difference ($p > 0.05$), whereas those with different letter indicate significant difference ($p < 0.05$)).

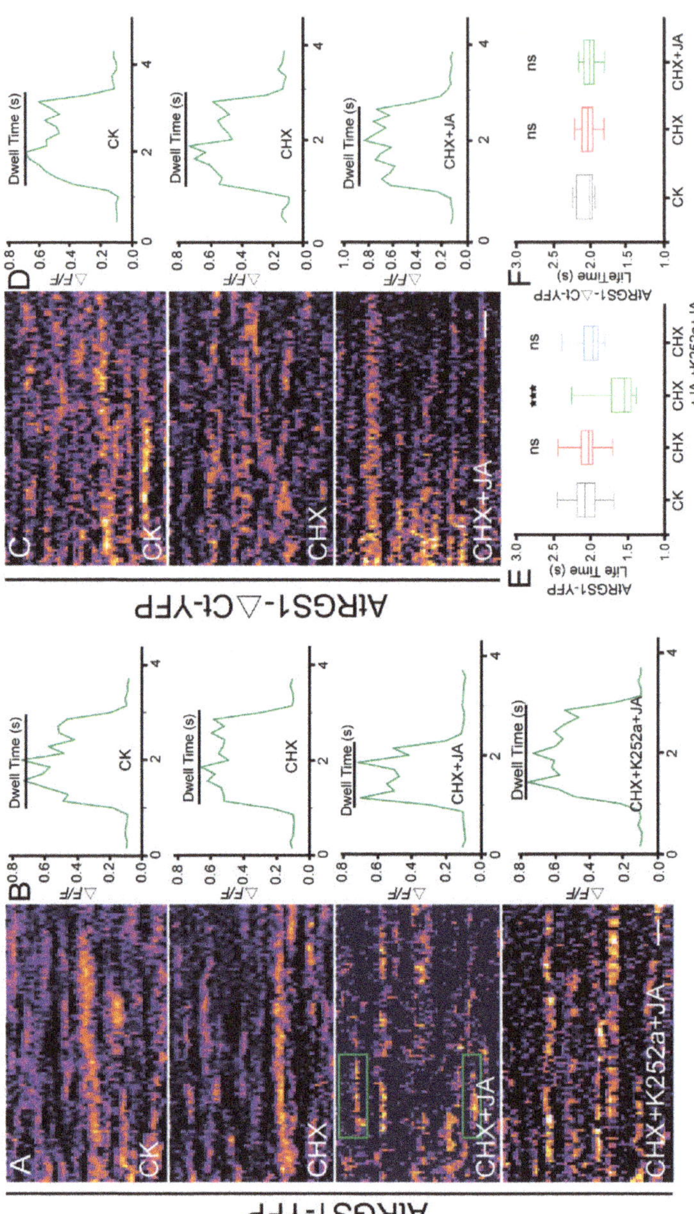

Figure 6. JA treatment affects the dwell time of AtRGS1. (**A**) and (**C**), Representative kymograph showing individual AtRGS1-YFP (**A**) and AtRGS1-ΔCt-YFP (**C**) dwell times without treatment (CK), with CHX treatment, with MeJA treatment for 8 h plus CHX pretreatment, and with MeJA treatment for 8 h plus CHX and K252a pretreatment. Bar = 0.5 s. (**B**) and (**D**), Representative traces of normalized fluorescence of AtRGS1-YFP (**B**) and AtRGS1-ΔCt-YFP (**D**) under different conditions as (**A**) and (**C**) by MATLAB analysis. (**E**,**F**), The average dwell time of AtRGS1-YFP (**E**) and AtRGS1-ΔCt-YFP (**F**) under the different treatment as (**A**) and (**C**). Error bars represent SD ($n = 10$). Significant differences are denoted by asterisks (*** $p < 0.001$, ns, no significant difference, Student's *t* test).

To investigate the role of the C-terminal domain of the AtRGS1 protein in its endocytosis upon JA stimulation, we used a C-terminal truncation mutant AtRGS1-ΔCt-YFP assay (Figure S1). As shown in Figure 5C, MeJA did not induce significant internalization of AtRGS1-YFP when the C-terminal domain was truncated, whereas the full-length AtRGS1 induced an increase in the number of fluorescent spots in the cytoplasm at the same time. Quantification assays confirmed that JA treatment induced a two-fold increment in AtRGS1 internalization but not in that of AtRGS1-ΔCt-YFP (Figure 5D). We next examined the effect of C-terminal truncation on its surface lifetime exposed to JA treatment. In contrast to the relative low stability of AtRGS1-YFP, the life time of AtRGS1-ΔCt-YFP was much longer ($t = 1.56$ s vs. $t = 2.07$ s) (Figure 6C,D,F). These results indicate that the C-terminal domain on AtRGS1 is necessary for JA-induced internalization.

Based on these data, we provide compelling evidence to support that both phosphorylation and C-terminal domain are prerequisites for JA-induced endocytosis of AtRGS1. AtRGS1 has a C terminus highly enriched in serine residues that are analogous to phosphorylated sequence of GPCRs [6], some of which are phosphorylated upon activation by D-glucose and flg22 for internalization [15,16]. These findings prompt us to propose that JA mediates endocytosis of AtRGS1 via the phosphorylation of its C-terminal region. However, identification of the candidate kinase and phosphorylated sites is warranted.

2.4. JA Treatment Affects PM Dynamics of AtRGS1 and AtRGS1-ΔCt

PM proteins are not static; instead, they exhibit dynamic behaviors. Except for subcellular trafficking, changes in the diffusion dynamics within the PM also occur when cells sense an extracellular signal. Single-particle tracking (SPT) techniques could trace the movements of single protein particles from sequential images by VA-TIRFM. Thus, various diffusion properties of PM proteins in plant cells, such as velocity, trajectory, diffusion coefficient, and mean square displacement, have been quantified by SPT analysis with a high spatiotemporal resolution [21]. Herein, using VA-TIRFM and SPT analysis, we quantitated the diffusion coefficient and motion range of AtRGS1-YFP and AtRGS1-ΔCt-YFP, which reflected their lateral mobility within the PM.

As shown in Figure 7A,C, the motion ranges and diffusion coefficients of AtRGS1-YFP always yielded a single population with different treatments. No distinct differences were found between seedlings without treatment (CK) and with CHX treatment only, whereas the dynamic behaviors of AtRGS1-YFP particles triggered by MeJA significantly changed (Figure 7B,D). For motion range, we found an overall increase after JA treatment, with $\hat{G}$ from 0.40 μm (CK) to 0.49 μm (Figure 7B). Moreover, the diffusion coefficient of AtRGS1-YFP upon JA stimulation was markedly higher than that under the non-treated condition (CK) ($\hat{G} = 2.97 \pm 0.18 \times 10^{-3}$ μm^2/s vs. $1.97 \pm 0.08 \times 10^{-3}$ μm^2/s) (Figure 7D). Taken together, AtRGS1 proteins exhibit comparable trends to increased lateral mobility within the PM in response to JA treatment.

Parallel experiments were also performed on AtRGS1-ΔCt-YFP. Consistent with AtRGS1-YFP, the motion ranges and diffusion coefficients of AtRGS1-ΔCt-YFP were also characterized by a single population (Figure 7E,G). Seedlings without treatment (CK) and treated with CHX only showed similar dynamic features (Figure 7F,H). Compared to the non-treated condition (CK), there was a distinct trend towards an increased motion range ($\hat{G} = 0.44$ μm vs. $\hat{G} = 0.49$ μm) and decreased diffusion coefficient ($\hat{G} = 2.59 \pm 0.08 \times 10^{-3}$ μm^2/s vs. $1.68 \pm 0.17 \times 10^{-3}$ μm^2/s) (Figure 7F,H). These observations suggest that JA treatment leads to a longer motion trajectory and lower diffusion coefficient of AtRGS1-ΔCt.

Collectively, full-length AtRGS1 and C-terminus-truncated AtRGS1 proteins had similar variations in motion range upon JA treatment and showed the opposite behavior for diffusion coefficients. Dynamic analysis of these molecules as individual entities will shed light on the underlying mechanisms of G protein signaling.

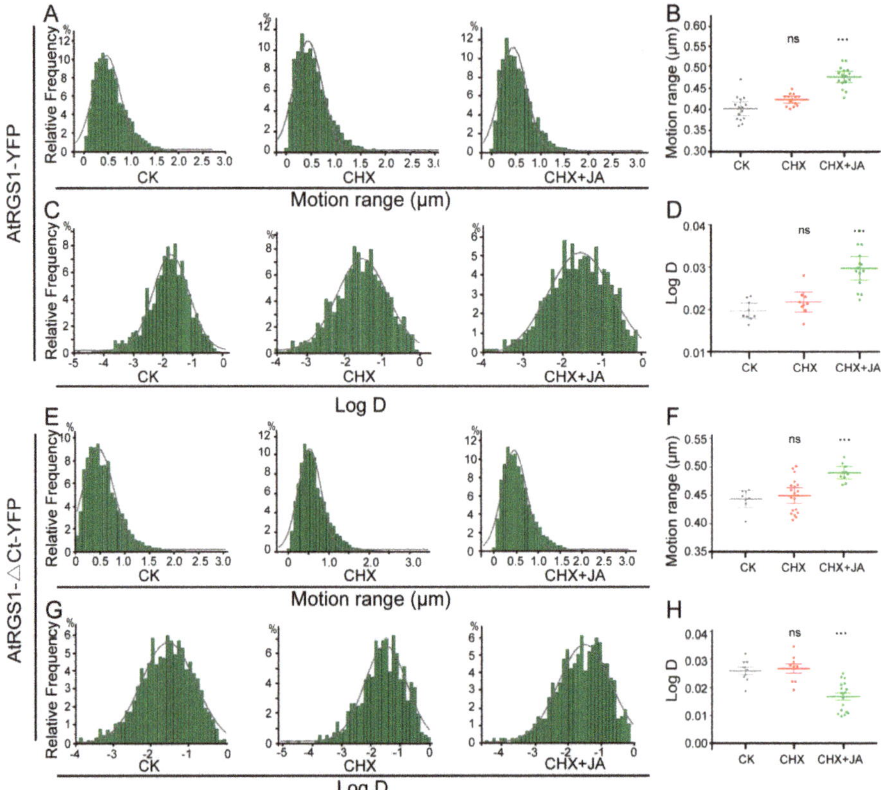

Figure 7. JA treatment affects PM dynamics of AtRGS1 and AtRGS1-ΔCt. (**A,E**), Distribution of the motion range of AtRGS1-YFP (**A**, *n*= 12589) and AtRGS1-ΔCt-YFP (**E**, *n*= 12776) without treatment (CK), with CHX treatment, and with MeJA treatment for 8 h plus CHX pretreatment. (**B,F**), The average frequency of the AtRGS1-YFP and AtRGS1-ΔCt-YFP motion range under different conditions as (**A,E**). Error bars represent SD (*n* = 8–20). Significant differences are denoted by asterisks (*** *p* < 0.001, ns, no significant difference, Student's *t* test). (**C,G**): Distribution of the diffusion coefficients of AtRGS1-YFP (**C**) and AtRGS1-ΔCt-YFP (**G**) without treatment (CK), with CHX treatment, and with MeJA treatment for 8 h plus CHX pretreatment. (**D,H**): The average frequency of AtRGS1-YFP and AtRGS1-ΔCt-YFP diffusion coefficients under different conditions as (**C,G**). Error bars represent SD (*n* = 8–20). Significant differences are denoted by asterisks (*** *p* < 0.001, ns, no significant difference, Student's *t* test).

3. Materials and Methods

3.1. Plant Materials and Growth Conditions

Arabidopsis thaliana ecotype Columbia-0 (Col-0) was used as a wild-type control. The *gpa1-4*, *agb1-2*, and *gpa1-4/agb1-2* mutants [15], and the 35S::AtRGS1-YFP [15] and 35S::AtRGS1-ΔCt-YFP [16] transgenic lines were kindly provided by Dr. Alan M. Jones, University of North Carolina at Chapel Hill (Chapel Hill, NC, USA). Seeds were surface-sterilized with ethanol and H_2O_2 mixture (70% ethanol:30% H_2O_2 = 4:1), and then stratified at 4 °C for 3 days in the dark. For root length measurements, anthocyanin measurement, and real-time PCR analysis, seeds were plated and grown on $\frac{1}{2}$ MS medium supplemented with 1% sucrose under a 16-h light (23–25 °C)/8-h dark (17–20 °C) photoperiod. For confocal and VA-TIRFM analyses, seeds were sown on liquid $\frac{1}{2}$ MS medium without sucrose, followed by 2 h light, and then grown in the dark for 3 days followed by drug treatment and imaging.

3.2. Anthocyanin Measurements

The anthocyanin quantification of 11-day-old *Arabidopsis* seedlings grown on MS medium with 0 or 25 mM MeJA was performed as previously described [22]. Anthocyanin content was reported as (A535–A650)/g fresh weight. For each treatment, the experiment was repeated thrice.

3.3. Root Length Measurement

The primary root length of 9-day-old *Arabidopsis* seedlings grown on MS medium with 0, 10, and 25 mM MeJA was measured by ImageJ software (Image high-energy version 1.4.3.67, National Institutes of Mental Health, Bethesda, MD, USA). For each treatment, the experiment was repeated thrice.

3.4. Quantitative Real-Time PCR Analysis

Three-week-old plants grown on MS medium were drenched in solution containing 100 mM MeJA or water for 8 h in the daytime and then harvested. Total RNA extraction and cDNA synthesis were conducted with an RNA prep Pure Plant Kit (Tiangen Biotech Co., Ltd., Beijing, China) and PrimeScript 1st Strand cDNA Synthesis Kit (Takara Bio Co., Ltd., Shiga, Japan), respectively. Quantitative real-time PCR analysis was performed with the Bio-Rad CFX Connect real-time PCR system (Bio-Rad, Hercules, CA, USA) by SYBR Premix Ex Taq II (Tli RNase H Plus) (Takara Bio Co., Ltd., Shiga, Japan). *ACTIN1* served as the internal control. Gene-specific primers are listed in Table S1. For each treatment, the experiment was repeated thrice.

3.5. Transient Infiltration of N. benthamiana Leaves

The AtRGS1-GFP and AtGPA1-mCherry plant expression constructs were developed by inserting the AtRGS1 and AtGPA1 coding sequence without a stop codon to pCAMBIA2300-35S-GFP and pCAMBIA1300-35S-mCherry vectors, respectively. The primers are shown in Table S1. Leaves of seven-week-old *N. benthamiana* were co-infiltrated with equal volumes of different combinations of *Agrobacterium* strains (AtRGS1-GFP with AtGPA1-mCherry). Plants were maintained at 22 °C for two days in the dark before imaging.

3.6. Drug Treatment

Dark-grown 3-day-old seedlings were incubated in deionized water containing the indicated compounds for the following times and concentrations: 4–10 h for 50 μm CHX; 2 h pretreatment with 50 μm CHX and then 2–8 h for 100 μm MeJA plus CHX; and 2 h pretreatment with 50 μm CHX, 30 min pretreatment with 10 μm K252a, and then 8 h for 100 μm MeJA plus CHX and K252a.

3.7. Confocal Microscopy

A Zeiss LSM710 confocal laser scanning microscope with a C-Apochromat 40/1.2 water immersion objective was used to image the vertical optical sections (Z stacks) of hypocotyl epidermal cells located 2 to 4 mm below the cotyledons. The YFP fluorescence signals were excited by a 514 nm laser line and were collected at 526 to 569 nm by the photomultiplier detector. A series of Z-section images were acquired from the top layer of cells with 0.5-μm steps. Images at 2 to 3 μm below the apical plasma membrane were used to quantify the internalization of AtRGS1-YFP or AtRGS1-ΔCt-YFP by the software ImageJ (Image high-energy version 1.4.3.67, National Institutes of Mental Health, Bethesda, MD, USA). For internalization analysis, the ratio between the fluorescence signal intensity in the cytoplasm and the whole cell represented the fraction of internalized AtRGS1-YFP or AtRGS1-ΔCt-YFP. For each treatment, the experiment was repeated thrice.

3.8. VA-TIRFM and Fluorescence Image Analysis

The epidermal cells of hypocotyl were imaged on an Olympus microscope using a 100× 1.45 NA oil immersion objective. AtRGS1-YFP or AtRGS1-ΔCt-YFP proteins were excited with a 488 nm laser line from a diode laser (Chang-chun New Industries Optoelectronics Technology), and the emission fluorescence was collected with an EM-CCD camera (ANDOR iXon DV8897D-CS0-VP, Andor Technology, Belfast, UK) using bandwidth filters ranging from 505 to 540 nm. Images were acquired with 150 ms exposure times and a time-lapse series of single particles of AtRGS1-YFP was taken with up to 100 images per sequence. The stand-alone MATLAB (R2014) Graphical User Interface program was used for SPT analysis according to the method described by Jaqaman [23]. Kymograph and lifetime analyses were performed as previously described [24]. After tracking the trajectories of individual particles, the mean square displacement (MSD) was computed [25]. The diffusion coefficient of a specific particle was calculated by linear fit to MSD versus time (MSD-t) plots [25]. The motion range was determined as the largest displacement during the lifetime [25]. The distribution of the diffusion coefficients or motion range was plotted in a histogram [25]. The data were fitted by a Gaussian function, and the position of the peak (denoted as $\hat{G}$) was considered as characteristic values for further analysis [25]. For each treatment, the experiment was repeated twice.

3.9. FLIM-FRET Measurement

FLIM was performed on an inverted Olympus FV1200 microscope equipped with a PicoQuant picoHarp300 controller (PicoQuant, Berlin, Germany). Excitation at 488 nm was conducted with a picosecond pulsed diode laser, and the emitted light was filtered with a 505/540 nm bandpass filter and detected with an MPD SPAD detector. Single-pixel fluorescence lifetimes were averaged across a representative region of interest (ROI). The resulting images showed the corresponding fluorescence lifetime, τ, for each pixel in a rainbow color code from navy ($\tau = 2.1$ ns) to red ($\tau = 2.7$ ns). FRET efficiency (%) was calculated as described previously [26]. For each treatment, the experiment was repeated twice.

3.10. Western Blot Analysis

Total proteins were extracted from the transgenic seedlings expressing AtRGS1-YFP and AtRGS1-ΔCt-YFP. The amount of AtRGS1 and AtRGS1-ΔCt proteins was determined using a specific anti-GFP antibody (Sigma, St.Louis, MI, USA).

3.11. Data Analysis

Significant differences are denoted by letters ($p < 0.05$; Duncan multiple-comparison test) and by asterisks (* $p < 0.05$, *** $p < 0.001$, ns, not significant; Student's *t* test). The standard deviation (SD) was used to calculate the error bars.

3.12. Accession Numbers

This data used in this study are available from the *Arabidopsis* Information Resource (TAIR) database under accession numbers AT3G26090.1 (AtRGS1), AT2G26300 (AtGPA1), AT4G03415 (AGB1), AT5G24780 (VSP1), and AT3G45140 (LOX2).

4. Conclusions

In conclusion, these results demonstrate that the G protein-mediated pathway is involved in JA signaling, which possibly requires G protein activation via phosphorylation and C-terminus-dependent AtRGS1 endocytosis indirectly by JA.

Supplementary Materials: Supplementary materials can be found at http://www.mdpi.com/1422-0067/20/15/3779/s1. Figure S1: Western blot analysis of the AtRGS1-YFP and AtRGS1-ΔCt-YFP; Figure S2: Co-transient expression of AtRGS1-GFP and AtGPA1-mCherry in *N. benthamiana* leaves; Table S1: Primers used in this study.

Author Contributions: L.L. and B.S. conducted experiments and edited figures; X.Q. assisted in performing experiments and collating data; X.Z. and S.S. gave valuable suggestions and modified the manuscript; and X.S. supervised the whole process and wrote the manuscript.

Funding: This research was funded by the the Fundamental Research Funds for the Central Universities (2018ZY37) and the National Natural Science Foundation of China (31871424).

Acknowledgments: We sincerely thank Alan M. Jones (University of North Carolina at Chapel Hill, USA) for providing the seeds of the *gpa1-4*, *agb1-2*, *gpa1-4/agb1-2* mutants and the 35S::AtRGS1-YFP, 35S::AtRGS1-ΔCt-YFP transgenic lines and kindly suggestions.

Conflicts of Interest: The authors declare no conflict of interest. The funders had no role in the design of the study; in the collection, analyses, or interpretation of data; in the writing of the manuscript, or in the decision to publish the results.

References

1. Wasternack, C.; Song, S. Jasmonates: Biosynthesis, metabolism and signalling by proteins activating and repressing transcription. *J. Exp. Bot.* **2017**, *6*, 1303–1321. [CrossRef] [PubMed]

2. Huang, H.; Liu, B.; Liu, L.; Song, S. Jasmonate action in plant growth and development. *J. Exp. Bot.* **2017**, *68*, 1349–1359. [CrossRef] [PubMed]

3. Shan, X.; Yan, J.; Xie, D. Comparison of phytohormone signalling mechanisms. *Curr. Opin. Plant Biol.* **2012**, *15*, 84–91. [CrossRef] [PubMed]

4. Gilman, A.G. G proteins: Transducers of receptor-generated signals. *Annu. Rev. Biochem.* **1987**, *56*, 615–649. [CrossRef] [PubMed]

5. Wettschureck, N.; Offermanns, S. Mammalian G proteins and their cell type specific functions. *Physiol. Rev.* **2005**, *85*, 1159–1204. [CrossRef] [PubMed]

6. Lefkowitz, R.J. Historical review: A brief history and personal retrospective of seven-transmembrane receptors. *Trends Pharmacol.* **2004**, *28*, 413–422. [CrossRef]

7. Ma, H.; Yanofsky, M.F.; Meyerowitz, E.M. Molecular cloning and characterization of GPA1, a G protein alpha subunit gene from *Arabidopsis thaliana*. *Proc. Natl. Acad. Sci. USA* **1990**, *87*, 3821–3825. [CrossRef]

8. Weiss, C.A.; Garnaat, C.W.; Mukai, K.; Hu, Y.; Ma, H. Isolation of cDNAs encoding guanine nucleotidebinding protein beta-subunit homologues from maize (ZGB1) and *Arabidopsis* (AGB1). *Proc. Natl. Acad. Sci. USA* **1994**, *91*, 9554–9558. [CrossRef]

9. Mason, M.G.; Botella, J.R. Completing the heterotrimer: Isolation and characterization of an *Arabidopsis thaliana* G protein gamma-subunit cDNA. *Proc. Natl. Acad. Sci. USA* **2000**, *97*, 14784–14788. [CrossRef]

10. Mason, M.G.; Botella, J.R. Isolation of a novel G-protein gamma-subunit from *Arabidopsis thaliana* and its interaction with Gβ. *Biochim. Biophys.* **2001**, *1520*, 147–153. [CrossRef]

11. Chen, J.G.; Willard, F.S.; Huang, J.; Liang, J.; Chasse, S.A.; Jones, A.M.; Siderovski, D.P. A seventransmembrane RGS protein that modulates plant cell proliferation. *Science* **2003**, *301*, 1728–1731. [CrossRef] [PubMed]

12. Johnston, C.A.; Taylor, J.P.; Gao, Y.; Kimple, A.J.; Grigston, J.C.; Chen, J.G.; Siderovski, D.P.; Jones, A.M.; Willard, F.S. GTPase acceleration as the rate-limiting step in *Arabidopsis* G protein-coupled sugar signalling. *Proc. Natl. Acad. Sci. USA* **2007**, *104*, 17317–17322. [CrossRef] [PubMed]

13. Urano, D.; Chen, J.G.; Botella, J.R.; Jones, A.M. Heterotrimeric G protein signalling in the plant kingdom. *Open Biol.* **2013**, *3*, 120186. [CrossRef] [PubMed]

14. Phan, N.; Urano, D.; Srba, M.; Fischer, L.; Jones, A.M. Sugar-induced endocytosis of plant 7TM-RGS proteins. *Plant Signal. Behav.* **2012**, *8*, e22814. [CrossRef] [PubMed]

15. Urano, D.; Phan, N.; Jones, J.C.; Yang, J.; Huang, J.; Grigston, J.; Taylor, J.P.; Jones, A.M. Endocytosis of the seven-transmembrane RGS1 protein activates G-protein-coupled signalling in *Arabidopsis*. *Nat. Cell Biol.* **2012**, *14*, 1079–1088. [CrossRef] [PubMed]

16. Tunc-Ozdemir, M.; Urano, D.; Jaiswal, D.K.; Clouse, S.D.; Jones, A.M. Direct modulation of heterotrimeric G protein-coupled signalling by a receptor kinase complex. *J. Biol. Chem.* **2016**, *291*, 13918–13925. [CrossRef]

17. Tunc-Ozdemir, M.; Jones, A.M. Ligand-induced dynamics of heterotrimeric G protein-coupled receptor-like kinase complexes. *PLoS ONE* **2017**, *12*, e0171854. [CrossRef]

18. Liang, X.; Ding, P.; Lian, K.; Wang, J.; Ma, M.; Li, L.; Li, L.; Li, M.; Zhang, X.; Chen, S.; et al. *Arabidopsis* heterotrimeric G proteins regulate immunity by directly coupling to the FLS2 receptor. *eLife* **2016**, *5*, e13568.

19. Liang, X.; Ma, M.; Zhou, Z.; Wang, J.; Yang, X.; Rao, S.; Zhou, J.M. Ligand-triggered de-repression of *Arabidopsis* heterotrimeric G proteins coupled to immune receptor kinases. *Cell Res.* **2018**, *28*, 529–543. [CrossRef]

20. Colaneri, A.C.; Tunc-Ozdemir, M.; Huang, J.; Jones, A.M. Growth attenuation under saline stress is mediated by the heterotrimeric G protein complex. *BMC Plant Biol.* **2014**, *14*, 129. [CrossRef]
21. Wang, L.; Xue, Y.; Xing, J.; Song, K.; Lin, J. Exploring the spatiotemporal organization of membrane proteins in living plant cells. *Annu. Rev. Plant Biol.* **2018**, *69*, 525–551. [CrossRef] [PubMed]
22. Shan, X.; Zhang, Y.; Peng, W.; Wang, Z.; Xie, D. Molecular mechanism for jasmonate-induction of anthocyanin accumulation in *Arabidopsis*. *J. Exp. Bot.* **2009**, *60*, 3849–3860. [CrossRef] [PubMed]
23. Jaqaman, K.; Loerke, D.; Mettlen, M.; Kuwata, H.; Grinstein, S.; Schmid, S.L. Danuser G Robust single-particle tracking in live-cell time-lapse sequences. *Nat. Methods* **2008**, *5*, 695–702. [CrossRef] [PubMed]
24. Eichel, K.; Jullié, D.; Zastrow, M.V. β-Arrestin drives MAP kinase signalling from clathrin-coated structures after GPCR dissociation. *Nat. Cell Biol.* **2016**, *18*, 303–310. [CrossRef] [PubMed]
25. Cui, Y.; Yu, M.; Yao, X.; Xing, J.; Lin, J.; Li, X. Single-particle tracking for the quantification of membrane protein dynamics in living plant cells. *Mol. Plant* **2018**, *11*, 1315–1321. [CrossRef] [PubMed]
26. Long, Y.; Stahl, Y.; Weidtkamp-Peters, S.; Postma, M.; Zhou, W.; Goedhart, J.; Blilou, I. In vivo FRET–FLIM reveals cell-type-specific protein interactions in *Arabidopsis* roots. *Nature* **2017**, *548*, 97–102. [CrossRef] [PubMed]

© 2019 by the authors. Licensee MDPI, Basel, Switzerland. This article is an open access article distributed under the terms and conditions of the Creative Commons Attribution (CC BY) license (http://creativecommons.org/licenses/by/4.0/).

International Journal of
Molecular Sciences

Article

Jasmonic Acid Methyl Ester Induces Xylogenesis and Modulates Auxin-Induced Xylary Cell Identity with NO Involvement

Federica Della Rovere [1], Laura Fattorini [1], Marilena Ronzan [1], Giuseppina Falasca [1], Maria Maddalena Altamura [1,*] and Camilla Betti [2,*]

[1] Department of Environmental Biology, Sapienza University of Rome, Piazzale Aldo Moro 5, 00185 Rome, Italy

[2] Department of Medicine, University of Perugia, Piazzale Menghini 8/9, 06132 Perugia, Italy

* Correspondence: mariamaddalena.altamura@uniroma1.it (M.M.A.); camilla.betti@unipg.it (C.B.); Tel.: +39-06-4991-2452 (M.M.A.); +39-07-5578-2402 (C.B.)

Received: 29 July 2019; Accepted: 6 September 2019; Published: 10 September 2019

Abstract: In *Arabidopsis* basal hypocotyls of dark-grown seedlings, xylary cells may form from the pericycle as an alternative to adventitious roots. Several hormones may induce xylogenesis, as Jasmonic acid (JA), as well as indole-3-acetic acid (IAA) and indole-3-butyric acid (IBA) auxins, which also affect xylary identity. Studies with the ethylene (ET)-perception mutant *ein3eil1* and the ET-precursor 1-aminocyclopropane-1-carboxylic acid (ACC), also demonstrate ET involvement in IBA-induced ectopic metaxylem. Moreover, nitric oxide (NO), produced after IBA/IAA-treatments, may affect JA signalling and interact positively/negatively with ET. To date, NO-involvement in ET/JA-mediated xylogenesis has never been investigated. To study this, and unravel JA-effects on xylary identity, xylogenesis was investigated in hypocotyls of seedlings treated with JA methyl-ester (JAMe) with/without ACC, IBA, IAA. Wild-type (wt) and *ein3eil1* responses to hormonal treatments were compared, and the NO signal was quantified and its role evaluated by using NO-donors/scavengers. Ectopic-protoxylem increased in the wt only after treatment with JAMe(10 µM), whereas in *ein3eil1* with any JAMe concentration. NO was detected in cells leading to either xylogenesis or adventitious rooting, and increased after treatment with JAMe(10 µM) combined or not with IBA(10 µM). Xylary identity changed when JAMe was applied with each auxin. Altogether, the results show that xylogenesis is induced by JA and NO positively regulates this process. In addition, NO also negatively interacts with ET-signalling and modulates auxin-induced xylary identity.

Keywords: adventitious rooting; auxin; ectopic metaxylem; ectopic protoxylem; ethylene; hypocotyl; jasmonates; nitric oxide; xylogenesis

---

1. Introduction

In vascular plants, the process leading to the differentiation of the xylem conducting elements (xylary elements, XEs) is still not fully understood. The XEs differentiate either from the procambium, and are named protoxylem and metaxylem, or from the vascular cambium (deuteroxylem). However, XEs may be produced also by other cell types, both in planta and, mainly, in vitro. Auxin is necessary for XE induction and early development [1], acting through the expression of the same genes both in planta and in vitro [2]. In in vitro cultured systems, XE formation occurs either via a direct trans-differentiation event, like in zinnia (*Zinnia elegans*) mesophyll cells [3], or via a cell de-differentiation process, followed by mitotic proliferation, as in *Arabidopsis thaliana* stem thin cell layers (TCLs) [4]. Additionally, other cell types than those of the procambium/cambium may be triggered in planta to either trans-differentiate into XEs, e.g., after a mechanical wounding [3], or to proliferate. The latter process causes the formation of cells that only subsequently differentiate into supernumerary XEs and has been observed in the hypocotyls of *Arabidopsis* dark-grown seedlings treated with naphthaleneacetic acid (NAA) [5]. The xylogenic response in planta, defined as ectopic xylary cell formation, occurs because vascular pattern formation is not cell clonal. XE-arrangement can be, in fact, altered by local signals, e.g., derived by phytohormones, or modified in response to environmental stimuli [6]. Xylogenesis is always preceded by an endogenous auxin (indole-3-acetic acid, IAA) accumulation [7–9]. In the basal hypocotyl of *Arabidopsis* dark-grown seedlings, ectopic XEs formation in the pericycle periclinal derivatives is controlled by either IAA, or indole-3-butyric acid (IBA), one of the possible IAA precursor ([10,11] and references therein), or the synthetic auxin NAA [5,7]. Moreover, in lettuce, pith cells trans-differentiate into XEs in the presence of NAA and either the ethylene (ET) precursor 1-aminocyclopropane-1-carboxylic acid (ACC), or the ET-releasing compound ethephon, thus demonstrating that ET and auxin are both required for xylogenesis in this species [12]. Xylem cells differentiation requires the realization of programmed cell death (PCD) [3]; it has been shown that ET is involved in PCD signalling during early and late stages of XE differentiation, at least in zinnia cell cultures [13,14]. Furthermore, exogenous ACC/ET stimulates xylem formation in stems/cuttings of gymnosperms and angiosperms [15–17]. In *Arabidopsis* dark-grown seedlings, it has been recently demonstrated that ET is involved in ectopic xylem formation through the activation of the ETHYLENE INSENSITIVE 3 (EIN3) and EIN3-LIKE 1 (EIL1) transcription factors (TFs) [11]. Both these TFs activate numerous ET-responsive genes [18] and when mutated cause complete ET-insensitivity ([19] and references therein) and persistent auxin-sensitivity [20]. In the basal hypocotyl of *Arabidopsis* dark-grown seedlings, the treatment with IAA (10 μM) induces exclusively ectopic protoxylem formation, whereas exogenous IBA (10 μM) only ectopic metaxylem, showing that the two auxins have different effects on XE identity. Both ectopic protoxylem and metaxylem are induced by ACC, when applied to the seedlings at 0.1 μM concentration together with IAA (10 μM), but only ectopic metaxylem is formed when seedlings are treated with ACC in combination with IBA (10 μM) [11].

It has been reported that both antagonistic and synergistic interactions occur also between ET and jasmonates (JAs) in numerous morphogenic and defense responses, mainly relying on EIN3/EIL1 functions [21].

In zinnia cell cultures genes involved in jasmonic acid (JA) synthesis and signalling are expressed at early stages of the xylogenic process [1]. In *Arabidopsis*, JAs trigger cambium cell division leading to deuteroxylem formation in shoots [22]. In addition, the early release of JA, after treatment with JA methyl ester (JAMe), and due to its demethylation [23], induces extra xylem in roots [24] modulating XE formation by controlling polar auxin transport in the vascular tissues [25]. However, the effect of JA seems to be dose-dependent, because low xylogenesis occurs after treatment with 1 μM JAMe, while high xylogenesis with 10 μM JAMe [24]. Moreover, when JAMe is applied at 10 μM concentration, xylogenesis is stimulated also in *Arabidopsis* TCLs cultured with IBA and Kinetin [26].

Nitric oxide (NO) is a reactive free radical molecule, which serves as a messenger in several cell differentiation events [27], including PCD [28]. In *Zinnia elegans*, a NO burst occurs before secondary cell wall formation and cell autolysis both in the differentiating xylem of the vascular bundles and in mesophyll cells undergoing trans-differentiation into XEs ([29] and references therein). The production of NO might be stimulated by auxin and be also spatially and temporally controlled ([27] and references therein). Moreover, NO can also affect organ elongation by regulating auxin transport, as seen in rice roots [30].

Exogenous NO may negatively impact ET synthesis; however, it may also stimulate ET production ([27] and references therein), and affect ET-signalling at the level of ETHYLENE INSENSITIVE 2 (EIN2) [31]. In the presence of ET, EIN2 activates a TF cascade, involving EIN3 and EIL1, which leads to specific plant developmental responses [32]. However, a possible interaction between NO and ET-signalling mediated by EIN3/EIL1 has been never demonstrated in the process of xylogenesis. Recent studies show that ET and NO may act as transducers of PCD signalling in early stages of xylogenesis, and as co-actors at later stages of cell death [33,34]. In addition, ET and NO are both involved in adventitious rooting in competition with xylogenesis in numerous plant species, including *Arabidopsis* ([11,35] and references therein).

A crosstalk between NO and JA may occur in some developmental processes possibly mediated by *OPR3*, a gene coding for OPDA-reductase 3 (OPR3), which converts 12-oxophytodienoic acid (OPDA) into the first precursor of JA ([26] and other references therein). In *Arabidopsis*, *OPR3* expression is increased by NO, resulting into an increased JA production ([36] and references therein), although an OPR3-independent JA biosynthetic pathway has also been recently discovered in this species [37]. Moreover, in *Taxus* cells, exogenous JAMe is known to stimulate NO synthase to induce NO production [38]. Therefore, it seems that JA and NO can modulate each other production, but their possible interaction in xylogenesis is not yet characterized.

The aim of the present research is to determine the involvement of JA, ET and NO in the control of endogenous/exogenous-auxin-induced xylogenesis through a possible crosstalk mediated by EIN3/EIL1. To this aim, ectopic xylem formation was investigated in the hypocotyl of dark-grown *Arabidopsis* seedlings exposed to various concentrations of JAMe with/without ACC, and exogenous IBA or IAA. The xylogenic response in the wild-type (wt) was compared with the *ein3eil1* mutant, the NO production monitored, and the NO putative role evaluated by measuring the effects of treatments with NO-donors/scavengers.

Results show that the ectopic formation of protoxylem was enhanced by JAMe when applied alone at a specific concentration (10 μM). The *ein3eil1* mutant was sensitive to JAMe. In fact, the stimulation of XEs mediated by JAMe occurred also in the mutant, suggesting that a negative interaction between JA and ET-signalling is involved in xylogenesis. This was strengthened by the reduction in xylogenesis observed in the wt after the combined application of JAMe with the ET-precursor ACC, in comparison with JAMe single treatments. Nitric oxide was detected at early stages of both xylogenesis and adventitious rooting in the hypocotyl pericycle cells, and its production was highly enhanced by JAMe (10 μM).

Surprisingly, xylogenesis did not increase quantitatively with any JAMe treatment combined with either IBA or IAA. Conversely, the xylary identity changed, in comparison with either auxin single treatments. In addition, the IBA/IAA-induced adventitious rooting was increased by the same JAMe concentration enhancing xylogenesis when applied alone. This suggests a role for JA in modulating adventitious rooting and xylogenesis programs in the same target cells through an interaction with NO.

The role of JA in inducing xylogenesis and in modifying xylary cell identity is further discussed.

2. Results

2.1. A Specific Concentration of Exogenous JAMe Enhances Ectopic Protoxylem Formation but its Combination with ACC Reduces it, without Affecting Xylary Identity

The hypocotyl base of seedlings grown in hormone free (HF) condition showed a sporadic cell proliferation in the pericycle (Figure 1), with occasional ectopic protoxylem formation (Figure 2A, and Table 1). Pericycle proliferation did not increase with 0.01 μM and 1 μM JAMe, but a significant increase was observed with 10 μM JAMe (Figure 1). Independently from the treatment, the proliferation was always associated with ectopic xylary cells (XEs) formation (Table 1).

In the mutant seedlings, the cell proliferation with XEs from the pericycle significantly increased compared to the wt with any JAMe treatment condition (Figure 1). Thus, in contrast with the JA-insensitivity observed in other morphogenic processes, e.g., root hair formation [39], the *ein3eil1* mutant perceived JA, giving rise to a pericycle xylogenic proliferation with significant increases even at the lowest JAMe dose tested. Independently from treatment and genotype, only ectopic protoxylem was formed (Figure 2B–E and Table 1). This showed that the JA deriving from the exogenous JAMe did not change the xylary cell identity determined by endogenous hormones, even in the impaired ET-perception mutant.

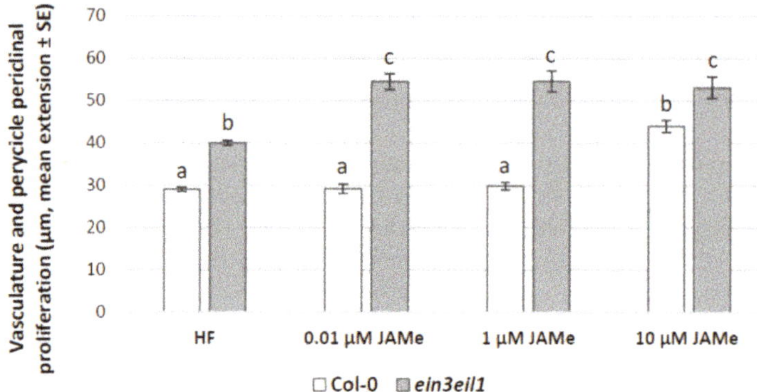

Figure 1. Pericycle periclinal proliferation in the basal hypocotyl of *Arabidopsis* Col-0 (wt) and *ein3eil1* double mutant seedlings either cultured with different JAMe concentrations (0.01, 1, 10 μM) or in hormone-free (HF) condition. Mean radial extension (± Standard error (SE)) of the vasculature and the pericycle periclinal proliferation in the basal 5 mm of the hypocotyl. Different letters among different treatments within the same genotype, or between the two genotypes within the same treatment, indicate significant differences at least at $p < 0.001$. The same letters indicate no significant difference. Two-way ANOVA followed by Tukey's post-test. $n = 30$.

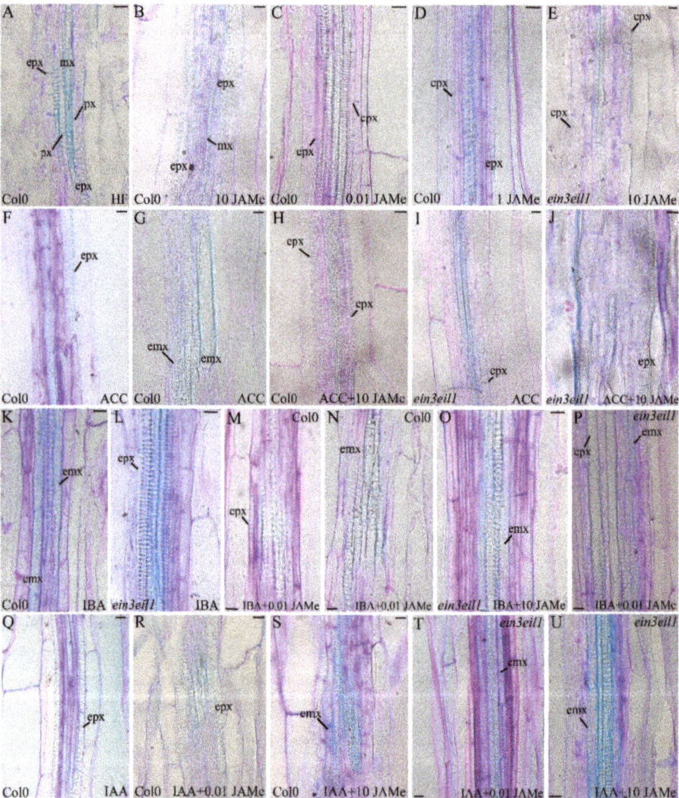

Figure 2. Xylogenesis in the basal hypocotyl of *Arabidopsis* Col-0 (wt) and *ein3eil1* double mutant seedlings dark-grown for 22 DAS with different treatments.

(**A**) Ectopic protoxylem elements occasionally formed by periclinal pericycle cell divisions in the basal 5 mm of Col-0 hypocotyl in the absence of exogenous hormones (hormone-free, HF). (**B–U**) Ectopic protoxylem or ectopic metaxylem elements formed in Col-0 (**B–D,F–H,K,M,N,Q–S**) or *ein3eil1* (**E,I,J,L,O,P,T,U**) basal hypocotyl treated with: 0.01 µM JAMe (**C**), 1 µM JAMe (**D**), 10 µM JAMe (**B,E**), 0.1 µM ACC (**F,G,I**), 0.1 µM ACC + 10 µM JAMe (**H,J**), 10 µM IBA (**K,L**), 10 µM IBA + 0.01 µM JAMe (**M,N,P**), 10 µM IBA + 10 µM JAMe (**O**), 10 µM IAA (**Q**), 10 µM IAA + 0.01 µM JAMe (**R,T**), 10 µM IAA + 10 µM JAMe (**S,U**). emx, ectopic metaxylem; epx, ectopic protoxylem; mx, metaxylem; px, protoxylem. Radial longitudinal sections stained with toluidine blue. Scale bars = 10 µm (**B–U**), 20 µm (**A**).

Table 1. Quantification of ectopic xylary elements (XEs) detected in the basal hypocotyl of Col-0 and *ein3eil1* seedlings. Mean number (± standard error (SE)) of XEs and percentage of ectopic protoxylem and metaxylem elements detected in an area of 150×150 µm^2 in the middle portion of the basal hypocotyl with different treatments. ACC, IBA and IAA were always used at 0.1 µM, 10 µM and 10 µM, respectively. Different letters among values within the same column indicate significant differences at least at $p < 0.05$. The same letter indicates no statistical difference. One-way ANOVA followed by Tukey's post-test. $n = 30$.

Treatment	Col-0 Ectopic XEs Mean Number (± SE)	*ein3eil1* Ectopic XEs Mean Number (± SE)	Col-0 Ectopic Protoxylem %	Col-0 Ectopic Metaxylem %	*ein3eil1* Ectopic Protoxylem %	*ein3eil1* Ectopic Metaxylem %
HF (control)	0.5 ± 0.2 [a]	0.6 ± 0.3 [a]	100	0	100	0
0.01 µM JAMe	1.3 ± 0.3 [a]	2.3 ± 0.7 [b]	100	0	100	0
1 µM JAMe	1.0 ± 0.3 [a]	2.4 ± 0.7 [b]	100	0	100	0
10 µM JAMe	2.8 ± 0.3 [b]	2.5 ± 0.4 [b]	100	0	100	0
0.1 µM ACC	1.1 ± 0.3 [a]	0.6 ± 0.2 [a]	45.5	54.5	100	0
10 µM JAMe + ACC	0.8 ± 0.2 [a]	2.4 ± 0.3 [b]	100	0	100	0
10 µM IBA	4.4 ± 0.6 [b]	2.0 ± 0.1 [b]	0	100	100	0
IBA + 0.01 µM JAMe	3.5 ± 0.4 [b]	2.4 ± 0.3 [b]	17.6	82.4	20.7	79.3
IBA + 1 µM JAMe	3.7 ± 0.8 [b]	2.3 ± 0.4 [b]	22.7	77.3	22.2	77.8
IBA + 10 µM JAMe	4.0 ± 0.6 [b]	3.6 ± 0.2 [b]	61.9	38.1	50	50
10 µM IAA	2.8 ± 0.5 [b]	2.0 ± 0.4 [b]	100	0	100	0
IAA + 0.01 µM JAMe	2.5 ± 0.4 [b]	1.8 ± 0.4 [b]	100	0	35	65
IAA + 1 µM JAMe	2.5 ± 0.2 [b]	2.2 ± 0.3 [b]	33.3	66.7	25	75
IAA + 10 µM JAMe	2.6 ± 0.5 [b]	3.8 ± 0.8 [b]	7.7	92.3	9	91

In the basal portion of wt seedlings treated with ACC a pericycle periclinal proliferation with xylogenesis comparable to HF condition was observed (Figure 3). However, both protoxylem and metaxylem elements were formed (Figure 2F,G, and Table 1). The extension of the vasculature, including the portion with pericycle proliferation, and the number of XEs formed were significantly reduced after ACC treatment in comparison with 10 µM JAMe (Figure 3 and Table 1). Moreover, there was also a significant decrease in proliferation with xylogenesis when the two compounds were combined in comparison with both ACC or JAMe alone (Figure 3). Interestingly, only protoxylem was formed in the combined treatment, as with JAMe (Figure 2B,H, in comparison, and Table 1).

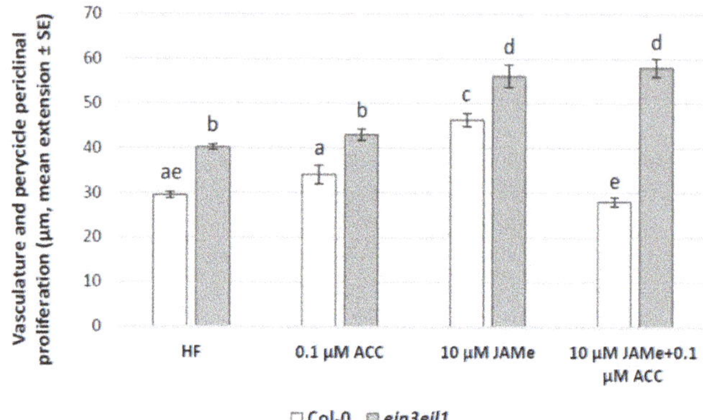

Figure 3. Pericycle periclinal proliferation in the basal hypocotyl of *Arabidopsis* Col-0 (wt) and *ein3eil1* double mutant seedlings either cultured with ACC (0.1 µM), or with JAMe (10 µM) alone or combined with ACC (0.1 µM), and without exogenous hormones (hormone-free, HF). Mean radial extension (± Standard Error (SE)) of the vasculature plus pericycle periclinal proliferation in the basal 5 mm of the hypocotyl. Different letters among different treatments within the same genotype, or between the two genotypes within the same treatment, indicate significant differences at least at $p < 0.05$. The same letters indicate no significant difference. Two-way ANOVA followed by Tukey's post-test. $n = 30$.

In the *ein3eil1* mutant, the periclinal proliferation with xylogenesis observed in the presence of ACC was comparable to that obtained in HF condition (Figure 3), in accordance with the mutant insensitivity to the ACC-derived ET. However, the xylogenic proliferation observed in the ACC and JAMe combined treatment was higher than the wt, reaching values not significantly different from those of JAMe alone (Figure 3). This confirmed the mutant responsiveness to JAMe and the stimulating effect of the latter on xylogenesis (Figures 1 and 3, and Table 1). In addition, only ectopic protoxylem was formed by the mutant (Figure 2I,J and Table 1).

Taken together, the results supported the hypothesis that JA positive role on xylogenesis was negatively affected by ET perception, with no effect on JAMe-determined xylary cell identity.

2.2. JAMe Changes the Xylary Cell Identity Determined by Exogenous IBA or IAA and Promotes Adventitious Rooting when Applied with IAA/IBA

Experiments were carried out to identify possible roles of JAMe on xylary cell identity in cooperation/antagonism with either IAA or IBA control, and in the presence/absence of ET perception through EIN3 and EIL1.

The pericycle periclinal proliferation with xylogenesis and the number of XEs formed (Table 1) in the wt basal hypocotyl were comparable after IBA, IAA or 10 µM JAMe treatments, and noticeably higher than in HF condition (Figures 1, 4 and 5).

In the *ein3eil1* mutant, the pericycle proliferation was significantly lower than the wt for IBA treatment (Figure 4), but higher for IAA (Figure 5). Additionally, the treatment with JAMe, in combination with either IBA or IAA, did not cause significant increases in the number of XEs in comparison with each auxin alone (Table 1).

The presence of JAMe in combination with either auxin resulted into changes in the xylary identity.

In fact, both in the wt and in the mutant, each JAMe concentration caused the formation of both protoxylem and metaxylem when combined with IBA (Figure 2M–P, and Table 1). Conversely, treatments with IBA induced only metaxylem in the wt, and only protoxylem in the mutant (Figure 2K-L, and Table 1). In the wt, when 0.01 µM JAMe was combined with IAA, only protoxylem was formed, as for IAA single treatment (Figure 2R,Q in comparison, and Table 1). For treatments with 1 µM JAMe, metaxylem could be detected and became more abundant than protoxylem after treatment with

10 µM JAMe (Figure 2S, and Table 1). Interestingly, in the mutant, where only protoxylem was formed after IAA treatment (Table 1), metaxylem was instead abundantly and increasingly produced, after combined treatments with IAA and increasing concentrations of JAMe (Figure 2T,U, and Table 1).

Collectively, the results showed that JAMe was able to change auxin-induced ectopic xylary cell identity, with an enhancing effect with higher concentrations.

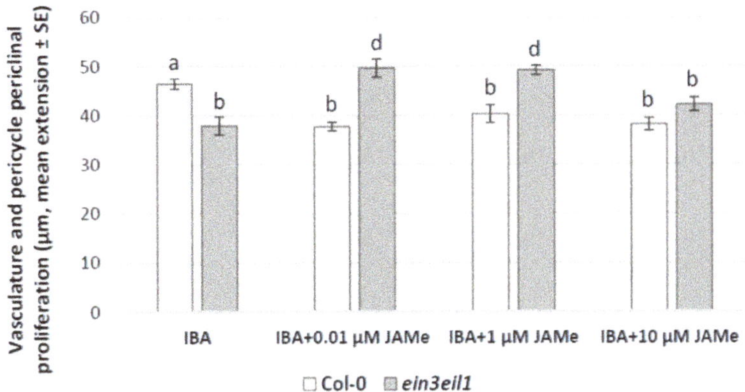

Figure 4. Pericycle periclinal proliferation in the basal hypocotyl of *Arabidopsis* Col-0 (wt) and *ein3eil1* double mutant seedlings either cultured with IBA (10 µM) or with IBA combined with 0.01, 1 or 10 µM JAMe. Mean radial extension (± standard error (SE)) of the vasculature plus pericycle periclinal proliferation in the basal 5 mm of the hypocotyls. Different letters among different treatments within the same genotype, or between the two genotypes within the same treatment, indicate significant differences at least at $p < 0.05$. The same letters indicate no significant difference. Two-way ANOVA followed by Tukey's post-test. $n = 30$.

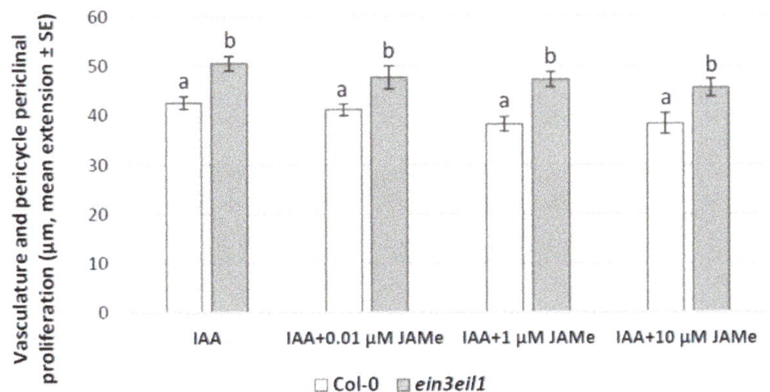

Figure 5. Pericycle periclinal proliferation in the basal hypocotyl of *Arabidopsis* Col-0 (wt) and *ein3eil1* double mutant seedlings either cultured with IAA (10 µM) or with IAA combined with 0.01, 1 or 10 µM JAMe. Mean radial extension (± standard error (SE)) of the vasculature plus pericycle periclinal proliferation in the basal 5 mm of the hypocotyl. Different letters among different treatments within the same genotype, or between the two genotypes within the same treatment, indicate significant differences at least at $p < 0.05$. The same letters indicate no significant difference. Two-way ANOVA followed by Tukey's post-test. $n = 30$.

As adventitious roots (ARs) and adventitious root primordia (ARPs) are induced by auxin in competition with xylogenesis in the basal hypocotyl of *Arabidopsis* [11], the density of the ARs/ARPs was also evaluated. In the wt, the ARP/AR formation caused by IBA application did not change when also JAMe was present in the medium at 0.01 and 1 µM concentration. Nevertheless, the AR-response increased many folds ($p < 0.0001$) in hypocotyls of seedlings treated with 10 µM JAMe (Figure 6). Similar results were also obtained when IAA was applied in combination with JAMe. However, the AR/ARP production in the combined treatment with IAA and 10 µM JAMe was significantly lower than with 10 µM JAMe + IBA (Figure 6). A similar trend was observed in the mutant treated with the same hormone conditions, even if a significant reduction in AR/ARP density in comparison with the wt was noticed (Figure 6).

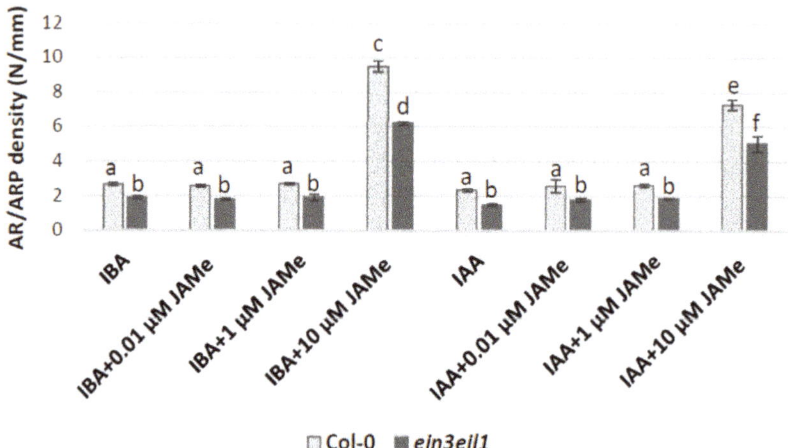

Figure 6. Mean density of adventitious roots (ARs) and adventitious root primordia (ARPs), evaluated as N/mm (± standard error (SE)) in the basal 5 mm of the hypocotyl of Col-0 (wt) and *ein3eil1* double-mutant seedlings cultured with IBA (10 µM) or IAA (10 µM), alone or combined with 0.01, 1 or 10 µM JAMe. Different letters among different treatments within the same genotype, or between the two genotypes within the same treatment, indicate significant differences at least at $p < 0.05$. The same letters indicate no significant difference. Two-way ANOVA followed by Tukey's post-test. $n = 30$.

2.3. Nitric Oxide is an Early Marker of Cell Reactivation in the Hypocotyl Basal Pericycle Involved in Either Xylogenesis or Adventitious Rooting, and its Signal is Enhanced by JAMe

Wild type seedlings cultured in HF conditions or treated with either 10 µM JAMe or 0.1 µM ACC were grown for 22 days in the presence of either the NO donor sodium nitroprusside (SNP) or the NO scavenger 2-4-carboxyphenyl-4,4,5,5-tetramethylimidazoline-1-oxyl-3-oxide (cPTIO). The NO signal was detected with the NO-fluorescent indicator diaminofluorescein-FM diacetate (DAF-FMDA) (Figure 7 and Figure S1), and the derived epifluorescence signal quantified (Figure 8).

The basal hypocotyl of HF-grown seedlings showed a hardly detectable fluorescence signal without addition of SNP in the culture medium. This indicated a very low NO production in the poorly extended periclinal pericycle derivatives, leading to ectopic xylary cells, as well as in the rare anticlinal derivatives and protruding adventitious root primordia (ARPs) (Figure 7A,B, and insets, Figure 8 and Figure S1A). In the presence of SNP, the induced NO production was very low and not significantly different from the HF condition (Figure 7C,D, and insets, and Figure 8). The fluorescence signal was also almost absent in the presence of cPTIO alone (Figure S1B, and Figure 8). In contrast, when the seedlings were cultured with JAMe +/-SNP, an evident NO signal characterized the hypocotyl vasculature, being more intense in pericycle cells (Figure 7E, and inset, Figure 8) and in their first periclinal derivatives (Figure 7F, upper inset and arrows, Figure 8). The signal was instead highly reduced by the cPTIO application in combination with JAMe (Figure 7G and inset, and Figure 8). Moreover, for JAMe +/− SNP treatments, the ability to produce NO decreased with the progress of xylem differentiation becoming almost absent in the differentiated ectopic xylary elements (Figure 7H, inset, and arrows). The ectopic XEs formed were identified as protoxylem type (Figure 7F, lower inset and Figure 7H, inset and arrows), as seen with JAMe treatments in the absence of the NO-donor (Figure 2B, and Table 1). Interestingly, the NO signal was present also in the rarely formed anticlinal derivatives (Figure 7I, and inset), leading to ARP formation, and at the ARP base (Figure S1C), with the NO signal quenching at about ARP VII stage [40] (Figure S1D,E).

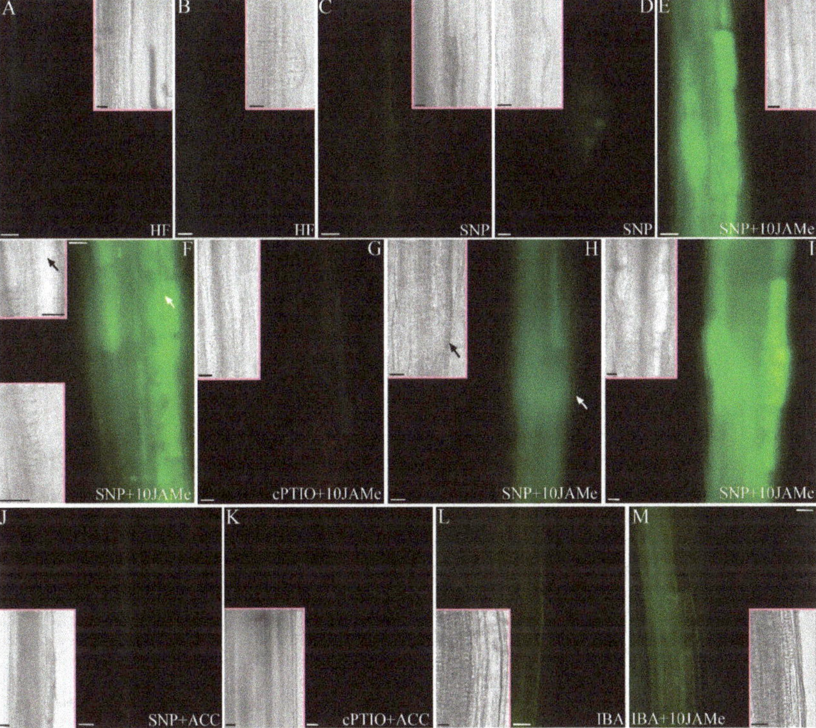

Figure 7. Analysis of NO epifluorescence signal, due to diaminofluorescein-FM diacetate (DAF-FMDA) treatment, in the basal hypocotyl of Col-0 (wt) seedlings dark-grown for 22 DAS. (**A**) Absence of epifluorescence signal in the pericycle periclinal derivatives occasionally formed in hormone-free (HF) condition. (**B**) Absence of epifluorescence signal in an adventitious root primordium derived from anticlinal pericycle divisions in HF condition. (**C,D**) Weak signal in correspondence of cells derived from periclinal (**C**) or anticlinal (**D**) divisions of the pericycle in seedlings treated with 50 µM sodium nitroprusside (SNP). (**E,F**) Strong NO signal in the pericycle cells (**E**), and their first periclinal derivatives (**F**; arrow in the inset) in 50 µM SNP + 10 µM JAMe (SNP + 10JAMe)–treated seedlings. (**G**) Absence of epifluorescence signal in the pericycle after 100 µM 2-4-carboxyphenyl-4,4,5,5-tetramethylimidazoline-1-oxyl-3-oxide (cPTIO) application to 10 µM JAMe-treated seedlings (cPTIO + 10JAMe). (**H**) Strong NO signal in pericycle periclinal derivatives in the hypocotyl of a seedling treated with SNP + 10JAMe. The signal disappears in the differentiated xylary elements (the arrow indicates an ectopic protoxylem element). (**I**) Strong NO signal in pericycle anticlinal derivatives occasionally formed after SNP + 10JAMe application.

(J) Weak NO signal in pericycle-derived cells in 50 μM SNP + 0.1 μM ACC (SNP + ACC)-treated seedlings. (**K**) Absence of NO signal in the hypocotyl of seedling cultured with 100 μM cPTIO + 0.1 μM ACC (cPTIO + ACC). (**L,M**) Epifluorescence signal in the hypocotyl of seedlings treated with 10 μM IBA (IBA; **L**) or 10 μM IBA + 10 μM JAMe (IBA + 10 JAMe; **M**). The signal intensity increases in the presence of JAMe. All the insets show bright field images of whole-mount seedling hypocotyls. Scale bars = 10 μm (**A,B,D–I,K–M**, Insets in **A–F,I,K,L**), 20 μm (**C,J**, Insets in **G,H,M**), 30 μm (Inset in **J**).

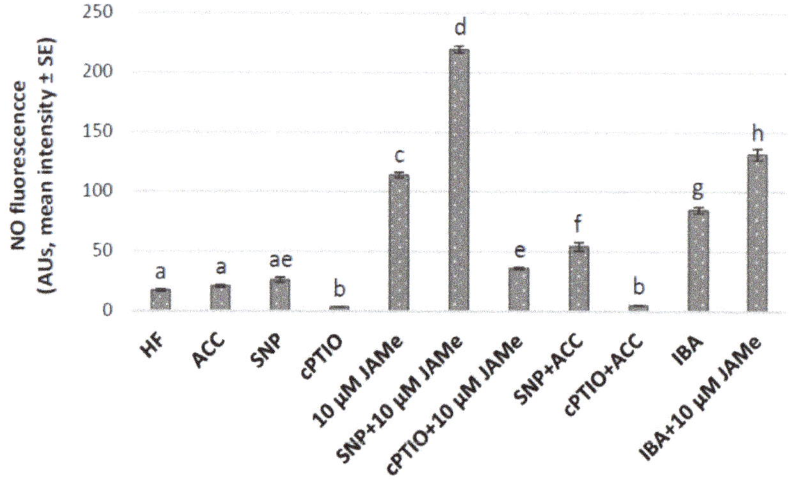

Figure 8. Quantification of the epifluorescence signal due to nitric oxide (NO) production, by means of DAF-FMDA probe, in the basal hypocotyl pericycle cells and in their early derivatives of Col-0 (wt) seedlings grown in darkness for 22 DAS with different treatments. Mean intensity (± standard error (SE)) in Arbitrary Units (AUs) in seedlings cultured in control condition (HF) or with 0.1 μM ACC (ACC), 50 μM SNP (SNP), 100 μM cPTIO (cPTIO), 10 μM JAMe, SNP + 10 μM JAMe, cPTIO + 10 μM JAMe, SNP + ACC, cPTIO + ACC, 10 μM IBA (IBA), 10 μM IBA + 10 μM JAMe. Different letters among treatments indicate significant differences at least at $p < 0.05$. The same letter indicates no significant difference. One-way ANOVA followed by Tukey's post-test. $n = 100$.

In addition, a very low fluorescence signal was observed for the ACC treatment, as measured for HF condition (Figure 8). A weak, but significant, increase appeared instead for ACC treatment combined with SNP (Figure 8). However, the signal intensity in the ACC and SNP combined treatment was much lower than for the 10 μM JAMe and SNP combined treatment (Figure 7E,F,J in comparison and Figure 8), and similarly lowered by cPTIO application (Figures 7K and 8). No changes in xylary cell identity were induced by NO production.

The DAF-FMDA fluorescence was also evaluated in basal hypocotyls of seedlings cultured with JAMe (10 μM) + IBA (10 μM) in comparison with IBA (10 μM) alone, because xylogenesis and AR-formation were more responsive to treatments with IBA than IAA (Figure 6 and Table 1).

The NO production was significantly higher when the seedlings were cultured with IBA and JAMe than only with IBA (Figure 7L,M) or JAMe (Figure 8). This was independent from the initiation of either xylogenesis or rhizogenesis, suggesting a program-independent NO involvement in modulating JAMe effects when combined with IBA.

3. Discussion

The data presented show that exogenous JAMe, applied at a specific concentration, induces xylogenesis in *Arabidopsis* seedlings, with a positive involvement of NO and a negative involvement of ET-signalling, and modulates IAA/IBA-determined xylary cell identity.

3.1. The Action of JAMe on Xylogenesis is Negatively Affected by ET Signalling by EIN3EIL1

The ectopic formation of protoxylem in the basal hypocotyl is enhanced by JAMe, when applied at 10 µM concentration. Interestingly, JA has been recently demonstrated to induce extra xylem also in the roots of the same ecotype of *Arabidopsis* here investigated, with high increases in xylogenesis with 10 µM JAMe [24], in accordance with the present results for the hypocotyl. Moreover, the JA signalling mutant *coronatine insensitive1-1* (*coi1-1*) does not form ectopic XEs in response to JA, whereas the JA biosynthesis mutant *oxophytodienoate-reductase3* (*opr3*) does form XEs, suggesting that the JA response, instead of synthesis, is responsible for xylogenesis induction [24]. It is, thus, possible that also in the hypocotyl the JA signalling, more than its biosynthesis, is positively involved in the control of xylogenesis. The JASMONATE ZIM-DOMAIN (JAZ) proteins are the target of CORONATINE INSENSITIVE1 (COI1) protein, and COI1–JAZ is the co-receptor of the JA-isoleucine conjugate (JA-Ile), active form of JA, in the responsive target cells, in our case in the hypocotyl pericycle cells. In accordance, *coi1* mutants are JA/JAMe-insensitive ([41] and references therein). It is widely known that the interaction of JA-Ile with COI1 induces the proteolysis of JAZs, releasing MYC2. MYC2 regulates the JA responses by controlling the expression of JA-responsive genes, including, in our hypothesis, those involved in xylogenesis. One of the JA-responsive xylogenic gene might be *ARF17*, belonging to the Auxin Response Factors (ARFs) mediating auxin-induced gene activation [42]. In fact, in *Arabidopsis* TCLs cultured with IBA and Kinetin, JA/JA-Ile is immunolocalized in the xylogenic cells, where *ARF17* is expressed and upregulated by 10 µM JAMe [26]. Moreover, in some reports cytokinin is considered important for xylogenesis to occur [3], but in others it is reported as a negative regulator of xylem development ([24] and references therein). JAMe treatments have been demonstrated to reduce cytokinin effects in the *Arabidopsis* root vasculature, and cytokinin treatments to nullify JAMe promotion of ectopic xylem formation [24]. In accordance, our previous [11] and present results on *Arabidopsis* hypocotyls of dark-grown seedlings show that xylogenesis occurs in the absence of exogenous cytokinin, and that this condition may favour the JA inductive action.

We also show that the *ein3eil1* mutant is JA-sensitive in xylogenesis, whereas it is known to be insensitive in other processes, e.g., the induction of pathogen-related genes and root hair development [39]. Moreover, the xylogenic effect of 10 µM JAMe is observed in the mutant as in the wt, showing that xylogenesis is negatively affected by the interaction of EIN3EIL1 with JA derived by JAMe application. The reduction of the xylary response observed in the wt, after exposition to JAMe combined with the ET-precursor ACC, supports the validity of this interpretation. The JA-sensitivity of *ein3eil1* in xylogenesis is in accordance with the observation that EIN3 and EIL1 are positive regulators only in a specific subset of JA responses [39,43].

3.2. Nitric Oxide is a Common Marker of JAMe-induced Xylogenesis and Auxin-Induced Adventitious Rooting Acting Downstream to Pericycle Cell Determination to either Program

Results showed that NO is early produced in the pericycle derivatives, from which either XEs or ARPs are formed, showing that NO is a common marker for both programs. Its presence in pericycle cells, either dividing periclinally to generate XEs or anticlinally to generate ARPs, demonstrates that it is an early marker for either xylogenesis or rhizogenesis. It is known that the exogenous treatment with the NO-donor SNP enhances initiation of root hairs in the root elongation zone through the reorientation of cortical microtubules [44]. Furthermore, cell plate formation is known to be very sensitive to perturbations of the microtubular cytoskeleton, and a peculiar target of nitrotyrosine, coming from NO-mediated post-translational modification ([45] and references therein). It is, thus, possible that a NO-guided rearrangement in the cytoskeleton and cell plate orientation may occur, resulting into either xylogenesis or rhizogenesis. However, NO seems to act downstream on factor(s) causing the pericycle cells to become committed to either program. Interestingly, NO production increases when 10 µM JAMe is applied. NO signalling is involved also in stomatal closure, with NO production highly enhanced by 10 µM JAMe [46]. Previous reports also showed that exogenous JAMe stimulates NO synthase to induce NO production in *Taxus* cells [38], and that JA and NO modulate each other production [47]. In addition, our quantitative data show that NO production, either with ACC single treatment, or when ACC was combined with SNP, was much lower than with 10 μM JAMe, combined or not with the same NO-donor compound. This suggests an antagonistic connection between JAs and ET in the control of xylogenesis, related to NO levels and, possibly, NO signal activity.

Interestingly, the disappearance of the NO signal occurs only later in xylogenesis, i.e., when the XEs become mature, while in rhizogenesis it occurs at an early stage (stage VII) of ARP formation. This means that a sustained NO production is needed until PCD and lignin deposition are concluded. On the other hand, the disappearance of the NO signal occurs in a still widely meristematic condition during rhizogenesis. In accordance, in *Zinnia elegans* a NO burst occurs before the processes of secondary cell wall formation and cell autolysis, in both differentiating xylem of vascular bundles and trans-differentiating xylary cells of leaf mesophyll ([29] and references therein). Moreover, in accordance with our results, in *Populus* roots, NO signalling contributes to the onset of xylary differentiation and to further stages of maturation, but it is not observed in the mature vessels [34].

It is known that IBA induces AR-formation by conversion into IAA, and this involves NO activity ([35] and references therein). Present and past results together show that NO is present in the ARPs meristematic domes in TCLs [35] and in planta (present results). It is possible that NO signalling is a long-lasting process in xylogenesis, in comparison with rhizogenesis, because it is necessary for consecutive events until xylary maturation. Conversely, ARPs do not need prolonged NO activity, being capable to sustain their growth after quiescent centre definition in their meristematic dome. This normally occurs at stage VII of ARP development [7,40], i.e., when the NO signal is here shown to be quenched.

Moreover, the DAF-FMDA fluorescence revealed that NO production was higher in the presence of JAMe (10 μM) combined with IBA (10 μM), than with IBA (10 μM) alone, independently of the initiation of either xylogenesis or rhizogenesis. This strengthens the idea of a program-independent NO involvement in both processes, as well as the concurrent role of JAMe combined with IBA in enhancing early NO formation.

Auxin (natural or synthetic) is necessary to trigger xylogenesis ([1] and present results), and IAA and its precursor IBA are present in the hypocotyls of dark-grown *Arabidopsis* seedlings [19]. Moreover, it is known that endogenous auxin accumulates before the formation of XEs in numerous plants/culture systems, and this occurs both after either a dedifferentiation in the target cells or their direct trans-differentiation ([11] and references therein). In particular, in *Arabidopsis* hypocotyls endogenous auxin accumulates in the basal pericycle before the first divisions occur [40]. Present data show that only a rare xylary formation is induced by the endogenous hormonal input (HF condition), whereas an enhancement is detected in the presence of 10 μM JAMe. Although a synergism of auxin and NO during the activation of cytokinesis has been reported [48], it still remains to be explained how the endogenous auxin with exogenous JAMe direct the basal pericycle cells towards the realization of the xylogenic program and not root formation, by using the same NO signalling molecule.

Surprisingly, xylogenesis does not increase with any JAMe concentration combined with either IBA or IAA, in comparison with the auxins single treatments. Additionally, the IBA/IAA-induced adventitious rooting is enhanced by the same JAMe concentration (10 μM) that enhances xylogenesis when applied alone. This suggests that JA acts as a common player between the two programs in the same target cells, with a possible different action depending on the endogenous auxin levels. Xylogenesis might need lower levels of auxin to be initiated, in accordance with the endogenous IAA levels monitored in the absence/presence of IBA in *Arabidopsis* dark-grown seedlings [19]. Moreover, this possibility has been also previously hypothesized based on the activity of the auxin influx carrier AUX1 in switching between the two developmental programmes [7]. In addition, it has been suggested that the JA-mediated activation of auxin response in *Arabidopsis* roots requires the function of AUX1, and JAMe upregulates AUX1 expression [49].

3.3. The Auxin-Determined Xylary Cell Identity is Modulated by JAMe

It is known that auxins are essential for determining xylem cell identity, with the formation of auxin maxima in the target cells mediating the process [50]. In various plant species, VASCULAR-RELATED NAC-DOMAIN6 and 7 (VND6 and 7) are master TFs in the control of XE identity, because *VND6* and *VND7* ectopic expression causes the trans-differentiation of either metaxylem or protoxylem elements ([51] and references therein). In fact, in *Arabidopsis* and poplar roots, these genes have been proposed as transcription switches for metaxylem and protoxylem formation [52]. VND7, in particular, exhibits a critical role in regulating protoxylem formation in the root, and metaxylem formation in both root and shoot [53]. In addition, SCARECROW (SCR) TF is induced by auxin [54], and forms with SHORT ROOT (SHR) TF a complex that produces the microRNA 165/166, involved in the destabilization of *HD-ZIP III* genes [55], determinant for metaxylem specification [7]. Both *Arabidopsis shr* and *scr* null mutants produce ectopic metaxylem in dark-grown seedlings [7], supporting a possible involvement of SHR-SCR complex activities in the specific realization of metaxylem. Moreover, it has been recently demonstrated that the metaxylem and protoxylem identity might be mutually repressed. Glycogen Synthase Kinase 3 (GSK3) proteins, like BRASSINOSTEROID INSENSITIVE 2 (BIN2), BIN2-LIKE 1 (BIL1), BIL2, SHAGGY-RELATED KINASE 11 (ATSK11) and ATSK13 have been demonstrated to interact with PHLOEM INTERCALATED WITH XYLEM (PXY), with this interaction seeming to specify repression of xylem identity [56].

The present results show that JAMe changes auxin-induced ectopic xylary cell identity, with an enhancing effect caused by the increase in its concentration. This is the first report of a role for JAs in changing xylary cell identity.

It is known, and here confirmed, that ectopic XE formation is stimulated by the exogenous application of IAA and IBA at the same concentration (10 µM) in dark-grown *Arabidopsis* hypocotyls. Nevertheless, the two auxins affect differently the xylary identity, with metaxylem induced by IBA and protoxylem by IAA [11 and present results].

We showed that the pericycle periclinal proliferation with xylogenesis in the wt was comparable for each auxin and JAMe (10 µM) individual treatments, and higher than HF or ACC, confirming the xylogenic role of the two exogenously applied auxins, but also of JAMe acting in combination with the endogenous hormonal pool. However, the presence of JAMe in combination with either IBA or IAA did not increase the number of the XEs formed in comparison with IBA/IAA alone, but affected xylary identity.

Independently of treatment and genotype, we observed only ectopic protoxylem formation after JAMe treatment, as observed for HF condition, thus showing that the applied JAMe did not change the xylary cell identity induced by the endogenous hormones. This was also demonstrated in the mutant impaired in ET-perception, but JA-sensitive, excluding a possible involvement of the EIN3EIL1 complex in this process. However, in the wt, JAMe, especially at the highest concentration, induced both ectopic protoxylem and metaxylem when combined with either IBA or IAA. Interestingly, in the mutant, forming only protoxylem after either IAA or IBA single treatments, metaxylem was instead abundantly formed when JAMe was applied, especially at 10 µM, combined with each auxin. Altogether, the results show that JAMe-modulation of ectopic xylem identity induced by exogenous IAA or IBA bypassed ET-perception by EIN3EIL1.

In conclusion, the basic findings of this research add new elements to the intricate switching of xylogenesis and adventitious rooting in the same cells. Here, JAs are demonstrated to be actors in both developmental programs, and in the modulation of xylary cell identity, with an important mediation by NO, as summarized in the model of Figure 9.

The present findings may be also exploited to implement practical aspects related to the creation of plants with improved xylem, favouring plant survival under water stress conditions, or for biotechnological purposes, e.g., for the optimization of biofuels and biomaterials production.

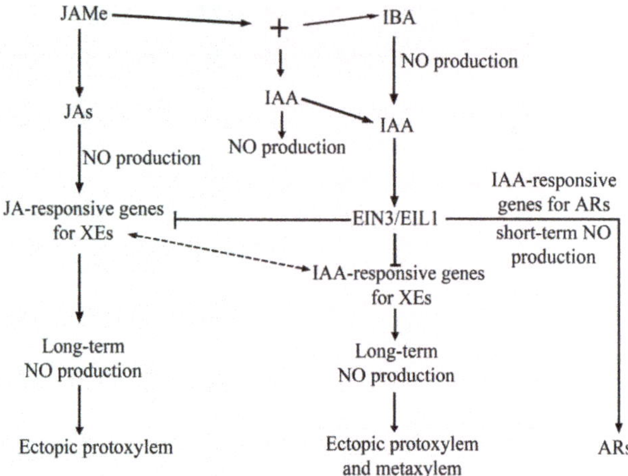

Figure 9. Proposed model for jasmonates (JAs) and auxin (IAA/IBA) roles in the formation of ectopic xylary cells (XEs) in the pericycle derivative cells of *Arabidopsis* basal hypocotyl of seedlings cultured with JAMe (10 µM) +/− IBA (10 µM) or IAA (10 µM). The pathway leading to the formation of adventitious roots (ARs) as an alternative to XE formation is also shown. The green part of the figure shows JA's effect in the absence of exogenous auxin. JAs derived by JAMe demethylation [23] induce NO production. The NO signal triggers the expression of JA-induced genes responsible for XE formation. This occurs through a negative interaction with ET-signalling by EIN3/EIL1. In the following steps of the differentiation process, a long-term NO production occurs and lasts until the maturation of ectopic protoxylem elements. When IBA is applied with JAMe (the pink part of the figure) is converted into IAA, also inducing NO-formation [35]. When IAA is alternatively applied with JAMe, NO-formation may also occur [27]. Cell perception of the endogenous IAA derived by the application of the two exogenous auxins is associated with a positive crosstalk with ET through EIN3EIL1 in AR-formation [11], and with a negative crosstalk in xylogenesis (present results). Different pathways of NO production occur in the two programs. A short-term NO production is associated with the expression of the IAA-induced genes involved in AR-formation, whereas a long-term NO production is proposed to be induced after the possible interaction of JA-responsive and IAA-responsive genes for XE-formation thereby leading to the ectopic formation of both protoxylem and metaxylem.

4. Materials and Methods

4.1. Plant Growth

One hundred seeds of Col-0 ecotype of *Arabidopsis thaliana* (L.) Heynh and of its homozygous double mutant *ein3eil1* [39] (provided by Hongwei Guo, Peking University, China) were sown, after sterilization, on square Petri plates (12 cm × 12 cm; 10 seeds per plate) containing full strength Murashige and Skoog (MS) [57] salts supplemented with 0.55 mM myo-inositol (Fluka, Buchs, Switzerland), 0.1 µM thiamine-HCl (Sigma-Aldrich, St. Louis, MO, USA), 1% (*w/v*) sucrose (Sigma-Aldrich) and 0.8% (*w/v*) agar (Sigma-Aldrich) (pH 5.7). As an alternative to this Hormone Free (HF) medium, either JAMe (0.01, 1 or 10 µM) (Duchefa Biochemie B.V, Haarlem, NH, USA), or ACC (0.1 µM) (Sigma-Aldrich), alone or combined with 10 µM JAMe, were added. The final JAMe concentrations in the media were obtained taking the exact amounts from a stock solution (10^{-3} M) in 10% ethanol. To dissolve the JAMe, diluted ethanol was preferred to pure ethanol, suggested by the producer (Duchefa Biochemie B.V, Haarlem, NH), in order to keep the final ethanol concentration in the media below 0.1% to avoid possible negative effects on germination [58].

Media containing the components of the HF medium and either IAA or IBA (both from Sigma-Aldrich), at 10 µM concentration, alone or combined with JAMe (0.01, 1 or 10 µM), were also prepared. Moreover, treatments with either 50 μM of the NO donor sodium nitroprusside (SNP; Merck, Darmstadt, Germany) or 100 μM of the NO scavenger 2-4-carboxyphenyl-4,4,5,5-tetramethylimidazoline-1-oxyl-3-oxide (cPTIO; Sigma-Aldrich) were performed either in HF condition or with JAMe (10 μM) or ACC (0.1 μM). The media were sterilized by autoclaving at 120 °C for 20 min. IBA and IAA were added before autoclaving taking the appropriate volume from stock-solutions (10^{-3} M). Sterile stock-solutions of JAMe (10^{-3} M), ACC (10^{-3} M), SNP (10^{-2} M) and cPTIO (10^{-2} M) were prepared by filtering (with a 0.22 μm pore filter), and the appropriate volume was taken to reach the final concentration in the already autoclaved medium, after the medium temperature decreased up to 45–50 °C.

Independently from the treatment, after sowing, the seeds were stratified for three days at 4 °C under continuous darkness and exposed to white light (intensity 100 $\mu E \cdot m^{-2} \cdot s^{-1}$) for about 6 h, to induce germination. The plates were then placed in vertical position under continuous darkness until 22 days after stratification (DAS), at 22 ± 2 °C.

4.2. Histological Analysis

After 22 DAS, 30 seedlings per genotype and per treatment were fixed, dehydrated, embedded in Technovit 7100 (Heraeus Kulzer, Germany), longitudinally sectioned (8 μm thickness) with the Microm HM 350 SV microtome (Microm, Walldorf, Germany), and stained with 0.05% toluidine blue (all procedures according to [40]). Sections were taken from the basal portion (5 mm in length) of the hypocotyl, according to Fattorini and co-workers [11], and observed with a Leica DMRB microscope; images were acquired with a DC500 camera (Leica, Wetzlar, Germany).

4.3. Nitric Oxide Detection

Intracellular NO content in 22d old Col-0 seedlings, cultured either in HF medium or treated with 10 μM JAMe, or 0.1 μM ACC, with/without SNP (50 μM) or cPTIO (100 μM), or with 10 μM IBA or 10 μM IBA + 10 μM JAMe, was monitored using the cell-permeable diacetate derivative diaminofluorescein-FM (DAF-FMDA; Sigma). Seedlings were incubated in 20 mM HEPES/NaOH buffer (pH 7.4) supplemented with 5 μM DAF-FMDA for 20 min [59] at 22 DAS, after having verified that no significant epifluorescence signal was detectable with the buffer alone (Figure S2). The excess of the fluorescent probe was removed by washing the seedlings for three times with the buffer, after which they were observed using a Leica DMRB microscope equipped with the specific set of filters (EX 450–490, DM 510, LP 515). The images were acquired with a Leica DC500 digital camera and analyzed with the IM1000 image-analysis software (Leica). The intensity of the green fluorescent signal was quantified in the basal hypocotyl of ten seedlings per treatment. Ten measures per seedling were randomly carried out in the pericycle cell derivatives using the ImageJ software (National Institute of Health, Bellevue, WA, USA), and expressed in arbitrary units (AUs; from 0 to 255). The values were averaged and normalized to those measured in hypocotyls incubated in the buffer without the fluorescent probe according to Fattorini and coworkers [35].

4.4. Measurement Procedures and Statistical Analysis

The hypocotyls of 30 seedlings per genotype and treatment were cut into three portions. Measures of the radial extension of the vascular system, including the de novo formed cells by the pericycle periclinal divisions, were carried at the middle of the hypocotyl basal portion, according to Fattorini and co-workers [11], and expressed as mean values (± standard error (SE)) In the same hypocotyl portion, the ectopic XEs present in an area of 150×150 μm^2 were counted and quantified as mean numbers (±SE). The protoxylem vs. metaxylem identity was determined and expressed as percentage on the total of the de novo formed XEs. Adventitious roots were counted in the same hypocotyl portions and evaluated as mean density (±SE).

One-way or two-way analysis of variance (ANOVA, $p < 0.05$) was used to compare the effects of different treatments or different treatments and genotypes, respectively, and, if ANOVA showed significant effects, Tukey's post-test was applied (GraphPad Prism 6.0, GraphPad Software, Inc., La Jolla, CA, USA). All the experiments were repeated three times, with very similar results.

Supplementary Materials: Supplementary materials can be found at http://www.mdpi.com/1422-0067/20/18/4469/s1.

Author Contributions: F.D.R. conceived and performed a large part of the experiments. L.F. performed the other part of the experiments and carried out the statistical evaluation of the data. M.R. prepared the histological materials and acquired the images. G.F. contributed to the statistical evaluation of data and the interpretation of the epifluorescence images. M.M.A. interpreted the data and contributed to write the manuscript. C.B. designed the experiments, interpreted the data and contributed to write the manuscript.

Funding: This research was funded by Sapienza University of Rome (grant number RM118164286B32D4 to LF).

Acknowledgments: We thank Hongwei Guo (Peking University, Beijing, China) for the kind donation of *ein3eil1* seeds.

Conflicts of Interest: The authors declare no conflict of interest.

Abbreviations

ACC	1-Aminocyclopropane-1-carboxylic acid
ANOVA	Analysis of variance
AR	Adventitious root
ARP	Adventitious root primordium
AU	Arbitrary unit
cPTIO	2-4-carboxyphenyl-4,4,5,5-tetramethylimidazoline-1-oxyl-3-oxide
DAF-FMDA	Diaminofluorescein-FM diacetate
DAS	Days after stratification
ET	Ethylene
HF	Hormone-free
IAA	Indole-3-acetic acid
IBA	Indole-3-butyric acid
JA	Jasmonic acid
JAMe	Jasmonic acid methyl ester
SE	Standard error
SNP	Sodium nitroprusside
TF	Transcription factor
XE	Xylary element

References

1. Yoshida, S.; Iwamoto, K.; Demura, T.; Fukuda, H. Comprehensive analysis of the regulatory roles of auxin in early transdifferentiation into xylem cells. *Plant Mol. Biol.* **2009**, *70*, 457–469. [CrossRef] [PubMed]

2. Miyashima, S.; Sebastian, J.; Lee, J.Y.; Helariutta, Y. Stem cell function during plant vascular development. *EMBO J.* **2013**, *32*, 178–193. [CrossRef]

3. Fukuda, H. Tracheary element differentiation. *Plant Cell* **1997**, *9*, 1147–1156. [CrossRef] [PubMed]

4. Falasca, G.; Zaghi, D.; Possenti, M.; Altamura, M.M. Adventitious root formation in *Arabidopsis thaliana* thin cell layers. *Plant Cell Rep.* **2004**, *23*, 17–25. [CrossRef] [PubMed]

5. Falasca, G.; Altamura, M.M. Histological analysis of adventitious rooting in *Arabidopsis thaliana* (L.) Heynh seedlings. *Plant Biosystems* **2003**, *137*, 265–273. [CrossRef]

6. Fukuda, H. Signals that control plant vascular cell differentiation. *Nat. Rev. Mol. Cell Biol.* **2004**, *5*, 379–391. [CrossRef] [PubMed]

7. Della Rovere, F.; Fattorini, L.; D'Angeli, S.; Veloccia, A.; Del Duca, S.; Cai, G.; Falasca, G.; Altamura, M.M. Arabidopsis SHR and SCR transcription factors and AUX1 auxin influx carrier control the switch between adventitious rooting and xylogenesis in planta and in in vitro cultured thin cell layers. *Ann. Bot.* **2015**, *115*, 617–628. [CrossRef] [PubMed]

8.	Sundberg, B.; Uggla, C.; Tuominen, H. Cambial growth and auxin gradients. In *Cell and Molecular Biology of Wood Formation*; Savidge, R.A., Barnett, J.R., Napier, R., Eds.; BIOS Scientific Publishers: Oxford, UK, 2000; pp. 169–188.

9.	McCann, M.C.; Domingo, C.; Stacey, N.J.; Milioni, D.; Roberts, K. Tracheary element formation in an in vitro system. In *Cell and Molecular Biology of Wood Formation*; Savidge, R.A., Barnett, J.R., Napier, R., Eds.; BIOS Scientific Publishers: Oxford, UK, 2000; pp. 457–470.

10.	Charton, L.; Plett, A.; Linka, N. Plant peroxisomal solute transporter proteins. *J. Integrative Plant Biol.* **2019**, *61*, 817–835. [CrossRef] [PubMed]

11.	Fattorini, L.; Della Rovere, F.; Andreini, E.; Ronzan, M.; Falasca, G.; Altamura, M.M. Indole-3-butyric acid induces ectopic formation of metaxylem in the hypocotyl of *Arabidopsis thaliana* without conversion into indole-3-acetic acid and with a positive interaction with ethylene. *Int. J. Mol. Sci.* **2017**, *18*, 2474. [CrossRef]

12.	Miller, A.R.; Pengelly, W.L.; Roberts, L.W. Introduction of xylem differentiation in *Lactuca* by ethylene. *Plant Physiol.* **1984**, *75*, 1165–1166. [CrossRef] [PubMed]

13.	Pesquet, E.; Tuominen, H. Ethylene stimulates tracheary element differentiation in *Zinnia elegans* cell cultures. *New Phytol.* **2011**, *190*, 138–149. [CrossRef] [PubMed]

14.	Pesquet, E.; Zhang, B.; Gorzsás, A.; Puhakainen, T.; Serk, H.; Escamez, S.; Barbier, O.; Gerber, L.; Courtois-Moreau, C.; Alatalo, E.; et al. Non-cell-autonomous postmortem lignification of tracheary elements in *Zinnia elegans*. *Plant Cell* **2013**, *25*, 1314–1328. [CrossRef] [PubMed]

15.	Biondi, S.; Scaramagli, S.; Capitani, F.; Marino, G.; Altamura, M.M.; Torrigiani, P. Ethylene involvement in vegetative bud formation in tobacco thin layers. *Protoplasma* **1998**, *202*, 134–144. [CrossRef]

16.	Eklund, L.; Tiltu, A. Cambial activity in 'normal' spruce *Picea abies* Karst (L.) and snake spruce *Picea abies* (L.) Karst f. virgata (Jacq.) Rehd. in response to ethylene. *J. Exp. Bot.* **1999**, *50*, 1489–1493. [CrossRef]

17.	Love, J.; Björklund, S.; Vahala, J.; Hertzberg, M.; Kangasjärvi, J.; Sundberg, B. Ethylene is an endogenous stimulator of cell division in the cambial meristem of *Populus*. *Proc. Natl. Acad. Sci. USA* **2009**, *106*, 5984–5989. [CrossRef] [PubMed]

18.	Zhang, J.; Yu, J.; Wen, C.K. An alternate route of ethylene receptor signaling. *Front. Plant Sci.* **2014**, *5*, 648. [CrossRef] [PubMed]

19.	Veloccia, A.; Fattorini, L.; Della Rovere, F.; Sofo, A.; D'Angeli, S.; Betti, C.; Falasca, G.; Altamura, M.M. Ethylene and auxin interaction in the control of adventitious rooting in *Arabidopsis thaliana*. *J. Exp. Bot.* **2016**, *67*, 6445–6458. [CrossRef] [PubMed]

20.	Růžička, K.; Ljung, K.; Vanneste, S.; Podhorsk, R.; Beeckman, T.; Friml, J.; Benková, E. Ethylene regulates root growth through effects on auxin biosynthesis and transport-dependent auxin distribution. *Plant Cell.* **2007**, *19*, 2197–2212. [CrossRef] [PubMed]

21.	Zhu, Z. Molecular basis for jasmonate and ethylene signal interactions in Arabidopsis. *J. Exp. Bot.* **2014**, *65*, 5743–5748. [CrossRef] [PubMed]

22.	Sehr, E.M.; Agusti, J.; Lehner, R.; Farmer, E.E.; Schwarz, M.; Greb, T. Analysis of secondary growth in the Arabidopsis shoot reveals a positive role of jasmonate signalling in cambium formation. *Plant J.* **2010**, *63*, 811–822. [CrossRef] [PubMed]

23.	Fattorini, L.; Falasca, G.; Kevers, C.; Rocca, L.M.; Zadra, C.; Altamura, M.M. Adventitious rooting is enhanced by methyl jasmonate in tobacco thin cell layers. *Planta* **2009**, *231*, 155–168. [CrossRef] [PubMed]

24.	Jang, G.; Chang, S.H.; Um, T.Y.; Lee, S.; Kim, J.K.; Choi, Y.D. Antagonistic interaction between jasmonic acid and cytokinin in xylem development. *Sci. Rep.* **2017**, *7*, 10212. [CrossRef] [PubMed]

25.	Jang, G.; Lee, S.; Chang, S.H.; Kim, J.-K.; Choi, Y.D. Jasmonic acid modulates xylem development by controlling polar auxin transport in vascular tissues. *Plant Biotechnol. Rep.* **2018**, *12*, 265–271. [CrossRef]

26.	Fattorini, L.; Hause, B.; Gutierrez, L.; Veloccia, A.; Della Rovere, F.; Piacentini, D.; Falasca, G.; Altamura, M.M. Jasmonate promotes auxin-induced adventitious rooting in dark-grown *Arabidopsis thaliana* seedlings and stem thin cell layers by a cross-talk with ethylene signalling and a modulation of xylogenesis. *BMC Plant Biol.* **2018**, *18*, 182. [CrossRef] [PubMed]

27.	Freschi, L. Nitric oxide and phytohormone interactions: Current status and perspectives. *Front Plant Sci.* **2013**, *4*, 398. [CrossRef] [PubMed]

28.	Neill, S. NO way to die – nitric oxide, programmed cell death and xylogenesis. *New Phytol.* **2005**, *165*, 5–8. [CrossRef] [PubMed]

29. Gabaldón, C.; Gómez, R.L.V.; Pedreño, M.A.; Ros, B.A. Nitric oxide production by the differentiating xylem of *Zinnia elegans*. *New Phytol.* **2005**, *165*, 121–130. [CrossRef] [PubMed]

30. Sun, H.; Feng, F.; Liu, J.; Zhao, Q. Nitric oxide affects rice root growth by regulating auxin transport under nitrate supply. *Front. Plant Sci.* **2018**, *9*, 659. [CrossRef] [PubMed]

31. Niu, Y.H.; Guo, F.Q. Nitric oxide regulates dark-induced leaf senescence through *EIN2* in *Arabidopsis*. *J. Integr. Plant Biol.* **2012**, *54*, 516–525. [CrossRef] [PubMed]

32. Alonso, J.M.; Stepanova, A.N. The ethylene signalling pathway. *Science* **2004**, *306*, 1513–1515. [CrossRef]

33. Iakimova, E.T.; Woltering, E.J. Xylogenesis in zinnia (*Zinnia elegans*) cell cultures: Unravelling the regulatory steps in a complex developmental programmed cell death event. *Planta* **2017**, *245*, 681–705. [CrossRef] [PubMed]

34. Bagniewska-Zadworna, A.; Arasimowicz-Jelonek, M.; Smoliński, D.J.; Stelmasik, A. New insights into pioneer root xylem development: Evidence obtained from *Populus trichocarpa* plants grown under field conditions. *Ann. Bot.* **2014**, *113*, 1235–1247. [CrossRef] [PubMed]

35. Fattorini, L.; Veloccia, A.; Della Rovere, F.; D'Angeli, S.; Falasca, G.; Altamura, M.M. Indole-3-butyric acid promotes adventitious rooting in *Arabidopsis thaliana* thin cell layers by conversion into indole-3-acetic acid and stimulation of anthranilate synthase activity. *BMC Plant Biol.* **2017**, *17*, 121. [CrossRef] [PubMed]

36. Corpas, F.J.; Barroso, J.B. Peroxisomal plant metabolism – an update on nitric oxide, Ca2+ and the NADPH recycling network. *J. Cell Sci.* **2018**, *131*, jcs202978. [CrossRef]

37. Chini, A.; Monte, I.; Zamarreño, A.M.; Hamberg, M.; Lassueur, S.; Reymond, P.; Weiss, S.; Stintzi, A.; Schaller, A.; Porzel, A.; et al. An OPR3-independent pathway uses 4,5-didehydrojasmonate for jasmonate synthesis. *Nat. Chem. Biol.* **2018**, *14*, 171–178. [CrossRef]

38. Wang, J.W.; Wu, J.Y. Nitric oxide is involved in methyl jasmonate-induced defense responses and secondary metabolism activities of Taxus cells. *Plant Cell Physiol.* **2005**, *46*, 923–930. [CrossRef]

39. Zhu, Z.; An, F.; Feng, Y.; Li, P.; Xue, L.; Angelo, A.M.; Jiang, Z.; Kim, J.M.; To, T.K.; Li, W.; et al. Derepression of ethylene-stabilized transcription factors (EIN3/EIL1) mediates jasmonate and ethylene signaling synergy in Arabidopsis. *Proc. Natl. Acad. Sci. USA* **2011**, *108*, 12539–12544. [CrossRef]

40. Della Rovere, F.; Fattorini, L.; D'Angeli, S.; Veloccia, A.; Falasca, G.; Altamura, M.M. Auxin and cytokinin control formation of the quiescent centre in the adventitious root apex of Arabidopsis. *Ann. Bot.* **2013**, *112*, 1395–1407. [CrossRef]

41. Betti, C.; Della Rovere, F.; Ronzan, M.; Fattorini, L. EIN2 and COI1 control the antagonism between ethylene and jasmonate in adventitious rooting of *Arabidopsis thaliana* thin cell layers. *PCTOC* **2019**, *138*, 41–51. [CrossRef]

42. Tiwari, S.B.; Hagen, G.; Guilfoyle, T. The roles of auxin response factor domains in auxin-responsive transcription. *Plant Cell* **2003**, *15*, 533–543. [CrossRef]

43. Song, S.; Huang, H.; Gao, H.; Wang, J.; Wu, D.; Liu, X.; Yang, S.; Zhai, Q.; Li, C.; Qi, T.; et al. Interaction between MYC2 and ETHYLENE INSENSITIVE3 modulates antagonism between jasmonate and ethylene signaling in Arabidopsis. *Plant Cell* **2014**, *26*, 263–279. [CrossRef]

44. Yemets, A.I.; Krasylenko, Y.A.; Sheremet, Y.A.; Blume, Y.B. Microtubule reorganization as a response to implementation of NO signals in plant cells. *Cytol. Genet.* **2009**, *43*, 73–79. [CrossRef]

45. Jovanović, A.M.; Durst, S.; Nick, P. Plant cell division is specifically affected by nitrotyrosine. *J. Exp. Bot.* **2010**, *61*, 901–909. [CrossRef] [PubMed]

46. Saito, N.; Nakamura, Y.; Mori, I.C.; Murata, Y. Nitric oxide functions in both methyl jasmonate signalling and abscisic acid signalling in Arabidopsis guard cells. *Plant Signal. Behav.* **2009**, *4*, 119–120. [CrossRef]

47. Zhou, J.; Jia, F.; Shao, S.; Zhang, H.; Li, G.; Xia, X.; Zhou, Y.; Yu, J.; Shi, K. Involvement of nitric oxide in the jasmonate-dependent basal defense against root-knot nematode in tomato plants. *Front. Plant Sci.* **2015**, *6*, 193. [CrossRef] [PubMed]

48. Ötvös, K.; Pasternak, T.P.; Miskolczi, P.; Domoki, M.; Dorjgotov, D.; Szűcs, A.; Bottka, S.; Dudits, D.; Fehér, A. Nitric oxide is required for, and promotes auxin-mediated activation of cell division and embryogenic cell formation but does not influence cell cycle progression in alfalfa cell cultures. *Plant J.* **2005**, *43*, 849–860. [CrossRef]

49. Sun, J.; Xu, Y.; Ye, S.; Jiang, H.; Chen, Q.; Liu, F.; Zhou, W.; Chen, R.; Li, X.; Tietz, O.; et al. Arabidopsis *ASA1* is important for jasmonate-mediated regulation of auxin biosynthesis and transport during lateral root formation. *Plant Cell* **2009**, *21*, 1495–1511. [CrossRef] [PubMed]

50. Bishopp, A.; Help, H.; El-Showk, S.; Weijers, D.; Scheres, B.; Friml, J.; Benková, E.; Mähönen, A.P.; Helariutta, Y. A mutually inhibitory interaction between auxin and cytokinin specifies vascular pattern in roots. *Curr. Biol.* **2011**, *21*, 917–926. [CrossRef] [PubMed]

51. Fukuda, H. Signaling, transcriptional regulation, and asynchronous pattern formation governing plant xylem development. *Proc Jpn. Acad. Ser. B Phys. Biol. Sci.* **2016**, *92*, 98–107. [CrossRef] [PubMed]

52. Kubo, M.; Udagawa, M.; Nishikubo, N.; Horiguchi, G.; Yamaguchi, M.; Ito, J.; Mimura, T.; Fukuda, H.; Demura, T. Transcription switches for protoxylem and metaxylem vessel formation. *Gene Dev.* **2005**, *19*, 1855–1860. [CrossRef] [PubMed]

53. Yamaguchi, M.; Kubo, M.; Fukuda, H.; Demura, T. VASCULAR-RELATED NAC-DOMAIN7 is involved in the differentiation of all types of xylem vessels in Arabidopsis roots and shoots. *Plant J.* **2008**, *55*, 652–664. [CrossRef] [PubMed]

54. Moubayidin, L.; Di Mambro, R.; Sozzani, R.; Pacifici, E.; Salvi, E.; Terpstra, I.; Bao, D.; van Dijken, A.; Dello Ioio, R.; Perilli, S.; et al. Spatial coordination between stem cell activity and cell differentiation in the root meristem. *Dev. Cell* **2013**, *26*, 405–415. [CrossRef] [PubMed]

55. Carlsbecker, A.; Lee, J.Y.; Roberts, C.J.; Dettmer, J.; Lehesranta, S.; Zhou, J.; Lindgren, O.; Moreno-Risueno, M.A.; Vatén, A.; Thitamadee, S.; et al. Cell signalling by microRNA165/6 directs gene dose-dependent root cell fate. *Nature* **2010**, *465*, 316–321. [CrossRef] [PubMed]

56. Etchells, J.P.; Smit, M.E.; Gaudinier, A.; Williams, C.J.; Brady, S.M. A brief history of the TDIF-PXY signalling module: Balancing meristem identity and differentiation during vascular development. *New Phytol.* **2016**, *209*, 474–484. [CrossRef] [PubMed]

57. Murashige, T.; Skoog, F. A revised medium for rapid growth and bio assays with tobacco tissue cultures. *Physiol. Plant.* **1962**, *15*, 473–497. [CrossRef]

58. Hirayama, T.; Fujishige, N.; Kunii, T.; Nishimura, N.; Iuchi, S.; Shinozaki, K. A novel ethanol-hypersensitive mutant of Arabidopsis. *Plant Cell Physiol.* **2004**, *45*, 703–711. [CrossRef] [PubMed]

59. Chen, W.W.; Yang, J.L.; Qin, C.; Jin, C.W.; Mo, J.H.; Ye, T.; Zheng, S.J. Nitric oxide acts downstream of auxin to trigger root ferric-chelate reductase activity in response to iron deficiency in Arabidopsis. *Plant Physiol.* **2010**, *154*, 810–819. [CrossRef] [PubMed]

© 2019 by the authors. Licensee MDPI, Basel, Switzerland. This article is an open access article distributed under the terms and conditions of the Creative Commons Attribution (CC BY) license (http://creativecommons.org/licenses/by/4.0/).

International Journal of
Molecular Sciences

MDPI

Article

Evolutionary Analysis of JAZ Proteins in Plants: An Approach in Search of the Ancestral Sequence

Adrián Garrido-Bigotes [1], Felipe Valenzuela-Riffo [2] and Carlos R. Figueroa [2,*]

[1] Laboratorio de Epigenética Vegetal, Facultad de Ciencias Forestales, Universidad de Concepción; Concepción 4070386, Chile; adrigarrido@udec.cl
[2] Institute of Biological Sciences, Campus Talca, Universidad de Talca, Talca 34655488, Chile; felvalenzuela@utalca.cl
* Correspondence: cfigueroa@utalca.cl; Tel.: +56-71-2200277

Received: 13 September 2019; Accepted: 10 October 2019; Published: 12 October 2019

Abstract: Jasmonates are phytohormones that regulate development, metabolism and immunity. Signal transduction is critical to activate jasmonate responses, but the evolution of some key regulators such as jasmonate-ZIM domain (JAZ) repressors is not clear. Here, we identified 1065 JAZ sequence proteins in 66 lower and higher plants and analyzed their evolution by bioinformatics methods. We found that the TIFY and Jas domains are highly conserved along the evolutionary scale. Furthermore, the canonical degron sequence LPIAR(R/K) of the Jas domain is conserved in lower and higher plants. It is noteworthy that degron sequences showed a large number of alternatives from gymnosperms to dicots. In addition, ethylene-responsive element binding factor-associated amphiphilic repression (EAR) motifs are displayed in all plant lineages from liverworts to angiosperms. However, the cryptic MYC2-interacting domain (CMID) domain appeared in angiosperms for the first time. The phylogenetic analysis performed using the Maximum Likelihood method indicated that JAZ ortholog proteins are grouped according to their similarity and plant lineage. Moreover, ancestral JAZ sequences were constructed by PhyloBot software and showed specific changes in the TIFY and Jas domains during evolution from liverworts to dicots. Finally, we propose a model for the evolution of the ancestral sequences of the main eight JAZ protein subgroups. These findings contribute to the understanding of the JAZ family origin and expansion in land plants.

Keywords: Jasmonate-ZIM domain; JAZ repressors; Jas domain; TIFY; degron; phylogenetic analysis; ancestral sequences

---

1. Introduction

Jasmonates (JAs) are phytohormones that regulate the defense responses, growth and development, fertility and reproduction, as well as the biosynthesis of secondary metabolites in terrestrial plants [1]. Their biosynthesis and canonical signaling pathways have been well characterized in vascular plants [2]. Specifically, the JA-signaling pathway is critical for suitable responses to development or environmental stress. The activation of the JA-signaling pathway starts with the perception of the bioactive JA, jasmonoyl-isoleucine (JA-Ile), which is a necessary step for the activation of JA responses. The perception mechanism is mediated by the protein co-receptor complex CORONATINE INSENSITIVE1 (COI1)–jasmonate ZIM-domain (JAZ) in model plants of vascular plants groups such as *Arabidopsis thaliana* and *Fragaria vesca* [3,4], among others. The presence of JA-Ile leads to JAZ protein degradation by the proteasome, and the release of MYC transcription factors (TFs), which are master regulators of JA responses [5–8]. Besides, additional proteins of the JA-signaling pathway such as the adaptor protein Novel Interactor of JAZ (NINJA), co-repressor proteins, e.g., like TOPLESS (TPLs), histone deacetylases (HDAs), antagonistic TFs, e.g., JASMONATE-ASSOCIATED MYC2-like (JAMs), and MYC2-TARGETED BHLH (MTB) proteins establish a fine-tuning repressor mechanism on MYC

TFs [9–13]. In the case of lower plants such as the liverwort *Marchanthia polymorpha*, COI1, JAZ co-repressor and MYC TFs are conserved, although the ligand molecule is dinor-oxophytodienoic acid (dnOPDA) [14–16]. Thus, the machinery of perception of the JA-signaling pathway and JAZ repressors seems to be conserved from liverworts to angiosperms, although some structural and functional differences could exist because of the great phylogenetic distance of these lineages.

JAZ repressors belonging to the TIFY family have been extensively studied in higher plants [17–21]. In Arabidopsis, this subfamily comprises 13 JAZ proteins (Figure 1) [17,22], while in *M. polymorpha* it only contains a single JAZ protein [14,15]. Normally, these proteins contain the TIFY (ZIM) conserved domain (Figure 1) consisting of 28 amino acid (aa) residues with the highly conserved motif TIF(F/Y)XG in *A. thaliana* [23], which is involved in the JAZ–NINJA and JAZ–JAZ protein interactions (Figure 1) [9,24]. However, some JAZ proteins such as AtJAZ13 lack the TIFY domain (Figure 1) [22]. In addition, all JAZ proteins contain the Jas domain, which is constituted by 26 aa that mediate the COI1–JAZ and MYC2–JAZ interactions (Figure 1) [25]. Generally, the Jas domain contains the degron sequence LPIAR(R/K), which is crucial for the COI1–JA-Ile–JAZ complex formation [3], the Jas motif for interaction with TFs and the nuclear localization signal (NLS) (Figure 1) [25]. Nevertheless, the functionality of JAZ proteins can be different depending on specific protein domains or motifs. For instance, some JAZ proteins own an alternative degron sequence, e.g., JAZ7 and JAZ8 in Arabidopsis, which is not functional, and therefore cannot be degraded by the proteasome [26]. Besides, the additional cryptic MYC2-interacting domain (CMID) has been identified in some JAZ proteins such as *A. thaliana* JAZ1 and JAZ10 proteins [27] (Figure 1), although it is weakly conserved [27]. Finally, ethylene-responsive element binding factor-associated amphiphilic repression (EAR) motifs (LxLxL) are present in some JAZ proteins such as *A. thaliana* JAZ5, JAZ6, JAZ7 or JAZ8 (Figure 1), which are related with the interaction with TPL proteins and the resultant repression [26,28]. Therefore, the diversity of JAZ proteins generated by the absence/presence of these domains or by particular changes in structural and functional domains leads to the regulation of a wide set of responses in higher plants, although a functional redundancy of JAZ repressors has been suggested [29]. In turn, the degron is a key sequence for JA responses, and the canonical degron defined as LPIAR(R/K) has been well studied in *A. thaliana* [3]. In this sequence, the fifth and sixth amino acid residues (from the N- to the C-terminus) are the highest conserved [3,4] and are involved in the interaction with COI1 in the presence of JA-Ile [3]. Moreover, these two residues remain conserved in lower plants such as *M. polymorpha* [14,15], in monocots such as *Oryza sativa* [19], and in angiosperms such as *F. vesca* [18], *Solanum lycopersicum* [17] or *Vitis vinifera* [30]. However, information about degron sequences in other plant lineages is limited.

In general, the evolution of the JA-signaling pathway and JAZ proteins through the plant kingdom has been scarcely explored. However, some evolutionary analysis on the TIFY family and the F-box domain of COI1 receptors' plant hormone signaling has been performed [21,31]. The signal transduction of JA is not present in algae lineage and appeared in land plants for the first time [31], thus increasing the number and diversity of JAZ proteins. Several authors suggest that the JA-signaling pathway arose from a common ancestor with auxins because the receptors exhibit a high similarity, share TPL proteins such as co-repressors, and the repressors are degraded by the proteasome [2]. Bai et al. [21] reported that the number of proteins that are members of the JAZ family increases as we move along the phylogenetic scale, with lower plants such as bryophytes containing 1–9 members and angiosperms comprising more than 10 members in the JAZ subfamily. On one hand, genome-wide characterization of JAZ proteins has mainly been studied in several families of higher plants, both in monocots such as rice [19] and wheat [20], and in dicots such as cotton [32], strawberry [18], tomato [17] and grape [30], among others. On the other hand, in gymnosperms, as well as in mosses and lycophytes, this protein family has been less studied. Nevertheless, data about the origin and evolution of JAZ proteins through the plant kingdom are still unknown. The large amount of genomic data now available in the plant kingdom allows the study of the evolutionary history of JAZ proteins. The aim of this research was to analyze JAZ proteins' evolution through the structural characterization of domains and motifs, phylogenetic analysis and reconstruction of the ancestral sequence for JAZ proteins in different lineages of land plants.

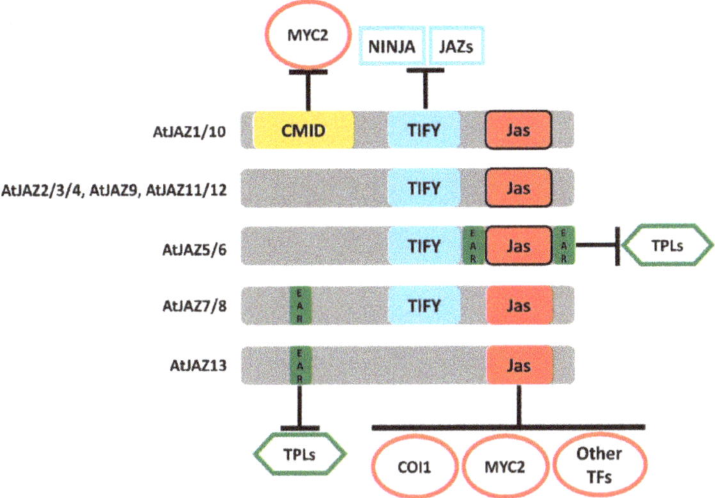

Figure 1. Schematic representation of *Arabidopsis thaliana* JAZ proteins showing the main motifs and domains through which they interact with other proteins (interactions indicated by T-bars and proteins by circles, rectangles and hexagons). Frame boxes of Jas domains indicate the conservation of functional degrons. CMID, Cryptic MYC2-interacting domain; COI1, CORONATINE INSENSITIVE1; EAR, ethylene-responsive element binding factor-associated amphiphilic repression; JAZ, jasmonate ZIM-domain; NINJA, Novel Interactor of JAZ; TFs, transcription factors; TPL, TOPLESS. This figure is based on References [3,9,22,24,25,27,29].

2. Results

2.1. Identification of JAZ Protein Family in the Plant Kingdom

In order to identify JAZ proteins in different lineages of the plant kingdom, *A. thaliana* JAZ proteins were used as queries by using blastp searches against the proteomes included in the PLAZA database. Previously reported JAZ sequences were directly obtained from their respective reports (Table 1). After removing partial sequences, sequences lacking the Jas domain, sequences with additional domains such as PEAPOD (PPD) or GATA zinc-finger (ZML), and redundant sequences, 1065 sequences were identified. In this sense, we identified 183, 6, 196 and 389 new JAZ genes corresponding to gymnosperms, *Amborella trichopoda*, monocots and dicots, respectively. In several organisms such as *Brachypodium dystachon*, *T. aestivum*, *Zea mays*, *Hevea brasiliensis*, *Populus trichocarpa*, *Malus* × *domestica*, *Prunus persica*, *Pyrus* × *bretschneideri* and *Solanum lycopersicum*, additional JAZ genes were identified (Supplementary Table S1).

The number of JAZ genes ranged from 1 to 50 in *M. polymorpha* and *T. aestivum*, respectively. However, 10–20 was the most common number of genes in most species (Table 1, Supplementary Table S1). In algae lineage, no JAZ sequences were detected (Table 1). The length of amino acidic sequences was highly variable through different plant lineages, but the normal range was between 100 and 300 aa (Supplementary Table S1). The longest JAZ protein corresponding to *Selaginella moellendorffii* showed 1346 aa, and the shortest exhibited 50 aa in *Erythranthe guttata*. These results indicated that in the different lineages and species, the number of genes and the protein length were variable, whereas these sequences retained the characteristic motifs and domains of the JAZ family.

Table 1. JAZ family genes in the plant kingdom.

Lineage	Organism	Number of Genes	Number of JAZ Genes	Reference
Algae	*Chlamydomonas reinhardtii*	17,741	0	[21], this research [1]
Moss	*Physcomitrella patens*	32,926	9	[21], this research
Liverwort	*Marchantia polymorpha*	19,278	1	[14]
Lycophyte	*Selaginella moellendorffii*	22,285	8	[21]
Gymnosperms	*Cycas micholitzii*	28,901	2	This research
	Ginkgo biloba	30,404	7	This research
	Gnetum montanum	32,549	2	This research
	Picea abies	66,632	30	This research
	Picea glauca	28,909	14	This research
	Picea sitchensis	20,434	12	This research
	Pinus pinaster	76,426	19	This research
	Pinus sylvestris	36,106	25	This research
	Pinus taeda	84,446	40	This research
	Pseudotsuga menziesii	149,717	30	This research
	Taxus baccata	32,062	5	This research
	Taxus chinensis	Unknown	9	[33]
Amborellales	*Amborella trichopoda*	26,846	6	This research
Monocots	*Spirodela polyrhiza*	19,623	7	This research
	Elaeis guineensis	29,808	14	This research
	Ananas comosus	270,240	8	This research
	Musa acuminata	37,582	34	This research
	Phalaenopsis equestris	29,431	9	This research
	Brachypodium distachyon	34,310	17	[21,34], this research
	Hordeum vulgare	25,780	8	This research
	Oropetium thomaeum	28,446	16	This research
	Oryza brachyantha	34,155	9	This research
	Oryza sativa	42,189	15	[19,21]
	Phyllostachis edulis	31,978	18	[35]
	Setaria italica	34,584	18	This research
	Sorghum bicolor	34,211	18	[21], this research
	Triticum aestivum	114,581	50	[20],this research
	Zoysia japonica	59,271	18	This research
	Zea mays	44,474	38	[36], this research
	Zostera marina	20,450	7	This research
Dicots	*Actinidia chinensis*	39,040	5	This research
	Amaranthus hypochondriacus	23,847	8	This research
	Beta vulgaris	26,920	7	This research
	Chenopodium quinoa	44,776	6	This research
	Daucus carota	32,113	10	This research
	Arabidopsis lyrata	31,073	13	This research
	Arabidopsis thaliana	69,810	13	[17,22]
	Brassica oleraceae	59,225	25	This research
	Brassica rapa	40,492	28	This research
	Capsella rubella	26,521	10	This research
	Schrenkiella parvula	26,313	12	This research
	Carica papaya	27,768	7	This research
	Tarenaya hassleriana	30,556	16	This research
	Citrullus lanatus	597,261	10	This research
	Erythranthe guttata	28,140	10	This research
	Cucumis melo	28,608	10	This research
	Cucumis sativus	21,503	11	This research
	Hevea brasiliensis	42,550	14	[37,38], this research
	Manihot esculenta	33,033	16	This research
	Ricinus communis	31,221	9	This research
	Arachis ipaensis	41,840	10	This research
	Cajanus cajan	48,680	11	[39]
	Cicer arietinum	170,274	9	This research
	Glycine max	56,044	24	This research
	Medicago truncatula	50,894	12	[21], this research
	Trifolium pratense	39,948	10	This research
	Vigna radiata	22,368	10	This research
	Utricularia gibba	25,930	11	This research
	Gossypium raimondi	37,505	15	[40]
	Theobroma cacao	29,232	8	This research
	Corchorus oliotorus	37,281	8	This research
	Eucalyptus grandis	36,349	11	This research
	Nelumbo nucifera	26,685	9	This research
	Ziziphus jujuba	31,701	10	This research

Table 1. *Cont.*

Lineage	Organism	Number of Genes	Number of JAZ Genes	Reference
Dicots	*Fragaria vesca*	32,831	12	[18]
	Malus × domestica	55,620	21	[41], this research
	Prunus persica	26,873	12	[42], this research
	Pyrus × bretschneideri	42,812	12	[43], this research
	Coffea canephora	469,604	7	This research
	Citrus clementina	24,533	7	This research
	Populus trichocarpa	42,950	13	[44], this research
	Capsicum annum	35,884	10	This research
	Solanum lycopersicum	34,725	13	[17], this research
	Solanum tuberosum	39,028	13	This research
	Petunia axillaris	35,812	13	This research
	Vitis vinifera	26,346	11	[21,30]

[1] 'This research' refers to new JAZ sequences found in the PLAZA 4.0 database. For more details, see the Materials and Methods section.

2.2. Conserved Motifs and Domains of JAZ Proteins in Land Plants

To gain insights into the evolution of domains and motifs in lineages of the plant kingdom, consensus sequences of the Jas and CMID domains and the TIFY and EAR motifs were constructed. The TIFY motif and the Jas domain showed stronger conservation in entire plant lineages. However, not all JAZ proteins maintained these sequences (Figure 2, Supplementary Table S1). The majority of JAZ proteins displayed the TIFYXG consensus sequences along the evolutionary history of the plant kingdom (Figure 2, Supplementary Table S1). Motifs of the Jas domain also indicated a high conservation in plant lineages (Figure 2). On one hand, the degron showed a LPQARK conserved sequence in lower plants (liverwort, moss and lycophyte) and LPIAR(R/K) in gymnosperms, *A. trichopoda*, monocots and dicots (Figure 2, Supplementary Table S1). The fifth and sixth amino acidic residues of the degron consensus sequence exhibited the highest conservation (Figure 2). On the other hand, logo sequences of the Jas motif showed higher similarity among lower plants than among higher plants (Figure 2). However, consensus sequences of the Jas motif could be defined as X-SL-X-R-FL-X-KRK-X-R. Finally, Nuclear Localization Signal (NLS) was the most variable motif between different species (Figure 2), defined by the consensus sequence X5-PY [25]. Moreover, the first five amino acidic residues displayed a stronger variability from gymnosperms to dicots (Figure 2, Supplementary Table S1). In the cases of moss and lycophyte, the fifth position was occupied by Pro amino acid (Figure 2, Supplementary Table S1). Therefore, the TIFY motif and the Jas domain were highly conserved in plant kingdom.

Otherwise, EAR motifs corresponding to both N- and C-terminal regions of JAZ proteins shared the basic sequence LxLxL in all plant lineages (Figure 3). However, the EAR motif defined as DLNEPT, which is located between the TIFY and Jas domains, was only conserved in dicot plants (Figure 3). So, both EAR N- and C-terminals were conserved from moss to dicots, while the EAR motif defined as DLNEPT appeared in dicot plants.

For a better understanding of the evolution corresponding to the N-terminal region of AtJAZ1 and AtJAZ10 orthologs involved with MYC2 interaction [27], a search was performed on 1065 JAZ protein sequences reported in this research (Supplementary Table S1). Regarding AtJAZ1 proteins, which contain the CMID domain and the adjacent EAR-like motif, gymnosperms and lower plants did not show this conserved region, but it appeared in *A. trichopoda* for the first time, and it was also conserved in angiosperms (Figure 4a). In all groups, the residues F, C, L and Y were the most conserved in the CMID domain. In the case of EAR-Like motif, it displayed a lower conservation in monocots and dicots (Figure 4a). On the other hand, the CMID domain of AtJAZ10 and its orthologs appeared only in the monocot *Elaeis guinensis* (Supplementary Table S1). However, this domain was extensively conserved in dicots (Figure 4b, Supplementary Table S1). In monocots, the CMID domain of the AtJAZ10 ortholog protein was shorter than in dicots, where the monocot had 17 aa against the 33 aa of the dicots (Figure 4b). Besides, the C-terminal region of the CMID domain was more conserved than the N-terminal region in dicots (Figure 4b). Hence, these results showed that the CMID domain of AtJAZ1 and AtJAZ10 orthologs emerged in basal angiosperms and in angiosperms, respectively.

It is important to highlight that some previously reported JAZ proteins showed additional domains by using the Conserved Domains Database (CDD). For instance, the wheat JAZs TaJAZ4-A, TaJAZ4-B, TaJAZ4-D, TaJAZ5-A, TaJAZ11-A, TaJAZ14-A, TaJAZ14-B and TaJAZ14-D [20] exhibited the ZML domain. In any case, these sequences were considered for subsequent analysis.

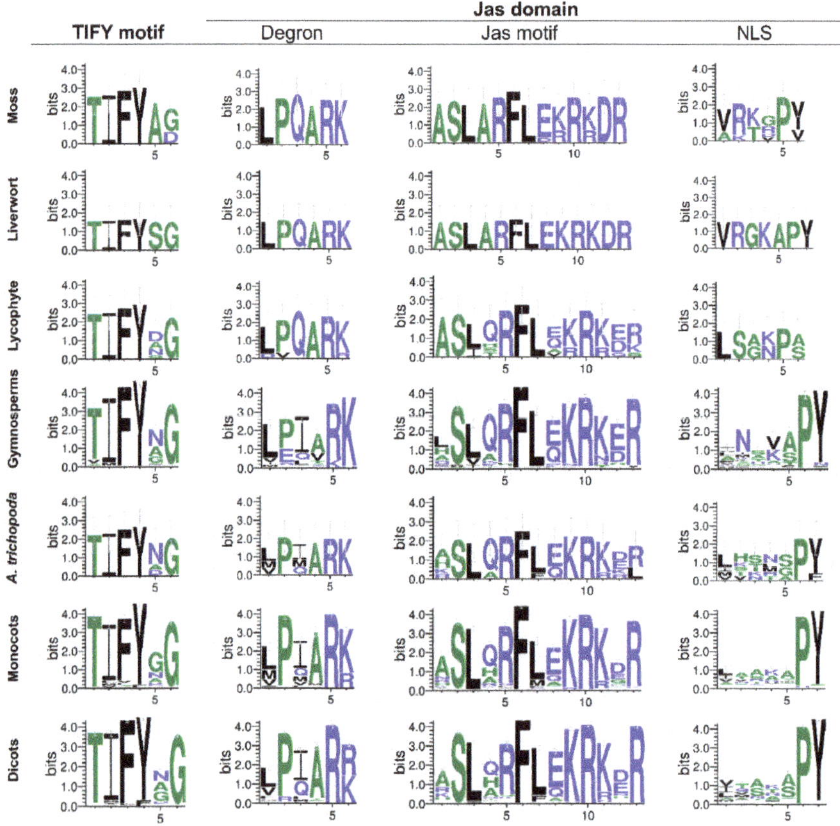

Figure 2. Consensus sequences of the TIFY motif and the Jas domain for different lineages in the plant kingdom. TIFY motif logo sequence and Jas domain showing degron, Jas motif and Nuclear Localization Signal (NLS) logo sequences. Logo sequences of the TIFY motif were obtained from 9, 1, 8, 159, 6, 287 and 515 sequences for moss, liverwort, lycophyte, gymnosperms, *Amborella trichopoda*, monocots and dicots, respectively. Logo sequences of the degron were obtained from 9, 1, 8, 167, 6, 287 and 516 sequences for moss, liverwort, lycophyte, gymnosperms, *A. trichopoda*, monocots and dicots, respectively. Logo sequences of the Jas motif were obtained from 9, 1, 8, 195, 6, 303 and 543 sequences for moss, liverwort, lycophyte, gymnosperms, *A. trichopoda*, monocots and dicots, respectively. Logo sequences of NLS were obtained from 7, 1, 2, 168, 6, 278 and 510 sequences for moss, liverwort, lycophyte, gymnosperms, *A. trichopoda*, monocots and dicots, respectively. JAZ proteins lacking the TIFY motif were not included in the analysis. For more details, see the Materials and Methods section.

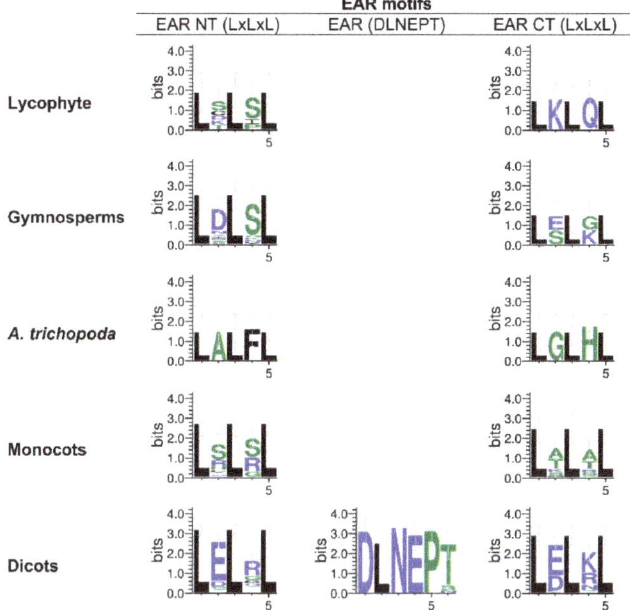

Figure 3. Consensus sequences of EAR motifs for different lineages in the plant kingdom. Logo sequences of the EAR NT motif were obtained from 5, 17, 1, 28 and 118 sequences for lycophyte, gymnosperms, *Amborella trichopoda*, monocots and dicots, respectively. Logo sequences of the EAR (DLNEPT) motif were obtained from 19 sequences for dicots. Logo sequences of the EAR CT motif were obtained from 1, 2, 2, 27 and 45 sequences for lycophyte, gymnosperms, *A. trichopoda*, monocots and dicots, respectively. CT, C-terminal region; EAR, ethylene-responsive element binding factor-associated amphiphilic repression; NT, N-terminal region. For more details, see the Materials and Methods section.

2.3. Degron Analysis in Land Plants

To explore different degron sequences in land plants, the canonical degron LPIAR(R/K) of *A. thaliana* was used as a reference because it is a critical motif for the co-receptor COI1–JAZ complex formation. We took into account that the amino acidic residues at the fifth and sixth position are the most important for the degron functionality [3,45], through all JAZ proteins in the plant kingdom. Next, we searched degron variants in entire species according to these criteria. As we moved through the evolutionary scale, from moss to dicots, we observed that the number of degron variants increased (Figure 5a, Supplementary Table S2). Moss and liverwort shared 100% frequency of the degron sequence LPQARK (Figure 5a, Supplementary Table S2). In lycophyte, LPQARK was the major degron and for first time, the new degron sequence DVQARK appeared (Figure 5a, Supplementary Table S2). In the case of gymnosperms, the majority of sequences displayed alternative degrons (Supplementary Table S2), and the canonical degron LPIAR(R/K) represented 18.46% (Figure 5a). Also, LEIV(R/K)K, VPQARK and LEIA(R/K)K degrons surged in this group (Figure 5a). Otherwise, the basal angiosperm *A. trichopoda* exhibited LPIARK with higher frequency, as well as VPQARK for gymnosperms (Figure 5a). We considered that monocots and dicots contained a similar percentage of the canonical degron LPIAR(R/K), and in these groups, the diversity of alternative degrons increased (Figure 5a). Finally, from lycophyte to dicots, JAZ sequences without degrons emerged (Figure 5a). Thus, the canonical sequence LPIAR(R/K) was the most represented degron from gymnosperms to dicots. However, a wide diversity of degron sequences was observed for gymnosperm, monocot and dicot lineages (Figure 5a). Besides this, JAZ proteins without degron sequences emerged in lycophytes and were maintained to dicots (Figure 5a).

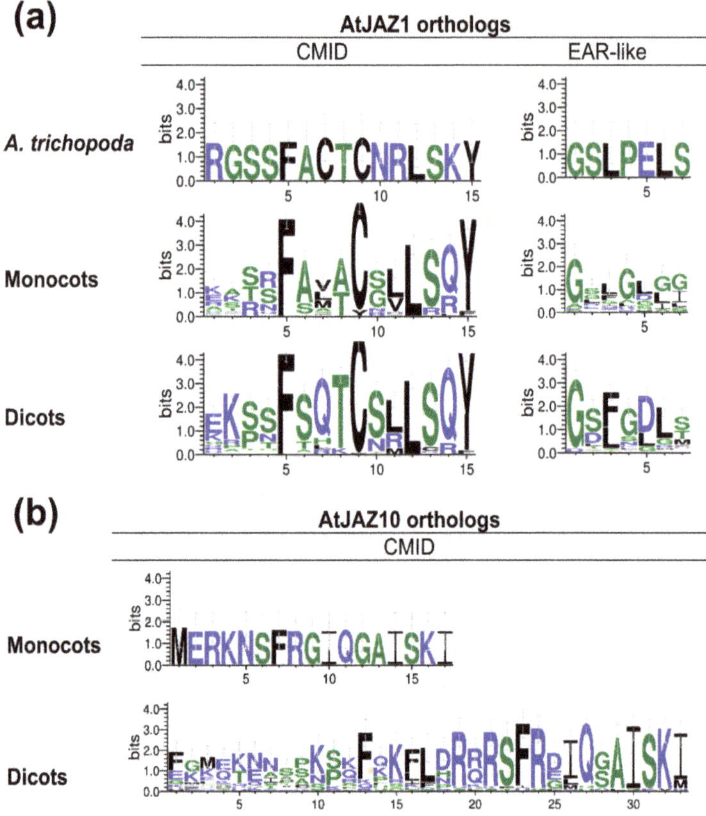

Figure 4. Consensus sequences of the N-terminal region for AtJAZ1 and AtJAZ10 orthologs in different plant kingdom lineages. (**a**) CMID domain and EAR-like logo sequences for AtJAZ1 orthologs in *Amborella trichopoda*, monocots and dicots. (**b**) CMID domain for AtJAZ10 orthologs in monocots and dicots. Logo sequences of the CMID domain were obtained from 1, 55 and 78 sequences for AtJAZ1 orthologs in *A. trichopoda*, monocots and dicots, respectively. Logo sequences of the EAR-like domain for AtJAZ1 orthologs were obtained from 1, 26 and 74 sequences for AtJAZ1 orthologs in *A. trichopoda*, monocots and dicots, respectively. Logo sequences of the CMID domain were obtained from 1 and 28 sequences for AtJAZ10 orthologs in monocots and dicots, respectively. CMID, cryptic MYC2-interacting domain; EAR, ethylene-responsive element binding factor-associated amphiphilic repression; JAZ, jasmonate-ZIM domain. For more details, see the Materials and Methods section.

To study the conservation of the amino acid biochemical features in degrons, we analyzed the percentage of each biochemical group at the different positions in degron sequences containing Arg/Lys (R/K) at the fifth and sixth positions, respectively. We observed differences depending on plant lineage. For instance, degrons in moss and liverwort maintained the same pattern of biochemical groups at the first four amino acid positions: the first, second and fourth positions corresponded to non-polar amino acids, and the third corresponded to amino acids with polar groups in their lateral chain (Figure 5b). In lycophyte, the first amino acid can also be a negatively charged amino acid (Figure 5b). Gymnosperms' degron was characterized by polar and non-polar amino acids at the first and third positions, while any biochemical group was present at the second and fourth positions (Figure 5b). In *A. trichopoda*, non-polar amino acids were established at the first four positions and the third position can be occupied by polar residues (Figure 5b). In monocots and dicots, non-polar residues appeared mostly at the first four positions. However, depending on the lineage, the resting positions showed higher variability (Figure 5b). Therefore, as we moved along the evolutionary scale of the plant kingdom, a diversity of amino acidic residues with different physicochemical characteristics at the first four positions of the degron sequence was observed.

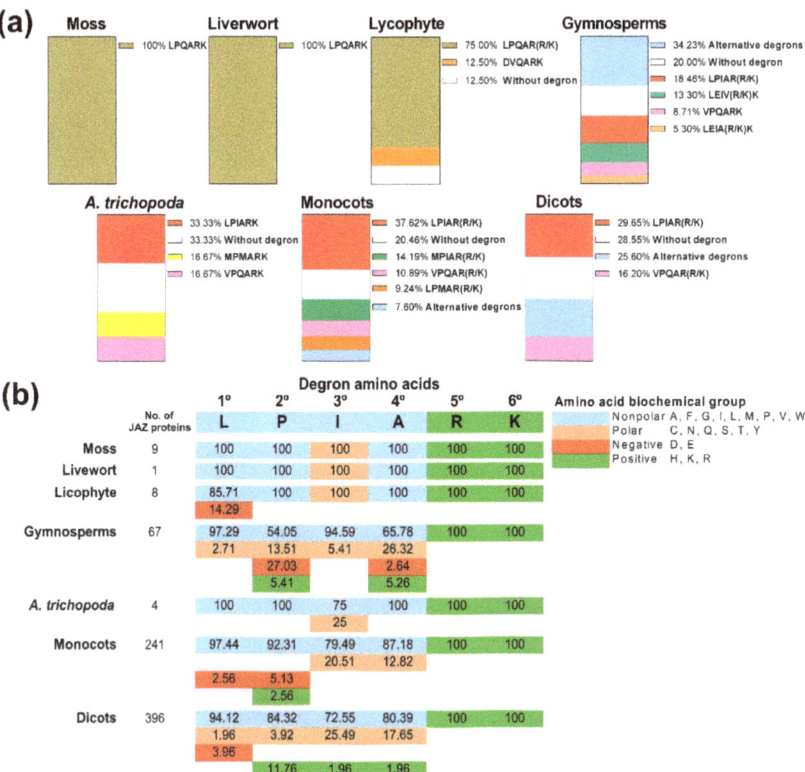

Figure 5. Analysis of degron sequences in different lineages of the plant kingdom. (**a**) Percentage (%) of specific degron sequences in each plant lineage. (**b**) Percentage (%) of different amino acid biochemical groups in each position of the degron sequence compared with the canonical degron LPIARK. The percentage (%) of specific degrons was obtained from 9, 1, 8, 195, 6, 303 and 543 sequences for moss, livewort, lycophyte, gymnosperms, *Amborella trichopoda*, monocots and dicots, respectively. The percentage (%) of different amino acid biochemical groups in each position of the degron sequence was obtained from 1, 1, 3, 37, 3, 39 and 51 different degron sequences for moss, livewort, lycophyte, gymnosperms, *A. trichopoda*, monocots and dicots, respectively.

2.4. Phylogenetic Analysis of JAZ Proteins in Land Plants

In order to study the phylogenetic relationships between different JAZ protein orthologs in different plant lineages, an unrooted phylogenetic tree was constructed by the Maximum Likelihood method. Globally, JAZ ortholog proteins were grouped in nine (I–IX) clades (Figure 6). *A. thaliana* JAZ1, JAZ2, JAZ5 and JAZ6 were grouped together in clade I, which contained ortholog sequences from monocots and dicots (Figure 6). Clade II was only constituted by JAZ orthologs of monocots and a JAZ ortholog of *M. × domestica* (Figure 6). Otherwise, clade III clustered AtJAZ7, AtJAZ8 and AtJAZ13 ortholog proteins in dicots, but also showed grouping along with JAZ ortholog proteins of some monocot species (Figure 6). In clade IV, JAZ proteins corresponding to liverwort, moss and lycophyte were grouped together (Figure 6). AtJAZ3, AtJAZ4 and AtJAZ9 were clustered near their ortholog proteins with some JAZ proteins of lycophyte and gymnosperms into clade V (Figure 6). On the other hand, clade VI included some JAZ proteins of dicots and monocots along with a lycophyte JAZ protein (Figure 6). Clade VII showed the clustering of AtJAZ11, AtJAZ12 and their respective orthologs in gymnosperms, monocots and dicots. Moreover, clade VIII only clustered gymnosperm and monocot JAZ sequences. Finally, AtJAZ10 orthologs were grouped in clade IX (Figure 6). These results indicated that different JAZ ortholog proteins are phylogenetically related based on their higher similarity and land plant lineage.

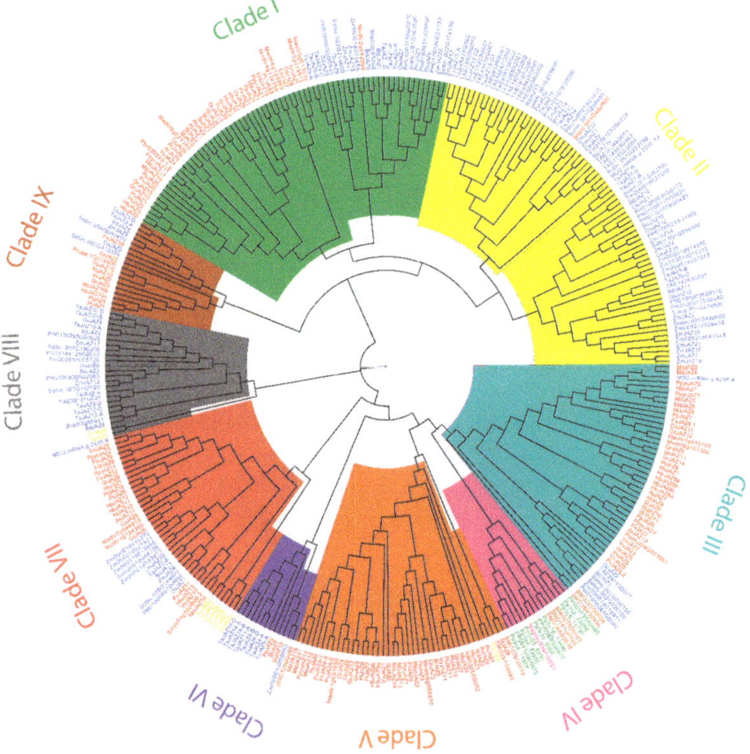

Figure 6. Phylogenetic analysis of JAZ proteins in land plants. The phylogenetic tree was constructed by using the full-length amino acid sequences of liverwort, moss, lycophyte, gymnosperms, monocots and dicots. Purple, green, brown, yellow, blue and red used in sequence names indicate liverwort, moss, lycophyte, gymnosperm, monocot and dicot JAZ proteins, respectively. Different colors indicate clades for different JAZ ortholog proteins. JAZ, jasmonate-ZIM domain.

2.5. Ancestral Sequences of JAZ Proteins in Land Plants

To understand the evolution of the JAZ protein family, a reconstruction of the JAZ ancestral sequences from liverwort to dicots was performed by the Maximum Likelihood method. For this analysis, only 308 JAZ sequences were considered, according to previously characterized JAZ sequences, including additional putative JAZs reported in this research (Table 1). As a result of this analysis, an ancestral tree grouping all JAZ proteins was obtained. The PhyloBot software showed a set of ancestors represented by numbers (Figure 7, Supplementary Table S3, http://www.phylobot.com/464456268, last accessed date: 19 August 2019). The JAZ protein of *M. polymorpha* was individually grouped in the first tree branch. Next, the ancestor 518 was common from moss to dicots (Figure 7).

Then, a series of ancestors originated different tree branches, which clustered different species of lycophyte, gymnosperms, monocots and dicots. First, AtJAZ3, AtJAZ4 and AtJAZ9 orthologs diverged in an independent manner from the common ancestor 450 (Figure 7). AtJAZ9 appeared first, and then AtJAZ3 and AtJAZ4 ortholog proteins diverged. The *Fragaria vesca* JAZs FvJAZ4-1, FvJAZ4-2, FvJAZ4-3 and FvJAZ9 were originated from the ancestor 471 (Figure 7). Second, the ancestor 329 and subsequent ancestors originated the groups of JAZ1, JAZ2, JAZ5 and JAZ6 orthologs. AtJAZ1, AtJAZ2 and FvJAZ1 arose from the ancestor 406, which was also the ancestor sequence of AtJAZ5, AtJAZ6 and FvJAZ5. An independent branch of the ancestor 330 gave rise to an exclusive group of monocot JAZ sequences (Figure 7). Third, the ancestors 327 and 326 originated some *Taxus chinensis* JAZ proteins. Fourth, the ancestors 325 and 541 were the common ancestors for AtJAZ10, AtJAZ7, AtJAZ8 and AtJAZ13 orthologs, respectively. These orthologs shared the ancestor sequence 541 and were separated in two subgroups. Fourth, the ancestor sequence 321 initiated the cluster of AtJAZ11 and AtJAZ12 orthologs (Figure 7). Finally, some ancestors, such as 346 and 406, were exclusively ancestors of JAZs for monocot and dicot plants, respectively. These results displayed an approach to the evolutionary history of different JAZ subgroups.

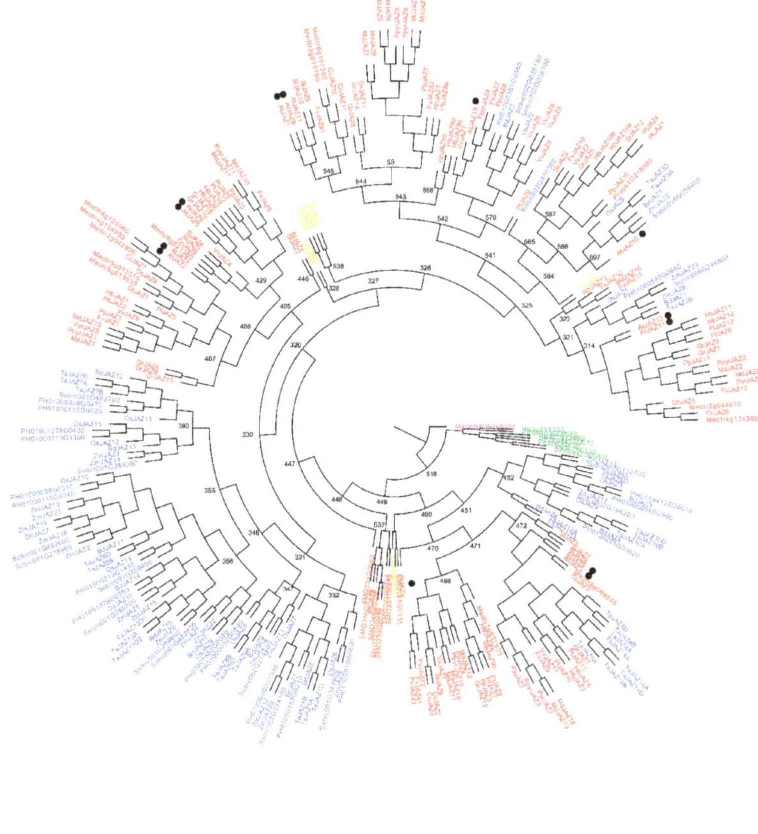

Figure 7. Reconstruction of JAZ ancestral sequences in land plants. The ancestral tree was performed using 308 full-length sequences corresponding to liverwort, moss, lycophyte, gymnosperms, monocots and dicots. Node numbers indicate some output ancestors obtained by using the PhyloBot software. The analysis of JAZ ancestral sequences is available to download at the following URL: http://www.phylobot.com/464456268 (last accessed date: 19 August 2019).

3. Discussion

JAZ proteins are key positive and negative regulators of the JA-signaling pathway [46], regulating different JA-responses such as tolerance against biotic and abiotic stresses, developmental processes, reproduction and secondary metabolism [29,47]. However, the evolutionary history of JAZ proteins and how new functionalities emerged along with the plant kingdom evolution are still unclear. In this research, the changes in JAZ protein sequences along the evolutionary scale of land plants are reported. Moreover, the ancestral sequences that gave rise to different JAZ groups are proposed by the reconstruction of JAZ ancestors.

The mechanism of JA-signal transduction is similar to that of auxins, which contain a receptor, the ubiquitin ligase complex for repressors' degradation, co-repressors, specific transcription factors and homologous repressors, leading to the hypothesis that JAs and auxins share a common ancestor [2]. The JA-signaling pathway, including JAZ proteins, is conserved in liverwort, moss, lycophyte, monocots and dicots [31]. However, JAZ protein members are more numerous in dicots versus liverwort [14,15,31], according to what was observed in the 82 organisms considered in this study (Table 1). The number of JAZ proteins increased during the evolution of vascular plants (Table 1), and this possibly boosted the colonization of new environments, enabling a higher tolerance against stress [1]. However, this expansion of the JAZ gene family was not related to an increase in gene number (Table 1), similar to that observed for the JAZ-homologous repressors, AUX/IAA proteins [48]. A high number of JAZ members involves functional redundancy, such as that reported in Arabidopsis [29] against what was observed in *M. polymorpha*, which only contains a single JAZ protein [14,15]. The expansion of JAZ proteins may be the consequence of gene duplication, alternative splicing and the emergence of new functional domains in proteins [27,49,50].

JAZ proteins in Arabidopsis and in other organisms such as *O. sativa, T. aestivum, S. lycopersicum* or *F. vesca* shape different groups regarding the conserved domains [17–20]. JAZ proteins, a subfamily within the TIFY family, are characterized by the TIFY domain [21,23] and the Jas domain [21,25]. In Arabidopsis, AtJAZ1–12 contain both conserved domains, while AtJAZ13 is non-TIFY domain [17,22]. JAZ proteins in non-vascular plants exhibited TIFY and Jas domains, but the gymnosperms showed some sequences without the TIFY domain (Supplementary Table S1), suggesting that this domain arose in the ancestor of this lineage. It is noteworthy that the TIFY and Jas domains were highly conserved from liverwort to dicots (Figure 2), according to what was previously reported in different species [17–20,41]. However, some repressors such as AtJAZ7, AtJAZ8 and AtJAZ13 contain an alternative degenerated Jas domain, which was observed from gymnosperms to dicots, ruling out the perception of JA-Ile and JAZ degradation [22,26]. Additional protein regions, such as EAR motifs, are conserved in AtJAZ5, AtJAZ6, AtJAZ7 and AtJAZ8 [26]. EAR motifs are involved in the recruiting of TPL proteins, which are present in land plants and in algae lineage [2,31], and proteins containing these motifs were detected in all plant lineages from moss to dicots (Figure 3). This suggests that JAZ proteins evolved from an ancestor sequence with EAR motifs. Otherwise, the CMID domain, involved in the interaction between JAZ repressors and MYC2 transcription factors, is weakly conserved through evolution and is different between AtJAZ1 and AtJAZ10 orthologs [27]. Although the logo sequences indicated a high conservation of this domain in monocots and dicots (Figure 4), the evolution for each one seems to be different because the amino acid composition is slightly variable and CMID is only present in AtJAZ10 orthologs of monocots and dicots, while for AtJAZ1 orthologs, it is represented from basal angiosperms (Figure 4).

The degron sequence of JAZ repressors is related with the JAZ degradation by the 26S proteasome after the formation of the complex JAZ–JA-Ile–COI1 [3], promoting the release of MYC2 TFs and the activation of JA responses [5,6]. The canonical degron LPIAR(R/K) was described in *A. thaliana* JAZ proteins for the first time, and then other species showed this conserved sequence (Supplementary Table S1) [17–19,30]. We also observed that the LPIAR(R/K) sequence is the most common degron, at least for gymnosperms, *A. trichopoda*, monocots and dicots (Figure 5a). In contrast, JAZ proteins of lower plants such as moss and lycophyte exhibited LPQARK (Figure 5a), the same sequence reported for the liverwort *M. polymorpha* [14,15]. The last amino acidic residues R and R/K are directly involved in the interaction with COI1 and JA-Ile, unlike the other residues that could participate in the three-dimensional structure and the binding affinity to the ligand, therefore regulating this interaction [3]. During the evolution of land plants, we observed that the number of alternative degrons rises, changing in the first four residues (Figure 5a). It is important to note that some JAZ proteins lacking the conserved degron emerged in the lycophyte lineage, according to that reported for the AtJAZ7, AtJAZ8 and AtJAZ13 orthologs [5,22]. On the other hand, the physicochemical characteristics of the different amino acidic residues are critical for the structure and function of the proteins [51,52]. We observed that all of the JAZ proteins containing the conserved degron maintained R and R/K at the fifth and sixth positions respectively (Figure 5b), as reported in Sheard et al. [3], so the function is determined by their physicochemical properties. This could be applied to the first four residues, detecting that in different lineages these positions can be occupied by different amino acids with similar physicochemical groups. Otherwise, the change of the fifth positively charged residues prevents the interaction with COI1 [26]. In summary, although the sequence LPIAR(R/K) is the canonical degron, some JAZ proteins changed to amino acid residues with similar or different biochemical properties, extending the diversity of degrons in the plant kingdom.

Therefore, all ortholog proteins belonging to different JAZ families in land plants showed similar conserved domains, and this is evidenced in the phylogenetic relationships. The phylogenetic tree and JAZ ancestral sequence reconstruction included previously reported and new putative JAZ sequences identified in the present research (Table 1). The more similar JAZ sequences are clustered closely (Figure 6), according to what was previously observed in different plant species such as *S. lycopersicum* [17], *V. vinifera* [30], *F. vesca* [18], *O. sativa* [19], and *T. aestivum* [20]. Specifically, AtJAZ1 and AtJAZ2 ortholog proteins clustered in the same subgroup (Figure 6), similar to that observed in *O. sativa* [19], *F. vesca* [18], and *M. × domestica* [41]. Otherwise, AtJAZ5 and AtJAZ6 orthologs, which contain a conserved EAR motif in the C-terminal region (Supplementary Table S1) [28], are grouped in the same clade (Figure 6), according to what was observed in *F. vesca* [18]. AtJAZ3, AtJAZ4 and AtJAZ9 proteins contain highly similar sequences and shape the same clade [17,18,41], in a comparable manner to what was observed in this research (Figure 6). Moreover, AtJAZ10 orthologs showing the CMID domain in a single monocot and in several dicots (Figure 4b, Supplementary Table S1), according to previous reports for Arabidopsis [27] and FvJAZ10 [18], formed an independent clade, as well as AtJAZ11 and AtJA12 ortholog proteins (Figure 6). Finally, a well differentiated clade is formed by JAZ lacking the conserved Jas (AtJAZ7 and AtJAZ8) and TIFY (AtJAZ13) domains. So, phylogenetic trees allow us to relate ortholog JAZ proteins regarding their sequence similarities and domain conservation. However, how the different JAZ protein subgroups evolved from an ancestral JAZ sequence in land plants cannot be elucidated.

The reconstruction of the ancestral sequence for a specific protein family is a powerful tool to understand the origin and the evolutionary process followed by protein sequences [53], but the current software can only provide the most conserved domains with higher fidelity [54]. JAZ proteins are represented in lower land plants by 1–9 members (Table 1, Supplementary Table S1) [14,15,21], compared to 40, 50 and 13 members in *Pinus taeda*, *Zea mays* and *A. thaliana*, respectively (Table 1) [22], showing a big expansion of the JAZ repertoire in vascular plants. Until now, *M. polymorpha* is the least evolved land plant with a sequenced genome [55], and its single JAZ protein has the TIFY and Jas domains highly conserved (Supplementary Table S1, Figure 2) [14,15], so it is the best model to study the evolution and functions of JAZ proteins [46]. The liverwort lineage is located in the first branch of the ancestral tree (Figure 7), suggesting that an ancestral sequence would be the origin of JAZ proteins. Being that the algae lineage contains TPL proteins [31], which are recruited by EAR motifs such as LxLxL [9], we suppose that the JAZ ancestor had this motif. However, because of proteins in lower plants, the conservation of these motifs is not represented in the ancestral sequences. In the reconstruction of the evolutionary history of JAZ ancestors, the other JAZ proteins in the different lineages were originated from the ancestor 518 (Figure 7). The group of the AtJAZ3, AtJAZ4 and

AtJAZ9 ortholog proteins experienced a divergence from the ancestor 449, constituting a well-defined group (Figure 7). Then, a set of ancestral JAZ proteins gave rise to the ancestor 330, which was separated into two branches: AtJAZ1, AtJAZ2, AtJAZ5 and AtJAZ6 orthologs grouping dicots in the same clade, while monocots were grouped in the other JAZ ortholog group (Figure 7). AtJAZ5 and AtJAZ6 contain the conserved motif LxLxL [28], while AtJAZ1 and AtJAZ2 lack it, suggesting that the ancestor 406 originator owned this conserved motif, but this one was lost during the evolution of AtJAZ1 and AtJAZ2 orthologs and maintained to the AtJAZ5 and AtJAZ6 origin (Figure 7). The last big group of JAZ orthologs showed the sequence 327 as the common ancestor for AtJAZ7, AtJAZ8, AtJAZ13, AtJAZ10, AtJAZ11 and AtJAZ12 (Figure 7), which diverged into three differentiated groups: the first containing AtJAZ7, AtJAZ8, and AtJAZ13, the second only with AtJAZ10, and the third with AtJAZ11 and AtJAZ12. These groups are characterized by the lack of the degron sequence for COI1 interaction in AtJAZ7, AtJAZ8 and AtJAZ13 [22,26], and the presence of the CMID conserved domain in the case of the AtJAZ10 orthologs [27]. An important point is that AtJAZ7 and AtJAZ8 ortholog proteins contain the EAR motif in the N-terminal region [18,26], suggesting that the previous ancestors should have contained this conserved sequence. In summary, the oldest ancestor would show the TIFY and Jas domains according to what was observed in the JAZ sequence of liverwort (Figure 7) [14,15].

Subsequently, we propose a general model of JAZ evolution in land plants, showing changes in ancestral sequences in eight different JAZ groups (Figure 8). During the evolutionary process, the putative ancestors (Supplementary Table S3) showed small changes in amino acidic residues of the TIFY and Jas domains, allowing the evolution of structures and functions. The TIFY motif is the most conserved sequence through evolution. In the case of the degron and Jas motif, some synonym substitutions are observed between the different ancestors (Figure 8), so this may possibly involve an adaptation to the environment [46]. It is noteworthy that the AtJAZ7, AtJAZ8 and AtJAZ13 orthologs' clade lost the Jas conserved motif during evolution, originating a new functional group of JAZ proteins. Besides that, the non-TIFY JAZ13 repressors lost the TIFY domain (Figures 7 and 8). Otherwise, the NLS region is the most variable between different JAZ groups, similar to what was observed for AUX/IAA proteins [48]. However, the oldest ancestor of the JAZ proteins of land plants is still unknown.

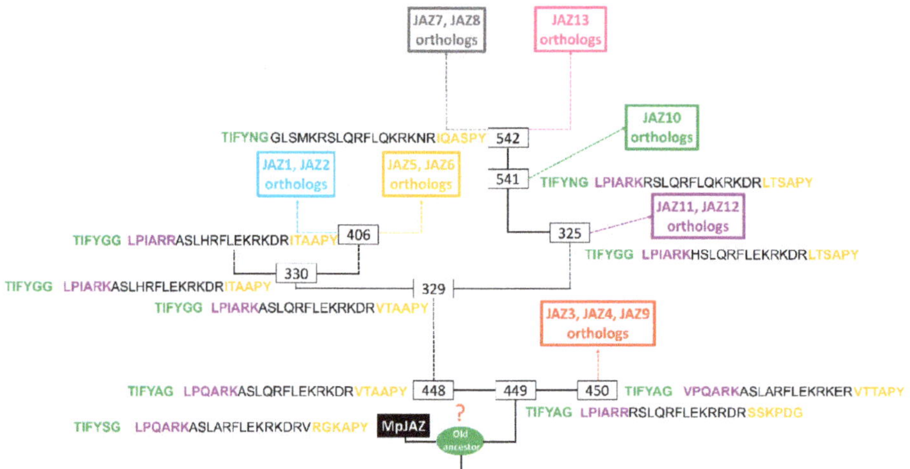

Figure 8. Proposed model of JAZ protein evolution. Numbers framed in boxes indicate the output name of the ancestor assigned by the PhyloBot software. The analysis of JAZ ancestral sequences is available to download at the following URL: http://www.phylobot.com/464456268, last accessed date: 19 August 2019. The TIFY motif and the Jas domain are displayed near to each ancestor. Different colors in boxes show different subgroups of eight JAZ ortholog proteins. Dashed lines indicate intermediate ancestors. The length of branches does not correspond to evolution time. JAZ, jasmonate-ZIM domain.

Regarding physiological implications, JAs are essential signaling phytohormones for stress tolerance, metabolism and developmental responses [29]. Thus, the evolution of key components of the signaling pathway, such as JAZ repressors, is necessary for overcoming environmental constraints [46]. In this sense, JAZ proteins are critical for the activation of JA responses, through the perception mechanism of JA-Ile and dinor-OPDA in *A. thaliana* and *M. polymorpha*, respectively [3,14]. The degron sequence is critical for interaction with COI1 and JA-Ile in *A. thaliana* [3], and a high diversity of degron sequences was detected in land plants (Supplementary Table S2, Figure 5). This could be related to different interaction affinities for the perception complex, because JAZs containing different degron sequences show diverse binding affinities in the interaction with COI1 mediated by JA-Ile [26], leading to fine-tuning of JA-response activation depending on the type of JAZ protein involved. In the case of JAZ7, JAZ8 and JAZ13 orthologs, they lack the conserved degron sequence (Supplementary Table S1) [22,26] and this prevents their degradation and maintains the repression over the transcription factors, which is related to pulsed responses and desensitization to JAs [5]. Besides this, JAZ proteins are involved in the formation of a co-repressor complex by the interaction with NINJA adaptor protein [27], and they recruit TPL proteins mediated by TIFY and EAR-like motifs, respectively [25]. Thus, JAZ5–8 and JAZ13 orthologs that contain EAR motifs in land plants (Supplementary Table S1, Figure 3) allow the formation of a repressor complex and the inhibition of JAZ responses mediated by TFs [22,26]. Some JAZ orthologs such as JAZ1 and JAZ10 show the N-terminal region containing the CMID domain and the EAR-like motif, involved in the attenuation of JA responses by the repression of MYC2 TF [27]. Moreover, JAZs are repressors of different TF families involved in defense, secondary metabolite biosynthesis, and tolerance to abiotic stress, among others [29]. In this case, JAZ proteins show a certain degree of functional redundancy, but it seems to be restrained to specific TF families [29]. The JAZ–TFs interaction is mediated by the Jas motif [25], which is highly conserved along the evolutionary scale (Supplementary Table S1, Figure 2), however, specific interactions are observed between JAZ proteins and some TFs [29]. In summary, the gain or loss of specific domains and motifs, along with changes in the amino acid composition observed in different lineages of land plants (Supplementary Table S1, Figures 1–4), may explain the higher ability for adaption under different environmental conditions and stresses.

In conclusion, all lineages of land plants contain JAZ proteins with the TIFY and Jas domains highly conserved. Besides this, some JAZs in subgroups such as JAZ1, JAZ5/6, JAZ7/8, and JAZ10 hold additional domains, e.g., EAR motifs or CMID domains. However, the origin of these domains is unknown, although it could be explained by some still unknown ancestor. New approaches for studying the conservation and evolution of protein sequences in each lineage could clarify the origin of JAZ proteins in the plant kingdom. These results establish a basis to understand the functional role of JAZ proteins during the evolution process in land plants, and they could be used to obtain crops more tolerant to environmental stresses through genetic breeding or gene modification.

4. Materials and Methods

4.1. Bioinformatic Identification of JAZ Proteins

JAZ protein sequences of *A. thaliana* were used for the searching of ortholog proteins in different plant genomes available in the PLAZA v4.0 database (Ghent University, Belgium, https://bioinformatics. psb.ugent.be/plaza/, last accessed date: 9 August 2019) by the BLASTP tool. Then, the resulting sequences were filtered. First, partial sequences and those with an e-value > 0.0 were removed. Second, NCBI's Conserved Domain Database (CDD) was used for the identification of proteins containing the Jas domain, and proteins lacking this sequence were removed. Third, sequences containing PPD or ZML domains were removed. Fourth, redundant sequences were removed after the multiple alignment with those previously reported by using Clustal Omega v2.0.12 (EMBL-EBI, Wellcome Genome Campus, Cambridgeshire, UK, https://www.ebi.ac.uk/Tools/msa/clustalo/, last accessed date:

15 April 2019). The resulting 1065 JAZ sequences used in the following analysis can be found in Supplementary Table S1.

4.2. Structural Analysis of JAZ Protein Domains

The TIFY and Jas domains of newly identified JAZ proteins in moss, lycophytes, gymnosperms, *A. trichopoda*, monocots and dicots were characterized by multiple alignment with respect to *A. thaliana* JAZ proteins and by using the Conserved Domains Database (CDD) v3.12 (National Center of Biotechnology Information, Maryland, USA; https://www.ncbi.nlm.nih.gov/Structure/cdd/cdd.shtml, last accessed date: 17 April 2019). Other conserved sequences, such as CMID domains and EARs motifs, were searched manually using *A. thaliana* sequences as a reference. The TIFY and Jas domains of previously reported JAZ proteins were obtained from the information collected in relative publications (Supplementary Table S1). Logo sequences of structural domains were obtained by the Weblogo v3.0 tool (University of California, Berkeley, California, USA; http://weblogo.threeplusone.com/create.cgi, last accessed date: 7 May 2019).

4.3. Analysis of Degron Sequences

For the degron analysis in the different plant lineages, only amino acidic sequences with Arg/Lys (R/K) at the fifth and sixth amino acidic positions from N- to C-terminus according to the canonical degron sequence LPIAR(R/K) [3] were considered as functional degrons. Then, the percentage (%) of each specific degron was calculated in the different lineages. In order to clearly display the more represented degrons, all sequences that individually showed a percentage lower than 5% were included together as alternative degrons. The percentage of each type of amino acidic residue at the different positions was calculated according to the physicochemical properties (non-polar, polar, negatively or positively charged) by using the Amino Acid Explorer tool (University of Maryland, Maryland, USA; https://www.ncbi.nlm.nih.gov/Class/Structure/aa/aa_explorer.cgi, last accessed date: 12 June 2019).

4.4. Phylogenetic Analysis

The evolutionary relationships between JAZ proteins were inferred using 308 full-length amino acidic sequences by using the Maximum Likelihood algorithm and a Jones, Taylor and Thornton (JTT) matrix-based model. The bootstrap consensus trees were inferred from 1000 replicates. Evolutionary analyses were conducted in Molecular Evolutionary Genetics Analysis X (MEGA X) (University of Pennsylvania, Pennsylvania, USA [56], last accessed date: 9 August 2019). An unrooted phylogenetic tree was constructed using 308 amino acidic JAZ sequences corresponding to previously identified and characterized JAZ proteins.

4.5. Reconstruction of JAZ Ancestral Sequences

For the ancestral JAZ reconstructions, different AtJAZ protein orthologs from different species were used. The amino acidic sequences were obtained from the PLAZA v4.0 database (Ghent University, Belgium, https://bioinformatics.psb.ugent.be/plaza/, last accessed date: 7 August 2019) and from previously reported JAZ sequences. Ancestral sequences were inferred using PhyloBot software (University of California, San Francisco, USA) [57] and using 308 JAZ sequences. Sequences were aligned by using MSAProbs [58] and MUltiple Sequence Comparison by Log-Expectation (MUSCLE) (California, USA) [59] with default settings. The results of the MSAProbs alignment and the tree drawn under the PROTCATJTT model were further analyzed and visualized by FigTree v1.4.3 (The University of Edinburgh, Scotland, UK, http://tree.bio.ed.ac.uk/software/figtree/, last accessed date: 16 August 2019). The analysis of JAZ ancestral sequences is available to download at the following URL: http://www.phylobot.com/464456268 (last accessed date: 19 August 2019).

Supplementary Materials: Supplementary materials can be found at http://www.mdpi.com/1422-0067/20/20/5060/s1. Table S1. Main amino acid sequence features of JAZ proteins of different plant lineages contained in the PLAZA database. Table S2. List of specific degron sequences for different plant lineages. Table S3. List of protein ancestor sequences obtained by PhyloBot software.

Author Contributions: Conceptualization: C.R.F. Methodology: A.G.-B., F.V.-R. and C.R.F. Validation: A.G.-B., F.V.-R. and C.R.F. Formal analysis: A.G.-B., F.V.-R. and C.R.F. Investigation: A.G.-B., F.V.-R. and C.R.F. Resources: A.G.-B., F.V.-R. and C.R.F. Data curation: A.G.-B. Writing—original draft preparation: A.G.-B. and C.R.F. Writing—review and editing: A.G.-B., F.V.-R. and C.R.F. Visualization: A.G.-B. and C.R.F. Supervision: C.R.F. Project administration: C.R.F. Funding acquisition: C.R.F.

Funding: This research was funded by the National Commission for Scientific and Technological Research (CONICYT, Chile) grant FONDECYT/Regular 1181310 to C.R.F.

Acknowledgments: A.G.-B. and F.V.-R. acknowledge CONICYT (grant FONDECYT/Postdoctorado 3190894) and Universidad de Talca for a doctoral scholarship, respectively.

Conflicts of Interest: The authors declare no conflict of interest.

References

1. Howe, G.A.; Major, I.T.; Koo, A.J. Modularity in Jasmonate Signaling for Multistress Resilience. *Annu. Rev. Plant Biol.* **2018**, *69*, 387–415. [CrossRef] [PubMed]

2. Han, G.-Z. Evolution of jasmonate biosynthesis and signaling mechanisms. *J. Exp. Bot.* **2017**, *68*, 1323–1331. [CrossRef] [PubMed]

3. Sheard, L.B.; Tan, X.; Mao, H.; Withers, J.; Ben-Nissan, G.; Hinds, T.R.; Kobayashi, Y.; Hsu, F.-F.; Sharon, M.; Browse, J.; et al. Jasmonate perception by inositol-phosphate-potentiated COI1-JAZ co-receptor. *Nature* **2010**, *468*, 400–405. [CrossRef] [PubMed]

4. Valenzuela-Riffo, F.; Garrido-Bigotes, A.; Figueroa, P.M.; Morales-Quintana, L.; Figueroa, C.R. Structural analysis of the woodland strawberry COI1-JAZ1 co-receptor for the plant hormone jasmonoyl-isoleucine. *J. Mol. Graph. Model.* **2018**, *85*, 250–261. [CrossRef]

5. Chini, A.; Fonseca, S.; Fernández, G.; Adie, B.; Chico, J.M.; Lorenzo, O.; García-Casado, G.; López-Vidriero, I.; Lozano, F.M.; Ponce, M.R.; et al. The JAZ family of repressors is the missing link in jasmonate signalling. *Nature* **2007**, *448*, 666–671. [CrossRef]

6. Thines, B.; Katsir, L.; Melotto, M.; Niu, Y.; Mandaokar, A.; Liu, G.; Nomura, K.; He, S.Y.; Howe, G.A.; Browse, J. JAZ repressor proteins are targets of the SCF(COI1) complex during jasmonate signalling. *Nature* **2007**, *448*, 661–665. [CrossRef]

7. Lorenzo, O.; Chico, J.M.; Sánchez-Serrano, J.J.; Solano, R. JASMONATE-INSENSITIVE1 Encodes a MYC Transcription Factor Essential to Discriminate between Different Jasmonate-Regulated Defense Responses in Arabidopsis. *Plant Cell Online* **2004**, *16*, 1938–1950. [CrossRef]

8. Fernández-Calvo, P.; Chini, A.; Fernández-Barbero, G.; Chico, J.-M.; Gimenez-Ibanez, S.; Geerinck, J.; Eeckhout, D.; Schweizer, F.; Godoy, M.; Franco-Zorrilla, J.M.; et al. The Arabidopsis bHLH transcription factors MYC3 and MYC4 are targets of JAZ repressors and act additively with MYC2 in the activation of jasmonate responses. *Plant Cell* **2011**, *23*, 701–715. [CrossRef]

9. Pauwels, L.; Barbero, G.F.; Geerinck, J.; Tilleman, S.; Grunewald, W.; Pérez, A.C.; Chico, J.M.; Bossche, R.V.; Sewell, J.; Gil, E.; et al. NINJA connects the co-repressor TOPLESS to jasmonate signalling. *Nature* **2010**, *464*, 788–791. [CrossRef]

10. De Geyter, N.; Gholami, A.; Goormachtig, S.; Goossens, A. Transcriptional machineries in jasmonate-elicited plant secondary metabolism. *Trends Plant Sci.* **2012**, *17*, 349–359. [CrossRef]

11. Fonseca, S.; Fernández-Calvo, P.; Fernández, G.M.; Díez-Díaz, M.; Gimenez-Ibanez, S.; López-Vidriero, I.; Godoy, M.; Fernández-Barbero, G.; Leene, J.V.; Jaeger, G.D.; et al. bHLH003, bHLH013 and bHLH017 Are New Targets of JAZ Repressors Negatively Regulating JA Responses. *PLoS ONE* **2014**, *9*, e86182. [CrossRef] [PubMed]

12. Sasaki-Sekimoto, Y.; Jikumaru, Y.; Obayashi, T.; Saito, H.; Masuda, S.; Kamiya, Y.; Ohta, H.; Shirasu, K. Basic helix-loop-helix transcription factors JASMONATE-ASSOCIATED MYC2-LIKE1 (JAM1), JAM2, and JAM3 are negative regulators of jasmonate responses in Arabidopsis. *Plant Physiol.* **2013**, *163*, 291–304. [CrossRef] [PubMed]

13. Liu, Y.; Du, M.; Deng, L.; Shen, J.; Fang, M.; Chen, Q.; Lu, Y.; Wang, Q.; Li, C.; Zhai, Q. MYC2 Regulates the Termination of Jasmonate Signaling via an Autoregulatory Negative Feedback Loop. *Plant Cell* **2019**, *31*, 106–127. [CrossRef] [PubMed]

14. Monte, I.; Ishida, S.; Zamarreño, A.M.; Hamberg, M.; Franco-Zorrilla, J.M.; García-Casado, G.; Gouhier-Darimont, C.; Reymond, P.; Takahashi, K.; García-Mina, J.M.; et al. Ligand-receptor co-evolution shaped the jasmonate pathway in land plants. *Nat. Chem. Biol.* **2018**, *14*, 480–488. [CrossRef] [PubMed]

15. Monte, I.; Franco-Zorrilla, J.M.; García-Casado, G.; Zamarreño, A.M.; García-Mina, J.M.; Nishihama, R.; Kohchi, T.; Solano, R. A Single JAZ Repressor Controls the Jasmonate Pathway in Marchantia polymorpha. *Mol. Plant* **2019**, *12*, 185–198. [CrossRef]

16. Peñuelas, M.; Monte, I.; Schweizer, F.; Vallat, A.; Reymond, P.; García-Casado, G.; Franco-Zorrilla, J.M.; Solano, R. Jasmonate-related MYC Transcription Factors are Functionally Conserved in Marchantia polymorpha. *Plant Cell* **2019**, *31*, 2491–2509.

17. Chini, A.; Ben-Romdhane, W.; Hassairi, A.; Aboul-Soud, M.A.M. Identification of TIFY/JAZ family genes in Solanum lycopersicum and their regulation in response to abiotic stresses. *PLoS ONE* **2017**, *12*, e0177381. [CrossRef]

18. Garrido-Bigotes, A.; Figueroa, N.E.; Figueroa, P.M.; Figueroa, C.R. Jasmonate signalling pathway in strawberry: Genome-wide identification, molecular characterization and expression of JAZs and MYCs during fruit development and ripening. *PLoS ONE* **2018**, *13*, e0197118. [CrossRef]

19. Ye, H.; Du, H.; Tang, N.; Li, X.; Xiong, L. Identification and expression profiling analysis of TIFY family genes involved in stress and phytohormone responses in rice. *Plant Mol. Biol.* **2009**, *71*, 291–305. [CrossRef]

20. Wang, Y.; Qiao, L.; Bai, J.; Wang, P.; Duan, W.; Yuan, S.; Yuan, G.; Zhang, F.; Zhang, L.; Zhao, C. Genome-wide characterization of JASMONATE-ZIM DOMAIN transcription repressors in wheat (Triticum aestivum L.). *BMC Genomics* **2017**, *18*, 152. [CrossRef]

21. Bai, Y.; Meng, Y.; Huang, D.; Qi, Y.; Chen, M. Origin and evolutionary analysis of the plant-specific TIFY transcription factor family. *Genomics* **2011**, *98*, 128–136. [CrossRef] [PubMed]

22. Thireault, C.; Shyu, C.; Yoshida, Y.; St Aubin, B.; Campos, M.L.; Howe, G.A. Repression of jasmonate signaling by a non-TIFY JAZ protein in Arabidopsis. *Plant J.* **2015**, *82*, 669–679. [CrossRef] [PubMed]

23. Vanholme, B.; Grunewald, W.; Bateman, A.; Kohchi, T.; Gheysen, G. The tify family previously known as ZIM. *Trends Plant Sci.* **2007**, *12*, 239–244. [CrossRef] [PubMed]

24. Chini, A.; Fonseca, S.; Chico, J.M.; Fernández-Calvo, P.; Solano, R. The ZIM domain mediates homo- and heteromeric interactions between Arabidopsis JAZ proteins. *Plant J.* **2009**, *59*, 77–87. [CrossRef] [PubMed]

25. Pauwels, L.; Goossens, A. The JAZ proteins: A crucial interface in the jasmonate signaling cascade. *Plant Cell* **2011**, *23*, 3089–3100. [CrossRef] [PubMed]

26. Shyu, C.; Figueroa, P.; Depew, C.L.; Cooke, T.F.; Sheard, L.B.; Moreno, J.E.; Katsir, L.; Zheng, N.; Browse, J.; Howe, G.A. JAZ8 lacks a canonical degron and has an EAR motif that mediates transcriptional repression of jasmonate responses in Arabidopsis. *Plant Cell* **2012**, *24*, 536–550. [CrossRef]

27. Goossens, J.; Swinnen, G.; Vanden Bossche, R.; Pauwels, L.; Goossens, A. Change of a conserved amino acid in the MYC2 and MYC3 transcription factors leads to release of JAZ repression and increased activity. *New Phytol.* **2015**, *206*, 1229–1237. [CrossRef]

28. Kagale, S.; Links, M.G.; Rozwadowski, K. Genome-Wide Analysis of Ethylene-Responsive Element Binding Factor-Associated Amphiphilic Repression Motif-Containing Transcriptional Regulators in Arabidopsis. *Plant Physiol.* **2010**, *152*, 1109–1134. [CrossRef]

29. Chini, A.; Gimenez-Ibanez, S.; Goossens, A.; Solano, R. Redundancy and specificity in jasmonate signalling. *Curr. Opin. Plant Biol.* **2016**, *33*, 147–156. [CrossRef]

30. Zhang, Y.; Gao, M.; Singer, S.D.; Fei, Z.; Wang, H.; Wang, X. Genome-wide identification and analysis of the TIFY gene family in grape. *PLoS ONE* **2012**, *7*, e44465. [CrossRef]

31. Wang, C.; Liu, Y.; Li, S.-S.; Han, G.-Z. Insights into the Origin and Evolution of the Plant Hormone Signaling Machinery. *Plant Physiol.* **2015**, *167*, 872–886. [CrossRef] [PubMed]

32. Sun, Q.; Wang, G.; Zhang, X.; Zhang, X.; Qiao, P.; Long, L.; Yuan, Y.; Cai, Y. Genome-wide identification of the TIFY gene family in three cultivated *Gossypium* species and the expression of JAZ genes. *Sci. Rep.* **2017**, *7*, 42418. [CrossRef] [PubMed]

33. Zhang, M.; Chen, Y.; Nie, L.; Jin, X.; Fu, C.; Yu, L. Molecular, structural, and phylogenetic analyses of Taxus chinensis JAZs. *Gene* **2017**, *620*, 66–74. [CrossRef] [PubMed]

34. Zhang, L.; You, J.; Chan, Z. Identification and characterization of TIFY family genes in Brachypodium distachyon. *J. Plant Res.* **2015**, *128*, 995–1005. [CrossRef]

35. Huang, Z.; Jin, S.-H.; Guo, H.-D.; Zhong, X.-J.; He, J.; Li, X.; Jiang, M.-Y.; Yu, X.-F.; Long, H.; Ma, M.-D.; et al. Genome-wide identification and characterization of TIFY family genes in Moso Bamboo (Phyllostachys edulis) and expression profiling analysis under dehydration and cold stresses. *PeerJ* **2016**, *4*, e2620. [CrossRef]

36. Zhang, Z.; Li, X.; Yu, R.; Han, M.; Wu, Z. Isolation, structural analysis, and expression characteristics of the maize TIFY gene family. *Mol. Genet. Genom.* **2015**, *290*, 1849–1858. [CrossRef]

37. Chao, J.; Zhao, Y.; Jin, J.; Wu, S.; Deng, X.; Chen, Y.; Tian, W.-M. Genome-Wide Identification and Characterization of the JAZ Gene Family in Rubber Tree (Hevea brasiliensis). *Front. Genet.* **2019**, *10*, 372. [CrossRef]

38. Hong, H.; Xiao, H.; Yuan, H.; Zhai, J.; Huang, X. Cloning and characterisation of JAZ gene family in Hevea brasiliensis. *Plant Biol. (Stuttg)* **2015**, *17*, 618–624. [CrossRef]

39. Sirhindi, G.; Sharma, P.; Arya, P.; Goel, P.; Kumar, G.; Acharya, V.; Singh, A.K. Genome-wide characterization and expression profiling of TIFY gene family in pigeonpea (Cajanus cajan (L.) Millsp.) under copper stress. *J. Plant Biochem. Biotechnol.* **2016**, *25*, 301–310. [CrossRef]

40. He, D.H.; Lei, Z.P.; Tang, B.S.; Xing, H.Y.; Zhao, J.X.; Jing, Y.L. Identification and analysis of the TIFY gene family in Gossypium raimondii. *Genet. Mol. Res.* **2015**, *14*, 10119–10138. [CrossRef]

41. Li, X.; Yin, X.; Wang, H.; Li, J.; Guo, C.; Gao, H.; Zheng, Y.; Fan, C.; Wang, X. Genome-wide identification and analysis of the apple (Malus × domestica Borkh.) TIFY gene family. *Tree Genet. Genomes* **2015**, *11*, 808. [CrossRef]

42. Sherif, S.; El-Sharkawy, I.; Mathur, J.; Ravindran, P.; Kumar, P.; Paliyath, G.; Jayasankar, S. A stable JAZ protein from peach mediates the transition from outcrossing to self-pollination. *BMC Biol.* **2015**, *13*, 11. [CrossRef] [PubMed]

43. Ma, Y.; Shu, S.; Bai, S.; Tao, R.; Qian, M.; Teng, Y. Genome-wide survey and analysis of the TIFY gene family and its potential role in anthocyanin synthesis in Chinese sand pear (Pyrus pyrifolia). *Tree Genet. Genomes* **2018**, *14*, 25. [CrossRef]

44. Wang, Y.; Pan, F.; Chen, D.; Chu, W.; Liu, H.; Xiang, Y. Genome-wide identification and analysis of the Populus trichocarpa TIFY gene family. *Plant Physiol. Biochem.* **2017**, *115*, 360–371. [CrossRef]

45. Melotto, M.; Mecey, C.; Niu, Y.; Chung, H.S.; Katsir, L.; Yao, J.; Zeng, W.; Thines, B.; Staswick, P.; Browse, J.; et al. A critical role of two positively charged amino acids in the Jas motif of Arabidopsis JAZ proteins in mediating coronatine- and jasmonoyl isoleucine-dependent interactions with the COI1 F-box protein. *Plant J.* **2008**, *55*, 979–988. [CrossRef]

46. Howe, G.A.; Yoshida, Y. Evolutionary Origin of JAZ Proteins and Jasmonate Signaling. *Molec. Plant* **2019**, *12*, 153–155. [CrossRef]

47. Wasternack, C.; Feussner, I. The Oxylipin Pathways: Biochemistry and Function. *Annu. Rev. Plant Biol.* **2018**, *69*, 363–386. [CrossRef]

48. Wu, W.; Liu, Y.; Wang, Y.; Li, H.; Liu, J.; Tan, J.; He, J.; Bai, J.; Ma, H. Evolution Analysis of the Aux/IAA Gene Family in Plants Shows Dual Origins and Variable Nuclear Localization Signals. *Int. J. Mol. Sci.* **2017**, *18*, 2107. [CrossRef]

49. Chung, H.S.; Howe, G.A. A Critical Role for the TIFY Motif in Repression of Jasmonate Signaling by a Stabilized Splice Variant of the JASMONATE ZIM-Domain Protein JAZ10 in Arabidopsis. *Plant Cell* **2009**, *21*, 131–145. [CrossRef]

50. Chaudhary, S.; Khokhar, W.; Jabre, I.; Reddy, A.S.N.; Byrne, L.J.; Wilson, C.M.; Syed, N.H. Alternative Splicing and Protein Diversity: Plants Versus Animals. *Front. Plant Sci.* **2019**, *10*, 708. [CrossRef]

51. Bowie, J.U.; Reidhaar-Olson, J.F.; Lim, W.A.; Sauer, R.T. Deciphering the message in protein sequences: Tolerance to amino acid substitutions. *Science* **1990**, *247*, 1306–1310. [CrossRef] [PubMed]

52. Zhang, X.-C.; Wang, Z.; Zhang, X.; Le, M.H.; Sun, J.; Xu, D.; Cheng, J.; Stacey, G. Evolutionary dynamics of protein domain architecture in plants. *BMC Evol. Biol.* **2012**, *12*, 6. [CrossRef] [PubMed]

53. Harms, M.J.; Thornton, J.W. Evolutionary biochemistry: Revealing the historical and physical causes of protein properties. *Nat. Rev. Genet.* **2013**, *14*, 559–571. [CrossRef] [PubMed]

54. Joy, J.B.; Liang, R.H.; McCloskey, R.M.; Nguyen, T.; Poon, A.F.Y. Ancestral Reconstruction. *PLoS Comput. Biol.* **2016**, *12*. [CrossRef]

55. Bowman, J.L.; Kohchi, T.; Yamato, K.T.; Jenkins, J.; Shu, S.; Ishizaki, K.; Yamaoka, S.; Nishihama, R.; Nakamura, Y.; Berger, F.; et al. Insights into Land Plant Evolution Garnered from the Marchantia polymorpha Genome. *Cell* **2017**, *171*, 287–304. [CrossRef]
56. Kumar, S.; Stecher, G.; Li, M.; Knyaz, C.; Tamura, K. MEGA X: Molecular Evolutionary Genetics Analysis across Computing Platforms. *Mol. Biol. Evol.* **2018**, *35*, 1547–1549. [CrossRef]
57. Hanson-Smith, V.; Johnson, A. PhyloBot: A Web Portal for Automated Phylogenetics, Ancestral Sequence Reconstruction, and Exploration of Mutational Trajectories. *PLoS Comput. Biol.* **2016**, *12*, e1004976. [CrossRef]
58. Liu, Y.; Schmidt, B.; Maskell, D.L. MSAProbs: Multiple sequence alignment based on pair hidden Markov models and partition function posterior probabilities. *Bioinformatics* **2010**, *26*, 1958–1964. [CrossRef]
59. Edgar, R.C. MUSCLE: Multiple sequence alignment with high accuracy and high throughput. *Nucleic Acids Res.* **2004**, *32*, 1792–1797. [CrossRef]

 © 2019 by the authors. Licensee MDPI, Basel, Switzerland. This article is an open access article distributed under the terms and conditions of the Creative Commons Attribution (CC BY) license (http://creativecommons.org/licenses/by/4.0/).

International Journal of
Molecular Sciences

Article

iTRAQ-Based Quantitative Proteomic Analysis of the Arabidopsis Mutant *opr3-1* in Response to Exogenous MeJA

Jiayu Qi, Xiaoyun Zhao and Zhen Li *

State Key Laboratory of Plant Physiology and Biochemistry, College of Biological Sciences, China Agricultural University, Beijing 100193, China; qijiayu97@163.com (J.Q.); xiaoyunzhao@cau.edu.cn (X.Z.)
* Correspondence: lizhenchem@cau.edu.cn

Received: 9 December 2019; Accepted: 14 January 2020; Published: 16 January 2020

Abstract: Jasmonates (JAs) regulate the defense of biotic and abiotic stresses, growth, development, and many other important biological processes in plants. The comprehensive proteomic profiling of plants under JAs treatment provides insights into the regulation mechanism of JAs. Isobaric tags for relative and absolute quantification (iTRAQ)-based quantitative proteomic analysis was performed on the Arabidopsis wild type (Ws) and JA synthesis deficiency mutant *opr3-1*. The effects of exogenous MeJA treatment on the proteome of *opr3-1*, which lacks endogenous JAs, were investigated. A total of 3683 proteins were identified and 126 proteins were differentially regulated between different genotypes and treatment groups. The functional classification of these differentially regulated proteins showed that they were involved in metabolic processes, responses to abiotic stress or biotic stress, the defense against pathogens and wounds, photosynthesis, protein synthesis, and developmental processes. Exogenous MeJA treatment induced the up-regulation of a large number of defense-related proteins and photosynthesis-related proteins, it also induced the down-regulation of many ribosomal proteins in *opr3-1*. These results were further verified by a quantitative real-time PCR (qRT-PCR) analysis of 15 selected genes. Our research provides the basis for further understanding the molecular mechanism of JAs' regulation of plant defense, photosynthesis, protein synthesis, and development.

Keywords: jasmonic acid; *opr3*; stress defense; quantitative proteomics

---

1. Introduction

Jasmonic acid (JA) is a plant hormone that plays an integral role in the regulation of plant growth and development as well as in plant defense against wounding, herbivory attack, and other biotic and abiotic stresses [1]. Jasmonates (JAs) are a class of compounds derived from jasmonic acid with varying biological activities, including the active form (3R,7S) jasmonoyl-isoleucine (JA-Ile) and the volatile methyl ester form methyl jasmonate (MeJA) [2,3]. As a stress-related hormone, JAs are involved in plant defense against insects and pathogens [4,5], the response to ultraviolet radiation [6], drought, and other abiotic stresses [7–10]. In unwounded plant tissues, JAs regulate plant growth and development [11], affect root growth [12], senescence, and stamen development [13–17]. However, when encountering insect herbivory attack, plants immediately undergo a series of physiological responses and initiate a rapid biosynthesis of JAs to trigger the transduction of a stress signal, which results in the activation of the wound defense mechanism [18–20]. Most plants derive their jasmonates from octadecanoid (18-carbon) fatty acid, while some plants also produce jasmonates from hexadecanoid (16-carbon) fatty acid [21]. The major JA synthesis pathway starts from α-linolenic acid (18:3) released from membrane lipids. The linolenic acid is oxygenated by 13-lipoxygenase (LOX) to form 13-hydroperoxylinolenic acid (13-HPOT) in the plasma membrane. The resulting fatty acid hydroperoxide is released from the plasma membrane to peroxisome, which is then dehydrated by allene oxide synthase (AOS) and cyclized by allene oxide cyclase (AOC) to form the cyclopentenone 12-oxo-phytodienoic acid (9S,13S-OPDA). The pentacyclic ring double bond in 9S,13S-OPDA is reduced by OPDA reductase 3 (OPR3) in the peroxisome to form 8-(3-oxo-2 (pent-2-enyl)-cyclopentyl) octanoic acid (OPC:8). Finally, 3R,7S-JA is generated from OPC:8 after three cycles of β-oxidation in the cytoplasm [22,23]. JA can also be produced from the OPDA derivative 4,5-didehydro-JA when OPR3 is completely knocked-out [24]. JA can be catalyzed by JA carboxyl methyltransferase (JMT) to form MeJA, while methyl jasmonate esterase (MJE) can convert MeJA to form JA. The active form of jasmonic acid, JA-Ile, is formed through the conjugation of jasmonic acid with isoleucine and is perceived by the COI1/JAZ co-receptor [25,26]. MeJA is volatile and can penetrate through plasma membrane readily, and it can be quickly converted to JA and eventually to JA-Ile to participate in systemic signaling during development and responses to stress [27]. MeJA has been widely used to study jasmonates signaling pathways and the mechanisms of plant defense.

Mutants defective in JA biosynthesis and response have been used to investigate the roles of JAs in defense and development [28–31]. These mutants included *dad1, fad3-3, fad7-2, fad8, dde2-2, dde1*, and *opr3*. OPR3 is one of the restrictive enzymes in the JA synthesis pathway. There are six OPR enzymes in Arabidopsis, but only OPR3 can effectively catalyze the reduction of 9S,13S-OPDA in plants [32–35]. Endogenous JA is nearly absent in *opr3-1* and the mutant showed three characteristic phenotypes: floral organs develop normally within the closed bud, but the anther filaments do not elongate enough to reach the locules above the stigma at the anthesis stage; the anther locules do not dehisce during flowering; the pollen grains are predominantly inviable [36]. However, *opr3-1* is not a null mutant, it can form mature full-length *OPR3* transcripts and synthesize JA under specific conditions like under B. cinerea infection [37]. In the OPR3 complete knock-out mutant (*opr3-3*), JA can still be synthesized through OPDA derivative 4,5-didehydro-JA under the catalysis of OPR2 [24]. Exogenous MeJA treatment can restore stamen development, the inhibition of root growth and the degradation of the jasmonate repressor JAZ1 in JA biosynthesis-deficient mutants [24], such as *opr3*, but not in JA signaling-deficient mutants [24]. Compared with other JA synthesis defect mutants, the *opr3-1* mutant is more resistant to necrotrophic fungus, Alternaria brassicicola, as well as to the soil gnat, Bradysia impatiensthus [37]. So, *opr3-1* is a valuable model to investigate the mechanism of the JA signaling pathway due to its nearly absent endogenous JAs [36,38]. It has been reported that exogenous JAs were involved in fertility regulation and root growth [39,40], whether the application of exogenous JAs on biosynthesis-deficient mutants, such as *opr3-1*, affects other functions of Arabidopsis remains an intriguing research topic.

The fast development of transcriptics and mass spectrometry-based quantitative proteomics approaches provide powerful tools to investigate the biological response of plants under external stress conditions or exogenous hormone stimulus. Pauwels L et al. investigated alterations in the transcriptome of the fast-dividing cell culture of Arabidopsis after exogenous MeJA treatment. The results showed early MeJA response genes encoded the JA biosynthesis pathway proteins and key regulators of MeJA responses, including most JA ZIM domain proteins and MYC2, meanwhile, in the second transcriptional wave, MeJA response transcripts were mainly involved in cellular metabolism and cell cycle progression [41]. Mata-Perez et al. used RNA-seq to study the profiles of Arabidopsis cell suspension cultures transcriptomes after linolenic acid treatment, identified 533 up- and 2501 down-regulated genes. RNA-seq data analysis showed that an important set of these genes was associated with the JA biosynthetic pathway, including LOX and AOC. In addition, several transcription factor families involved in the response to biotic stresses, such as pathogen attacks or herbivore feeding, were identified [42]. Guo et al. used an iTRAQ-based quantitative proteomics approach to analyze broccoli sprouts treated with exogenous jasmonic acid and found that photosynthesis and protein synthesis were inhibited after JA treatment, which was responsible for the slower growth of broccoli, but carbon metabolism and amino acid metabolism-related proteins were up-regulated [43]. Farooq et al. investigated MeJA-induced Arsenic tolerance in *Brassica napus* leaves using iTRAQ and 110 differentially regulated proteins were identified—proteins that were involved in stress and defense, photosynthesis, carbohydrates and energy production, protein metabolism, and secondary metabolites [44]. Alvarez et al. investigated the changes in protein redox regulation in response to oxidative stress induced by MeJA in Arabidopsis shoots and roots using quantitative proteomics approach and confirmed cysteine residues of proteins were involved in redox regulation, which provided a deeper understanding of the jasmonate signaling and regulation network [45].

Most of the reports investigated the effects of exogenous JAs on stress and defense responses in the presence of endogenous JA. There were very limited reports on the effects of exogenous JAs in the absence of endogenous JAs. The recovery of fertility in *opr3-1* after exogenous MeJA treatment indicated that exogenous JAs can, at least partially, replace the role of endogenous JA. Thus, we raise the following question: which signaling pathways and metabolic processes can be affected by exogenous JAs in the absence of endogenous JAs?

In this study, we used an iTRAQ-based quantitative proteomic method to investigate the effects of exogenous MeJA on JA synthesis deficient mutant *opr3-1* (Figure S1). A total of 126 differentially regulated proteins (DRPs) were identified between the control and treatment groups of both genotypes (Arabidopsis wild type (Ws) and *opr3-1*) after MeJA treatment. These DRPs were involved in metabolism processes, responses to stress, the defense against pathogens and wounds, photosynthesis, protein synthesis, as well as development processes. The transcriptional level of 15 selected genes from the DRPs was further validated by qRT-PCR analysis. Our work contributed to a better understanding of the molecular mechanisms of JAs regulating plant defense, photosynthesis, protein synthesis, and development.

2. Results

2.1. Overview of Protein Identified in Ws and opr3-1

A quantitative proteomics analysis of Arabidopsis wild type (Ws) and JA synthesis deficient mutant (*opr3-1*) treated with 0.25 mM of MeJA for 8 h was performed to identify differentially expressed proteins between these genotypes under exogenous MeJA treatment. A total of 45,691 unique spectra corresponding to 25,957 unique peptides and 3683 proteins were identified in this experiment. Among them, 3386 proteins can be identified in all three replicates and 3214 proteins can be quantified (Figure 1a). It can be seen that sequence coverages of the identified proteins were mostly below 30%, and most of the identified proteins were in the mass range 20–30 KD and 30–40 KD (Figure 1b,c).

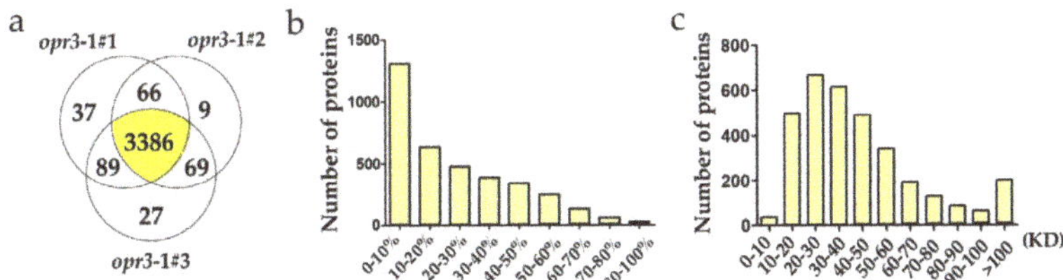

Figure 1. Information on the identified proteins. (**a**) A Venn diagram of the number of proteins identified in three replicates; (**b**) the distribution of sequence coverage; (**c**) the distribution of the mass of the identified proteins.

2.2. Identification of Differentially Regulated Proteins

In order to explore the effect of exogenous JAs on *opr3-1* at the proteome level, differentially regulated proteins (DRPs, fold change > 1.5, $p < 0.05$) were screened according to the intensity of the iTRAQ reporter ions. A total of 126 DRPs were identified between the control and treatment groups of both genotypes (Figure 2a). To further understand the effects of exogenous JAs on the proteome of Arabidopsis in the absence of endogenous JA, we screened DRPs between *opr3-1* and *opr3-1* after the MeJA treatment and removed the proteins that showed significant changes in abundance in Ws after

MeJA treatment. The remaining 97 DRPs were considered as proteins that were induced by exogenous JAs. Among them, 44 proteins were up-regulated and 53 proteins were down-regulated (Figure 2b). These DRPs were used for the following functional analysis.

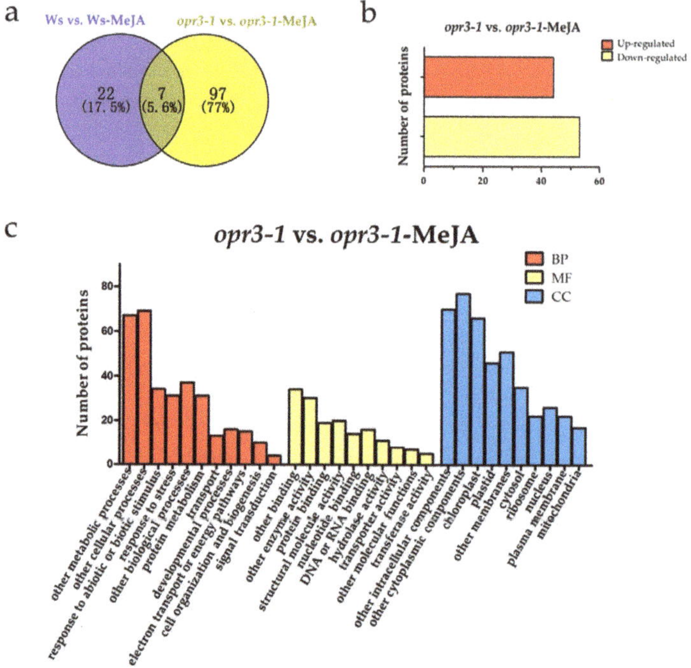

Figure 2. Functional classifications of differentially regulated proteins. (**a**) A Venn diagram of differentially regulated proteins; (**b**) the numbers of up-regulated and down-regulated proteins; (**c**) the GO assignment of DRPs in *opr3-1* in response to methyl jasmonate (MeJA) treatment (*opr3-1*-MeJA). BP: Biological Process; MF: Molecular Function; CC: Cellular Component.

2.3. Functional Analysis of Differentially Regulated Proteins

GO analysis of the DRPs showed that the DRPs responded to MeJA in *opr3-1* can be classified into 11 biological process categories: metabolic processes (20.48%), cellular processes (21.10%), the response to abiotic or biotic stimulus (10.39%), the response to stress (9.48%), other biological processes (11.31%), protein metabolism (9.48%), transport (3.97%), developmental processes (4.89%), electron transport or energy pathways (4.58%), cell organization and biogenesis (3.05%), and signal transduction (1.22%). For molecular functions, 20.73% of the proteins were related to binding activity, followed by enzyme activity (18.29%), structural molecule activity (12.19%), protein binding (11.58%), DNA or RNA binding (9.76%), nucleotide binding (8.54%), hydrolase activity (6.71%), transporter activity (4.88%), other molecular functions (4.27%), and transferase activity (3.05%). In the cellular components category, 17.82% of the DRPs were cytoplasmic components, followed by intracellular components (16.20%), chloroplast (15.28%), other membranes (11.81%), plastids (10.65%), cytosol (8.10%), nucleus (6.02%), ribosome (5.09%), plasma membrane (5.09%), and mitochondria (3.93%) (Figure 2c).

A Kyoto Encyclopedia of Genes and Genomes (KEGG) pathway analysis of the DRPs between *opr3-1* and *opr3-1*-MeJA showed 20 functional classes (Figure S2). Most of the proteins were enriched in metabolic pathways, protein synthesis, photosynthesis, the biosynthesis of secondary metabolites, carbon metabolism, and the biosynthesis of amino acids.

A STRING analysis was performed to investigate the interaction network among these DRPs. The DRPs can be divided into three groups (Figure 3). They were involved in protein synthesis (red group), energy metabolism (green group), and photosynthesis (blue group). Among these proteins, ATP synthase gamma chain 1 (AT4G04640.1, No.2) is involved in the regulation of ATPase activity, it catalyzes the conversion of ATP from ADP in the presence of a proton gradient across the membrane [46]. The abundance of this protein decreased by 0.59 fold in *opr3-1* after MeJA treatment, indicating that MeJA treatment reduced the synthesis of ATP and impaired the energy metabolism of *opr3-1*. An oxygen-evolving enhancer protein (AT4G05180.1, No.4) is required for photosystem II assembly/stability. The loss of the oxygen-evolving enhancer protein induces significant decreases in photosystem II function [47], and this protein was up-regulated by 1.80 folds after MeJA treatment in *opr3-1*. The expression of the Chlorophyll a–b binding protein (AT3G27690.1, No.3) was increased by 1.54 folds in *opr3-1* after MeJA treatment. This protein acts as a light receptor and is closely related to photosystems. The up-regulation of these two proteins in *opr3-1* after MeJA treatment indicated that MeJA treatment could enhance photosynthesis in *opr3-1*. When the plant is mechanically damaged, the JAs' content increases abruptly [48], while the application of exogenous MeJA simulates the process of pest or bacteria invasion and leaf damage, which results in the activation of the JAs' synthesis pathway, however, in the *opr3-1* mutant, the in vivo synthesis of JA is inhibited due to the lack of OPR3 enzyme, thus, the *opr3-1* mutant provides an excellent model to investigate the effects of exogenous JAs without background interference from endogenous JAs. We found that the abundance of pigment defective 334 (AT4G32260.1, No.1), which has hydrogen ion transmembrane transporter activity and is involved in the defense response to bacterium, was increased by 1.66 folds in *opr3-1* after MeJA treatment. The up-regulation of this protein suggested that the application of exogenous MeJA could induce the defense mechanism against bacterium invasion. Proteins in the red group (protein synthesis-related process) were closely interconnected, these proteins (AT2G41840.1, AT4G26230.1, AT3G05560.1, AT3G02080.1, AT3G02560.2, AT2G43030.1, No.5—10) mostly belonged to the ribosomal protein family and were involved in translation. The abundance of these proteins decreased by 0.66, 0.65, 0.61, 0.64, 0.66, and 0.65 folds in *opr3-1* after MeJA treatment, respectively. The down-regulation of these proteins indicated that MeJA treatment inhibited protein synthesis in *opr3-1*.

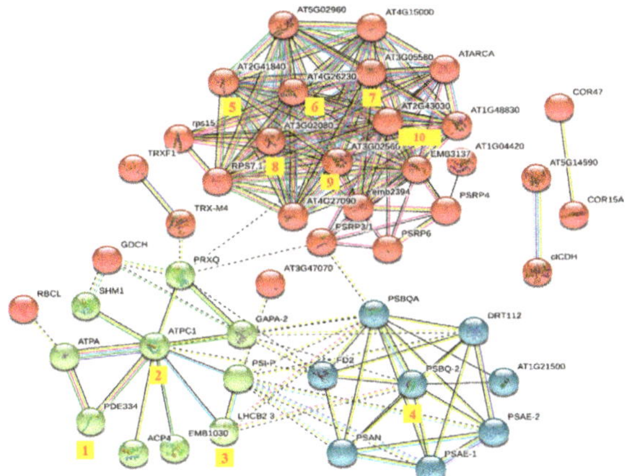

Figure 3. Protein–protein interaction network of differentially regulated proteins (DRPs) in *opr3-1* under MeJA-treatment. Red group: protein synthesis; green group: energy metabolism; blue group: photosynthesis.

2.4. Verification of the DRPs by qRT-PCR

To validate the iTRAQ results, the transcriptional levels of 15 candidate DRPs were analyzed using qRT-PCR (Figure 4). Among them, six DRPs showed similar trends of variation in their mRNA expression level compared with protein expression, including proteins involved in JA synthesis (OPR3, AOC), photosynthesis (PRXQ), protein domain specific binding (GRF5), and defense against pathogens and wounds (BG2, Thioredoxin M1) (Figure 4). *OPR3* and *AOC* are key genes in the synthesis of the JA pathway, *OPR3* was not expressed in the *opr3-1* mutant as expected, the expression of *AOC* was significantly increased after MeJA treatment in both genotypes, but the expression level was generally lower in the *opr3-1* mutant compared with the wild type (Figure 4). However, the alteration in protein expression levels did not always correlate well with the changes in mRNA expression. In this study, we found several genes with discrepancies in protein and mRNA abundances. For example, the abundance of RPS2C and RPL22B (protein synthesis-related) decreased in *opr3-1* after MeJA treatment, while their mRNA expression showed no significant changes in *opr3-1*. MeJA treatment resulted in accumulations of PDE334 and PR5 (related to defense against pathogen) in *opr3-1* but their mRNA expression decreased in *opr3-1* after treatment. Such discrepancies between qRT-PCR and iTRAQ results can be attributed to the post-transcriptional, translational, and post-translational regulation of gene expression [49].

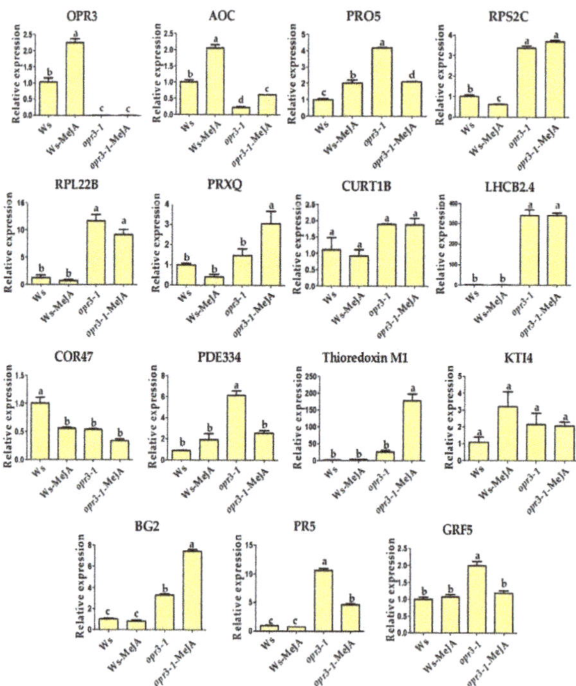

Figure 4. Relative mRNA expression levels of selected genes measured by qRT-PCR. OPR3: 12-oxophytodienoate reductase 3; AOC: Allene oxide cyclase 1; PRO5: Profilin-5; RPS2C: 40S ribosomal protein S2-3; RPL22B: 60S ribosomal protein L22-2; PRXQ: Peroxiredoxin Q; CURT1B: Curvature thylakoid 1B; LHCB2.4: Chlorophyll a-b binding protein 2.4; COR47: Dehydrin COR47; PDE334: Pigment defective 334; Thioredoxin M1: Arabidopsis thioredoxin m-type 1; KTI4: Kunitz trypsin inhibitor 4; BG2: Glucan endo-1,3-beta-glucosidase; PR5: Pathogenesis-related protein 5; GRF5: 14-3-3-like protein GF14 upsilon.

3. Discussion

Jasmonates, including jasmonic acid, methyl jasmonate, and jasmonoyl-isoleucine are crucial plant hormones widely present in higher plants. They play important roles in regulating seed germination, growth, pollen fertility, the response to external damage (mechanical, herbivore, insect damage) and pathogenic infections. The endogenous level of JA in *opr3-1* was only about 1/5 of the wild type (Figure S5). The deficiency of endogenous JAs in the *opr3-1* mutant results in the anther filaments not elongating enough to reach the stigma, anther locules not dehiscing, and inviable pollen grains, which eventually results in male sterility. The application of exogenous JAs can restore the male sterile phenotype. The nearly absence of endogenous JAs in *opr3-1* provide an excellent model to investigate the regulation mechanism of JA on plant development and defense response. Previous studies mostly focused on changes in the gene expression levels induced by exogenous JA, while changes at the proteomic levels were less explored [50,51]. Therefore, we used an iTRAQ-based quantitative proteomics approach to identify responsive proteins in JA synthesis deficient mutants, after exogenous MeJA treatment, as a means of exploring the regulatory roles of JAs. This study not only discovered the classic JA-induced proteins reported in previous studies [52,53] but also discovered some new proteins affected by exogenous MeJA, which are mainly involved in protein synthesis, photosynthesis, the response to stress, energy metabolism, and pollen development (Figure 5).

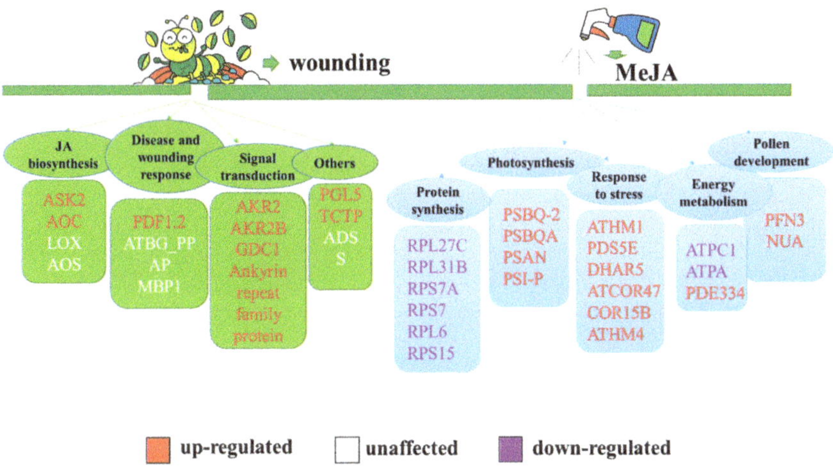

Figure 5. Summary of the biological processes affected by exogenous MeJA in the absence of endogenous jasmonic acid (JA). The red: up-regulation of protein expression; white: no significant change; purple: down-regulation of protein expression. Green group: proteins affected by JAs reported previously; blue group: proteins affected by jasmonates (JAs) discovered in this study. Orange group: plant hormone affected by JAs in this study.

3.1. MeJA-Induced Physiological Changes

The physiological assays showed that there was no significant difference in the content of H_2O_2 between Ws and *opr3-1* under normal condition (Figure S3a,b). After 8 h of MeJA treatment, the H_2O_2 content in Ws increased, while the opposite trend was observed in *opr3-1* (Figure S3c). In addition, *opr3-1* had higher Peroxidase (POD) content under normal conditions, and the content of POD in *opr3-1* was almost two times higher than that of Ws. The MeJA treatment led to decreases in POD content in both genotypes and a larger decrease in *opr3-1* was observed (Figure S3d), but *opr3-1* still managed to maintain a higher POD content than Ws. These results indicate that *opr3-1* can maintain a better reactive oxygen species (ROS) scavenging capability under exogenous MeJA treatment. The higher POD content in *opr3-1* may contribute to its lower ROS level, even though the POD content decreased in both genotypes after MeJA treatment. In the proteomics data, it was found that the abundances of ROS scavenge-related proteins (AT1G19570.1, AT3G26060.1, AT1G03680.1) were up-regulated in *opr3-1* after MeJA treatment, which may also explain the decrease in ROS content in *opr3-1*.

The contents of the free amino acid of the two genotypes were determined by liquid chromatograph-mass spectrometer (LC-MS) after derivatization with the AccQ tag reagent and the effect of exogenous MeJA treatment on free amino acids was revealed by principal component analysis (Figure S4). There were significant differences in free amino acid contents between Ws and *opr3-1* under normal conditions, and MeJA treatment did not show a significant effect on amino acids' contents in both genotypes. These data show that there was a significant difference in endogenous free amino acids' contents between Ws and *opr3-1*. Such a discrepancy cannot be compensated by the exogenous application of MeJA, indicating that exogenous JAs cannot fully replace the functions of endogenous JAs.

The contents of three major plant hormones (ABA, JA, SA) were measured using LC-MS in Ws and *opr3-1* before and after MeJA treatment (Figure S5). SA, JA, and ABA play important roles in plant defense and stress response. SA is best known for its central role in the plant defense response against pathogens and as an inducer of systemic acquired resistance. It is synthesized from chorismate via isochorismate. The infection of plants by pathogens results in an increase in SA levels both at the site of infection and in distant tissues [54]. ABA is an isoprenoid compound associated with seed dormancy, drought responses, and other growth processes [55]. ABA levels are regulated by a variety of environmental factors, including drought, cold, and other biotic or abiotic stresses [56]. In response to abiotic stress, the crosstalk between plant hormonal signaling pathways prioritizes defense over other cellular functions [57]. SA and JA-mediated signaling pathways are closely related in plant stress resistance, and they crosstalk through certain regulatory factors, such as *NPR1* [58]. The antagonism between SA and JA signaling pathways results in the downregulation of a large set of JA-responsive genes, including the marker genes PDF1.2 and VSP2 in the presence of SA [59]. In this study, the significantly reduced JA content in *opr3-1* confirmed the mutation of the *OPR3* gene, and the spike in JA content after exogenous MeJA treatment indicated a quick absorption and transformation of MeJA (Figure S5). Moreover, the ABA and SA contents showed similar trends after MeJA treatment, i.e., their contents were both significantly higher in *opr3-1* than in Ws under normal conditions, and they were both significantly decreased after MeJA treatment. Also, the plant hormone contents in MeJA-treated *opr3-1* was similar to those of the untreated wild type. These data indicate that there was antagonism between JA and ABA/SA in both Ws and *opr3-1*, and the lack of endogenous JA resulted in higher ABA and SA levels in untreated *opr3-1*, while the application of MeJA reduced the levels of these two hormones.

3.2. Proteins Response to MeJA Treatment in opr3-1

3.2.1. Stress-Related Proteins

When wounded or under insect or pathogen attack, plants initiate defensive mechanisms and activate the JA synthesis pathway, which results in a sharp increase in JA content [60]. In this experiment, the application of exogenous MeJA simulated such stress process in the plants. We found four wound-related proteins, DHAR1 (AT1G19570.1), ATCOR47 (AT1G20440.1), PDE334 (AT4G32260.1), and ATHM1 (AT1G03680.1). The abundance of these proteins were up-regulated by 1.62, 1.70, 1.66, and 1.51 folds, respectively, in *opr3-1*, after MeJA treatment. DHAR1 is a key component of the ascorbate recycling system; it is involved in ROS scavenging under oxidative stresses [61]. ATHM1 is the key enzyme of the oxidative pentose phosphate pathway, which supplies reducing power (as NADPH) in non-photosynthesizing cells, and is involved in the response to oxidative stress and regulates the carbohydrate metabolic process [62]. The up-regulated expression of DHAR1 and ATHM1 indicates the enhanced ROS scavenging capability of *opr3-1* in the presence of exogenous MeJA, which is evidenced by reduced H_2O_2 level in MeJA-treated *opr3-1* (Figure S3c). The accumulation of ATCOR47 is triggered in response to the presence of fungus and PDE334 and is involved in the response to an invasion of bacterium [63,64]. The accumulation of these two proteins indicates that the application of exogenous MeJA could also induce the expression of proteins involved in the response to fungus and bacterium invasion.

3.2.2. Pollen Development-Related Proteins

In *opr3-1*, the pollen grains are inviable, the anthers are abnormally dehydrated, and the anther filaments do not elongate, resulting in male sterility. These defects can be remedied by the application of exogenous MeJA, indicating that JA is required for male gamete development [65]. In the proteomic results, we found several up-regulated proteins in response to MeJA in *opr3-1* that were involved in anther and pollen development.

Nuclear pore anchor (NUA, AT1G79280.2) is a component of the nuclear pore complex, it mediates the transportation of RNA and other cargoes between the nucleus and the cytoplasm. Nuclear pore anchor mutants *nua-1* and *nua-4* showed diverse developmental phenotypes, including early flowering, stunted growth, and shortened anther filament [66], indicating that NUA is required for filament elongation. Profilin-3 (AT5G56600.1) is a ubiquitous eukaryotic protein that regulates the actin cytoskeleton, which is essential for pollen development. Profilin-3 can rearrange the actin cytoskeleton during pollen germination, and recently, it has been identified as a potent regulatory factor in pollen development [67]. The up-regulation of these two proteins (1.66 and 1.32 folds) in *opr3-1*, after MeJA treatment, indicates that these two proteins are induced by exogenous MeJA and that they may be curial components for restoring *opr3-1*'s fertility.

3.2.3. Protein Synthesis-Related Proteins

Ribosomes contain a large number of ribosomal proteins, which can catalyze the peptidyl transfer reaction for polypeptide synthesis. They are responsible for protein synthesis and play a major role in regulating cell growth, differentiation, and development [68]. In this study, we found five 40S ribosomal proteins (AT2G41840.1, AT3G02080.1, AT1G48830.1, AT3G02560.1, AT5G02960.1), four 60S ribosomal proteins (AT4G15000.1, AT4G26230.1, AT3G05560.1, AT4G27090.1), two 30S ribosomal proteins (ATCG00900.1, AT5G14320.1), and one 50S ribosomal protein (AT2G43030.1) that were dramatically down-regulated by MeJA in *opr3-1* (Table S1). The decreased abundance of these ribosomal proteins suggests that MeJA treatment inhibited protein synthesis in *opr3-1*.

3.2.4. Photosynthesis-Related Proteins

In plants, photosynthesis is an important metabolic process and is susceptible to environmental stress. It has been reported that photosynthesis rate was promoted in Arabidopsis under drought stress [69]. In this study, iTRAQ data show that MeJA application could remarkably enhance the expression of photosynthesis-related proteins. The photosystem I reaction center subunit N (AT5G64040.2) may function in mediating the binding of the antenna complexes to the PSI reaction center and core antenna. It plays an important role in docking plastocyanin to the PSI complex [47]. Photosystem I protein P (AT2G46820.1) is a part of the photosystem I complex [70]. Photosystem II subunit Q-1 (AT4G05180.1) and photosystem II subunit Q-2 (AT4G21280.2) encode the PsbQ subunit of the oxygen evolving complex of photosystem II. They are required for photosystem II assembly/stability [71]. These proteins were up-regulated by 1.51, 1.95, 1.70, and 1.55 folds in *opr3-1* after MeJA treatment, respectively. The up-regulation of these photosynthesis-related proteins suggests that exogenous MeJA treatment enhances plant photosynthesis processes.

4. Materials and Methods

4.1. Plant Materials and Growth Conditions

The seeds of *Arabidopsis thaliana* ecotype Wassileskija (Ws) and mutant *opr3-1* were sterilized with 1% NaClO for 10 min, followed by washing with distilled water and sowing onto a 1/2 MS medium for 10 days. Afterwards the seedlings were transferred to pots to a climate chamber (22 °C; 8/16 h light/dark cycle, 65% rh).

Four-week-old Arabidopsis (bolting but not flowering) plants were sprayed with 250 μM MeJA in 0.05% Tween-20, and the control groups were sprayed with 0.05% Tween-20 without MeJA. After 8 h of treatment, the leaves of the plants were collected, ground into a fine powder in liquid nitrogen and stored at −80 °C until protein extraction.

4.2. Protein Extraction and Digestion

Arabidopsis leaf proteins were extracted by a modified phenol extraction method [72]. In brief, 0.5 g of leaves were ground into fine powder in liquid nitrogen, and a 3 mL protein extraction buffer (500 mM Tris–HCl, 700 mM sucrose, 500 mM EDTA,100 mM KCl, 1% protease inhibitor cocktail, 1% phospho-STOP, pH 8.0) was added and ground for 10 min. Then, 3 mL of Tris-saturated phenol was added and ground for another 10 min. The phenol layer was collected after centrifugation and the proteins were precipitated with 0.1 M ammonium acetate in methanol overnight at −20 °C. The protein pellet was washed three times with pre-cooled acetone and dried. The protein pellet was dissolved with 7 M urea/2 M thiourea, and the protein concentration was measured by Bradford assay.

The protein was digested with a modified filter-aided sample preparation (FASP) workflow [73]. In short, 200 μg of protein was loaded onto an ultrafiltration device (10 KDa, MWCO, 500 μL, Sartorius, Gottingen, Germany), reduced with 50 mM DTT at 56 °C and alkylated with 200 mM IAM for 30 min, in the dark, at room temperature. The protein was digested with trypsin with a protein:enzyme ratio of 50:1 at 37 °C for 16 h.

4.3. iTRAQ Labeling, High pH Reversed-Phase Fractionation, and NanoLC-MS Analysis

An iTRAQ 8-plex kit was used to label peptides from Ws and *opr3-1*, with or without MeJA treatment, following the manufacturer's instructions. The details of the iTRAQ channels used for each sample are listed in Table S2. Three biological replicates were analyzed.

The labeled peptides were pooled and fractionated with a C18 column (2.1 mm × 100 mm, 2.6 μm, Kinetex, Phenomenex) using a gradient elution program of 20 mM ammonium acetate in water (pH 10.0) and 20 mM ammonium acetate in 90% acetonitrile (pH 10.0) on a High Performance Liquid Chromatography (HPLC) system (H-Class bio, Waters, Milford, MA, USA). The peptides were pooled into 12 fractions, dried in a vacuum concentrator and resuspended with 0.1% formic acid.

Protein identification was performed with a Q-Exactive high resolution mass spectrometer (Thermo Fisher Scientific, Waltham, MA, United States) coupled with nanoAcquity HPLC (Waters, Milford, MA, USA). The labeled peptides were loaded on an Acclaim PepMap C18 trap column (75 μm × 2 cm, 3 μm, C18, 100Å, Thermo Fisher Scientific) and separated by a home-made C18 column (100 μm × 15 cm, 3 μm, C18, 125Å, Phenomenex) at a flow rate of 400 nL/min. Peptide elution was achieved through a linear gradient of Buffer B (0.1% formic acid in acetonitrile) in 120 min. An MS survey scan was performed between 300–1800 m/z with a resolution of 70,000. Higher energy collisional dissociation (HCD) fragmentation was performed for the 10 most intensive precursor ions with a resolution of 17,500, and the dynamic exclusion time was 30 s.

The MS raw files were processed with a Mascot distiller and searched with Mascot (version 2.6.0, Matrix Science, London, United Kingdom) against the TAIR10 database. Scaffold Q+ (version 4.8.7, Proteome Software, Portland, OR, United States) was used for quantitative analysis. The search parameters were as follows: enzyme specificity was set as trypsin with two missed cleavages; precursor ion mass tolerance was set at 10 ppm and MS/MS fragment ion mass tolerance was at 0.02 Da; the fixed modification was carbamidomethyl (C) and variable modification was oxidation (M); iTRAQ 8-plex was selected for quantification; only peptides with a false discovery rate (FDR) less than 1% were used for subsequent data analysis.

4.4. Bioinformatics Analysis

The identified proteins were annotated using the TAIR database (https://www.arabidopsis.org/, Fremont, CA, USA). The Kyoto Encyclopedia of Genes and Genomes (KEGG) pathway enrichment was performed using an online searching tool (http://www.omicsolution.org/wu-kong-beta-linux/passwd/KEGGEnrich/, Shanghai, China). The protein interaction analysis was performed using the String program (version 11.0, http://www.stringdb.org/, Hinxton, UK).

4.5. Plant Hormone Assay

Plant hormone contents were assayed using a published method [74].

4.6. Quantitative RT-PCR Analyses

The total RNA was extracted individually using 1mL of TRI reagent (Thermo Fisher Scientific, Waltham, MA, USA). For all samples, 2 µg of total RNA was converted to cDNA using M-MLV reverse transcriptase (Promega, Madison, WI, USA). Quantitative RT-PCR was performed with the Applied Biosystems 7500 RT-PCR system with SYBR Premix Ex *Taq* (Takara, Tokyo, Japan). The gene-specific primers (a single peak in qPCR melting curve products) used are listed in Table S4, and *ACTIN* was used as control. The relative quantification of RNA expression was calibrated using the formula $2^{-\Delta\Delta Ct}$ method.

5. Conclusions

In this study, we investigated the effects of exogenous MeJA on Arabidopsis using the JA synthesis deficient mutant *opr3-1*. The differential defense against stress, photosynthesis, and development-related proteins were up-regulated in *opr3-1* after MeJA treatment, meanwhile, MeJA could also down regulate the expression of a large number of ribosomal proteins. Our study shows that, in the absent of endogenous JA, exogenous MeJA enhances Arabidopsis' defense against stress, photosynthesis, and developmental processes, whilst also inhibiting protein synthesis process. For plant hormones, a trace level of JA could still be detected in *opr3-1*, indicating that the JA synthesis capability of the *opr3-1* mutant was significantly blocked but not completely inhibited, and we also found antagonism between JA and SA/ABA. The presented results provide a new framework and candidate protein list for further understanding the molecular mechanisms of exogenous JAs-regulated plant defense, photosynthesis, protein synthesis, and development process.

Supplementary Materials: Supplementary materials can be found at http://www.mdpi.com/1422-0067/21/2/571/s1. Figure S1. Workflow of the experiment. Figure S2. KEGG pathway enrichments of DRPs in MeJA-treated *opr3-1* compared with the *opr3-1* control group. Figure S3. Effects of MeJA treatment for 8 h on physiological parameters assays of two genotypes. (a)Water content; (b) Chlorophyll content; (c) H_2O_2 content; (d) POD content. Figure S4. Principle component analysis score plot of free amino acid contents of the two genotypes under MeJA treatment. Figure S5. Plant hormone contents in the two genotypes under MeJA treatment. (a) Jasmonic acid; (b) Abscisic acid; (c) Salicylic acid. Table S1. Details of differentially regulated proteins related to protein synthesis in *opr3-1*. Table S2. Details about the iTRAQ channels. Table S3. DRPs detected in Ws and *opr3-1* under MeJA treatment. Table S4. Primer sequences for gene expression analysis by RT-qPCR.

Author Contributions: Conceptualization, Z.L.; methodology, J.Q. and Z.L.; formal analysis, Z.L. and J.Q.; investigation, J.Q.; resources, Z.L.; data curation, Z.L. and J.Q.; writing—original draft preparation, J.Q.; writing—review and editing, X.Z.; Z.L.; supervision, Z.L.; project administration, Z.L.; funding acquisition, Z.L. All authors have read and agreed to the published version of the manuscript.

Funding: This research was funded by The State Key Laboratory of Plant Physiology and Biochemistry at China Agricultural University.

Acknowledgments: We are very thankful to Pei Liu (Professor, College of Resources and Environmental Sciences, China Agricultural University) for providing experimental materials, and we would like to thank Yiting Shi (Associate Professor, State Key Laboratory of Plant Physiology and Biochemistry, College of Biological Sciences, China Agricultural University) for the help in qRT-PCR experiments and providing thoughtful insights.

Conflicts of Interest: The authors declare no conflict of interest.

Abbreviations

ABA	Abscisic acid
AOC	Allene oxide cyclase
AOS	Allene oxide synthase
DRPs	Differential regulated proteins
FDR	false discovery rate
GO	Gene ontology
H_2O_2	Hydrogen peroxide
IAA	Auxin
JA	Jasmonic acid
JA-Ile	Jasmonoyl-isoleucine
JAs	Jasmonates
JMT	Jasmonic acid carboxyl methyltransferase
KEGG	Kyoto Encyclopedia of Genes and Genomes
LOX	13-lipoxygenase
MeJA	Methyl jasmonate
OPR3	12-oxophytodienoic acid reductase 3
PCA	Principal component analysis
POD	Peroxidase
ROS	Reactive oxygen species
SA	Salicylic acid
STRING	Search Tool for the Retrieval of Interacting Genes

References

1. Creelman, R.A.; Tierney, M.L.; Mullet, J.E. Jasmonic acid/methyl jasmonate accumulate in wounded soybean hypocotyls and modulate wound gene expression. *Proc. Natl. Acad. Sci. USA* **1992**, *89*, 4938–4941. [CrossRef]

2. Wasternack, C.; Hause, B. Jasmonates: Biosynthesis, perception, signal transduction and action in plant stress response, growth and development. An update to the 2007 review in Annals of Botany. *Ann. Bot.* **2013**, *111*, 1021–1058. [CrossRef]

3. Goossens, J.; Fernández-Calvo, P.; Schweizer, F.; Goossens, A. Jasmonates: Signal transduction components and their roles in environmental stress responses. *Plant Mol. Biol.* **2016**, *91*, 673–689. [CrossRef]

4. Devoto, A. The Jasmonate Signal Pathway. *Plant Cell* **2002**, *14*, S153–S164.

5. Kessler, A.; Baldwin, I.T. Plant responses to insect herbivory: The emerging molecular analysis. *Annu. Rev. Plant Biol.* **2002**, *53*, 299–328. [CrossRef] [PubMed]

6. Conconi, A.; Smerdon, M.J.; Howe, G.A.; Ryan, C.A. The octadecanoid signalling pathway in plants mediates a response to ultraviolet radiation. *Nature* **1996**, *383*, 826–829. [CrossRef]

7. Koo, A.J.K.; Howe, G.A.; Hause, B.; Wasternack, C.; Strack, D. The wound hormone jasmonate. *Phytochemistry* **2009**, *69*, 1571–1580. [CrossRef] [PubMed]

8. Huang, H.; Liu, B.; Liu, L.; Song, S. Jasmonate action in plant growth and development. *J. Exp. Bot.* **2017**, *68*, 1349–1359. [CrossRef] [PubMed]

9. Cipollini, D. Interactive effects of lateral shading and jasmonic acid on morphology, phenology, seed production, and defense traits in Arabidopsis thaliana. *Int. J. Plant Sci.* **2005**, *166*, 955–959. [CrossRef]

10. Fujita, M.; Fujita, Y.; Maruyama, K.; Seki, M.; Hiratsu, K.; Ohme-Takagi, M.; Tran, L.S.; Yamaguchi-Shinozaki, K.; Shinozaki, K. A dehydration-induced NAC protein, RD26, is involved in a novel ABA-dependent stress-signaling pathway. *Plant J.* **2004**, *39*, 863–876. [CrossRef] [PubMed]

11. Koda, Y. The Role of Jasmonic Acid and Related Compounds in the Regulation of Plant Development. *Int. Rev. Cytol.* **1992**, *135*, 155–199. [PubMed]

12. Staswick, P.E.; Su, W.; Howell, S.H. Methyl Jasmonate Inhibition of Root Growth and Induction of a Leaf Protein are Decreased in an Arabidopsis thaliana Mutant. *Proc. Natl. Acad. Sci. USA* **1992**, *89*, 6837–6840. [CrossRef] [PubMed]

13. Ajin, M.; Bryan, T.; Byongchul, S.; BMarkus, L.; Goh, C.; Koo, Y.J.; Yoo, Y.J.; Yang, D.C.; Giltsu, C.; John, B. Transcriptional regulators of stamen development in Arabidopsis identified by transcriptional profiling. *Plant J. Cell Mol. Biol.* **2010**, *46*, 984–1008.

14. Ajin, M.; John, B. MYB108 acts together with MYB24 to regulate jasmonate-mediated stamen maturation in Arabidopsis. *Plant Physiol.* **2009**, *149*, 851–862.

15. Susheng, S.; Tiancong, Q.; Huang, H.; Qingcuo, R.; Dewei, W.; Changqing, C.; Wen, P.; Yule, L.; Jinrong, P.; Daoxin, X. The Jasmonate-ZIM domain proteins interact with the R2R3-MYB transcription factors MYB21 and MYB24 to affect Jasmonate-regulated stamen development in Arabidopsis. *Plant Cell* **2011**, *23*, 1000–1013.

16. Reeves, P.H.; Ellis, C.M.; Ploense, S.E.; Miin-Feng, W.; Vandana, Y.; Dorothea, T.; Aurore, C.; Ina, H.; Kennerley, B.J.; Charles, H. A regulatory network for coordinated flower maturation. *PLoS Genet.* **2012**, *8*, e1002506. [CrossRef]

17. Xiao, S.; Dai, L.; Liu, F.; Wang, Z.; Peng, W.; Xie, D. COS1: An Arabidopsis coronatine insensitive1 suppressor essential for regulation of jasmonate-mediated plant defense and senescence. *Plant Cell* **2004**, *16*, 1132–1142. [CrossRef]

18. Kemal, K. Diverse roles of jasmonates and ethylene in abiotic stress tolerance. *Trends Plant Sci.* **2015**, *20*, 219–229.

19. Campos, M.L.; Jin-Ho, K.; Howe, G.A. Jasmonate-triggered plant immunity. *J. Chem. Ecol.* **2014**, *40*, 657–675. [CrossRef]

20. Howe, G.A.; Jander, G. Plant Immunity to Insect Herbivores. *Annu. Rev. Plant Biol.* **2008**, *59*, 41–66. [CrossRef]

21. Farmer, E.E.; Weber, H.; Vollenweider, S. Fatty acid signaling in Arabidopsis. *Planta* **1998**, *206*, 167–174. [CrossRef] [PubMed]

22. Farmer, E.E.; Ryan, C.A. Octadecanoid Precursors of Jasmonic Acid Activate the Synthesis of Wound-Inducible Proteinase Inhibitors. *Plant Cell* **1992**, *4*, 129–134. [CrossRef] [PubMed]

23. Sanders, P.M.; Lee, P.Y.; Biesgen, C.; Boone, J.D.; Beals, T.P.; Weiler, E.W.; Goldberg, R.B. The arabidopsis DELAYED DEHISCENCE1 gene encodes an enzyme in the jasmonic acid synthesis pathway. *Plant Cell* **2000**, *12*, 1041–1061. [CrossRef] [PubMed]

24. Chini, A.; Monte, I.; Zamarreno, A.M.; Hamberg, M.; Lassueur, S.; Reymond, P.; Weiss, S.; Stintzi, A.; Schaller, A.; Porzel, A.; et al. An OPR3-independent pathway uses 4,5-didehydrojasmonate for jasmonate synthesis. *Nat. Chem. Biol.* **2018**, *14*, 171–178. [CrossRef] [PubMed]

25. Fonseca, S.; Chini, A.; Hamberg, M.; Adie, B.; Porzel, A.; Kramell, R.; Miersch, O.; Wasternack, C.; Solano, R. (+)-7-iso-Jasmonoyl-L-isoleucine is the endogenous bioactive jasmonate. *Nat. Chem. Biol.* **2009**, *5*, 344–350. [CrossRef] [PubMed]

26. Sheard, L.B.; Tan, X.; Mao, H.; Withers, J.; Ben-Nissan, G.; Hinds, T.R.; Kobayashi, Y.; Hsu, F.-F.; Sharon, M.; Browse, J. Jasmonate perception by inositol-phosphate-potentiated COI1–JAZ co-receptor. *Nature* **2010**, *468*, 400–405. [CrossRef]

27. Robson, F.; Okamoto, H.; Patrick, E.; Harris, S.R.; Wasternack, C.; Brearley, C.; Turner, J.G. Jasmonate and Phytochrome A Signaling in Arabidopsis Wound and Shade Responses Are Integrated through JAZ1 Stability. *Plant Cell* **2010**, *22*, 1143–1160. [CrossRef]

28. Kachroo, A.; Lapchyk, L.; Fukushige, H.; Hildebrand, D.; Klessig, D.; Kachroo, P. Plastidial fatty acid signaling modulates salicylic acid- and jasmonic acid-mediated defense pathways in the Arabidopsis ssi2 mutant. *Plant Cell* **2003**, *15*, 2952–2965. [CrossRef]

29. Ishiguro, S.; Kawai-Oda, A.; Ueda, J.; Nishida, I.; Okada, K. The DEFECTIVE IN ANTHER DEHISCIENCE gene encodes a novel phospholipase A1 catalyzing the initial step of jasmonic acid biosynthesis, which synchronizes pollen maturation, anther dehiscence, and flower opening in Arabidopsis. *Plant Cell* **2001**, *13*, 2191–2209. [CrossRef]

30. Stintzi, A.; Weber, H.; Reymond, P.; Browse, J.; Farmer, E.E. Plant defense in the absence of jasmonic acid: The role of cyclopentenones. *Proc. Natl. Acad. Sci. USA* **2001**, *98*, 12837–12842. [CrossRef]

31. Rudus, I.; Terai, H.; Shimizu, T.; Kojima, H.; Hattori, K.; Nishimori, Y.; Tsukagoshi, H.; Kamiya, Y.; Seo, M.; Nakamura, K.; et al. Wound-induced expression of DEFECTIVE IN ANTHER DEHISCENCE1 and DAD1-like lipase genes is mediated by both CORONATINE INSENSITIVE1-dependent and independent pathways in Arabidopsis thaliana. *Plant Cell Rep.* **2014**, *33*, 849–860. [CrossRef] [PubMed]

32. Fragoso, V.; Rothe, E.; Baldwin, I.T.; Kim, S.G. Root jasmonic acid synthesis and perception regulate folivore-induced shoot metabolites and increase Nicotiana attenuata resistance. *New Phytol.* **2014**, *202*, 1335–1345. [CrossRef] [PubMed]

33. Hu, X.; Li, W.; Chen, Q.; Yang, Y. Early signals transduction linking the synthesis of jasmonic acid in plant. *Plant Signal. Behav.* **2009**, *4*, 696–697. [CrossRef] [PubMed]

34. Zhou, J.; Guo, L. Jasmonic Acid (JA) Acts as a Signal Molecule in LaCl-Induced Baicalin Synthesis in Scutellaria baicalensis Seedlings. *Biol. Trace Elem. Res.* **2012**, *148*, 392–395. [CrossRef]

35. Fang, L.; Su, L.; Sun, X.; Li, X.; Sun, M.; Karungo, S.K.; Fang, S.; Chu, J.; Li, S.; Xin, H. Expression of Vitis amurensis NAC26 in Arabidopsis enhances drought tolerance by modulating jasmonic acid synthesis. *J. Exp. Bot.* **2016**, *67*, 2829–2845. [CrossRef]

36. Stintzi, A.; Browse, J. The Arabidopsis male-sterile mutant, opr3, lacks the 12-oxophytodienoic acid reductase required for jasmonate synthesis. *Proc. Natl. Acad. Sci. USA* **2000**, *97*, 10625–10630. [CrossRef]

37. Chehab, E.W.; Kim, S.; Savchenko, T.; Kliebenstein, D.; Dehesh, K.; Braam, J. Intronic T-DNA insertion renders Arabidopsis opr3 a conditional jasmonic acid-producing mutant. *Plant Physiol.* **2011**, *156*, 770–778. [CrossRef]

38. Stenzel, I.; Hause, B.; Miersch, O.; Kurz, T.; Maucher, H.; Weichert, H.; Ziegler, J.; Feussner, I.; Wasternack, C. Jasmonate biosynthesis and the allene oxide cyclase family of Arabidopsis thaliana. *Plant Mol. Biol.* **2003**, *51*, 895–911. [CrossRef]

39. Nagpal, P.; Ellis, C.M.; Weber, H.; Ploense, S.E.; Barkawi, L.S.; Guilfoyle, T.J.; Hagen, G.; Alonso, J.M.; Cohen, J.D.; Farmer, E.E. Auxin response factors ARF6 and ARF8 promote jasmonic acid production and flower maturation. *Development* **2005**, *132*, 4107–4118. [CrossRef]

40. Yamane, H.; Abe, H.; Takahashi, N. Jasmonic Acid and Methyl Jasmonate in Pollens and Anthers of Three Camellia Species. *Plant Cell Physiol.* **1982**, *23*, 1125–1127.

41. Pauwels, L.; Morreel, K.; Witte, E.D.; Lammertyn, F.; Montagu, M.V.; Boerjan, W.; Inzé, D.; Goossens, A. Mapping methyl jasmonate-mediated transcriptional reprogramming of metabolism and cell cycle progression in cultured Arabidopsis cells. *Proc. Natl. Acad. Sci. USA* **2007**, *105*, 1380–1385. [CrossRef]

42. Mata-Perez, C.; Sanchez-Calvo, B.; Begara-Morales, J.C.; Luque, F.; Jimenez-Ruiz, J.; Padilla, M.N.; Fierro-Risco, J.; Valderrama, R.; Fernandez-Ocana, A.; Corpas, F.J.; et al. Transcriptomic profiling of linolenic acid-responsive genes in ROS signaling from RNA-seq data in Arabidopsis. *Front. Plant Sci.* **2015**, *6*, 122. [CrossRef] [PubMed]

43. Guo, L.; Wang, P.; Gu, Z.; Jin, X.; Yang, R. Proteomic analysis of broccoli sprouts by iTRAQ in response to jasmonic acid. *J. Plant Physiol.* **2017**, *218*, 16–25. [CrossRef] [PubMed]

44. Farooq, M.A.; Zhang, K.; Islam, F.; Wang, J.; Athar, H.U.R.; Nawaz, A.; Ullah Zafar, Z.; Xu, J.; Zhou, W. Physiological and iTRAQ-Based Quantitative Proteomics Analysis of Methyl Jasmonate-Induced Tolerance in Brassica napus Under Arsenic Stress. *Proteomics* **2018**, *18*, e1700290. [CrossRef] [PubMed]

45. Alvarez, S.; Zhu, M.; Chen, S. Proteomics of Arabidopsis redox proteins in response to methyl jasmonate. *J. Proteom.* **2009**, *73*, 30–40. [CrossRef]

46. Inohara, N.; Iwamoto, A.; Moriyama, Y. Two genes, atpC1 and atpC2, for the gamma subunit of Arabidopsis thaliana chloroplast ATP synthase. *J. Biol. Chem.* **1991**, *266*, 7333–7338.

47. Yi, X.; Hargett, S.R.; Frankel, L.K.; Bricker, T.M. The PsbQ protein is required in Arabidopsis for photosystem II assembly/stability and photoautotrophy under low light conditions. *J. Biol. Chem.* **2006**, *281*, 26260–26267. [CrossRef]

48. Xu, L. The SCFCOI1 Ubiquitin-Ligase Complexes Are Required for Jasmonate Response in Arabidopsis. *Plant Cell Online* **2002**, *14*, 1919–1935. [CrossRef]

49. Jiang, Y.; Sun, A.; Zhao, Y.; Ying, W.; Sun, H.; Yang, X.; Xing, B.; Sun, W.; Ren, L.; Hu, B.; et al. Proteomics identifies new therapeutic targets of early-stage hepatocellular carcinoma. *Nature* **2019**, *567*, 257–261. [CrossRef]

50. Hossain, M.A.; Munemasa, S.; Uraji, M.; Nakamura, Y.; Mori, I.C.; Murata, Y. Involvement of Endogenous Abscisic Acid in Methyl Jasmonate-Induced Stomatal Closure in Arabidopsis. *Plant Physiol.* **2011**, *156*, 430–438. [CrossRef]

51. Ghuge, S.A.; Andrea, C.; Rodrigues-Pousada, R.A.; Alessandra, T.; Stefano, F.; Paraskevi, T.; Riccardo, A.; Alessandra, C. The MeJA-inducible copper amine oxidase AtAO1 is expressed in xylem tissue and guard cells. *Plant Signal. Behav.* **2015**, *10*, e1073872. [CrossRef] [PubMed]

52. Sasaki, Y.; Asamizu, E.; Shibata, D.; Nakamura, Y.; Kaneko, T.; Awai, K.; Amagai, M.; Kuwata, C.; Tsugane, T.; Masuda, T. Monitoring of methyl jasmonate-responsive genes in Arabidopsis by cDNA macroarray: Self-activation of jasmonic acid biosynthesis and crosstalk with other phytohormone signaling pathways. *DNA Res.* **2001**, *8*, 153–161. [CrossRef] [PubMed]

53. Ajin, M.; Kumar, V.D.; Matt, A.; John, B. Microarray and differential display identify genes involved in jasmonate-dependent anther development. *Plant Mol. Biol.* **2003**, *52*, 775.

54. Verma, V.; Ravindran, P.; Kumar, P.P. Plant hormone-mediated regulation of stress responses. *BMC Plant Biol.* **2016**, *16*, 86. [CrossRef] [PubMed]

55. Koornneef, M.; Reuling, G.; Karssen, C.M. The isolation and characterization of abscisic acid-insensitive mutants of Arabidopsis thaliana. *Physiol. Plant* **2010**, *61*, 377–383. [CrossRef]

56. Hirayama, T.; Shinozaki, K. Perception and transduction of abscisic acid signals: Keys to the function of the versatile plant hormone ABA. *Trends Plant Sci.* **2007**, *12*, 343–351. [CrossRef]

57. Spoel, S.H.; Dong, X. Making sense of hormone crosstalk during plant immune responses. *Cell Host Microbe* **2008**, *3*, 348–351. [CrossRef]

58. Spoel, H.S. NPR1 Modulates Cross-Talk between Salicylate- and Jasmonate-Dependent Defense Pathways through a Novel Function in the Cytosol. *Plant Cell* **2003**, *15*, 760–770. [CrossRef]

59. Leon-Reyes, A.; Van der Does, D.V.; Lange, E.S.D.; Delker, C.; Wasternack, C.; Wees, S.C.M.V.; Ritsema, T.; Pieterse, C.M.J. Salicylate-mediated suppression of jasmonate-responsive gene expression in Arabidopsis is targeted downstream of the jasmonate biosynthesis pathway. *Planta* **2010**, *232*, 1423–1432. [CrossRef]

60. Mcconn, M.; Creelman, R.A.; Bell, E.; Mullet, J.E.; Browse, J. Jasmonate is essential for insect defense in Arabidopsis. *Proc. Natl. Acad. Sci. USA* **1997**, *94*, 5473–5477. [CrossRef]

61. Dixon, D.P.; Davis, B.G.; Edwards, R. Functional divergence in the glutathione transferase superfamily in plants. Identification of two classes with putative functions in redox homeostasis in Arabidopsis thaliana. *J. Biol. Chem.* **2002**, *277*, 30859–30869. [CrossRef] [PubMed]

62. Nee, G.; Zaffagnini, M.; Trost, P.; Issakidis-Bourguet, E. Redox regulation of chloroplastic glucose-6-phosphate dehydrogenase: A new role for f-type thioredoxin. *FEBS Lett.* **2009**, *583*, 2827–2832. [CrossRef] [PubMed]

63. Newman, T.; De Bruijn, F.J.; Green, P.; Keegstra, K.; Kende, H.; Mcintosh, L.; Ohlrogge, J.; Raikhel, N.; Somerville, S.; Thomashow, M. Genes galore: A summary of methods for accessing results from large-scale partial sequencing of anonymous Arabidopsis cDNA clones. *Plant Physiol.* **1994**, *106*, 1241–1255. [CrossRef] [PubMed]

64. Vladan, B.; Jan, S.; Jürgen, S.; Carsten, K. Dehydrins (LTI29, LTI30, COR47) from Arabidopsis thaliana expressed in Escherichia coli protect Thylakoid membrane during freezing. *J. Serb. Chem. Soc.* **2013**, *78*, 1149–1160.

65. Sanders, P.M.; Bui, A.Q.; Weterings, K.; Mcintire, K.N.; Hsu, Y.C.; Pei, Y.L.; Mai, T.T.; Beals, T.P.; Goldberg, R.B. Anther developmental defects in Arabidopsis thaliana male-sterile mutants. *Sex. Plant Reprod.* **1999**, *11*, 297–322. [CrossRef]

66. Xu, X.M.; Rose, A.; Muthuswamy, S.; Jeong, S.Y.; Venkatakrishnan, S.; Zhao, Q.; Meier, I. NUCLEAR PORE ANCHOR, the Arabidopsis homolog of Tpr/Mlp1/Mlp2/megator, is involved in mRNA export and SUMO homeostasis and affects diverse aspects of plant development. *Plant Cell* **2007**, *19*, 1537–1548. [CrossRef]

67. Huang, S.; Mcdowell, J.M.; Weise, M.J.; Meagher, R.B. The Arabidopsis profilin gene family. Evidence for an ancient split between constitutive and pollen-specific profilin genes. *Plant Physiol.* **1996**, *111*, 115–126. [CrossRef]

68. Barakat, A.; Szick-Miranda, K.; Chang, I.F.; Guyot, R.; Blanc, G.; Cooke, R.; Delseny, M.; Bailey-Serres, J. The Organization of Cytoplasmic Ribosomal Protein Genes in the Arabidopsis Genome. *Plant Physiol.* **2001**, *127*, 398–415. [CrossRef]

69. Li, G.; Wu, Y.; Liu, G.; Xiao, X.; Wang, P.; Gao, T.; Xu, M.; Han, Q.; Wang, Y.; Guo, T.; et al. Large-scale Proteomics Combined with Transgenic Experiments Demonstrates An Important Role of Jasmonic Acid in Potassium Deficiency Response in Wheat and Rice. *Mol. Cell. Proteom.* **2017**, *16*, 1889–1905. [CrossRef]

70. Anastassia, K.; Maria, H.; Virpi, P.; Vainonen, J.P.; Suping, Z.; Poul Erik, J.; Henrik Vibe, S.; Vener, A.V.; Eva-Mari, A.; Anna, H. A previously found thylakoid membrane protein of 14kDa (TMP14) is a novel subunit of plant photosystem I and is designated PSI-P. *FEBS Lett.* **2005**, *579*, 4808–4812.

71. Haldrup, A.; Naver, H.; Scheller, H.V. The interaction between plastocyanin and photosystem I is inefficient in transgenic Arabidopsis plants lacking the PSI-N subunit of photosystem I. *Plant J.* **1999**, *17*, 689–698. [CrossRef] [PubMed]

72. Cao, J.; Li, M.; Chen, J.; Liu, P.; Li, Z. Effects of MeJA on Arabidopsis metabolome under endogenous JA deficiency. *Sci. Rep.* **2016**, *6*, 37674. [CrossRef]

73. Wisniewski, J.R.; Zougman, A.; Nagaraj, N.; Mann, M. Universal sample preparation method for proteome analysis. *Nat. Methods* **2009**, *6*, 359–362. [CrossRef] [PubMed]

74. Zhao, X.; Bai, X.; Jiang, C.; Li, Z. Phosphoproteomic Analysis of Two Contrasting Maize Inbred Lines Provides Insights into the Mechanism of Salt-Stress Tolerance. *Int. J. Mol. Sci.* **2019**, *20*, 1886. [CrossRef] [PubMed]

© 2020 by the authors. Licensee MDPI, Basel, Switzerland. This article is an open access article distributed under the terms and conditions of the Creative Commons Attribution (CC BY) license (http://creativecommons.org/licenses/by/4.0/).

MDPI

St. Alban-Anlage 66

4052 Basel

Switzerland

Tel. +41 61 683 77 34

Fax +41 61 302 89 18

www.mdpi.com

International Journal of Molecular Sciences Editorial Office

E-mail: ijms@mdpi.com

www.mdpi.com/journal/ijms

www.ingramcontent.com/pod-product-compliance
Lightning Source LLC
LaVergne TN
LVHW071506180726
843512LV00014B/1027